Evolution from molecules
to men

EVOLUTION FROM MOLECULES
TO MEN

Edited by D. S. Bendall
on behalf of Darwin College, Cambridge

CAMBRIDGE UNIVERSITY PRESS

Cambridge
London   New York   New Rochelle
Sydney   Melbourne

Published by the Press Syndicate of the University of Cambridge
The Pitt Building, Trumpington Street, Cambridge CB2 1RP
32 East 57th Street, New York, NY 10022, USA
296 Beaconsfield Parade, Middle Park, Melbourne 3206, Australia

First published 1983

Printed in Great Britain by the University Press, Cambridge

Library of Congress catalogue card number: 82-22020

British Library Cataloguing in Publication Data
Evolution from molecules to men
1. Human evolution
I. Bendall, D. S.
573.2   GN 281
ISBN 0 521 24753 5

U P

Contents

Contributors

Ms S. M. Adams, Department of Genetics, University of Leicester, Leicester LE1 7RH, UK

Professor G. E. Allen, Washington University, St Louis, Missouri 63130, USA

Professor F. J. Ayala, Department of Genetics, University of California at Davis, Davis, CA 95616, USA

Dr P. A. Barrie, Department of Genetics, University of Leicester, Leicester LE1 7RH, UK

Dr P. P. G. Bateson, Sub-Department of Animal Behaviour, High Street, Madingley, Cambridge CB3 8AA, UK

Dr A. Blanchetot, Department of Genetics, University of Leicester, Leicester LE1 7RH, UK

Dr W. F. Bodmer, Imperial Cancer Research Fund Laboratories, Lincoln's Inn Fields, London WC2A 3PX, UK

Professor R. W. Burkhardt, Department of History, University of Illinois, Urbana, ILL 61801, USA

Professor P. H. Clarke, Department of Biochemistry, University College London, Gower Street, London WC1E 6ET, UK

Dr T. H. Clutton-Brock, Large Animal Research Group, Department of Zoology, 34a Storey's Way, Cambridge CB3 0DT, UK

Dr R. Dawkins, Department of Zoology, South Parks Road, Oxford OX1 3PS, UK

Professor Dr M. Eigen, Max-Planck Institut für biophysikalische Chemie, Am Fassberg, D-3400 Göttingen, W. Germany

Professor S. J. Gould, Museum of Comparative Zoology, Harvard University, Cambridge, MA 02138, USA

Professor A. Hallam, Department of Geological Sciences, University of Birmingham, Birmingham B15 2TT, UK

Professor J. L. Harper, School of Plant Biology, University College of North Wales, Bangor, Wales

Mr S. Harris, Department of Genetics, University of Leicester, Leicester LE1 7RH, UK

Dr M. J. S. Hodge, Department of Philosophy, University of Leeds, Leeds LS2 9JT, UK

Professor D. L. Hull, University of Wisconsin, Milwaukee, Wisconsin 53201, USA

Professor Sir Andrew Huxley, Department of Physiology, University College London, Gower Street, London WC1E 6BT, UK

Professor G. Isaac, Department of Anthropology, University of California at Berkeley, Berkeley, CA 94720, USA

Dr F. Jacob, Institut Pasteur, 25 rue du Dr Roux, 75017 Paris, France

Dr A. Jeffreys, Department of Genetics, University of Leicester, Leicester LE1 7RH, UK

Professor R. C. Lewontin, Museum of Comparative Zoology, Harvard University, Cambridge MA 02138, USA

Professor J. Maynard Smith, School of Biological Sciences, University of Sussex, Falmer, Brighton, Sussex BN1 9QG, UK

Professor E. Mayr, Museum of Comparative Zoology, Harvard University, Cambridge, Mass 02138, USA

Professor E. Nevo, Institute of Evolution, Haifa University, Haifa, Israel

Professor J. Passmore, Department of Philosophy, Australian National University, History of Ideas Unit, Research School of Social Sciences, P.O. Box 4, Canberra ACT 2600, Australia

Professor Sir David Phillips, Molecular Biophysics Laboratory, Zoology Department, South Parks Road, Oxford OX1 3PS, UK

Professor J. Shapiro, Department of Microbiology, University of Chicago, 920 E 38 Street, Chicago, Ill 60637, USA

Dr M. J. E. Sternberg, Molecular Biophysics Laboratory, Zoology Department, South Parks Road, Oxford OX1 3PS, UK

Dr B. J. Sutton, Molecular Biophysics Laboratory, Zoology Department, South Parks Road, Oxford, OX1 3PS, UK

Dr B. A. O. Williams, King's College, Cambridge CB2 1ST, UK

Professor E. O. Wilson, Museum of Comparative Zoology, Harvard University, Cambridge, MA 02138, USA

Professor C. R. Woese, Department of Genetics and Development, University of Illinois, Urbana, Illinois 61801, USA

Mr D. Wood, Department of Genetics, University of Leicester, Leicester LE1 7RH, UK

Preface

The conference, entitled *Evolution of Molecules and Men*, which Darwin College organised from 27 June until 2 July 1982 has been called in *Science* 'the most official of the multiplicity of conferences to commemorate the centenary of Darwin's death'. Although there is, of course, no body with the authority to declare a conference on Darwin official, there is an obvious sense in which the label is not unjustified. No other conference took as wide-ranging a view of the present state of Darwinism, as this volume shows. This is not a volume of conference proceedings in the strict sense. However, with the exception of Sir Andrew Huxley's prologue, every contribution started as a paper at the conference, which has been more or less revised for publication. Only three participants felt that they preferred not to join in the publication for one reason or another.

A brief explanation of how the conference came about may be of interest to readers of this volume. The explanation actually goes back almost two decades. Darwin College was founded by three other Cambridge colleges – Gonville & Caius College, St John's College and Trinity College – to be the first exclusively postgraduate college in Cambridge. The move was taken soon after the death at the end of 1962 of Sir Charles Darwin, grandson of *the* Charles Darwin. After some negotiation, the family agreed to make their property available to the new college – a considerable site on an arm of the river acquired by Charles' son George in 1885, that included the family home, Newnham Grange, and the Old Granary. The college was also given permission to take the name Darwin. Although it was never the intention that the college should concentrate on biological studies, the name has inevitably created a special association, and it is a fact that research in biology occupies a large proportion of the research students and Fellows of the college.

As the centenary of Charles Darwin's death approached, the college felt that it wished to do something substantial to commemorate it. After considerable preliminary discussion with officers of the Royal Society and others, the college decided that the only really appropriate celebration would be one that looked at the whole field in order to assess the current status and prospects of Darwinian theory. The response was overwhelming, not only from various societies and institutions that helped with the finance, but from biologists all over the western world. The range of the papers, selected by a small college committee after very extensive discussion with experts in various fields, reflects the fact that Darwinian theory has implications for all aspects of biology and also for sociology and philosophy.

An early decision was taken to restrict the conference to about 300 participants in order to allow for serious formal and informal discussion. This was a difficult decision to make, and a harder one to enforce. It became increasingly obvious that a large number of people other than specialists have a serious interest in the subject and this volume is directed to them as much, or perhaps more, than to professional evolutionists and molecular biologists. Although the papers were all written by professionals in the strict sense of the word, they proved to appeal to a wider audience as well. That, we like to think, is a measure of the range and breadth of Darwinism, of Darwin's own work and of the avenues of inquiry that he opened up. As Professor Passmore said in his 'epilogue', a remarkable tour de force in summation: if the authors, 'active scientists' in most cases, 'usually begin with a bow in Darwin's direction, and an appropriate quotation,... they are not, in general, talking *about* Darwin. Rather, they are presenting us with independent investigation conducted in a Darwinian spirit, testimonies to the fact that Darwin is still a living force...There is something very special about a scientist who is commemorated in that way.'

The prologue by Sir Andrew Huxley actually results from a different activity of the college, which in 1977 initiated a series of annual public lectures to be given to a large academic audience by distinguished figures in all branches of the world of learning. This lecture was given on 4 May 1982 by Sir Andrew Huxley, currently the President of the Royal Society, on the subject of Charles Darwin. The committee felt that it would make an appropriate prologue to this volume and we thank Sir Andrew for agreeing to revise it for this purpose.

The college was able to run the conference with its own limited personnel and I believe it is not invidious if I single out for special thanks three people – Dr D. H. Mellor who acted as Conference Secretary, Dr D. S. Bendall who played a major role in the planning stage and has taken the responsibility for this volume, and my secretary Mrs Joyce Graham who was responsible for the administrative side.

Moses Finley
Master of Darwin College

September 1982

PROLOGUE

1

How far will Darwin take us?

ANDREW HUXLEY

The Origin of Species worked a revolution by bringing a very wide range of biological phenomena within the scope of natural explanation. In the nature of things, evolution, which has to do with events millions or hundreds of millions of years ago, is an area where detailed certainty is unattainable and will remain so. But a centenary of this kind is a suitable occasion not only for taking stock and asking ourselves how successful Darwin's ideas now appear to have been within the range to which he devoted himself, but also for going further and asking what are the prospects for extending natural explanations to aspects of living things to which Darwin did not address himself because in his time they were completely out of range.

Darwin was, of course, not the originator of the idea of evolution. If you mention evolution to a Frenchman the first name that will come to his lips is not Darwin but Lamarck, who lived and wrote half a century or so before Darwin. In Britain there was Darwin's own grandfather Erasmus Darwin. In the nineteenth century, Herbert Spencer (1852) was an ardent evolutionist well before the publication of the *Origin*, as was Robert Chambers, author of the *Vestiges of Creation*, published anonymously in 1844. The chief grounds in those days for thinking of evolution, or transmutation as it was commonly called, were the obvious resemblances between animals of any one group, and the fact that organisms could be classified in a hierarchical system – species within genera, genera within families and so on – so that it was possible to draw an evolutionary tree in which each branch point might correspond to the splitting into two of an ancestral species.

In the first half of the nineteenth century almost all biologists remained unconvinced, and fixity of species was the standard and orthodox position.

My grandfather, Thomas Henry Huxley, adopted an agnostic position, as
he did in many things, and writing much later (1888) he describes his
discussions in the 1850s with Herbert Spencer in the following words: 'I
took my stand upon two grounds: firstly, that up to that time, the evidence
in favour of transmutation was wholly insufficient; and, secondly, that no
suggestion respecting the causes of the transmutation assumed, which had
been made, was in any way adequate to explain the phenomena. Looking
back at the state of knowledge at that time, I really do not see that any
other conclusion was justifiable...So I took refuge in that 'thätige Skepsis',
which Goethe has so well defined; and reversing the apostolic precept to
be all things to all men, I usually defended the tenability of the received
doctrines, when I had to do with the transmutationists; and stood up for
the possibility of transmutation among the orthodox – thereby, no doubt,
increasing an already current, but quite undeserved, reputation for
needless combativeness.'

The situation was transformed by the joint publication in July 1858 of
the papers of Darwin and Alfred Russel Wallace proposing natural
selection as a mechanism for the transmutation of species. My grandfather
(1888) records that his reaction was: 'How extremely stupid not to have
thought of that'. Darwin then got down with extraordinary speed to
writing *The Origin of Species*, which he described as an abstract of the much
longer work that he had already begun, and the *Origin* was actually
published in November 1859.

Darwin had for twenty years been amassing evidence both for the
general proposition that the diversity of present-day living organisms had
been generated by 'descent with modification' from one or perhaps a few
primeval forms, and for the efficacy of natural selection as the principal
mechanism by which this modification had been brought about. He
presented this evidence in so persuasive and readable a form in the *Origin*
that the violent opposition which it generated lasted only for a few years,
at least among the educated public in Britain. In the *Origin*, Darwin held
back from explicitly applying his ideas to Man, and went no further than
saying in his concluding chaper that 'Light will be thrown on the origin
of man and his history', but by 1871 the climate of opinion had changed
so much that he felt able to publish *The Descent of Man*.

But even in *The Descent of Man*, Darwin (1871) excludes two great areas
of biology. He says 'In what manner the mental powers were
first developed in the lowest organisms, is as hopeless an enquiry as
how life itself first originated. These are problems for the distant future,
if they are ever to be solved by man'. So, in asking 'How far will Darwin

take us?' we ask in the first place whether we can now accept the two propositions to which he addressed himself – descent with modification, and modification by natural selection. In the second place, we ask what prospect there now is of solving the problems of the origin of life and of the nature of consciousness.

Descent with modification

The first sentence of *The Origin of Species* reads: 'When on board H.M.S. "Beagle", as naturalist, I was much struck with certain facts in the distribution of the organic beings inhabiting South America, and in the geological relations of the present to the past inhabitants of that continent'. As regards extinct forms, the point which directly suggested common ancestry was that many South American fossils were clearly related to present-day species which are found only in South America: for example, *Megatherium*, the giant ground-sloth, was related to present-day sloths, the giant armadillo *Glyptodon* was related to present-day armadillos, and fossil rodents resembled present-day South American rodents rather than those of North America or the Old World. As regards living organisms, the most striking instances were in the Galapagos Islands, where, for example, there are thirteen species of ground finches, closely related to one another, adapted to different ways of feeding, and none found anywhere else in the world – just the situation to be expected if they had been derived by descent with modification from a few individuals of a single species which had found their way to the islands. Most of Darwin's evidence is morphological but in *The Descent of Man* he also emphasizes the resemblance of the behaviour of the higher mammals to that of Man. He draws his examples equally from the plant and animal kingdoms, from developing as well as adult forms, and from fossils as well as recent organisms.

Darwin readily admitted the gaps that existed in sequences of fossils, but many of these gaps have since his time been at least partially filled, and this is especially true in the case of Man. When the *Origin* was published, the only known human fossil was the Neanderthal skeleton, whereas now there must, I suppose, be fossil remains as old as or older than the Neanderthal skeleton from thousands of individuals intermediate between present-day apes and man. The closeness of relationship between animals of widely different sorts has been immensely strengthened by the discoveries of biochemistry, the community of biochemical systems and indeed of the genetic system between widely different kinds of animals and plants. Finally, the relationships between animals traced by recent methods such as protein sequencing agree closely with the relationships traced by methods available in Darwin's time.

Among scientists, the proposition that existing species have arisen by descent with modification is now agreed practically universally. There is, however, vigorous opposition by creationists, especially in the United States. I do not regard this as a scientific disagreement: almost all creationists base themselves on authority which is not admitted by scientists, and they are attacking not so much evolution as science itself. From a scientific point of view one does not need to take them seriously, though from a sociological point of view, and from the point of view of the harm their doctrines may do, one does indeed have to take them seriously. One has to admit, of course, that there are still major gaps in the fossil record, notably between the different invertebrate phyla which seem to appear as distinct entities in the early Palaeozoic, but all biologists suppose that the fossil record is incomplete at that point because the missing organisms were soft-bodied, and have not been preserved; however, this is something that we have to bear in mind.

Natural selection

It is often forgotten that Darwin himself made no claim that natural selection is the only mechanism in evolution. In a letter to *Nature* in 1880 he asks whether it is possible to 'name any one who has said that the evolution of species depends only on natural selection? As far as concerns myself, I believe that no one has brought forward so many observations on the effects of the use and disuse of parts, as I have done in my "Variation of Animals and Plants under Domestication"; and these observations were made for this special object. I have likewise there adduced a considerable body of facts, showing the direct action of external conditions on organisms...'. Passages like this, which is not by any means unique, read oddly to the present-day biologist, but the reason for this is that we have all become accustomed to Mendelian inheritance, and we forget that a great difficulty facing the idea of natural selection in Darwin's time was the complete ignorance of the laws of heredity and the general assumption that most inherited differences were subject to some degree of blending inheritance or 'regression towards the mean', so that the variability that is obvious within any species would disappear in a few generations unless restored by external influences.

Darwin's own hypothesis (1868) for the mechanism of inheritance, which he called pangenesis, was conceived about 1840. It is a mechanism in which blending inheritance, and influences of use and disuse and so on, would occur. He supposed that the germ cells received 'gemmules' from all the cells of the body, so that influences of parts remote from the germ cells would be capable of influencing them and the hereditary material that they would then transmit. Of course, as we all know, pangenesis has received no support, and Mendelism and modern biochemical genetics do not allow for any influence from other parts of the body onto the genetic information which is passed on to the next generation. When I say this I must remind myself, and I hope everyone will remind themselves, that there is at any time a great danger of assuming that we already know the complete answer to a problem; we must beware that we do not make fools of ourselves, as so many people have done before, by assuming that what we now know is the whole truth – in this case, that there is no possibility of a reverse influence of remote parts of the body on the germ cells. But conversely, one need have no reservations in saying that Mendelian inheritance removes the need for postulating influences of external conditions, such as Darwin invoked for the maintenance of variability, since Mendelian inheritance is not subject to blending and maintains variability (except in very small populations where genes get lost), because, of course, the hereditary factors are merely reshuffled at each generation.

Mendel's theory did not become generally known within Darwin's lifetime. Although published in 1865 and 1866, it was known only to a few biologists, such as Nägeli, until it was rediscovered in 1900. It is paradoxical that when it was rediscovered it failed to be recognized as being the very thing that Darwinian selection required. The rediscoverers, Correns, Tschermak and De Vries, and their followers, such as Bateson in this country and Morgan in the United States, all thought in terms only of mutations with large effect and believed that ordinary small-range variation, apparently continuous, was not of a Mendelian kind. It was generally believed that wild species were more-or-less pure lines, that is to say the individuals were genetically almost identical, and that the variability within each species was not inherited and therefore could not be an effective basis for natural selection.

It was not until the 1930s and 1940s that it was recognized that most of the ordinary variability within a species obeys Mendel's laws, though in most cases this is not obvious because any one character is determined by several genes which are inherited independently. The history of this slow appreciation of the significance of Mendelian inheritance for evolution has been well set out in a recent collection of essays (Mayr & Provine, 1980).

The other great advance at that time was the demonstration by

R. A. Fisher (1930) of the power of natural selection. It is enormously more powerful than is intuitively obvious to an untutored person or to someone whose knowledge is based only on reading Darwin.

This synthesis, referred to as neo-Darwinism, was set out by my brother Julian Huxley in 1942 in his book *Evolution, the Modern Synthesis*. One can describe it as being much harder than the original Darwinism, in the sense that many of the ill-defined concepts in *The Origin of Species* – effects of use and disuse, and direct effects of environment on heredity – were dropped, partly for lack of evidence and partly because they ceased to be necessary, since ordinary variation was of a Mendelian kind. Apart from the fact that these concepts became superfluous for natural selection, one can regard them as not being important because, even if they did exist, they would not help as regards the aspects of evolution which are difficult to explain by natural selection, particularly the question whether what Darwin referred to as organs of extreme perfection and complication, like the eye or the ear, could have originated by such a crude process as mutation and selection.

So we can say that the modern equivalent of Darwin's own position is that the constructive force in evolution is natural selection acting on variability which is inherited according to Mendel's laws and which is maintained on the long time-scale by random mutations. The latter will no doubt have some direct influence on the course of evolution, but their importance is still a controversial question and in any case, without natural selection, they do not help to account for the adaptedness of organisms or the coordination of their parts. There are other questions about the mechanism of evolution that are much discussed at present. Does evolution go progressively with the whole of a species changing more or less simultaneously and continuously or does it go by the process referred to by Eldredge & Gould (1972) as 'punctuated equilibria', in which relatively rapid and local change is followed by a period of stagnation? And is cladistic taxonomy anti-Darwinian? These are interesting and important questions but I regard them as being questions within the Darwinian framework – they do not give rise to doubts as to whether Darwinism is right or not. They are really cases where it is not a question of available explanations being inadequate, it is a question of there being too many explanations: there are alternative explanations, and the question is which one is right, or rather, how great are their respective contributions? I will not say any more about this and will concentrate on those questions where doubt is expressed about the essentials of the neo-Darwinian position.

The chief of these is the adequacy of mutation and selection. In the first place, given Mendelian inheritance, natural selection must work to some extent. The question is whether its extent is limited to minor adaptations of a species to changes of environment, or perhaps to evolution of different species within a genus (admitted even by Linnaeus), or whether mutation and selection are able to account for everything: the development of new types of organisms, the extreme complexity that exists in living systems, and the marvellous coordination of different parts and different systems within one organism. This is a crucial question because there is at present no alternative to mutation and selection short of admitting some external, and one would have to say supernatural, intelligence, outside present-day science, but not for that reason necessarily inconceivable; Hoyle and others find themselves driven to that position.

If one asks 'Can something as crude as random mutation followed by natural selection account for the perfection of living things that we are all so conscious of?', the commonsense answer is 'No! That is much too crude a process.' With his usual characteristic honesty Darwin admits this, and in *The Origin of Species* under the heading *Organs of extreme perfection and complication* he begins: 'To suppose that the eye, with all its inimitable contrivances for adjusting the focus to different distances, for admitting different amounts of light, and for the correcting of spherical and chromatic aberration, could have been formed by natural selection, seems, I freely confess, absurd in the highest degree.'* Note that he said 'seems'; he does not say 'is', and he goes on after that passage to draw attention to intermediate stages of visual systems in various animals that might correspond to stages in the evolution of the eye.

Now the *Origin* is a very persuasive book; it brings together an extraordinary range of knowledge – it is difficult to imagine anyone nowadays writing so comprehensive a book – and if a critic of Darwin asserts that natural selection is incapable of producing the complexity that we know, the first reply to make is to ask if he has read *The Origin of Species*, and if he says 'no', I think you are entitled to disregard anything that he may say. Few, even among biologists, appreciated the power of natural selection until the publication in 1930 of R. A. Fisher's *Genetical Theory of Natural Selection*, one of the key works in establishing the neo-Darwinian synthesis. Fisher showed, for example, that if a mutation gives to animals that possess it a 1% advantage over those that do not, that mutation is likely to be incorporated in a large population within something of the order of 100 generations, that is to say, an advantage so slight that it would not

* *The Origin of Species*, 1st edn, p. 186; 4th edn (1866), p. 215.

be noticeable and would be extremely difficult even to measure by field experiments, becomes incorporated in a species in a time which is very short in geological terms. This power is far beyond what one ordinarily visualizes when thinking of natural selection.

So the second question to ask a critic is 'Have you read Fisher's *Genetical Theory of Natural Selection?*', and again, if the answer is 'no', you are entitled to disregard anything he says. To dismiss natural selection without having studied these questions to some serious extent is the equivalent of a biologist saying to a physicist that he does not accept relativity or quantum mechanics when he has not made any attempt to understand them. But some physicists as well as laymen do express doubts. They will say that biologists have been brainwashed, that we have all been told throughout our school and university training that natural selection is the way in which evolution has worked, and so we all believe it. There may be something in this: it is indeed a kind of thing that can happen, and I myself draw attention to examples where it has happened. There is undoubtedly an advantage in coming fresh to a field, so that one is free from such assumptions. On the other hand a little learning is a useful as well as a dangerous thing.

It is a common fallacy among critics of Darwinism to assume that each complex system that we know in living organisms was achieved in a single step, and to point out how improbable it is that such an event could occur by chance. No Darwinian supposes any such thing. The real question is: Can one think out a sequence of stages of evolution of an organ (or an enzyme or whatever it may be), such that each stage will be more valuable to the whole organism than the previous one? For example, both Hoyle & Wickramasinghe (1981) and the X-ray crystallographer Lipson (1980) mention haemoglobin, the oxygen carrier of the blood, as something so complex and subtle that there would be a vanishingly small chance of its being produced by random mutation so that natural selection could act so as to preserve it. They do not deny descent with modification but they postulate some supernatural intervention to govern the direction of the hereditary changes which are produced. I would agree if the question were the production *de novo* of the complete haemoglobin molecule, which contains some 600 amino acids and four haem groups, the latter being a moderately complicated organic molecule containing an iron atom, but that is not the question. The haemoglobin molecule is composed of four subunits, each one of which is similar to another well-known protein, myoglobin, of which there are large quantities in our muscles, and which is also an oxygen carrier, working in essentially the same way as haemoglobin. Haemoglobin has some rather minor, but quite important, different characteristics in its mode of carrying oxygen which give it an advantage as the carrier in the blood: it gets rid of its oxygen at a relatively high partial pressure so that it will diffuse quickly into the tissue. That is a relatively minor step, and I do not think that the majority of objectors to Darwinian evolution would say that the development of haemoglobin from myoglobin-like molecules was incredible. Still, the evolution of myoglobin, with about 150 amino acids and one haem group, is a formidable enough problem. But we can go back a step further. The haem group is identical to the essential group in the cytochromes. These are enzymes used in the oxidative processes within cells, very much more widespread than the oxygen carriers myoglobin and haemoglobin, and present in an enormous range of organisms, including some of the most primitive. It is not difficult to imagine that once cytochromes had evolved, the haem group might associate with some other protein and thus acquire its ability to carry oxygen molecules. But then this takes the problem back to the evolution of cytochrome. Cytochromes are in the oxidative chain and work not by carrying oxygen molecules but by actually undergoing oxidation and reduction: the iron atom changes between the ferrous and the ferric state. This process is something that happens even in the simplest iron compounds, unlike oxygen carriage, which is not imitated by any known simple compound. So one can visualize primitive semi-living systems in which there were oxidation–reduction reactions involving iron; then more complex iron molecules being evolved, leading ultimately to cyto-chromes; then the haem group combining with some other protein to produce a primitive myoglobin; and finally four myoglobins associating to form haemoglobin. So one can split the process of evolution of haemoglobin into many steps, each of which is intelligible from the point of view of Darwinian evolution.

Another example which illustrates the way in which plausible inter-mediate steps have turned up unexpectedly through research which was not in any way aimed directly at working out evolutionary questions is provided by the electric organs of fishes. In the *Origin* Darwin says 'It is impossible to conceive by what steps these wondrous organs have been produced'. He says 'In Gymnotus [the electric eel] and Torpedo [the electric ray] they no doubt serve as powerful means of defence and perhaps for securing prey; yet in the Ray, an analogous organ in the tail, even when greatly irritated, manifests, as lately observed by Matteucci, but little electricity; so little that it can hardly be of much use for these ends'.* The position remained equally puzzling until about 30 years ago when Hans

* *The Origin of Species*, 4th edn (1866), p. 224.

Lissmann (1951) showed that these weak electric organs are used as a kind of submarine radar, both for navigation and for finding prey, and that the fishes which possess them have a highly-developed electric sense. Later it was found that this electric sense is shared by many fishes without electric organs (Kalmijn, 1974). In all cases the receptors which detect changes in electric field are derived from the neuromasts of the lateral-line system; the function of the latter is to detect water movements but, like most sense organs, they are also moderately sensitive to electric fields. So it is probable that the powerful electric organ of *Gymnotus* was evolved through the following steps:

(1) Electric fields in the water (due to prey or to the Earth's magnetic field) caused small variations in the continuous discharge in sensory nerve fibres from neuromasts; natural selection was therefore able to develop an electric sense by increasing the sensitivity of some of these receptors to electric fields.

(2) These sensitive electroreceptors were activated by the action potentials from the fish's own muscles; natural selection was therefore able to develop some of the muscle into a 'weak electric organ' with an 'active detector' function.

(3) The 'weak electric organ' gave discharges strong enough that they occasionally activated nerve fibres in prey animals and thus made their movements uncoordinated; natural selection was therefore able to develop part of the 'weak electric organ' into a powerful organ capable of incapacitating any prey species.

A useful thought in relation to natural selection is that it imitates purpose, so that the 'argument from design' in favour of the existence of a deity transforms almost immediately into an argument for natural selection, though of course natural selection is immensely slower than rational purpose. There are also useful analogies between natural selection and the traditional crafts. I have been repeatedly impressed by the technical achievements that were made before there was any understanding of the processes that were involved – smelting of iron or making musical instruments, or the invention of the boomerang by Australian aborigines. These developments must have happened in a way that has analogies with natural selection: they must have depended on chance events with consequences that were noticed by the craftsman and were passed on by imitation to his children and their children. This is analogous to trans-mutation of species by mutation, selection and inheritance, but in the latter case the effect is picked out by differential survival, a process immensely slower than imitation by a human successor, and correspondingly evolution is an immensely slower process than even traditional development of human skills.

There remain great areas of uncertainty about natural selection. Direct evidence is almost unavailable on the evolution of highly organized systems, most of which are matters of biochemistry, or of soft structures, which are not preserved in fossils. The relevant genes for most of these things have not yet been identified. The only genes that are well known at present are those which actually produce enzymes, but those which determine the structure and shape of an organ like the eye have not yet been identified, and we therefore do not know their mutation rates and cannot calculate likely rates of evolution. For these reasons it seems to me that the neo-Darwinian position does remain a hypothesis, but undoubtedly a very powerful one which has not yet been found wanting and to which there is at present no alternative of a scientific kind. I see no reason to think of giving it up at this stage of the game.

The origin of life

The next of the questions to which I propose to address myself is whether the origin of life on earth is explicable in terms of physics and chemistry. I am not going to spend long on this. As you will know, there is a standard story that physical processes, including lightning and so forth, on the primitive earth gave rise to chemical compounds which formed a 'primeval soup' containing some at least of the precursors of substances for living systems – amino acids, sugars and purines have been shown to be produced in these ways. It is then supposed that random chemical reactions produced some unidentified substance, far simpler than anything known in present-day living things, which was capable of self-replication. From this point on, mutation and selection could give rise to further development of complexity. This idea can be regarded as an extension of Darwinian principles to the origin of life; indeed, Darwin himself evidently thought along these lines (F. Darwin, 1888).

In 1981, two books were published by very well known scientists, which discussed as an alternative the arrival of living matter from outside the solar system: Hoyle & Wickramasinghe's *Evolution from Space* and Francis Crick's *Life Itself*. Hoyle makes a calculation which purports to show that the spontaneous origin of life on earth is so improbable as to be impossible, and he takes this as grounds for postulating an extra-terrestrial source. But his calculation is not a fair one. In the first place the probability that he calculates is the chance of the spontaneous production simultaneously of

2000 different working enzyme molecules, coordinated so as to form a living system, whereas he ought to have calculated the chance of the production of some immensely simpler system capable of crude evolution which could lead, through many stages, to present-day living things. In the second place, what he calculates is the chance of this event taking place at one specified place and at one specified moment of time. This is of no interest: what we are interested in is the chance of an event of this kind occurring at *any* place on the surface of the earth and at *any* time within several hundred million years. Either of these two factors is enough to invalidate his argument.

Crick approaches the question in a very different way. After a discussion with which I do not at all disagree, he reaches what I regard as the only possible conclusion at this time, which is one of total uncertainty. We have no idea at present if there was a primeval soup; we do not know what its composition was; if some primitive self-replicating system was set up we do not know its chemical nature; and the degree of uncertainty is such that both in Crick's view and in mine it may turn out, when knowledge has advanced further, either that this kind of origin of life was virtually impossible, or that it was virtually inevitable. Crick goes on, not unreasonably, to discuss what to think if the outcome were that it was virtually impossible, but my own reaction is to wait and see. There is certain to be great progress; study of replication processes in advanced living things will, I am sure, have a great spin-off in the direction of ideas about the origin of life, and much progress has already been made in laboratory studies of simple self-replicating systems (Orgel, 1979).

Consciousness

My final question was: can natural explanations be extended to mental processes, to the fact that we are conscious of our surroundings, our awareness, our sensation and volition? Darwin kept off Man altogether in *The Origin of Species* but the public was perfectly clear that his proposals for the evolution of animals, if accepted, would apply also to the human body. Likewise he kept off mental phenomena, not only in the *Origin* but, as I have already said, also in his later book *The Descent of Man*. But the public reaction to the *Origin* must reflect a general feeling that acceptance of Darwinism in relation to our bodies implied some sort of demotion of mental processes to a status where they were governed in a crude way by physical or chemical events in the brain. I will admit that I used to have this kind of feeling, but for a good many years I have recognized that the possibilities of a physical system as complex as the human brain are so vast that there is no need for a human being to feel that he is reduced in stature if his thoughts and feelings are governed by physical events in his brain cells. If you just look at a human brain it is a not very interesting lump of greyish material, but if you stain it to show the nervous elements and look at a tiny piece of it under the microscope, the complexity is quite fantastic: it is such a tangle that you cannot begin to make head or tail of it. If you use a procedure which selectively stains an occasional nerve cell, you find that it has hundreds of processes connecting it to other cells, that connections are made to it from hundreds of other cells, and that there are something like 10^{11} of these cells in the brain. Even a simple computer is capable of a performance which is impressive to humans, and the human brain is immensely more complex than any computer either existing or thought of at the present time. So I am no longer distressed at the idea that my thoughts and actions may be governed by physical events in this complex thing that we all have inside our skulls.

Also, I used to be upset at the idea of possibly not having free will, but it now seems to me that even if we do not have free will, the events which govern our movements are so unpredictable that there is no need to be worried about it. They are unpredictable many times over. Even if we knew the complete sequence of nucleic acids in our chromosomes, there would still be random irregularities of growth, so that the detailed structures of our brains would be unknown short of examining them in immense detail with the electron microscope, which could only be done after we are dead. A second level of uncertainty is that every individual has a different experience throughout his life; there is no possibility of keeping track of what is recorded in either his conscious or his unconscious memory. Yet another level is that the events recorded by electrophysiologists in nerve cells are irregular – they are affected by Brownian motion, and this gives an indeterminacy of response even without invoking indeterminacy at the quantum level. I used to say that there is yet another more basic level of uncertainty, in the impossibility of knowing the precise genetic make-up of an individual, but geneticists have said to me that there has been such rapid progress in sequencing nucleic acids that complete sequencing of an individual's genome may be practicable in the relatively near future. But still, the other three levels of unpredictability are quite enough to protect us against feeling constrained even if free will in the ordinary sense does not exist. These are matters of one's emotional attitudes: it seems to me that the disparaging aspects of a Darwinian approach to the brain disappear if one thinks of one's brain as an essential part of one's conscious self, rather than as something separate from the conscious self which limits its independence. One might go a step further and adopt a position stated very nicely in a letter by Howard Florey: 'I prefer to look on the brain as the instrument of the mind, and not on mind as a mere secretion of the cortical cells.' (Macfarlane, 1979).

What about the actual relationship of mental states to physical processes in the brain? This is the long-standing mind–body problem which has exercised philosophers, I suppose since philosophy began. *Prima facie* it seems to me that this is an insoluble problem. Present-day physics and chemistry do not contain even the ingredients that might be expected to produce a conscious experience. An extra-terrestrial super-intelligence could understand the existence of automata which behave in every way like human beings, composed of matter of the kind that we are familiar with, but he would suppose that these were unconscious automata. As a rule scientists say little or nothing about this problem, and many brush it completely under the carpet and do not even think about it. Darwinism as expressed in the *Origin* and the *Descent of Man*, relating only to bodily features, was quickly accepted, and it led to an optimism that there would soon be a complete understanding of Nature, including mental phenomena. A striking contrast to this was expressed by the Swiss physiologist du Bois-Reymond (1872), who recognized the kind of difficulties that I have just been speaking of; he went to the opposite extreme and stated that there were two problems which were insoluble and would remain so: first the ultimate nature of matter and force – I am not going to say anything about that – and the second was conscious awareness in the mind. Of these problems he used the word *ignorabimus*, and received much obloquy from fellow scientists for so saying. All scientists will admit that we do not at present know about these things, but he went further and said that we never shall know. He was right to the extent that now, more than a hundred years later, we are no further forward.

How will things be in another hundred years? I prefer to regard the present situation as a challenge rather than a defeat, though I readily admit that there are only very general grounds for optimism. Early last century any physico-chemical explanation of the ordinary processes of life must have seemed impossible; in many respects it is now complete through the immense triumphs of biochemistry. Early this century even Bateson (1916), the well-known geneticist, said 'The supposition that particles of chromatin, indistinguishable from each other and indeed almost homogeneous under any known test, can by their material nature confer all the properties of life surpasses the range of even the most convinced materialism'. These problems which seemed insoluble to advanced biologists of their time have yielded, or are yielding, to gradual progress. Evolution seemed equally insoluble until there came the sudden illumination by the papers of Darwin and Wallace. The mind–body problem has not yet made even gradual progress, nor has it yet had its Darwin, but these past successes give one hope.

A passage written by the physiologist J. S. Haldane in 1935 is an example of a commonly held opinion: 'When we attribute conscious experience to a living organism we do so for a very good reason, which is that the behaviour of the organism exhibits retrospect and foresight'. I regard that as both wrong and misleading. One can imagine an unconscious automaton, with the complexity of the brain of even a small mammal, behaving in an apparently intelligent way and exhibiting what would look like retrospect and foresight. Linking consciousness to these particular aspects of behaviour begs a very important question, since it implies that conscious mind, independent of its connection with the brain, is capable of independent rational processes, and has to be postulated in order that an animal may perform its movements in these more-or-less rational ways. I regard this as a totally open question. The extreme opposite position would be to say that conscious experience is a mere epiphenomenon, that we are all really automata, and somehow we are aware of what is being done within ourselves. I regard this equally as an open position. One argument against this position is that, on the whole, sensations which correspond to biologically advantageous situations are pleasant, as when we satisfy hunger or thirst or other instincts, while sensations which correspond to biologically damaging situations – pain resulting from injury, for example – are unpleasant. This situation would be explained by natural selection if the sensation itself were part of the chain of events leading to action; but if sensations were side effects, their quality would be irrelevant, and there would be no explanation for this degree of appropriateness of the quality of the sensation. I readily admit, however, that this is far from being a strong argument.

Very few biologists, even very few neurobiologists, who might be expected to devote themselves to these questions, have had the courage to express clear opinions on these topics. A notable exception, of course, is Sir John Eccles in his joint book with Sir Karl Popper (Popper & Eccles, 1977). Eccles comes out squarely in favour of a largely independent conscious mind that interacts with specified regions in the brain. According to him, the mind is capable of reading out the state of groups of nerve cells, so that it becomes aware of sensations brought to it through the sense organs and nervous system of the animal, and it also interacts in the opposite direction, influencing the activities of other groups of cells and thus the movements of the animal. Eccles's arguments leave me in an agnostic state, as my grandfather was left in relation to evolution until reading the Darwin–Wallace papers and *The Origin of Species*. It may be that my attitude simply reflects the conservatism of approaching old age, just as many of the older biologists of last century – Owen, Sedgwick and von Baer, for example – were never converted to Darwinism, but I do not believe that younger people have responded to Eccles's proposition in the way that their grandfathers or great-grandfathers responded to Darwin.

Can one see any way into this problem? I do not see a way through neurophysiology: however much we may know about what nerve cells are doing, I do not see how it can tell us how their activities relate to conscious phenomena, although clearly, of course, they do relate; nor do I see how progress can come through introspection. The only possibility that comes to mind is the study of paranormal phenomena, such as thought trans-ference. At present it seems to me that it is difficult to dismiss the numerous and apparently convincing reports of such phenomena, but such reports are anecdotal, they are sporadic, they are unpredictable, no one has yet got them into a condition where they are repeatable, and it has to be admitted that they have been muddied by fraud. If genuine, they certainly imply that mental processes have some degree of existence independent of brains and independent of the now-known properties of matter. I have no idea whether or not one should hope that someone will make a break through in this field. If it were achieved, it would be the greatest advance in human understanding ever made, but it would open alarming prospects of mind control.

Well, I have posed some questions; I have not answered them, but I am quite sure that none of you expected that I would.

References

Bateson, W. (1916). Review of *The Mechanism of Mendelian Heredity*,
    T. H. Morgan, A. H. Sturtevant, H. J. Muller & C. B. Bridges. *Science*, **44**,
    536–43.
Crick, F. (1981). *Life Itself*. New York: Simon & Schuster; London; Macdonald.
Darwin, C. (1859). (4th edn, 1866). *On the Origin of Species by means of Natural
    Selection*. London: John Murray.
Darwin, C. (1868). *The Variation of Animals and Plants under Domestication*.
    London: John Murray.
Darwin, C. (1871). *The Descent of Man and Selection in Relation to Sex*, p. 36 (2nd
    edn, 1874, p. 66). London: John Murray.

Darwin, C. (1880). Sir Wyville Thomson and Natural Selection. *Nature*, **23**, 32.

Darwin, F. (1888). *The Life and Letters of Charles Darwin*, vol. 3, p. 18. London: John Murray.

du Bois-Reymond, E. (1872). *Über die Grenzen des Naturerkennens*. Leipzig: Veit. Reprinted in *Reden von Emil du Bois-Reymond*, 2nd edn (1912), vol. 1, pp. 441–73. Leipzig: Veit.

Eldredge, N. & Gould, S. J. (1972). Punctuated equilibria: an alternative to phyletic gradualism. In *Models in Paleobiology*, ed. T. J. M. Schopf, pp. 82–115. San Francisco: Freeman, Cooper.

Fisher, R. A. (1930). *The Genetical Theory of Natural Selection*. Oxford: Clarendon Press.

Haldane, J. S. (1935). *The Philosophy of a Biologist*, p. 79. Oxford: Clarendon Press.

Hoyle, F. & Wickramasinghe, N. C. (1981). *Evolution from Space*. London: Dent.

Huxley, T. H. (1888). On the reception of the 'Origin of Species'. In *Life and Letters of Charles Darwin*, ed. F. Darwin, vol. 2, pp. 188, 196, 197. London: John Murray.

Kalmijn, Ad. J. (1974). The detection of electric fields from inanimate and animate sources other than electric organs. In *Handbook of Sensory Physiology*, vol. III/3, ed. A. Fessard, pp. 147–200. Berlin & New York: Springer-Verlag.

Lipson, H. S. (1980). A physicist looks at evolution. *Physics Bulletin*, **31**, 138.

Lissmann, H. W. (1951). Continuous electrical signals from the tail of a fish, *Gymnarchus niloticus* Cuv. *Nature*, **167**, 201–2.

Macfarlane, R. G. (1979). *Howard Florey. The Making of a Great Scientist*, p. 63. Oxford: Oxford University Press.

Mayr, E. & Provine, W. B. (1980). *The Evolutionary Synthesis*. Cambridge, Mass.: Harvard University Press.

Orgel, L. E. (1979). Selection *in vitro*. *Proceedings of the Royal Society of London, Series B*, **205**, 435–42.

Popper, K. R. & Eccles, J. C. (1977). *The Self and its Brain*, chapter E7. Berlin, New York & London: Springer.

Spencer, H. (1852). The development hypothesis. *The Leader* 20 March 1852. Reprinted in *Essays: Scientific, Political, and Speculative*, 1891, vol. i, pp. 1–7. London: Williams & Norgate.

EVOLUTIONARY HISTORY

2

Darwin, intellectual revolutionary

ERNST MAYR

We live in a time of short memories. Only few papers nowadays cited in the scientific literature are more than three years old. And many of the younger generation are regrettably unaware of the revolutionary contributions made by the great men of the past. One can hardly find a better illustration for this than the work of Charles Darwin. Nearly all of his great innovations have become to such an extent an integral component of Western thinking that only the historians appreciate Darwin's pioneering role.

We speak rather glibly of the Darwinian revolution, but if one challenges someone to specify precisely just what he means by this term, one will invariably get an answer that is, at best, incomplete but more often partly wrong. I admit that it is almost impossible for a modern person to project himself back to the early half of the last century, to reconstruct the thinking of the pre-Darwinian period, that is the framework of ideas that was so thoroughly destroyed by Darwin. What we must, therefore, do is to ask what were the widely or universally held views that were challenged by Darwin in 1859 in the *Origin?* I will endeavour to show that the intellectual revolution generated by Darwin produced a far more fundamental and far-reaching change in the thinking of the Western World than is usually appreciated. And I shall also make it clear that this was not merely an incidental by-product of some biological speculations, but that Darwin had a comprehensive programme, to review and, where necessary, to revolutionize traditional ideas. I shall show that Darwin was a bold and often quite radical revolutionary.

Let me preface my analysis by the comment that I shall try to avoid as much as possible analysing by what pathway Darwin reached his novel ideas, because such an analysis is presented by Dr Hodge in the next chapter. We must begin our analysis by asking whether there was a prevailing world view in the Western World prior to 1859. The answer unfortunately is negative. Indeed, there was perhaps no other period before 1800 and after 1860 in which the prevailing thinking in England, France, and Germany was as different from that of the other two countries as it was in the first half of the nineteenth century. Natural theology was still dominant in England, but already obsolete in France and Germany. The conservative and rational France of Cuvier, as well as the romantic Germany of the *Naturphilosophen* were worlds apart from the England of Lyell, Sedgwick, and John Herschel. Darwin's intellectual milieu, of course, was that of England and his arguments were directed against his peers and their thinking.

Let me now take up the major concepts or ideas which Darwin encountered and which he tried to modify or replace. By far the most comprehensive of these, in fact an all-pervading ideology, was that of creationism. By creationism is understood the belief that the world must be interpreted as reflecting the mind of the Creator. But physicists and naturalists interpreted the implementation of the mind of the Creator in different ways. The order and harmony of the universe made the physical scientists search for laws, for wise institutions in the running of the universe installed by the Creator. Everything in nature was caused, but the causes were secondary causes, regulated by the laws instituted by the primary cause, the Creator. To serve his Creator best a physicist studied His laws and their working. Most naturalists, by contrast, concentrated on the wonderful adaptations of living creatures. These could not be explained readily as the result of general laws such as those of gravity, heat, light, or movements. Nearly all the marvellous adaptations of living creatures are so unique that it seemed vacuous to claim that they were due to 'laws'. It rather seemed that these aspects of nature were so special and unique that they could be interpreted only as caused by the direct intervention of the Creator, by His specific design. Consequently the functioning of organisms, their instincts, and their manifold interactions provided the naturalist with abundant evidence for design, and seemed to constitute irrefutable proof for the existence of a Creator. Darwin, even though a naturalist, tried valiantly to apply the thinking of the physicists, by looking for laws.

It was the task of natural theology to study the design of creation, and natural theology was thus as much science as it was theology. The two endeavours, theology and science, were indeed inseparable. Consequently, most of the greater scientific works of this period, as exemplified by Lyell's *Principles of Geology* (1830–33) or Louis Agassiz's *Essay on Classification* (1857) were simultaneously treatises of natural theology. Science and theology were fused into a single system and, as is obvious with hindsight, there could not be any truly objective and uncommitted science until science and theology had been cleanly and completely divorced from each other. The publication of Darwin's *The Origin of Species* was to a greater extent responsible for this divorce than anything else. Although Darwin was not yet a complete agnostic in 1859, he argued throughout the *Origin* against creationism, the intrusion of theology into science. And when Darwin said 'This whole volume is one long argument', he was clearly referring to his endeavour to 'get theology out of science', as Gillespie (1979) has put it. In this Darwin was completely successful, and the historians accept the year 1859 as the end of creditable natural theology.

I used to believe that it was a loss of Christian faith that induced Darwin to seek a complete autonomy of science from theology (Mayr, 1977). However, the recent analyses of Gillespie (1979), Ospovat (1981), and others have persuaded me that scientific findings had been primarily responsible for Darwin's change of mind. In particular it was Darwin's realization of the invalidity of three prominent doctrines among the numerous beliefs of creationism, that was of crucial importance for Darwin's change of mind:

(1) that of an unchanging world of short duration,
(2) that of the constancy and sharp delimitation of created species, and
(3) that of a perfect world explicable only by the postulate of an omnipotent and beneficent Creator.

As we shall presently discuss, Darwin's conclusions were reinforced by his conviction of the invalidity of several other tenets of creationism, such as the special creation of Man, but the three stated dogmas were clearly of primary importance.

Even though Bishop Ussher's calculation that the world had been created as recently as 4004 years B.C. was still widely accepted among the pious, the men of science had long become aware of the great age of the earth. The researches of the geologists, in particular, left no doubt of the immense age of the earth, thus providing all the time needed for abundant organic evolution. Lamarck was the first to draw the necessary conclusions from this.

Among the various discoveries of geology the one that was most important and most disturbing for the creationist was the discovery of extinction. Already in the eighteenth century Blumenbach and others had accepted extinction of formerly existing types like ammonites, belemnites, and trilobites, and of entire faunas, but it was not until Cuvier had worked out the extinction of a whole sequence of mammalian faunas in the Tertiary of the Paris Basin that the acceptance of extinction became inevitable. The ultimate proof for it was the discovery of fossil mastodons and mammoths, living survivors of which could not have possibly remained undiscovered in some remote part of the globe.

Three solutions were offered for the explanation of extinction. According to Lamarck no organism ever became extinct, there simply was such drastic transformation that the formerly existing types had changed beyond recognition. According to the progressionists, like Hugh Miller, Murchison, and Louis Agassiz, each former fauna had become extinct as a whole and was replaced by a newly created more progressive fauna. Such catastrophism was unpalatable to Charles Lyell, who produced a third theory consistent with his uniformitarianism. He believed that individual species became extinct one by one as conditions were changing and that the gaps thus created in nature were filled by the introduction of new species. It was an attempt at a reconciliation between the recognition of a changing world of long duration and the tenets of creationism.

The question of how these new species were 'introduced' was left unanswered by Lyell. He bequeathed this problem to Darwin for whom in due time it became the most important research programme. Darwin thus approached the problem of evolution in an entirely different manner from Lamarck. For Lamarck evolution was a strictly vertical phenomenon, more or less a temporalization of the *scala naturae*, proceeding in a single dimension, that of time. Evolution for him was a movement from less perfect to more perfect, an endeavour to establish a continuity among the major types of organisms, from the most primitive infusorians up to the mammals and Man. Lamarck's *Philosophie Zoologique* was the paradigm of what I designate as vertical evolutionism. Species played no role in Lamarck's thinking. New species originated all the time by spontaneous generation from inanimate matter, but this produced only the simplest infusorians. Each such newly established evolutionary line gradually moved up to ever greater perfection. Lamarck rightly called his work a 'philosophy', because he did not present testable scientific theories.

Darwin was unable to build on this foundation but rather started from the fundamental question that Lyell had bequeathed to him. Although Lyell had appealed to 'intermediate causes' as the source of the new species, his description of the process was that of special creation. 'Species may have been created in succession at such times and at such places as to enable them to multiply and endure for an appointed period and occupy an appointed space on the globe'. For Lyell each creation was a carefully planned event. The reason why Lyell, like Henslow, Sedgwick, and all the others of Darwin's scientific friends and correspondents in the middle of the 1830s, accepted the unalterable constancy of species, was ultimately a philosophical one. The constancy of species was the one piece of the old dogma of a created world that remained inviolate after the concepts of the recency and constancy of the physical world had been abandoned.

Another ideology which prevented Lyell from recognizing the change of species was *essentialism*, which had dominated Western thinking for more than two thousand years after Plato. According to the essentialist, the changing variety of things in nature is a reflection of a limited number of constant and sharply delimited underlying *eide*, or essences. Variation is merely the manifestation of imperfect reflections of the constant essences. For Lyell all nature consisted of constant types, each created at a definite time. 'There are fixed limits beyond which the descendants from common parents can never deviate from a certain type'. And he added emphatically: 'It is idle...to dispute about the abstract possibility of the conversion of one species into another, when there are known causes, so much more active in their nature, which must always intervene and prevent the actual accomplishment of such conversions' (Lyell, 1835 III: 162). For an essentialist there can be no evolution, there can only be a sudden origin of a new essence by a major mutation or saltation.

To the historian it is quite evident that no genuine and testable theory of evolution could develop until the possibility was accepted that species have the capacity to change, to become transformed into new species, and to multiply into several species. For Darwin to accept this possibility required a fundamental break with Lyell's thinking. The question which we must ask ourselves and which Dr Hodge discusses in detail is, how Darwin was able to emancipate himself from Lyell's thinking, and what observations or conceptual changes permitted Darwin to adopt the theory of a transforming capacity of species.

As Darwin tells us in his autobiography, he encountered many phenomena during his visit to South America on the *Beagle* which any modern biologist would unhesitatingly explain as clear evidence for evolution. Furthermore, when sorting his collections on the homeward voyage, his observations in the Galapagos Islands made him pen the memorable sentences (approximately July 1836): 'When I see these islands in sight of each other and possessed of but a scanty stock of animals, tenanted by these birds but slightly differing in structure and filling the same place in nature, I must suspect they are varieties...if there is the slightest foundation for these remarks the zoology of the archipelagoes will be well worth examining: for such facts would undermine the stability of species' (Barlow, 1963). Yet, it is evident that Darwin at that date had not yet consciously abandoned the concept of constant species. This Darwin apparently did in two stages. The discovery of a second, smaller, species of *Rhea* (South American ostrich), led him to the theory, consistent with essentialism, that an existing species could give rise to a new species, by a sudden leap. Such an origin of new species had been postulated, scores of times before (Osborn, 1894) from the Greeks to Robinet and Maupertuis. Typological new origins, however, are not evolution (Mayr, 1982: 301–42). The diagnostic criterion of evolutionary transformation is gradualness. The concept of gradualism, the second step in Darwin's conversion, was, apparently, first adopted by Darwin, as Sulloway (1982) has demonstrated so convincingly, when the ornithologist John Gould, who prepared the scientific report on Darwin's bird collections, pointed out to him that there were three different endemic species of mockingbirds on three different islands in the Galapagos. Darwin had thought they were only varieties.

The mockingbird episode was of particular importance to Darwin for two reasons. The Galapagos endemics were quite similar to a species of mockingbirds on the South American mainland and clearly derived from it. What was important was that the Galapagos birds were not the result of a single saltation, as Darwin had postulated for the new species of *Rhea* in Patagonia, but had gradually evolved into three different although very similar species on three different islands. This fact helped to convert Darwin to the concept of gradual evolution. Even more important was the fact that these three different species had branched off from a single parental species, the mainland mockingbird, an observation which gave Darwin the solution to the problem of the multiplication of species.

The problem of the introduction of new species posed by Lyell was thus solved by Darwin. New species can originate by what we now call geographical or allopatric speciation. This theory of speciation says that new species may originate by the gradual genetic transformation of isolated populations. By this thought Darwin founded a branch of evolutionism which, for short, we might designate as horizontal evolutionism in contrast to the strictly vertical evolutionism of Lamarck. The two kinds of evolutionism deal with two entirely different aspects of evolution even though the processes responsible for these aspects proceed simultaneously. Vertical evolutionism deals with adaptive changes in the time dimension, while horizontal evolutionism deals with the origin of new diversity, that is with the origin of new populations, incipient species, and new species, which enrich the diversity of the organic world and which are the potential founders of new evolutionary departures, of new higher taxa, of the occupants of new adaptive zones.

From 1837 on, when Darwin first recognized and solved the problem of the origin of diversity, this duality of the evolutionary process has been with us. Unfortunately, there were only few authors with the breadth of thought and experience of Darwin to deal simultaneously with both aspects of evolution. As it was, palaeontologists and geneticists concentrated on or devoted themselves exclusively to vertical evolution, while the majority of the naturalists studied the origin of diversity as reflected in the process of speciation and the origin of higher taxa.

The first author to have proposed geographic speciation was the geologist L. von Buch (1825) in a short statement which he failed to develop any further. M. Wagner (1841) and Alfred Russel Wallace (1855) also proposed it independently of Darwin. The discovery of the divergence of contemporary, geographically isolated, populations made it possible to incorporate the origin of organic diversity within the compass of evolution. For Darwin this was of crucial importance, because horizontal thinking permitted the solution of three important evolutionary problems.

(1) The problem of the multiplication of species.

(2) The resolution of the seeming conflict between the observed discontinuities in nature and the concept of gradual evolution.

(3) The problem of the evolution of the higher taxa owing to common descent.

Perhaps the most decisive consequence of the discovery of geographic speciation was that it led Darwin automatically to a branching concept of evolution. This is why branching entered Darwin's notebooks at such an early stage.

How are relatives connected?

For those who accepted the concept of the *scala naturae*, and in the eighteenth century this included most naturalists to a lesser or greater extent, all organisms were part of a single linear scale of ever-growing perfection. Lamarck still adhered, in principle, to this concept even though he allowed for some branching in his classification of the major phyla. Pallas and others had also published branching diagrams but it required the categorical rejection of the *scala naturae* by Cuvier in the first and second decades of the nineteenth century before the need for a new way to represent organic diversity became crucial. The Quinarians experimented with indicating relationship by osculating circles, but their diagrams did not fit reality at all well. The archetypes of Owen and the Naturphilosophen strengthened the recognition of discrete groups in nature but the use of the term affinity in relation to these groups remained meaningless prior to the acceptance of the theory of evolution.

Apparently very soon after Darwin had understood that a single species of South American mockingbird had given rise to three daughter species in the Galapagos Islands, he seemed to have realized that such a process of multiplication of species, combined with their continuing divergence, could give rise in due time to different genera and still higher categories. The members of a higher taxon then would be united by descent from a common ancestor. The best way to represent such common descent would be a branching diagram. Already in the summer of 1837 Darwin clearly stated that 'organized beings represent an irregularly branched tree' (*Notebook B*, 21) and he drew several tree diagrams in which he even distinguished living and extinct species by different symbols. By the time he wrote the *Origin* the theory of common descent had become the backbone of his evolutionary theory, not surprisingly so because it had extraordinary explanatory powers. Indeed the manifestations of common descent as revealed by comparative anatomy, comparative embryology, systematics ('natural system'), and biogeography became the main evidence for the occurrence of evolution in the years after 1859. Reciprocally the stated biological disciplines, which up to 1859 had been primarily descriptive, now became causal sciences, with common descent providing explanation for nearly everything that had previously been puzzling.

In these studies the comparative method played an important role. To be sure, the practitioners of idealistic morphology and the Naturphilosophen had also practised comparison with excellent results. But the archetypes which they had reconstructed had no causal explanation until they were reinterpreted by Darwin as reflecting the putative common ancestors.

The theory of common descent, once proposed, is so simple and so obvious that it is hard to believe that Darwin was the first to have adopted it consistently. Its importance was not only that it had such great explanatory powers but also that it provided for the living world a unity which had been previously missing. Up to 1859 one had been impressed primarily by the enormous diversity from the lowest plants to the highest vertebrates, but this diversity took on an entirely different complexion when it was realized that it all could be traced back to a common origin.

The final proof of this, of course, was not supplied until our time, when it was demonstrated that even the prokaryotes have the same genetic code as animals and plants.

Perhaps the most important consequence of the theory of common descent was the change in the position of Man. For theologians and philosophers alike Man was a creature apart from all other living nature. Aristotle, Descartes, and Kant agreed in this, no matter how much they disagreed in other aspects of their philosophies. Darwin, in the *Origin*, confined himself to the cautiously cryptic remark 'light will be thrown on the origin of Man and his history' (p. 488). But Ernst Haeckel, T. H. Huxley, and in 1871 Darwin himself, demonstrated conclusively that Man must have evolved from an ape-like ancestor, thus putting him right into the phylogenetic tree of the animal kingdom. This was the end of the traditional anthropocentrism of the Bible and of the philosophers. To be sure, the claim that 'man is nothing but an animal' was quite rightly rejected by all the more perceptive students of Man, yet one cannot question that Darwin was responsible for a fundamental reevaluation of the nature of Man and his role in the universe.

Gradualism

Darwin's solution for the problem of the multiplication of species and his discovery of the theory of common descent were accompanied by a number of other conceptual shifts. The most important one was his partial abandoning of essentialism in favour of gradualism and population thinking. The literature on the history of evolutionary biology lists numerous authors, stated to have been forerunners of Darwin by having adopted evolutionism. When more closely examined nearly all of these claims turn out to be invalid. Changes in the living world were ascribed to new origins by these earlier authors. But the sudden origin of a new species or still higher taxon is not evolution. Indeed, as has been rightly said by Reiser, if one is an essentialist, one cannot conceive of gradual evolution. Since an essence is constant and sharply delimited against other essences, it cannot possibly evolve. Change for an essentialist occurs only by the introduction of new essences. This was precisely Charles Lyell's interpretation of the introduction of new species throughout geological history, or Darwin's early explanation of the origin of the lesser ostrich of Patagonia.

The observation of nature seemed to give powerful support to the essentialist's claims. Wherever one looked one saw discontinuities, between species, between genera, between orders and even higher taxa. Quite naturally, the gaps between the higher taxa, like birds and mammals, or beetles and butterflies, were mentioned particularly often by Darwin's critics. And yet, as far back as Aristotle and his principle of plenitude (Lovejoy, 1936) there had been an opposing trend. It was expressed in the *scala naturae*, and even such an arch-essentialist as Linnaeus stated that the orders of plants were touching each other like countries on a map.

Nevertheless Lamarck was the first person to apply the principle of gradualism to explain the changes in organic life that could be inferred from the geological record. But there is no evidence that Darwin derived his gradualistic thinking from Lamarck. References to gradual changes are scattered through Darwin's notebooks from an early time (Kohn, 1980). Yet it is still somewhat uncertain what the exact sources of Darwin's gradualistic thinking were. One of the intellectual sources surely was Lyell's uniformitarianism, which Darwin had adopted quite early. There is also the fact that Darwin considered the changes of organisms either to be produced directly by the environment or to be at least an answer to the changes in the environment. Hence 'the changes in species must be very slow, owing to physical changes slow...' (*Notebook C*, 17). Gradualness was also favoured by Darwin's conclusion that changes in habit or behaviour may precede changes in structure (*Notebook C*, 57, 199). At that time Darwin still believed in a principle called Yarrell's Law, according to which it takes many generations of impact for the effects of the environment or of use and disuse to become strongly hereditary. As Darwin stated 'Variety when long in blood, gets stronger and stronger' (*Notebook C*, 136). Various other sources for Darwin's gradualistic thinking have been suggested in the recent literature, such as the writings of J. B. Sumner (Gruber, 1974: 125), or Leibniz's principle of plenitude (Stanley, 1981), but to me it seems more likely that Darwin had arrived at his gradualism empirically.

At least three observations may have been influential. First, the slightness of the differences among the mockingbird populations on the three Galapagos islands and the South American mainland, as well as a similarly slight difference among many varieties and species of animals. Second, the barnacle researches, where Darwin complained constantly to what great extent species and varieties were intergrading. And third, Darwin's work with the races of domestic pigeons, where he convinced himself that even the most extreme races which, if found in nature, would be unhesitatingly placed by taxonomists in different genera, were nevertheless the product of painstaking, long-continued, gradual, artificial selection. In his *Essay* of 1844, Darwin argues in favour of gradual evolution by analogy with what is found in domesticated animals and plants. And he postulates therefore that 'there must have existed intermediate forms between all the species of the same group, not differing more than recognized varieties differ' (p. 157). The abundant evidence which Darwin adduces in support of gradualism was to a considerable extent instrumental in the weakening and eventual refutation of essentialism. It was only by adopting a theory of gradual evolution that Darwin was able to escape from the constraints of essentialism.

Natural selection

The concept for which Darwin is better known than for any other is that of natural selection. When we speak of Darwinism today we mean evolution by natural selection. It is a concept of extraordinary complexity and Dr Hodge discusses in the next chapter how Darwin found the various components and how he pieced them together. Even though later in life Darwin always claimed that the analogy with the artificial selection practised by animal and plant breeders had given him the idea for the principle of natural selection, this later recollection of Darwin's is refuted by his notebook entry when he read Malthus on 28 September 1838.

One question that is still controversial today is whether Darwin came to the idea of natural selection exclusively on the basis of his natural history studies or whether he also used general contemporary ideas and applied them to natural populations. Social historians, in particular, have claimed that the socio-economic conditions of contemporary England and the social theories of Malthus had been the main source of Darwin's inspiration. To be sure, Darwin was not a recluse and he read widely in the contemporary literature; he surely was fully aware of what went on around him. Yet his notebooks demonstrate conclusively that the major impact on his thinking was his realization of the uniqueness of individuals, of the superfecundity of reproducing individuals, and of the struggle for existence among animals and plants. All these are natural history concepts. This I consider to be the case also for the competition between European settlers and primitive natives. Furthermore, if natural selection had really been the logical consequence of contemporary sociological and economic thought, everybody should have been able to see at once the logic of this new theory and adopted it speedily. Actually, except for a few naturalists, natural selection was almost universally rejected by Darwin's contemporaries. I consider all this as powerful indication that the theory of natural selection was primarily a natural history theory and not a socio-economic theory, as has been claimed by some social historians.

The process of natural selection, as clearly recognized by the modern evolutionist, is a two-step process: the first is the production in each generation of new, genetically different, individuals; and the second is the survival and successful reproduction of a minimal fraction of the numerous individuals produced by reproduction of the previous generation.

Of these two steps Darwin understood the second, the actual selection, far better than the first. Selection was a necessity because natural populations are stable, natural resources likewise are stable and limited, but all species have an enormous reproductive surplus. Therefore, on 28 September 1838, when he encountered a particular sentence in Malthus's essay, it set in motion an intellectual avalanche in Darwin's mind. The sentence read 'It may safely be pronounced, therefore, that population, when unchecked, goes on doubling itself every 25 years, or increases in a geometrical ratio'. This sentence gave an entirely new meaning to the concept of 'struggle for existence', a concept well known since the eighteenth century. It made Darwin understand two important facts, first that the struggle for existence is not between species but between individuals, and second that the struggle is far fiercer than either he or any of his forerunners had ever previously imagined. If Darwin at that time had still been a typologist, the fact that only one in a hundred, or one in a thousand, or one in a million, of the offspring of a parent survived, would have been irrelevant. If they were all merely reflections of the same essence it would not matter which one survived. It required the realization that each individual is uniquely different from every other one, a conceptualization which we now refer to as population thinking, before the realization of the intensity of the struggle for existence could be appreciated as important. And it was this new insight which led Darwin to his important conclusion that individuals that differ from each other in their attributes must also differ in their capacity for survival. As Darwin stated in his *Essay* of 1844 (p. 119): 'How can it be doubted from the struggle each individual has, to obtain subsistence that any minute variation in structure, habits, or instincts, adapting that individual better to the new conditions, would tell upon its vigour and health?' Even when exposed to the same struggle for existence different individuals have a different chance to survive and to reproduce. And this partly deterministic, partly probabilistic, process is the basic meaning of the term *natural selection*.

For many of the earlier evolutionists selection was mortality selection. It was a case of 'survival of the fittest'. Darwin himself saw things more clearly and specifically mentioned 'success in leaving progeny' (*Origin*, p. 62) as part of natural selection.

Since the rise of genetics it has become important to emphasize that the phenotype of the individual as a whole is the target of selection. The emphasis on single genes in the work of most mathematical population geneticists and their definition of evolution as 'a change in gene frequencies' has led to the unfortunate misunderstanding by certain outsiders, that the selection of individual genes is the basic thesis of neo-Darwinism. It is not!

The phenotype of the individual is the result of developmental processes and these therefore contribute importantly to the fitness of an individual. The praiseworthy recent emphasis on the importance of developmental processes does not however imply a refutation of Darwinism as has been claimed by some recent authors.

As perceptive as Darwin's analysis of the second step in the process of natural selection was, as confused was he about the first step, the production of the variation which is the raw material of natural selection. Let us remember that Darwin's major ideas on variation were formed between 1837 and 1858, in other words in a period well before the rise of genetics, before an understanding of nucleus and chromosomes, and well before anyone had made a distinction between genotype and phenotype. Not surprisingly, in spite of his continuing preoccupation with variation, it was a subject that utterly frustrated Darwin. He frequently changed his mind, often contradicted himself, and ultimately admitted that he did not really understand what caused variation and what its real nature was.

When one reads the nonsense about variation and evolution written by de Vries, Bateson, and Johannsen, the first biologists who did understand genetics, one will certainly have to judge Darwin's groping in the dark rather charitably. But there was one thing the naturalist and pigeon breeder Darwin understood very well, and it was the one thing that is most important for the student of evolution, the fact that abundant variation is always available. Hence there is always an opportunity for natural selection to act.

Variation and its causation is a subject on which more has been learned since the publication of the *Origin* than on any other aspect of evolution. And much of this book is devoted to reports on these findings and on the still-open frontiers. Most important for the evolutionist are perhaps two facts. The first is that each new genotype is the product of recombination and that therefore the factors responsible for recombination, and more generally for the production of new genetic systems, are of far greater importance to the evolutionist than mutation, even though admittedly, ultimately all genetic changes are due to mutation. This is why it was so misleading when the early geneticists attributed evolution to mutation pressure. The second fact is that each new individual is the product of a long and complicated process beginning with crossing-over and the reassortment of chromosomes during meiosis, and the random meeting of gametes during fertilization, to mention only a few of the series of steps, all of them with a strong stochastic component. What is often forgotten is that evolution is not a smooth continuous process, but consists of the formation of a brand-new gene pool in every generation, and the extraction from this gene pool of unique individuals, which are then exposed to the probabilistic process of natural selection. It is this remarkable mixture of chance and necessity which makes natural selection so different from anything philosophers had considered previously, and which makes it evident why the question of whether evolution is due to chance or necessity is so misleading.

The fundamental importance of the theory of natural selection when proposed in 1858/59 was that it challenged two of the most universally held ideologies of the period, that of cosmic teleology, and that of natural theology. According to the teleologists the world either was established toward an end or was still moving toward an objective, either by the guiding hand of a Creator or by secondary causes, that is, by laws that were guiding the course of events toward an ultimate goal. All this was categorically denied by Darwin, for whom each evolutionary change was a singular event controlled by the temporary constellation of selection forces. Indeed Darwin even stated that there would not be any change at all, when there was no change in the environment. A non-deterministic process like natural selection was quite unintelligible for any philosopher thinking in terms of Newtonian laws, and this is why Herschel referred contemptuously to natural selection as the law of the higgledy-piggledy.

Devastating as the denial of teleology was for many of Darwin's contemporaries such as Sedgwick and von Baer, the denial of design was even more sweeping. To explain all the beautiful adaptations of organisms, their adjustment to each other, their well-organized interdependence, and indeed the whole harmony of nature, as the result of such a capricious process as natural selection, was quite unacceptable to almost all of Darwin's contemporaries. The theory of natural selection amounted to the proposal to replace the hand of the Creator by a purely material and mechanical process, at that by one not deterministic and not predictive. As one critic put it, it dethroned God. Although an accommodation between religion and Darwinism was eventually reached, their mutual relation had first to go through a rather traumatic period. The current renaissance of the creationist movement shows that this relationship is still a precarious one.

Critical dates

An historian likes to cite dates for events in intellectual history as much as for political history. When one of the pioneers of intellectual history, like Darwin, introduces revolutionary new ideas, one would like to pinpoint the moment when he formulated the new concepts. The fixing of a definite date often helps in the determination of the factors that were instrumental in the development of the new idea. In line with this reasoning we have come to say that Darwin 'became an evolutionist' in March 1837 when Gould persuaded him that the varieties of mockingbirds on the Galapagos were good species, and we say that Darwin discovered the principle of natural selection on 28 September 1838 when reading Malthus. In both cases this dating is strongly reinforced by assertions made by Darwin himself.

A thorough study of the literature has convinced me that a major shift, virtually amounting to a conversion, did indeed occur on the mentioned dates. However, in both cases the crucial new information that brought about the conversion struck a 'prepared mind'. Darwin had already been vaguely thinking of a non-Lyellian change in species sometime before March 1837, a thought foreshadowed by his famous 'islands in sight of each other' statement, penned a year earlier while sailing from the Cape of Good Hope to Ascension (Sulloway, 1982) and by his saltational theory of the origin of Darwin's *Rhea*.

The same is true for natural selection; although clearly 28 September 1838 was decisive, yet Darwin, even prior to this date, made in his notebooks a number of 'proto-selectionist' statements, as Kohn (1980) calls them. Most interesting among these is Darwin's reference to sexual selection (*Notebook C*, 61). 'Whether species may not be made by a little more vigour being given to the chance offspring who have any slight peculiarity of structure. Hence seals take victorious seals, hence deer victorious deer, hence males armed and pugnacious all order; cocks all war-like...' And after the date of the Malthus episode his understanding of natural selection continued to mature (Hodge, this volume).

To repeat, even though the shifts both to gradual evolutionism and to natural selection were clearly rather sudden and drastic processes, nevertheless there are indications for a prior erosion of opposing views, and a 'preparing of the mind' for the final, decisive shift.

Darwin's uncertainties

Darwin, like all great pioneers, removed old uncertainties and created new ones. He was unable to find satisfying solutions for several of the problems he tried to solve. Some of the questions he posed are still unsolved or at least still the subject of controversy. As I have mentioned already, there was no other subject on which he was as uncertain as he was on variation. The science of genetics has removed some of these uncertainties, but the recent discovery of different kinds of DNA has reopened the subject.

Uncertainty still exists concerning speciation. At first Darwin was entirely convinced of the importance of isolation, even though apparently for the wrong reasons. Later, particularly after he thought that varieties of plants were as much incipient species as were varieties of animals, he downgraded the importance of isolation and attempted to replace it by his principle of character divergence. As the current writings on sympatric speciation, parapatric speciation, and stasipatric speciation show, the question of the importance of geographical isolation is still controversial today.

Darwin often commented on the steady action of natural selection and it is usually assumed that he favoured a steady evolutionary progression. Actually, Darwin was fully aware of the potential for evolutionary stasis. In his *Essay* of 1844 he said 'There is no necessity of modification in a species, when it reaches a new and isolated country. If it be able to survive and if slight variations better adapted to the new conditions are not selected, it might retain...its old form for an indefinite time' (p. 202). He also said 'As we see some species at present adapted to a wide range of conditions, so we may suppose that such species would survive unchanged and unexterminated for a long time' (p. 165).

Darwin was remarkably well aware that the size of a population might be an important evolutionary factor. However, since he considered variation to be an intermittent phenomenon, occurring mostly under special circumstances, he thought that the larger a population was, the greater the chance that it would produce a variation that would be favoured by selection. This opinion was reinforced by his observation that species from large continents were usually competitively superior to species from restricted areas. The genetic discoveries of the last 80 years make it possible to claim by contradistinction that it is precisely very small founder populations that have the best opportunity to initiate new evolutionary departures. However, this also is still a controversial area in which the last word has not yet been said.

Again and again when we examine a modern controversy carefully we find that the problem was already known to Darwin and that he had already made tentative suggestions as to its solution. There seem to have been no limits to the fertility of his mind.

Darwin's life work

Darwin would be remembered as an outstanding scientist even if he had never written a word about evolution. Indeed J. B. S. Haldane has gone so far as to say: 'In my opinion, Darwin's most original contribution to biology is not the theory of evolution, but his great series of books on experimental botany published in the latter part of his life' (Haldane, 1959; 358). This achievement is little known among non-biologists, nor is Darwin's equally outstanding work on the adaptation of flowers and on animal psychology, or his competent work on the barnacles and his imaginative work on earthworms. In all these areas Darwin was a pioneer, and although in some areas it took more than half a century before others continued to build on the foundations which Darwin had laid, it is now clear that he had attacked important problems with extraordinary originality, thereby becoming the founder of several now well-recognized separate disciplines. Darwin was the first to work out a sound theory of classification, one which is still adopted by the majority of taxonomists. His approach to biogeography, in which so much emphasis was placed on the behaviour and the ecology of organisms as factors of distribution, is much closer to modern biogeography than the purely descriptive–geographical approach that dominated biogeography for more than half a century after Darwin's death.

This volume illustrates that Darwin was a pioneer also in other areas: Professor Hull reports on Darwin's impact on philosophy. I have already mentioned Darwin's contribution to the end of anthropomorphism, and to the recognition of population thinking. It was clearly Darwin who established the philosophical foundations of historical biology, a branch of science which in my opinion is as scientific as are the physical sciences, even though dominated by rather different methods of discovery and explanation from those considered valid and necessary in the physical sciences. This includes limits to prediction, the importance of the observational–comparative method (as compared to the experimental one), the role of historical narratives, and much else. I am mentioning this only in order to illustrate the extraordinary richness and originality of Darwin's thought and the impact of his writings right up to the present day.

Conclusion

The students of the history of ideas are quite unanimous that the Darwinian revolution was the most fundamental of all intellectual revolutions in the history of mankind. While such revolutions as those brought about by Copernicus, Newton, Lavoisier, or Einstein affected only one particular branch of science, or the methodology of science as such, the Darwinian revolution affected every thinking man. A world view developed by anyone after 1859 was by necessity quite different from any world view formed prior to 1859. It is therefore eminently proper that we pay tribute to the memory of this great man. But how can we explain Darwin's greatness?

That he was a genius is hardly any longer questioned, some of his earlier detractors notwithstanding. But there must have been a score of other biologists of equal intelligence who failed to match Darwin's achievement. What then is it that distinguishes Darwin from all the others?

Perhaps we can answer this question by investigating what kind of a scientist Darwin was. As he himself has said, he was first and foremost a naturalist. He was a splendid observer and like all other naturalists he was interested in organic diversity and in adaptation. Naturalists are, on the whole, describers and particularists, but Darwin was also a great theoretician, something only very few naturalists have ever been. In that respect Darwin resembles much more some of the leading physical scientists. But Darwin differed from the run-of-the-mill naturalists also in another way. He was not only an observer but he was also a gifted and indefatigable experimenter whenever he dealt with a problem, the solution of which could be advanced by an experiment.

I think this suggests some of the sources of Darwin's greatness. The universality of his talents and interests had preadapted him to become a bridge builder between fields. It enabled him to use his background as a naturalist to theorize about some of the most challenging problems with which man's curiosity is faced. And contrary to widespread beliefs, Darwin was utterly bold in his theorizing.

There may be other ingredients to Darwin's greatness, but those that are apparent are these:

A brilliant mind, great intellectual boldness, and an ability to combine the best qualities of a naturalist-observer, of a philosophical theoretician, and of an experimentalist. The world has so far seen such a combination only once and this accounts for Darwin's unique greatness.

Further reading

Maynard Smith, J. (1975). *The theory of evolution.* 3rd edn. Harmondsworth: Penguin Books.
Mayr, E. (1977). Darwin and natural selection. *American Scientist,* **65**, 321–7.
Mayr, E. (1982). *The growth of biological thought.* Cambridge, Mass.: Harvard University Press.

References

Barlow, N. (ed.) (1963). Darwin's Ornithological Notes. *Bulletin British Museum (Natural History),* History Series, **2**, 201–78.
Bowler, P. J. (1976). *Fossils and Progress.* New York: Science History Publications.
Buch, L. von (1825). *Physicalische Beschreibung der Canarischen Inseln,* pp. 132–3. Berlin: Kgl. Akad. Wiss.
Darwin, C. (1844). Essay. In *The Foundations of the Origin of Species,* ed. F. Darwin (1909). Cambridge: Cambridge University Press.
Darwin, C. (1859). *On the Origin of Species,* 1964 facsimile edition, ed. E. Mayr. Cambridge: Harvard University Press.
Gillespie, N. C. (1979). *Charles Darwin and the Problem of Creation.* Chicago and London: University of Chicago Press.
Gruber, H. E. (1974). *Darwin on Man.* New York: Dutton.
Haldane, J. B. S. (1959). An Indian perspective of Darwin. *Centennial Review of Arts and Sciences, Michigan State University,* **3**, 357.
Kohn, D. (1980). Theories to work by: rejected theories, reproduction, and Darwin's path to natural selection. *Studies in the History of Biology,* **4**, 67–170.
Lamarck, J.-B. (1809). *Philosophie Zoologique.* Paris.
Lovejoy, A. O. (1936). *The Great Chain of Being.* Cambridge: Harvard University Press.
Lyell, C. (1830–1833). *Principles of Geology,* 3 vols. London: John Murray.
Lyell, C. (1835). *Principles of Geology,* 4th edn, III, 162. London: John Murray.
Mayr, E. (1977). Darwin and natural selection. *American Scientist,* **65**, 321–7.
Osborn, H. F. (1894). *From the Greeks to Darwin.* New York: Columbia University Press.
Ospovat, D. (1981). *The development of Darwin's theory.* Cambridge: Cambridge University Press.
Stanley, S. M. (1981). *The New Evolutionary Timetable.* New York: Basic Books.
Sulloway, F. J. (1982). Darwin and his finches: the evolution of a legend. *Journal of the History of Biology,* **15**, 1–53.
Wagner, M. (1841). *Reisen in der Regentschaft Algier in den Jahren 1836, 1837 und 1838.* Leipzig: Leopold Voss.
Wallace, A. R. (1855). On the law which has regulated the introduction of new species. *Annals and Magazine of Natural History,* ser. 2, **16**, 184–96.

3

The development of Darwin's general biological theorizing

M. J. S. HODGE

Charles Darwin (1809–1882) was only twenty seven when he returned, in 1836, from the five-year voyage of HMS *Beagle*. What is more, within the next five years he reached all those conclusions in general biological theory that his books were to expound three decades later. For, by 1841, he had worked out not only his theory of the origin of species, *natural selection*, but also, it seems, his theory of generation (or reproduction, including heredity, variation and so on), *pangenesis*.

This chapter surveys the development of his general biological theorizing over that remarkable early period. The analysis draws throughout on the work done in the last decade by Gruber (1974), Herbert (1974, 1977), Ghiselin (1975), Ruse (1975a, b; 1979), Schweber (1977, 1980), Kottler (1978), Manier (1978), Sulloway (1979, 1982a, b), Kohn (1980), Ospovat (1981), and Sloan (1983a, b) and is derived from four studies by the present writer (Hodge, 1982, 1983a, b; Kohn & Hodge, 1983) where full reference is made to the documentary sources and secondary literature.

Such a survey can serve more than mere biographical curiosity, and a final section will suggest how it may clarify some issues of current interest to historians, to philosophers and to biologists.

It can also free us from many mistaken myths about Darwin himself. These myths mostly trace to his own misleading reminiscences later in life, and have been relentlessly reaffirmed since, at the 1959 centennial symposia for example and in the 1978 BBC-TV series on Darwin; but they are nonetheless discredited by the scholarly industry now grown up around the rich manuscript archive from Darwin's early years (Kohn, 1983).

One is the romantic, really Wordsworthian, individualist myth so dear to the literary guardians of English national cultural stereotypes. It depicts the young Darwin as a lone, sporting gentleman, an amateur beetle-collector seeing nature as she really is by simply looking with the clear gaze of genius, unimpeded by any scientific training, theological prejudice, professional ambition and so on. Another is the Whiggish, anachronistic myth that Darwin's general biological thought consists of a molecule comprising just two atoms: the idea of evolution and the idea of natural selection. It depicts his early intellectual development as reducing to two moments of discovery, whereby he moves from having no coherent ideas to having just those ideas.

Fortunately, there is a single antidote effective against both these myths; and that is to start all over again with the most decisive source of Darwin's new identity, on the voyage, as a committed man of science: his zealous discipleship of Charles Lyell's (1797–1875) views in geology (including biogeography and ecology).

This antidote is effective against the romantic–individualist myth, because, as a protégé of Lyell, the young Darwin of the *Beagle* is at once invested with all the intellectual and institutional context that that myth would suppress. It is effective against the evolution-plus-natural-selection myth, because it forces us to reconstruct the narrative of his subsequent theorizing not as so many unknowing steps towards his final positions, but as so many deliberate departures from positions initially shared with his mentor.

Darwin as protégé of Lyell (1834–7)

Darwin's acceptance of Lyell's views was complete by mid-1834 when he sailed around the Horn. Having now studied all three volumes of the *Principles of Geology* (1830–3), he was applying Lyell's entire system, physical and organic worlds alike, to South America and beyond that the whole earth. Henceforth this system provided the framework for his preoccupation with the problems of the extinction and origin of species.

All the causes of change are presumed, in Lyell's system, to persist undiminished into the present, the human period, and on into the future. Now, as at all times, habitable dry land is being destroyed by subsidence and erosion in some regions, while it is being produced by sediment consolidation, lava eruption and elevationary earthquake action in others. Equally, Lyell has the long succession of faunas and floras brought about by a continual, one-by-one extinction of species and their replacement by new ones.

The epistemological rationale for his presumption, of the persistence of all such causes of change into the present and future, is the ideal of explanation by real or existing causes, *verae causae*. Like his friend (and the undergraduate Darwin's scientific hero) the physicist John Herschel (1792–1871), Lyell followed earlier writers, most notably the Scottish philosopher Thomas Reid (1710–1796), who had drawn this moral from the superior evidential credentials of the Newtonian gravitational force over the Cartesian ethereal vortices: any causes invoked in an explanatory theory should, ideally, be known to exist through direct observation independently of the facts they are supposed to explain. That force, unlike those vortices, was a well-evidenced explanation for the planets' orbits because, the argument went, the orbits themselves were not the sole evidence for its existence. It was a real not a conjectural cause; for it was known to exist from observations of swinging pendulums and falling stones down here on earth.

Lyell's system was, therefore, to exemplify an epistemological analogy. In geology, only causes active in the present, human period are accessible in principle, although often not in practice, to direct observation. So, in this science, the present is to the past as the terrestrial is to the celestial in Newtonian physical astronomy.

Such, then, was the context, at once systematic and epistemological wherein Darwin, from 1834 on, was theorizing about species extinctions and origins. His own thinking over the next three years was developed, accordingly, through successive disagreements with Lyell's views on the organic world, while he continued to accept his mentor's teaching on the physical world of land, sea and climate changes.

Lyell had made adaptational considerations alone completely decisive in determining the timing and placing of both species extinctions and species origins. Any species must eventually become extinct; for changes in local conditions will sooner or later allow other species better adapted to the changed conditions to invade and conquer in the struggle for existence. As for species origins, Lyell had argued against spontaneous generation and against new species arising by the modification ('transmutation') of older ones. But, in offering no positive alternative account, he had left the means whereby new species originate quite mysterious if not miraculous. He had held explicitly, however, that when and where any given species is created is determined by the conditions it needs to flourish. Conversely, the character and so the supraspecific group membership of the species that have originated in any area is determined by conditions there. So, on Lyell's account, if two areas are very similar in conditions they will have congeneric or cofamilial endemic species; if they are very dissimilar, then their endemic species will be of distinct families or orders.

Now, by early spring 1837, Darwin had decided that such purely adaptational explanations could not account adequately for the timing and placing of either the extinctions of old species or the origins of new ones. And he was emphasizing, in his *Red Notebook*, the parallel in the explanatory inadequacy of adaptation in the two cases (Herbert, 1980).

Already, in 1835, he had concluded that several species of large mammals, formerly flourishing on the eastern plains of South America, had later become extinct while no change in conditions physical or ecological had occurred. And he had adopted another theory of extinction, discussed but rejected by Lyell, wherein species are like individuals and die of old age. A species, as a succession of organisms produced sexually, might have only a limited total lifetime, Darwin argued; just as a succession of apple trees propagated by grafts was supposed to last only so long before degenerating and dying as if it were merely the extension of a single limited life.

As for the origin of species, reflection on many palaeontological and biogeographical facts, including those established by the expert judgments of Richard Owen and John Gould on the voyage material, had convinced Darwin, by early spring 1837, that the close similarity seen between any congeneric endemic species in two areas was not always explicable as a common adaptation to common conditions; for often the two sets of species were endemic in areas with very different climate, soil and so on.

However, if more recent species could descend from earlier ones, ancestry could explain what adaptation could not. Why did the species on younger land – the Galapagos islands or the southern Patagonian plains, for instance – often resemble closely other species on the nearest older land? Because they were descended from them, many sometimes descending from a single ancestral species. Thus did Darwin conclude that resemblances between species are often not due to adaptation but to inheritance from common ancestors; while differences are often adaptive and are due to differing, multiple divergences from those common ancestors. It is not, as he saw it, that species are not exquisitely adapted to their respective places in the economy of nature. They are and any theory of species origins must explain why. It is simply that adaptation and ancestry can explain what adaptation alone can not. On such grounds Darwin had decided, by that spring, for transmutation and common descent.

Always the bold 'philosopher' as much as the cautious 'naturalist', he soon went far beyond these disagreements with Lyell. He did so in a new sequence of theoretical notebooks, *B–E* (1837–9) and *M–N* (1838–9) – *M* for 'metaphysics', meaning mind, man, materialism, morals and so on. (See a note at the end of this chapter on editions of these texts.) He did so, moreover, in two ways that were consciously conditioned from the start by his immediate context and resources in 1837 as a biological theorist.

First, from way back in his Edinburgh days and his apprenticeship to Robert Grant in invertebrate zoology, he had been much pre-occupied with comprehensive generalizations about sexual and asexual modes of generation (Sloan, 1983*a*). His extinction theorizing had thus been developed, since 1835, at the first intersection of that old pre-occupation with his newer devotion to Lyellian geology.

Now, in the summer of 1837, he would understand the origins no less than the extinctions of species through appropriate comparisons and contrasts between sexual and asexual generation. Having read again a book he had admired when studying at Edinburgh, a book much concerned with precisely such comparisons and contrasts – his grandfather Erasmus Darwin's *Zoönomia* (1794–6) – he was soon taking its title for the opening heading of his *Notebook B*, where he was now to pursue his own inquiry into 'the laws of life'. He was soon making explicit, too, a fundamental teleological analogy, in which changing conditions are to sexual generation as sexual crossing is to asexual generation. Although it can extend an individual life, continued grafting eventually brings death without issue; and likewise sexual generation in unchanging conditions brings death, extinction of the species, without issue. With sexual crossing, however, an individual whose life has been extended by grafting, although not enabled to go on itself, can give rise to a new individual with a new lease on life. Likewise, then, in changing conditions, sexual generation is the providential means whereby a species can change into another, a new species adapted to the new conditions, so avoiding the extinction without issue that would otherwise occur.

Second, Lyell had outlined very fully – and of course rejected – the comprehensive transmutationist system of biology developed by Jean Lamarck, praising the French zoologist for his courage in extending his repugnant ideas to our own species, man. Darwin accordingly opened his *Notebook B*, in July 1837, with an integrated sequence of entries whose twenty-seven pages of argumentation were to match the structure of Lyell's exposition of Lamarck. Darwin, like Lamarck – as presented (and misrepresented) by Lyell – would now trace any degrees of difference, no matter how wide, to long-run divergences from common stocks. And he would trace any higher levels of bodily organization and mental faculties, explicitly including those in man, to long-run progress from remote starting points in the simplest organisms of all, infusorian monads.

With these decisions Darwin became more than the protégé of Lyell. He would be ever hereafter his own successor in developing sequels to the steps he had taken in the spring and summer of 1837.

The new programme pursued (1837–8)

Darwin himself often reflected in his notebooks on his new programme's presuppositions concerning God, nature, man and science.

God was, for Darwin then, still the traditional good and wise creator, but one never working in so many separate acts of miraculous interference, always through the natural consequences of a few initial enactments of general causal laws: as with planetary orbits and the law of gravitation, so, Darwin insisted, with species origins and the laws of generation. As for man, he is a species produced like any other, lawfully, his mental faculties the causal consequences of his bodily organization and not miraculously superadded; while science is a quest for lawful causes that are evidenced both directly and independently of the many, diverse facts they can explain, and indirectly and dependently, by the very multitude and diversity of those facts.

The place in the programme of his generational theory of species origins Darwin understood through the analogy of a tree of life. By the autumn of 1837, he had developed this analogy so as to understand all long-run trends in diversification and progress through an arboriform extrapolation, on a changing but stable Lyellian earth's surface, of successive species propagations; these being analogous to the successive bud propagations whereby any tree grows, with many buds ending without branching, in species extinctions, while other buds branch without ending, in species multiplications.

His species propagation theorizing itself was accordingly constructed, from the very opening of *Notebook B*, as an argument that starts with the sexual generation of one individual organism from another and ends with the propagation of one species from another. Moreover, this species propagation is ultimately made possible by the two features that Darwin sees distinguishing sexual from asexual generation: namely, maturation in the offspring and the interaction of two parents in their production. Thanks to the impressionability of immature organization, hereditary adaptive variation accompanies sexual generation in changing conditions. But how then can any species be constant in character across its entire range? Because crossing with the blending of parental characters keeps the species constant as long as the conditions are constant overall, only changing temporarily and locally. Conversely, then, a new variety can be

formed if this conservative action of crossing is circumvented by reproductive isolation of a few individuals in new conditions, whether that
isolation arises with or without geographical segregation.

And how may this variety formation proceed to new species formation?
As Darwin knew, the usual criteria for specific rather than mere varietal
distinctions were those that Swedes and Italians do not satisfy, but lions
and tigers do: namely, true breeding, lack of intermediate forms and
unwillingness or inability to produce fertile hybrids on crossing. So, the
final stage in Darwin's argumentation concerns how a species meeting
those three criteria would eventually arise with prolonged isolation and
divergence.

Over the next year, through the summer of 1838, it is the variety and
species formation stages of this argument that Darwin develops most fully
and explicitly.

The character gaps between good species he explains by continued
divergence and by the extinction of intermediate varieties. The true
breeding he explains by the law that in crossing the characters of an older
domestic race dominate those of a younger one. This dominance shows,
Darwin reasons, that older characters are more permanent, more deeply
embedded in the hereditary constitution and so more resistant to the
influence either of mates in crossing or of changing external conditions.
Such permanent divergences in character have often arisen in the distinct
races of a domesticated species, and in the wild they would be accompanied
by an unwillingness or inability to interbreed; for there, Darwin argues,
the reproductive system, with the associated instincts, is not disrupted as
it is in domesticated species.

So, for Darwin, in the early summer of 1838, the races of domesticated
species are providing positive analogies, especially for the final, species
formation, steps in his argument. But at the other, the opening end of that
argument, concerning the initiation and transmission of adaptive change,
they are a source of contrasts and not of comparison. Thus he says that
many varieties of domesticated species, fancy pigeon breeds for example,
are monstrous not adaptive; they can only be maintained by artificial
feeding and breeding, including selective breeding; they are quite unlike
wild, natural and adaptive varieties and even more unlike wild species.

Now, as Darwin sees it, at this time, for a change to be an adaptation
it must be more than merely hereditary and advantageous; it must be
necessary rather than accidental, elicited, that is, by the very conditions
that make it advantageous, as albinism sometimes seems to be by cold. And
it must be acquired by the whole of the race faced with those conditions,

so as not to be lost in crossing. Then, as with reproductive isolation, changed habits can initiate permanent adaptive changes in structure, especially in higher animals by entailing changes in the conditions of foetal maturation.

All these developments, in Darwin's comprehensive species propagation argumentation, are reinforced and not rejected in further developments, consummated in mid-September, 1838, in his *Notebook D* conjectures about sexual and asexual generation, more particularly about sexual buds (or ova) and asexual buds.

A sexual bud or ovum has started life, he supposes, as a bud from the mother when she was herself a newly fertilized egg in her mother, the grandmother. Why then does it not eventually become an exact facsimile of that mother? Because, he answers, of two lots of differences. During its maturation it is subject to conditions not exactly like those the mother matured in; and at fertilization it is acted on by a mate with a constitution unlike its own and unlike its mother's.

The whole object of sex is to have unlike acting on unlike so as to make possible the production of unlike offspring, thus allowing for adaptive change, Darwin argues. But where does the constitutional unlikeness of the two mates come from? Well, since each has the same ancestry, their constitutional difference must trace to the cumulative influence of different conditions in the two lines of descent from those ancestors. So, sexual generation is the means whereby the past influences of changing conditions can be accumulated and combined with present ones to ensure continued variability. If conditions change only locally and temporarily then only individual differences will result, but, Darwin concludes, if conditions change overall and permanently then a new species will eventually be formed.

Darwin is pleased with this novel analysis of how sex enables new species to arise from old. But he admits that it leaves as mysterious as in 1837, how the maturation of the individual fertilized ovum ensures that the changes be adaptive in each generation. He can only conjecture that additional maturational innovations will not become hereditary unless they harmonize with the previous ones that are already being recapitulated in maturation. Adaptive innovations could thus be separated, as the ones that are hereditarily transmitted, from the maladaptive, as the ones that are not; although Darwin notes that hereditary diseases show that this separation is often fallible.

Such is Darwin's species origin theorizing in mid-September 1838. It compounds still further those two legacies so actively conjoined since the previous summer of 1837: the historical, biogeographical (including ecological) concerns that he had inherited from Lyell, and the generational concerns deriving from his study with Grant and subsequent reading in Erasmus Darwin.

Consider, then, what geography and generation together have done for Darwin's understanding of the problems of organic diversity and the origin of species. First, his explanation of resemblances and differences among species contrasts directly with any developmental stage theorizing as found most famously in Robert Chambers, Herbert Spencer or Karl Marx. A developmental stage theory refers similarities and differences to more-or-less-equal advances made from the lowest point on a universal scale of advance. Thus Chambers explains the similarities between Old and New World monkeys as due to life having advanced to the same level of organization from quite independent origins of life in the two hemispheres (Hodge, 1972).

By contrast, for Darwin as a genealogical (ancestry and descent) theorist of historical biogeography, resemblances and differences are not traced to developmental advances, but to ramifying migrations and adaptive divergences from common ancestors that are more or less remote from their diverse descendants in time, place and character.

Now, Darwin at this time is explicitly taking each organism's ontogeny to recapitulate its phylogeny. But his very assumption that many descendant species may diverge from a single common ancestry precludes his construing descendants as grown-up ancestors and ancestors as descendants in embryo; for any given immature, embryonic life already has but one determinate mature adult future; if a puppy as a dog, if a tadpole as a frog.

It is, then, because he is explaining differences and resemblances as he is that Darwin, in 1838, needs a theory of purely opportunistic adaptive change in changing conditions, a theory making no developmentalist assumption as to a preferred direction that life will take provided it can go on at all. And his thinking is indeed knowingly premised on the assumption that in the complete absence of change in conditions there would be no changes in organization; so that whatever different changes in organization have occurred in the many lines descending from some common stock are due to differences in the conditions in those lines. Darwin, as a genealogical biogeographer, thinking horizontally as well as vertically, to use Ernst Mayr's (1982) terms, would explain change generationally but not developmentally.

This integration of horizontal, genealogical, geographical and

generational constraints is also the source for what Ernst Mayr (1982)
calls Darwin's populationist thinking about species. For it has led Darwin
to think of each species as spreading out into varying conditions, over a
range, over time. And only the extension of a species, not its intension, as
philosophers say – only the collective membership of a species, not the
properties earning the members their membership in it – can have a
geographical range and a historical career. Now, recall his account of
individual adaptations to conditions as arising in individual maturations
in sexual generation; and recall his account of hereditary variation as
embedded constitutionally through successive individual matings. And
then it is clear how Darwin has come to be thinking of a species as a
population of individuals, with each member differing from every other
because arising in a unique sequence of influences exerted by a unique
succession of conditions and mates.

The origins of natural selection and pangenesis (1838–42)

Darwin's mid-September 1838 theorizing, as developed in his *Notebook D*,
provided the immediate context for his most celebrated innovation of all:
the theory of natural selection. For – contrary to the legend that it was all
thought through in a day – this theory was worked out in three main stages
over the next half year.

The first stage involved only the opening steps in Darwin's overall
argumentation from individual generation to species formation. For it sees
him changing his mind about the adaptiveness of structural change rather
than about species formation as such.

This first stage did come on reading Thomas Malthus's (1766–1834)
Essay on the Principle of Population, on 28 September 1838; after a
generational conjecture – that higher animal foetuses are initially
hermaphrodite – had led him to Adolphe Quetelet's findings on the sex ratio
at birth and so, it seems, to Malthus as an author linked by a reviewer with
the Belgian social statistician (Schweber, 1977). Now, what Darwin's
reflections on Malthus did was to move him away from a prenatal,
maturational sorting of adaptive from maladaptive variation. For he moved
at once to a post-natal, ecological sorting wherein individual adaptive
variants are retained, while the maladaptive are eliminated in the Malthusian crush of population.

These initial reflections did not include, then, any analogy between
artificially and naturally selective breeding. But Darwin did draw an
explicit analogy between this teleology – of population pressure and

sorting, as ensuring adaptation of plant and animal structure to changing conditions – and Malthus's theistic teleology of superfecundity, as ensuring the energetic dispersal of ancient tribes beyond the original Asian seat of the human species. Man, naturally slothful according to Malthus, only spread into more adverse climes thanks to the local scarcities of food entailed by his excess fertility; with later settlers always eventually victorious because made doubly energetic in struggling with both rigorous conditions and previous settlers.

Alien species beating natives on their home ground had been decisive for Lyell's species extinction theorising, where the analogy with the European human conquests over American Indians was explicit. Colonizing species of alien genera doing likewise had long been decisive for Darwin's species origin theorizing, where the analogy with the European human conquests over Australasian natives was hardly less explicit. Throughout 1838, Darwin had been allowing for Lyellian competitive defeats to extinguish some species before their predetermined ageing overcame them. Now, his Malthusian reflections prompted both a new ecological under-standing of adaptive sorting and a final return to Lyellian ecology for all extinctions. Thus did ecological explanations regain ground earlier lost to generational ones.

The second stage began around the end of November 1838. For Darwin then drew his first explicit analogies between adaptation in wild species and the fitness for human ends of domestic races, as both due to selective breeding. To make this new comparison Darwin had to drop his old contrast between monstrous, selected, domestic races and adaptive, not selected, wild species. He dropped it first, it seems, on considering sporting-dog breeds, notably greyhounds; for these, although formed by the human artifice of selective breeding, had been given both structures and instincts useful in the wild. (Hence the carnivorous canines in Darwin's later thought-experiments on natural selection.) Up to now, domestic race formation had provided Darwin only with analogies for the formation of wild species as ancient, true breeding and intersterile races. With this new selective breeding analogy, domestic races, as adaptations, are also providing analogies for the formation of wild species as ancient and perfectly adapted races.

The third stage came seemingly early in 1839, when this new analogy prompted a further revision to the opening steps of the overall. argu-mentation. In artificial selection, the chosen end, for which the race is being fitted, does not elicit the variation being selected. So, Darwin reasoned, natural selection likewise could work with variation that is accidentally rather than necessarily adaptive. If changes in conditions disrupt the precise replication of parental characters so as to yield hereditary variation, then, providing only that some of it happens to be adaptive, this will suffice in the long run for selection as a cause of adaptive species formations.

With these developments beyond his mid-September 1838 positions, Darwin had reached the theory of natural selection much as he would publish it later. In March 1839, he even outlined the argumentative structure that opens his *Sketch* of 1842 (F. Darwin, 1909), the first, manuscript, version of *The Origin of Species* (1859).

By 1841, he had very probably worked out, also, his later theory of individual organism generation: pangenesis. The theory, as eventually set forth (1868), was constituted by two theses: that the generative material comes from all over the parent body or bodies, and that it consists of minute 'gemmules' budded off from every part.

Darwin's preoccupation with generation went back to Edinburgh and to consequent inquiry, when on the voyage, into modes of reproduction common to various invertebrate groups (Sloan, 1983*a*). In early 1837, he had supported the senescence analogy between sexual and asexual generation by interpreting all generation as division, whether artificial or natural, complete or incomplete, simultaneous or successive.

This thesis, that all generation is division, continued throughout the *B*, *C* and *D* notebooks and was reinforced by an explicit equating, in September 1838, of division and gemmation or budding. The equation made all generation, sexual or otherwise, a form of budding. However, in his theorizing about sexual generation, as being unique in allowing variation and adaptation, Darwin made a fundamental distinction between sexual buds, such as ova, and asexual buds. According to this distinction, all and only sexual buds are involved in maturation and fertilization. Consequently all and only sexual buds are impressionable, whether by the action of changed conditions or by a sexual element from a mate of unlike constitution. Conversely, while lacking those powers an asexual bud, or even a severed flatworm fragment, can do what an unfertilized ovum can not: namely, produce a whole organism without interactive collaboration with any other part; this power being credited by Darwin to the presence in such a bud or fragment, indeed in any healing flesh, of material determining growth for all the parts of the whole organism.

Switching now from 1838 to 1868, one sees, in direct contrast with this earlier view, that a principal object of pangenesis is to explain how in all generation, sexual and asexual, the powers are the same and so, too, the material. To this end, Darwin adduces, most especially, those phenomena that would be anomalous for any exclusive correlation of maturation, fertilization and impressionability with sexual rather than asexual modes of generation. Thus aphid parthenogenesis shows an unfertilized ovum producing a maturing organism with no prior interaction with a male element; again, so-called graft hybrids and the effects of pollen on non-germinal tissue in a female plant both show impressionability without fertilization and maturation; while sporting and reversion in asexual plant buds show variation without fertilization or maturation.

So, as contrasted with his September 1838 position, Darwin sees pangenesis as an evening up, on both sides, of all the powers of the sexual and asexual parts of any organism. Pangenesis is itself presented as a theory of how this identity of powers arises in development. In a developing organism, starting as a fertilized egg, each cell or tissue, throughout its own maturation, is budding off miniature facsimiles of itself, the 'gemmules'. The whole organism can then have, as an adult, the same powers in its asexual buds and its sexual organs, because the same material is there: namely, 'gemmules' collected from all over the body.

These gemmules have, accordingly, been invested by Darwin with two sorts of properties: those credited to every asexual part of the body, in 1838, to explain its generative and regenerative powers, and those invoked then to explain the impressionability and variability of immature ova. So, pangenesis could have been derived from the 1838 position, by pandynamic extension to the ova of powers previously denied to them, and by a panovulational extension to all other parts of powers and matters formerly reserved for the ovary.

But is that, in fact, how Darwin arrived at pangenesis? No known document confirms that it was. But all the indirect evidence, including the records of his reading in such writers on generation as Erasmus Darwin, Johannes Müller and Giorgio Gallesio, makes it most probable that he came to pangenesis in such a revision of his 1838 position, and that he did so in the years 1840–1.

However and whenever it was first formulated, this pangenetic reduction of every mode of generation to micro-ovulo-gemmation could take inheritance, in so far as it was completely conservative, to be effected by an exact replication of a whole in all its parts; so that variation, reversion and so on are explicable as disturbances, suspensions and complications of that fundamental replicative tendency. Thus could pangenesis unify all Darwin's generation theorizing from gemmules to organisms and on to species, and beyond them the whole tree of life.

Pangenesis was never, however, to meet the *vera causa* evidential ideal, as Darwin himself was keenly aware, and he published it only after natural selection had been launched unencumbered by any such conjectural causation for generation.

The argument of the 1842 *Sketch* (and so, too, of the *Origin*) was knowingly structured to accord with that *vera causa* ideal (Hodge, 1977). The existence of natural selection as a causal agency is evidenced first; then its adequacy to produce new species from old is argued by analogy from the ability of artificial selection, although much less precise and prolonged, to produce domestic races. Finally, Darwin displays the indirect evidence for the theory: the explanations it provides for a wide array of facts in biogeography, geology, embryology and so on.

With the 1842 *Sketch* written, Darwin's most creative period was ended. That year he moved out of London to the Kent countryside and was henceforth mainly writing books, raising children and nursing his health. He never had another fundamentally novel idea in general biological theory. But then he had already had enough to keep him and many others occupied for a very long time.

The nature of Darwin's science

Even this brief analysis of Darwin's biological theorizing in these early years suggests the following conclusions about the nature of his science in its formative phase.

The social context. The traditional way to connect Darwin's science with his society is through Malthus's *laissez faire* political economy. But this cannot be adequate. A capitalism connection is relevant in the 1840s and 1850s when Darwin is applying division of labour theory to the problems of divergence (Schweber, 1980; Ospovat, 1981). But in 1838 it was Malthus's theodicy of ancient empires, not his political economy of the modern state that bore decisively on Darwin's biogeography and ecology (Bowler, 1976). More generally, any *laissez faire* connection helps very little in understanding how Darwin came to take up the problems his theorizing was to solve. Here it is his family, especially his grandfather, and his mentor Lyell, that indicate connections with movements of thought directly linked with fundamental social change. Erasmus Darwin, a Birmingham Lunar Society man, belonged to a provincial movement of 'radical' dissent from national and metropolitan orthodoxies in politics and religion. Lyell was a 'liberal' Scots Whig very much in the *Edinburgh Review* tradition of John Playfair and his own father-in-law Leonard Horner, a tradition bent on using electoral and educational, including university, reforms to break the national and metropolitan hegemony of the Tory, Oxonian, Anglican establishment. It is in such mediating contexts, rather than in any direct tie to capital through *laissez faire*, that the social history of Darwin's science should be sought.

Laws, causes and chances. To Darwin, natural selection, as a causal agency and lawful process, was akin, in its *vera causa* credentials, to gravitational force in celestial mechanics. But his earliest critics often judged his theorizing not to match the standards set by Newtonian physics (Hull, 1973). Was Darwin, then, mistaken in so relating his own science to Newton's?

He was, in so far as he underestimated the implications of one major disanalogy: he had no law that was to natural selection as the Newtonian inverse square law (with proportionality to mass products) was to gravitational attraction. For this disanalogy arose from another: Darwin's natural selection as a lawful process was complex, being compounded from heredity, variation and superfecundity, each of those processes having its own laws; while Newton's gravitational force was not compound and had a single law of its own.

Now, from these disanalogies arose a further one. In simple cases, the consequences in certain conditions of Newtonian gravitational attraction could be deduced and the adequacy of this cause for certain phenomena, most notably elliptical planetary orbits, thereby established. By contrast, the consequences and so competences, and hence the adequacies or inadequacies, of natural selection, especially for long-run effects, were practically impossible to decide, at least for a finite intellect; although the young Darwin himself could consistently suppose that God in choosing this means for adapting life to a changing earth had foreseen all its consequences.

However, although these disanalogies were fundamental, Darwin's theorizing had not taken him out of the causal, lawful, deterministic Newtonian universe, into one as irreducibly acausal and absolutely probabilistic as is sometimes thought implicit in quantum mechanics.

Any natural selection involves differential reproductive success that is nonfortuitous because determined by the way variant organisms are interacting with an environment that is causally sensitive to those particular physical differences in the organisms. If more red than green members of a species of moth living on green foliage are being killed by predators, that may be selection; but only if the predators are not colour-blind; if they are it is random sampling error. But even when killed by colour-blind predators the moths are not victims of any capricious cosmic indeterminacy; for each of their deaths is a causally determined event. It is only as deaths of red or green moths in that environment that these are fortuitous events; this fortuitousness being, as philosophers say, description relative.

Again, as Darwin saw it, the chanciness of hereditary variation is relative, not absolute. The processes generating this variation he supposed to be lawful, causal and so determinate; but not causally sensitive to environmental conditions in such a way that any particular change in conditions elicits an increased supply of those variants that are adaptive in the new conditions.

So, the distinctions decisive for Darwin's account of the generation and the fate of variation are distinctions drawn within the presuppositions of a deterministic universe. In its dependence on those presuppositions his biology was more like statistical than either celestial or quantum mechanics (Hull, 1974).

The non-tautologousness of natural selection. Darwin's *vera causa* argumentation shows that natural selection is not tautologous for one reason, ultimately. The definitional question, of what natural selection is, can be answered by specifying necessary and sufficient conditions for its occurrence, namely hereditary variation that is causally relevant to reproductive success thanks to organism–environment interactions; it can, then, be answered without begging in advance of empirical inquiry all those further questions as to whether these conditions are ever met: whether, that is, any natural selection exists; and, if so, how it is distributed, what it can do now and what it has been responsible for in the past.

It is, then, a mistake to defend natural selection against the tautology objection by proposing criteria of fitness independent of reproductive success. Fitness differences are best understood as reproductive expectancy differences analogous to normalized life expectancy differences. And as such they have no causal or explanatory power of their own. If Jones has outlived Smith this cannot be explained by showing that he earlier had the higher life expectancy, and then arguing that this duly caused him to live the longer life. It can be explained by citing earlier physical differences; perhaps Jones jogged while Smith smoked. Likewise, in natural selection it is physical differences, not the differences in reproductive expectancies estimated from them, that can cause and can explain subsequent reproductive performance differences.

As differing expectancies, fitnesses can neither cause reproductive performance differences nor be definitionally equated with them. So, natural selection is no untestable tautology, but not because fitness measurements, as expectancy estimates, can sometimes be falsified by later performance measurements. Natural selection is no tautology, because there is no *a priori* proof, from its definition alone, for its existence, nor then for its prevalence, or its adequacy or its responsibility for evolution. If there were, then the last decade and a half of selectionist–neutralist controversies over all these different non-definitional issues could have been settled in advance without recourse to empirical data, in an armchair with a scientist's glossary and a logician's truth table. As Michael Ruse observes (1981) if selectionism is a tautology, neutralism is a contradiction. Darwin's strategy in structuring his argumentation to conform to the *vera causa* ideal shows why it is not.

Evolution, cytology, genetics and the unification of nineteenth-century biological theory. Early in 1900, the doyen of Columbia University biology, E. B. Wilson, introduced the new second edition of his treatise, *The Cell*, by arguing that the main challenge then facing biological science was to integrate its two greatest achievements in the past century: the theory of evolution and the theory of cells. And he explained why he saw August Weismann's theory of the continuity of the germ plasm as the most promising foundation for any such integration.

Wilson's position makes sense of a great deal in the history of general biological theory before and since 1900. It has often been said that the decisive development since Darwin was a new synthesis, in the 1920s and 1930s, of Mendel on heredity and Darwin on selection. Wilson brings out the importance of that earlier and no less fundamental post-Darwinian synthesis, which this later one presupposed: the synthesis of evolution and cytology. Since Wilson's teachings were a principal inspiration for the Morgan school, the twentieth century science of genetics may be said to have arisen within his Weismannist programme for the integration of evolutionary and cytological theorizing (Mayr, 1982).

Darwin's integration of evolutionary and physiological biology had been attempted, in the 1840s, through pangenesis. But pangenesis was not reconcilable with the cytological consensus just emerging when Darwin published it a quarter of a century later. Physiologists were then increasingly agreeing that every cellular organism is either a single cell or a cell colony arising from the successive divisions of a single cell, and that two cells come together to form one at fertilization, each having arisen by the division of one cell in the respective parent body.

In Darwin's version pangenesis could not be squared with these cytological generalizations; for, if each of the two masses of gemmules coming together at fertilization is taken to be one cell, then it has not arisen in the division of one cell in that parent; while, if each is taken to be a myriad of cells, then far too many are coming together at fertilization.

The problem of transforming Darwin's pangenetic theory to square it with cytology in general, and with Weismann's theory in particular, was taken up most systematically by De Vries; and later developments leading to the theory of the gene were to owe much to his solution: 'intracellular pangenesis'. So, Darwin's attempt to integrate evolutionary and physiological biology contributed indirectly to the unification that was called for in Wilson's 1900 programme and that genetics was to secure. That Darwin's ideas could have such manifold influence throughout the entire structure of modern biological theory should not now be surprising. Compulsive, selfconscious intellectual, 'philosopher' no less than 'naturalist', he had worked from his earliest years on a very broad canvas indeed.

A Note about Darwin's early notebooks

Darwin's *B–E* notebooks were published by De Beer, Rowlands & Skramovsky (1960–7) and again, in computer print-out form, by Barrett (1972). The *M–N* notebooks, in Barrett's (1974) edition, are available in Gruber (1974). All these six notebooks, together with the *Red Notebook* (1836–7), *Notebook A* (1837–8), the *Torn Apart Notebooks* (1839–42) and the *Questions and Experiments Notebook* (1839–44) are to be published, in 1983, in colour-film form, and, in 1984, in a complete printed text edited by P. H. Barrett, P. J. Gautrey, S. Herbert, D. Kohn and S. Smith. The publisher of the film and printed text will be the Cambridge University Library in association with the British Museum (Natural History).

Further reading

Hodge, M. J. S. (1983). *Darwin and the Theory of Natural Selection: Roles for Epistemological and Methodological Ideals in a Scientific Innovation*. Boston and Dordrecht: D. Reidel.

Kohn, D. (ed.) (1983). *The Darwinian Heritage: A Centennial Retrospect*. Princeton: Princeton University Press.

Mayr, E. (1982). *The Growth of Biological Thought. Diversity, Evolution and Inheritance*. Cambridge, Mass.: Harvard University Press.

Ruse, M. (1979). *The Darwinian Revolution*. Chicago: University of Chicago Press.

References

Barrett, P. H. (1972). *Computerized Print-out: Darwin's transmutation notebooks – B, C, D, E.* Lansing, Michigan: Michigan State University.

Barrett, P. H. (1974). Darwin's Early and Unpublished Notebooks. In *Darwin on Man*, ed. H. E. Gruber. New York: E. P. Dutton.

Bowler, P. J. (1976). Malthus, Darwin and the concept of struggle. *Journal of the History of Ideas*, **37**, 631–50.

Darwin, C. (1859). *On the Origin of Species by Means of Natural Selection.* London: John Murray.

Darwin, C. (1868). *The Variation of Animals and Plants under Domestication.* 2 vols. London: John Murray.

Darwin, E. (1794–6). *Zoönomia*, 2 vols., London: Johnson.

Darwin, F. (1909). *The Foundations of the Origin of Species: Two Essays written in 1842 and 1844 by Charles Darwin.* Cambridge: Cambridge University Press.

De Beer, G., Rowlands, M. J. & Skramovsky, B. M. (1960–7). Darwin's notebooks on the transmutation of species. *Bulletin of the British Museum (Natural History). Historical Series*, **2**, 27–200; **3**, 129–76.

Ghiselin, M. (1975). The rationale of pangenesis. *Genetics*, **79**, 47–57.

Gruber, H. E. (1974) *Darwin on Man.* New York: E. P. Dutton.

Herbert, S. (1974). The place of man in the development of Darwin's theory of transmutation. Part I. To July 1837. *Journal of the History of Biology*, **7**, 217–58.

Herbert, S. (1977). The place of man in the development of Darwin's theory of transmutation. Part 2. *Journal of the History of Biology*. **10**, 155–227.

Herbert, S. (1980). The Red Notebook of Charles Darwin. Edited with an introduction and notes. *Bulletin of the British Museum (Natural History). Historical Series*, **7**, 1–164.

Hodge, M. J. S. (1972). The universal gestation of nature: Chambers' *Vestiges* and *Explanations. Journal of the History of Biology*, **5**, 127–51.

Hodge, M. J. S. (1977). The structure and strategy of Darwin's 'long argument'. *British Journal for the History of Science*, **10**, 237–46.

Hodge, M. J. S. (1982). Darwin and the laws of the animate part of the terrestrial system (1835–1837): on the Lyellian origins of his zoonomical explanatory program. *Studies in History of Biology*, **6**, 1–106.

Hodge, M. J. S. (1983a). Darwin as lifelong generation theorist. In *The Darwinian Heritage: A Centennial Retrospect*, ed. D. Kohn. Princeton: Princeton University Press.

Hodge, M. J. S. (1983b). *Darwin and the Theory of Natural Selection. Roles for Epistemological and Methodological Ideals in a Scientific Innovation.* Boston and Dordrecht: D. Reidel.

Hull, D. (1973). *Darwin and his Critics.* Cambridge, Mass.: Harvard University Press.

Hull, D. (1974). *Philosophy of Biological Science.* Englewood Cliffs, N.J.: Prentice Hall.

Kohn, S. (1980). Theories to work by: Rejected theories, reproduction and Darwin's path to natural selection. *Studies in the History of Biology*, **4**, 67–170.

Kohn, D. (ed.) (1983). *The Darwinian Heritage: A Centennial Retrospect.* Princeton: Princeton University Press.

Kohn, D. & Hodge, M. J. S. (1983). The immediate origins of the theory of natural selection (1838–9). In *The Darwinian Heritage: A Centennial Retrospect*, ed. D. Kohn. Princeton: Princeton University Press.

Kottler, M. J. (1978). Charles Darwin's biological species concept and theory of speciation: the transmutation notebooks. *Annals of Science*, **35**, 275–97.

Lyell, C. (1830–33). *The Principles of Geology*, 3 vols., London: John Murray.

Mayr, E. (1982). *The Growth of Biological Thought. Diversity, Evolution and Inheritance*. Cambridge, Mass.: Harvard University Press.

Manier, E. (1978). *The Young Darwin and His Cultural Circle*. Dordrecht and Boston: D. Reidel.

Ospovat, D. (1981). *The Development of Darwin's Theory*. Cambridge: Cambridge University Press.

Ruse, M. (1975a). Darwin's debt to philosophy: an examination of the influence of the philosophical ideas of John F. W. Herschel and William Whewell in the development of Charles Darwin's theory of evolution. *Studies in History and Philosophy of Science*, **6**, 159–81.

Ruse, M. (1975b). Charles Darwin and artificial selection. *Journal of the History of Ideas*, **36**, 339–50.

Ruse, M. (1979). *The Darwinian Revolution*. Chicago: University of Chicago Press.

Ruse, M. (1981). Darwin's theory: an exercise in science. *New Scientist*, 25 June.

Schweber, S. S. (1977). The Origin of the *Origin* revisited. *Journal of the History of Biology*, **10**, 229–316.

Schweber, S. S. (1980). Darwin and the political economists: divergence of character. *Journal of the History of Biology*, **13**, 195–289.

Sloan, P. (1983a). Darwin, vital matter and the unity of nature. *Journal of the History of Biology*, **16**.

Sloan, P. (1983b). Darwin's early work on invertebrate reproduction. In *The Darwinian Heritage: A Centennial Retrospect*, ed. D. Kohn. Princeton: Princeton University Press.

Sulloway, F. J. (1979). Geographic isolation in Darwin's thinking: the vicissitudes of a crucial idea. *Studies in the History of Biology*, **3**, 23–65.

Sulloway, F. J. (1982a). Darwin and his finches: the evolution of a legend. *Journal of the History of Biology*, **15**, 1–53.

Sulloway, F. J. (1982b). Darwin's conversion: the *Beagle* voyage and its aftermath. *Journal of the History of Biology*, **15**, 325–96.

Wilson, E. B. (1900). *The Cell in Development and Inheritance*, 2nd edn. New York: Macmillan.

4

Darwin and the nature of science

DAVID L. HULL

Darwin constructed *The Origin of Species* (1859) as an argument *for* the gradual evolution of species primarily by means of chance variation and natural selection and *against* special creation, the belief that 'at innumerable periods in the earth's history certain elemental atoms have been commanded suddenly to flash into living tissues' (Darwin, 1859: 483). Critics both at the time and since agree with Rudwick (1972: 222) that in arguing against the special creationists, Darwin 'presented his theory with only a straw man to oppose it: *either* slow trans-specific evolution by means of "natural selection", *or* direct divine creation of new species from the inorganic dust of the Earth'. Rudwick (1972: 207) goes on to suggest that Darwin was carefully avoiding a third, more viable alternative:

On the contrary, although Darwin later suggested that the only alternative to his evolutionary theory was a naive creationism, in fact there was another explanation available, with intellectual credentials quite as high as Darwin's, and with considerably more credibility to the mind of the time. This alternative was well developed by one who started as Darwin's collaborator but who later became one of his most implacable opponents – the anatomist and paleontologist Richard Owen (1804–1892).

For want of a better term, I will call this third alternative to evolution and special creation 'idealism', and would add a fourth alternative as well – reverent silence. Given the highly inductive philosophies of science current at the time, scientists felt perfectly justified in remaining silent on those issues for which they lacked sufficient data. In the *Origin* Darwin (1859: 310) lists a dozen or so palaeontologists and geologists who 'unanimously, often vehemently, maintained the immutability of species' and in retrospect could not recall coming upon a single person, save R. E. Grant, 'who seemed to doubt about the permanence of species'

(Darwin, 1899: 1: 71). Lyell (1889: 2: 274) confirms Darwin's recollections and adds his own position to the list. 'But, speaking generally, it may be said that all the most influential teachers of geology, palaeontology, zoology, and botany continued till near the middle of this century either to assume the independent creation and immutability of species, or carefully to avoid expressing any opinion on this important subject.'

It is extremely difficult to gauge the prevalence of a belief at a particular time in history. The barriers confronting the type of statistical analysis necessary are all but immobilizing. Rudwick (1972: 208) rightly maintains that natural history was 'not old-fashioned in Owen's time' and goes on to add that 'Owen himself was not a kind of living fossil epistemologically: his view of nature was that of most of his contemporaries'. Paradis (1978: 120) claims, to the contrary, that the 'nature-philosophy' tradition exemplified by Owen's work 'had never been strong' in England, while Yeo (1979) argues that the Whewell–Owen concept of science at the time was struggling against great opposition to become accepted. G. H. Lewes (1852: 263), for one, viewed the sort of science that he preferred as a minority position in England: 'Although Germany and France have applied Goethe's morphological ideas with great success, yet England – true to her anti-metaphysical instinct, unhappily no more than an instinct with the majority – has been very chary of giving them admission, because the real philosophic method which underlies them is not appreciated.'

I have no idea how strong each of these traditions was in Great Britain at the time. I do know, however, that leading Victorian scientists can be found explicitly espousing each of the four alternatives mentioned and that Darwin had the same opinion of all but his own – they were not 'scientific'.

In the *Origin* Darwin chose to argue against the creationists. All he had to do was to extend the notion of science exemplified in Lyell's *Principles of Geology* (1830–3) to include the origin of species as well as their extinction. By the very act of publishing the *Origin*, Darwin was breaking the gentlemen's agreement that Chambers (1844) had broken in such an ungentlemanly fashion before him. But idealism was quite another matter. He had no idea of how to confront idealistic explanations in terms of Platonic ideas and polarizing forces. Instead he used the scientist's most potent weapon – silence. Although Lyell was Darwin's implicit opponent in the *Origin*, he never published a rebuttal. Although Owen is hardly mentioned, he published one of the most acrimonious critiques of Darwin and Wallace's theory, while at the same time claiming priority (Owen, 1860). After all, had he not repeatedly referred to the 'continuous operation of Creative power, or the ordained becoming of living things' (Owen, 1858: 314, 1860: 258)?

I think that Darwin's different reaction to Lyell and Owen can be explained by differences in their views on the nature of science. Darwin agreed with Lyell about the nature of science and wanted only to extend it a bit further. With respect to Owen, the differences were so profound that Darwin could not find sufficient grounds even for disagreement. To combat Owen, Darwin would have to engage in 'philosophy', an indulgence he tended to confine to his personal correspondence and private notebooks (Gruber & Barrett, 1974). The Darwinian revolution was as much concerned with the promotion of a particular view of science as it was with the introduction of a theory on the transmutation of species. However, before embarking on a discussion of the interconnections between philosophy and evolutionary theory, a few words must be said about terminology.

One perennial problem in the history of ideas is that no two people ever mean precisely the same thing by such terms as 'creation', 'idealism', or 'science'. The idealists were as mixed a lot as were the evolutionists. One solution is never to use the same term for any two authors, as self-defeating a suggestion as I have ever heard. But as soon as one claims that both Darwin and Huxley came to believe in the evolution of species while Sedgwick and Agassiz did not, one invites the specialist to list the indefinitely many ways in which the views of these workers differed. I think that such careful distinctions are absolutely necessary in the history of ideas, but I think that more general works, attempts to compare views and trace them through time, also have a function. In this paper I contrast evolutionism, creationism, and idealism, and try to explain why Darwin thought that his own concept of the origin of species was scientific while the explanations provided by the creationists and idealists were not. In order to circumvent the difficulties inherent in the use of general terms, I use specific scientists to illustrate general positions – Darwin for evolutionism, Owen for idealism, Whewell for creationism, and Lyell for reverent silence. In doing so, I do not mean to imply that the views of any of these scientists were somehow 'typical' of the general position. As I have argued elsewhere, it is as misleading to think of a particular exposition as being typical of a more general system of ideas as it is to think of a particular organism being typical of its species (Hull, 1976, 1978, 1983). In some cases, there is a point to such 'exemplifications'; in most cases not. The idealists present a special problem because they did not form a group. Claims about 'Darwinism' in the middle of the nineteenth century sound sensible because the Darwinians formed a fairly cohesive social group at the time. Talk of 'idealism' seems less appropriate because of the absence of such a group.

Philosophy of science

Darwin had the mixed fortune of attempting to solve the mystery of mysteries at the very time that philosophy of science was becoming a self-conscious discipline in the English-speaking world. John Herschel (1830), William Whewell (1837, 1840, 1849), John Stuart Mill (1843), and Lyell (1830–3) carried on at great length about the nature of science and proper scientific method. Darwin read the works of nearly all of these philosopher-scientists: Herschel before leaving on his voyage, Lyell on route, and Whewell upon his return. Darwin found himself largely in agreement with the ideas of these men and thought of himself as pursuing science in the ways they dictated.

Two problems arise, however. The first is that Herschel, Lyell, and Mill disagreed fundamentally with Whewell over the nature of the human mind and our knowledge of the empirical world. As Cannon (1976) and Hodge (1983) see the issue, one instructive way of viewing the Darwinian revolution is to interpret it as a conflict between Herschel, Mill, Lyell, and Darwin on the one side and people like Whewell, Owen (1846, 1848, 1849, 1851), and Edward Forbes (1854) on the other side – the 'empiricists' against the 'idealists' (see also Ellegård, 1957, 1958; and Hull, 1972).

The second problem is that the 'empiricists' and 'idealists', as much as they might disagree with each other about the fundamental nature of science, agreed that in *The Origin of Species* Darwin had failed to meet accepted standards of scientific method. Darwin's reasoning in the *Origin* was too hypothetical, too speculative, not sufficiently inductive. Although all of the philosophers mentioned acknowledged the role of hypotheses in science, they also thought that scientific theories could somehow be 'proved' in the sense that Newton's theories of gravitation and light had been proved. As Hodge (1983) shows, Darwin began by thinking he had presented a *vera causa* in the strong sense exemplified by Newton's gravitational theory, but in the face of the negative reactions of the leading philosophers of science of his day, he retreated to a weaker exemplar – contemporary theories of light.

According to one long list of commentators, the sort of reasoning exhibited by evolutionary biologists from Darwin to the present is sorely deficient (Popper, 1957, 1974; Himmelfarb, 1959). According to another equally long list of commentators, there is nothing whatsoever wrong with the methodology of evolutionary biologists (Ruse, 1971, 1975, 1977, 1979; Hull, 1973). In fact Ghiselin (1969) has argued that Darwin's scientific method is a paradigm of good scientific practice. I do not intend to go over these issues once again here. Instead I intend to pursue a somewhat more fundamental issue. Too often the content of scientific theories and our beliefs about the nature of science are treated as if they change in relative independence of each other, when in actual fact their development is closely interlaced. Just as our methodological beliefs influence the content of scientific theories, the content of these theories influences what we take to be proper method.

As Cohen (1981) argues, Newton not only transformed the science of mechanics but also changed the way subsequent scientists went about doing science. Similarly, as Rudwick (1972) has documented, Lyell was as concerned to redefine the science of geology as he was to support his own particular view of geological phenomena. As Laudan (1981: 9) puts the general position, 'it is shifting *scientific* beliefs which have been chiefly responsible for the major doctrinal shifts within the philosophy of science'.

I think that numerous puzzles about the reception of Darwin's *Origin* can be resolved if sufficient attention is paid to the influence that the introduction of such fundamental theories as evolutionary theory have on our understanding of science itself. Darwin thought of himself as continuing in the Herschel–Lyell tradition of 'inductive science'. Thus, he was bewildered by Herschel's characterizing his theory as the 'law of higgledy-piggledy' (Darwin, 1899: 2: 37), Mill's (1874) opting for divine plan over evolution, and Lyell's continued reluctance to come out in favour of his theory. That Whewell would reject Darwin's theory was understandable. Darwin's conception of science departed in significant respects from Whewell's modified Kantian views, but why were Herschel, Mill and Lyell so reluctant?

I propose to argue that Darwin's theory struck at the very foundations of the philosophical views held by these philosopher-scientists as well. As far as I can tell, Darwin was unaware of these implications of his theory. Several of his contemporaries perceived these conflicts through a glass darkly, but these dim perceptions were soon brushed aside, just as the triumph of Newton's theory stilled philosophical doubts about action-at-a-distance. However, just as these doubts were exhumed with the advent of relativity theory, conflicts between evolutionary theory and traditional philosophical concepts have surfaced again (Ghiselin, 1974; Hull, 1976, 1978; Wiley, 1981; Mayr, 1982; Gould, 1982).

Reverent silence

Not only did Lyell's *Principles of Geology* (1830–3) present a new theory of the formation of the earth's crust, but also 'it will endeavour to establish the *principles of reasoning* in the science' (Lyell, 1881: 1: 234). According to Lyell, geologists should explain past geological phenomena in the same way that they explain present-day phenomena, by means of the same kinds of causes acting at approximately the same rates as they are observed to operate today. Lyell extended his naturalistic view of science to include the extinction of species but not their origin. 'Whether new species are substituted from time to time for those which die out, is a point on which no decided opinion is offered; that data hitherto obtained being considered insufficient to determine the question' (Lyell, 1830: xii).

In his private correspondence, Lyell was a good deal less circumspect. In response to a letter from Herschel in 1836 acknowledging that a naturalistic explanation of the origin of species is scientifically permissible, Lyell (1881: 1: 467) agrees, complaining that the 'German critics have attacked me vigorously, saying that by the impugning of the doctrine of spontaneous generation, and substituting nothing in its place, I have left them nothing but the direct and miraculous intervention of the First Cause, as often as a new species is introduced, and hence I have overthrown my own doctrine of revolutions, carried on by a regular system of secondary causes'.

Lyell's German critics had a point. By arguing in print against Lamarck's theory of the transmutation of species, dismissing spontaneous generation as a 'fanciful notion left over from Aristotle' (Lyell, 1830: 1: 59), and couching his own agnosticism in terms of 'creation', Lyell could not help but realize the impression that he would have on his readers. It is certainly true that initially Darwin took Lyell's *Principles of Geology* as supporting his own Christian creationist belief. In his 1858 Presidential Address to the British Association for the Advancement of Science, Owen (see Basalla, Coleman & Kargon, 1970: 326) warns his audience that 'it may be well to bear in mind that by the word "creation", the zoologist means "a process he knows not what"'. Once again, even though 'creation' might well have been a code word for unknown natural processes, the use of this word, especially when it was coupled with all sorts of additional theistic references, was guaranteed to give just the opposite impression.

Although Darwin's dilemma – either miracles or evolution – was not fair to many of his contemporaries, it was certainly appropriate for some, including no less an authority than William Whewell. Whewell (1847: 3: 624–5) saw the dilemma in precisely these stark terms:

...either we must accept the doctrine of the transmutation of species, and must suppose that the organized species of one geological epoch were transmuted into those of another by some long-continued agency of natural causes; or else, we must believe in many successive acts of creation and extinction of species, out of the common course of nature; acts which, therefore we may properly call miraculous.

For his part, Whewell (1847: 3: 638–9) opts for the second alternative. Huxley, in his contribution to *The Life and Letters of Charles Darwin* (F. Darwin (ed.) 1899, 1, 548), poses Darwin's dilemma once again, while ridiculing Whewell's idealistic 'conceivability' criterion for knowledge, remarking, 'No doubt the sudden concurrence of half-a-ton of inorganic molecules into a live rhinoceros is conceivable, and therefore may be possible', but for his own part, he refused 'to run the risk of insulting any sane man by supposing that he seriously holds such a notion' (Huxley, 1870: 375).

Special creationists applied a double standard to scientific and theistic explanations of the origin of species. The standards for scientific explanations were extremely exacting while those for theistic explanations were left unspoken. In 1852, prior to the appearance of the *Origin*, Herbert Spencer (1868: 1: 1) objected to this double standard, complaining that those 'who cavalierly reject the Theory of Evolution as not being adequately supported by facts, seem to forget that their own theory is supported by no facts at all'. In the *Origin* itself, Darwin (1859: 483) echoed Spencer's complaint: 'Although naturalists very properly demand a full explanation of every difficulty from those who believe in the mutability of species, on their own side they ignore the whole subject of the first appearance of species in what they consider reverent silence.'

After the appearance of the *Origin*, William Hopkins (1860: 87) explicitly defended precisely this asymmetric view of natural and supernatural explanations. Hopkins justifies applying high standards of criticism to Darwin's theory while requiring nothing of its creationist alternative by claiming that the 'doctrine of successive creations, any more than that of final causes, does not pretend to be a *physical theory*'. To the contrary, it 'professes to be a negation of other theories rather than a theory of itself, and therefore cannot be called upon to account for phenomena at all in the physical sense in which we necessarily call upon a definite physical theory to account for them'.

Darwin not only shattered the reverent silence of men like Lyell but also directly confronted the legitimacy of explaining the origin of species by means of supernatural agencies. Darwin scholars are in wide agreement that Lyell posed the species problem that Darwin set about solving. Lyell also provided the general rules of reasoning that Darwin was to employ in solving it. Darwin was willing to push Lyell's line of reasoning into areas that Lyell himself was reluctant even to approach. By including the origin of species within the province of natural science, Darwin threatened the last citadel protecting mind, soul, and morals from the encroachment of science. The conflict continues to the present. In the nineteenth century, however, Darwin played Joshua to Lyell's Moses.

But this is not a scientific explanation

Cannon (1976: 382) encourages historians of Victorian science to give 'sympathetic attention to various schemes of that period – mostly Ideal Type schemes – which were not then obscurantist but were serious scientific proposals'. This is easier to encourage than to accomplish because the Ideal Type world view is so different from our own. To make matters worse, the world view which idealists were combatting is very similar to our own. Hence, even direct quotations have an air of ridicule about them. One thing is certain, however: in the struggle for existence, the Herschel– Lyell–Darwin concept beat the Whewell–Owen–Agassiz concepts hollow. Although these two views of nature, treated this broadly, have wide areas of overlap, where they differed most strongly, the idealists lost out. Should they have lost? Was the resulting science better because they lost? These are the sorts of questions that historians are not supposed to ask and, if asked, are all but impossible to answer.

In 1868 Owen proclaimed that if he were given the alternative, 'species by miracle or by law', he for one would opt for the latter (Owen, 1868: 3: 793). Louis Agassiz, in the margins of the copy of the *Origin* that Darwin sent him, wrote, 'What is the great difference between supposing that God makes variable species or that he makes laws by which species vary?' Later he asks again, 'What does this prove except an ideal unity holding all parts of one plan together?' (Lurie, 1960: 255). The point at issue was the nature of scientific laws and the relation between them and ideal plans. How could an ideal unity hold a plan together?

Because the issues under investigation are metaphysical, they are more than a little difficult to characterize adequately in the space of a few sentences, especially since no two of the authors under discussion held the same views. A good point of entry is the distinction made by Hopkins (1860: 740) between geometrical and physical laws. According to Hopkins, patterns exist out there in nature. Geometrical laws are deduced from the phenomena themselves and are 'entirely independent of any theory respecting the *physical causes* to which the phenomena are referable'. Physical laws to the contrary present appropriate physical causes.

According to Hopkins, Kepler's laws are geometrical while Newton's law of universal gravitation is physical. Mill (1843) designated much the same distinction by his contrast between uniformities of co-existence and causal laws. In biological contexts, the covariation of traits discerned by naturalists and comparative anatomists were examples of geometrical laws or uniformities of co-existence. None of the scientists under discussion doubted that the regularities referred to by the 'unity of type' existed. They differed with respect to their physical or causal explanation.

The preponderance of Owen's work was descriptive. Although his idealistic philosophy may have 'informed' his descriptive work, he tended to raise these issues explicitly only in the beginning or the conclusion of his works, usually in the high-flown theistic language common at the time. Other anatomists might object to the particular patterns that Owen discerned, but no one objected to his searching for them. As geometric laws they were as unproblematic as the relations between conic sections. It was the physical cause that Owen proposed that gave some of his contemporaries pause. Owen's (1848: 172) classic statement of the cause of the interrelationships that seemed so apparent in organic forms runs as follows: 'The Platonic *eidos*, or specific organizing principle or force, would seem to be in antagonism with the general polarizing force, and to subdue and mould it in subserviency to the exigencies of the resulting specific form.'

I wish I could go on at this juncture to expand on the preceding characterization, but I have been unable to discover such an amplification in Owen's work. He restates the preceding view in a variety of ways but does not present much in the way of detailed explication. What were these polarizing forces? According to what formulae did they vary? Owen never says, but in referring to 'polarizing forces', Owen thought of himself as carrying on in the Newtonian tradition. Just as Whewell (1840: 1: 331) explained the growth of crystals in terms of polarization, Owen explained the repetition of vertebrae in a backbone in these same terms. As Hall (1968) has shown, Newton's work let loose a tidal wave of forces and subtle fluids in science. In most cases, little came of them.

In the first edition of the *Origin* (1859: 435), Darwin mentions Owen's *On the Nature of Limbs* (1849) in connection with the law of the Unity of Type, finding it 'interesting'. In later editions, once Owen had come out strongly against Darwin's and Wallace's theory, Darwin was more candid, concluding his discussion of Owen with 'but this is not a scientific explanation'. In retrospect, Huxley recalls sharing Darwin's opinions of Owen's view. Huxley could not see what Agassiz's explanation of the coming and going of species as God thought of them actually explained. 'Neither did it help me to be told by an eminent anatomist that species had succeeded one another in time, in virtue of "a continuously operative creational law"' (Darwin, 1899: 1: 549).

One might complain that these comments by Darwin and Huxley are merely a sign of their partisanship. However, the Darwinians expressed precisely these same views long before the appearance of the *Origin*. The occasion was a paper by Edward Forbes (1854) in which Forbes characterized the distribution of organized beings through time as forming a figure eight with its constriction at the boundary between the Permian and Triassic epochs. As a description of the distribution of fossils, Forbes' claim was straightforward enough. It could be tested by discovering additional fossils. As it turned out, Forbes' distributional claim was soon refuted by J. Barrande's discovery of Cambrian fossils (Darwin, 1903: 2: 230). However, like Owen, Forbes did not stop with stating a geometrical law. He (Forbes, 1854: 428) went on to explain this distribution as a 'manifestation of force of development at opposite poles of an ideal sphere'.

Forbes' paper hardly went unnoticed by future evolutionists. Wallace was so incensed by Forbes' 'ideal absurdity' that he was led to write his 1855 paper (Marchant, 1916: 54). Darwin was just as irate. In a letter to Hooker in 1854, he remarks, 'It is very strange, but I think Forbes is often rather fanciful; his "Polarity" makes me sick – it is like "magnetism" turning a table' (Darwin, 1903: 1: 77). Wallace and Darwin were hardly disinterested observers, but Hopkins (1860: 749) raises exactly the same objections to Forbes' reference to polarity, and he can hardly be accused of being biased in favour of evolutionary theory. Hopkins does not object to Forbes' distributional claim as a geometrical law: 'If, however, the term *polarity* is intended, on the contrary, to convey the idea of some particular physical cause to which the phenomena are due, the theory becomes a physical theory, which we estimate, as such, at the lowest value, since so far from tracing the action of some definite physical cause, it does not even assign such causes in any comprehensible terms.'

Huxley's views on these issues are a good deal more equivocal. In his early writings, Huxley referred to such things as plans and archetypes as freely as Owen did. As Huxley (1893: 1: 7) recalls these early years, 'species work was always a burden to me; what I cared for was the architectural and engineering part of the business, the working out the wonderful unity of plan in the thousands and thousands of diverse living constructions, and the modifications of similar apparatuses to serve diverse ends'. However, Huxley (1853: 50) claims that for his part, all he means by the 'plan' of a particular invertebrate is a 'conception of a form embodying the most general propositions that can be affirmed respecting the Cephalus Mollusca, standing in the same relation to them as the diagram to a geometrical theorem, and like it, at once, imaginary and true'.

In one of his first letters to Huxley, Darwin (1903: 1: 73) applauds Huxley's work in comparative anatomy: 'The discovery of the type or "idea" (in your sense, for I detest the word as used by Owen, Agassiz & Co.) of each great class, I cannot doubt, is one of the very highest ends of Natural History...' Exactly how candid Darwin is being is difficult to judge because Huxley's views on the nature and role of archetypes in science are anything but transparent. For instance, in his review of the tenth edition of Chambers' *Vestiges of the Natural History of Creation* (1853), Huxley argues that he has nothing against archetypes as long as they are not supposed to *do* anything, as long as they are not supposed to be active agents. According to Huxley (1854: 427), the main message of the *Vestiges* is simply 'in all its naked crudeness, the belief *that a law is an entity* – a Logos intermediate between the Creator and his works'; (see also Huxley, 1871: for further discussion see Di Gregorio, 1981).

A very similar objection was to be made of Darwin's use of the notion of natural selection in analogy to artificial selection, as if he intended natural selection 'as an agent' (Darwin, 1903: 1: 126). He explained that he used 'natural selection' only as shorthand for the 'tendency to the preservation (owing to the severe struggle for life to which all organic beings at some time or generation are exposed) of any, the slightest, variation in any part, which is of the slightest use or favourable to the life of the individual which has thus varied; together with the tendency to its inheritance'. As misleading as the phrase 'natural selection' has been to many (Young, 1971), one can only be thankful that Darwin used it instead of his more exacting description. The point of the description is, however, that natural selection is not an agent over and above the natural processes to which Darwin refers.

Darwin (1859: 206) calls the Unity of Type one of the two commonly acknowledged 'great laws' upon which all organic beings have been formed, but he explains the unity of type in terms of unity of descent, not in terms of the action of some force or 'idea'. Neither sort of explanation can be ruled out *a priori*. The claim that certain limitations of structure inherent in nature explain the relatively few basic patterns that anatomists find in the structure of living organisms was far from implausible at the time. Explanations in terms of common descent were also plausible. And of course nothing prevents an anatomist from combining the two (Gould,

1977). However, explanations in terms of structural constraints are not of the same sort as those in terms of common descent. The former are characteristics of the inherent make-up of the universe; the latter are the result of historical contingencies. Unless one postulates an 'historical law' in the sense of a programmed sequence of events inherent in nature, explanations in terms of common descent are highly contingent. According to Darwin, the patterns so apparent in the living world do not reflect the workings of any underlying *Baupläne* but common history.

A metaphysical head

In reaction to a paper by James Dwight Dana (1857), Darwin remarked in a letter to Lyell, 'I could make nothing of Dana's idealistic notions about species; but then, as Wollaston says, I have not a metaphysical head'. The reason is that neither side fully understood the other's position on issues that can only be termed 'metaphysical'. For over two thousand years, species had been considered paradigm examples of natural kinds, like geometric figures and the physical elements. The point of Dana's paper was to show that all three could be treated in parallel ways. Throughout most of the history of Western thought, natural kinds were viewed as being eternal, immutable, and discrete. All Darwin claimed was that species are not eternal but temporary, not immutable but quite changeable, and not discrete but gradating imperceptibly through time one into another. Nevertheless, Darwin and his fellow Darwinians continued to treat species as natural kinds albeit very peculiar ones. Continuing to view species as akin to geometric figures and the physical elements even though they lack *all* the defining characteristics of this metaphysical category is akin to claiming that something is a triangle even though it has neither three sides, nor three angles, and is not a geometric figure anyway.

In response to the *Origin*, Louis Agassiz (1860: 143) objected:

It seems to me that there is much confusion of ideas in the general statement of the variability of species so often repeated lately. If species do not exist at all, as the supporters of the transmutation theory maintain, how can they vary? And if individuals alone exist, how can the differences which may be observed among them prove the variability of species?

As Beatty (1982) has argued, in part, all Darwin is doing is redefining a theoretical term, the sort of activity that goes on all the time in science. Just as the gene concept has varied throughout the history of genetics, as new theories of the gene were introduced and accepted, the species concept has varied. Darwin (1899: 2: 123) dismissed Agassiz's 'weak metaphysical and theological attack', but Agassiz had a point. Darwin was not merely redefining a term but redefining it in such a way that it could no longer belong to the traditional metaphysical category in which it had always been placed. It is one thing to redefine 'species' in terms of reproductive gaps, morphological gaps, etc. It was quite another thing to redefine 'species' so that species could no longer be the *sort* of thing that they had always been viewed as. Dana (1857) wrote his paper to show that natural kinds or 'species' in the generic sense are not the sort of thing that can evolve. According to Dana, the physical elements *qua* elements cannot possibly blend into each other. Either a substance is a multiple of a fixed number, or it is not. If it is, it is an element; if not, not. The same can be said for geometric figures. Conic sections can vary continuously, but triangularity cannot evolve into rectilinearity. Either a geometric figure has three sides, or it does not. Nothing in between can possibly exist (see Rudwick, 1972; Winsor, 1976; Manier, 1978; Ruse, 1979; Ospovat, 1981; Ridley, 1982).

According to such idealists as Agassiz and Dana, structural plans belong in the category 'unchangeable kind', not temporary manifestation. As Agassiz (1860: 143) expressed this conviction 'I have attempted to show that branches of the animal kingdom are founded upon different plans of structure, and for that very reason have embraced from the beginning representatives between which there could be no community of origin'. One cannot legitimately argue that species are immutable because that is how 'species' is defined. Definitions are themselves far from immutable. However, one can legitimately object to the assertion that species are mutable *and* that they are instances of a metaphysical category that is defined in terms of immutability. If species evolve *and* are instances of natural kinds, then both 'species' and 'natural kind' must be redefined. Perhaps someone could set out a notion of 'archetype' or 'plan' in which these entities can change through time, but to my knowledge, no one in Darwin's day attempted such a reformulation of traditional metaphysics.

In this regard, the metaphysics implicit in Darwin's theory conflicted with idealistic metaphysics, but it also conflicted with the metaphysics implicit in the Herschel–Lyell philosophy of science. According to these philosopher-scientists, statements of the covariation of the traits that characterize organisms are laws of nature – *geometric* laws but laws nonetheless. If species evolve, then it follows that laws of nature are evolving, the very state of affairs that Lyell was so concerned to avoid. Although early in his theorizing Darwin toyed with the idea that species might be very much like organisms with a definite life–span somehow built into their make-up, by the time he published the *Origin*, he had abandoned this idea (Hodge, 1983). As far as I can tell, no matter how much various Darwinians and anti-Darwinians might disagree with each other, none of them suggested that species belong in the metaphysical category 'individual'. A detailed discussion of this issue had to await the work of Michael Ghiselin (1969, 1974).

I have no intention of reading Ghiselin's conclusion back into the past. To the contrary, I do not think that anyone at the time saw the conflict in these terms. It is for this very reason that the issues were never joined, and such Darwinians as Huxley were able to remain 'pre-Darwinian' as long as they lived (see also Bartholomew, 1975). This state of affairs persists to the present. Even though comparative anatomists clearly acknowledge that species evolve, they insist that they can go about their business as if they did not (Huxley, 1874; Zangerl, 1948; Jardine, 1969; Nelson & Platnick, 1981).

The implications of moving species from the metaphysical category that can appropriately be characterized in terms of 'natures' to a category for which such characterizations are inappropriate are extensive and fundamental. If species evolve in anything like the way that Darwin thought they did, then they cannot possibly have the sort of natures that traditional philosophers claimed they did. If species in general lack natures, then so does *Homo sapiens* as a biological species. If *Homo sapiens* lacks a nature, then no reference to biology can be made to support one's claims about 'human nature'. Perhaps all people are 'persons', share the same 'personhood', etc., but such claims must be explicated and defended *with no reference to biology*. Because so many moral, ethical, and political theories depend on some notion or other of human nature, Darwin's theory brought into question all these theories. The implications are not entailments. One can always dissociate '*Homo sapiens*' from 'human being', but the result is a much less plausible position.

Dana (1857: 488) echoed a common view when he claimed that the kingdoms of life are made up of units properly termed 'species'. 'Were these units capable of blending with one another indefinitely, they would no longer be units, and species could not be recognized. The system of life would be a maze of complexities; and whatever its grandeur to a being that could comprehend the infinite, it would be unintelligible chaos to man'. Darwin (1859: 490) concluded his *Origin of Species* with the observation that there is a grandeur to his view of life, and I personally can attest to the fact that one need not be an infinite being to appreciate the grandeur of this view. Darwin (1859: 485) thought that his theory entailed that species must be treated in the same way as naturalists had treated genera.

as 'merely artificial combinations made for convenience'. Ghiselin (1969, 1974), building on the work of Mayr (1942, 1969), has shown that this inference does not necessarily follow. Just as Darwin carried Lyell's views somewhat deeper into the Promised Land, Ghiselin has played Joshua to Darwin's Moses.

I owe more than the usual note of thanks to David Roos for reading and criticizing an earlier draft of this paper, and the traditional demurrer that he cannot be held responsible for the results is also more appropriate than usual.

Further reading

Ghiselin, M. (1969). *The Triumph of the Darwinian Method*. Berkeley: University of California Press.
Rudwick, M. J. S. (1972). *The Meaning of Fossils*. New York: Science History Publications.
Ruse, M. (1979). *The Darwinian Revolution*. Chicago and London: University of Chicago Press.

References

Agassiz, L. (1860). Prof. Agassiz on the origin of species. *American Journal of Science*, 30, 142–54; also in *Annals and Magazine of Natural History*, 6, 219–32.
Bartholomew, M. (1975). Huxley's defense of Darwinism. *Annals of Science*, 32, 525–36.
Basalla, G., Coleman, W. & Kargon, R. H. (eds). (1970). *Victorian Science*, pp. 312–34. New York: Doubleday.
Beatty, J. (1982). What's in a word? In *Nature Animated*, ed. M. Ruse. Dordrecht: D. Reidel.
Cannon, W. Faye, (1976). The Whewell-Darwin controversy. *Journal of the Geological Society of London*, 132, 377–84.
Chambers, R. (1844). *Vestiges of the Natural History of Creation*. London: John Churchill.
Chambers, R. (1853). *Vestiges of the Natural History of Creation*, 10th edn. London: John Churchill.
Cohen, I. B. (1981). *The Newtonian Revolution*. New York: Cambridge University Press.
Dana, J. D. (1857). Thoughts on species. *Annals and Magazine of Natural History*, 20, 485–97.
Darwin, C. (1859). *On the Origin of Species* (1966). Cambridge, Mass.: Harvard University Press.
Darwin, F. (ed.). (1899). *The Life and Letters of Charles Darwin*. New York: D. Appleton.
Darwin, F. (ed.). (1903). *More Letters of Charles Darwin* (1972). New York: Johnson Reprint Corporation.
Di Gregorio, M. (1981). Order or process of nature: Huxley's and Darwin's different approaches to natural sciences. *History and Philosophy of the Life Sciences*, 2, 217–41.

Ellegård, A. (1957). The Darwinian theory and nineteenth century philosophies of science. *Journal of the History of Ideas*, **18**, 362–93.

Ellegård, A. (1958). *Darwin and the General Reader*. Göteborg: Göteborgs Universitets Ärsskrift.

Forbes, E. (1854). On the manifestation of polarity in the distribution of organized beings in time. *Royal Institution of Great Britain Proceedings*, **1**, 428–33.

Ghiselin, M. (1969). *The Triumph of the Darwinian Method*. Berkeley: University of California Press.

Ghiselin, M. (1974). A radical solution to the species problem. *Systematic Zoology*, **23**, 536–44.

Gould, S. J. (1977). *Ever Since Darwin*. New York: Norton.

Gould, S. J. (1982). Darwinism and the expansion of evolutionary theory. *Science*, **216**, 380–87.

Gruber, H. E. & Barrett, P. H. (1974). *Darwin on Man*. New York: Dutton.

Hall, T. (1968). On biological analogs of Newtonian paradigms. *Philosophy of Science*, **35**, 6–27.

Herschel, J. (1830). *Preliminary Discourse on the Study of Natural Philosophy* (1966). New York: Johnson Reprint Corporation.

Himmelfarb, G. (1959). *Darwin and the Darwinian Revolution*. Garden City, New York: Doubleday.

Hodge, M. J. S. (1983). *Darwin and Natural Selection: His Methods and His Methodology*. Dordrecht: D. Reidel.

Hopkins, W. (1860). Physical theories of the phenomena of life. *Fraser's Magazine*, **61**, 739–52; **62**, 74–90.

Hull, D. L. (1972). Charles Darwin and nineteenth-century philosophies of science. In *Foundations of Scientific Method: The Nineteenth Century*, ed. R. N. Giere & R. S. Westfall, pp. 115–32. Bloomington: Indiana University Press.

Hull, D. L. (1973). *Darwin and His Critics*. Cambridge, Massachusetts: Harvard University Press.

Hull, D. L. (1976). Are species really individuals? *Systematic Zoology*, **25**, 174–91.

Hull, D. L. (1978). A matter of individuality. *Philosophy of Science*, **45**, 335–60.

Hull, D. L. (1983). Darwinism as an historical entity. In *The Darwinian Heritage*, ed. D. Kohn. Princeton: Princeton University Press.

Huxley, T. H. (1853). On the morphology of the Cephalous Mollusca. *Philosophical Transactions of the Royal Society*, **143**, 29–66.

Huxley, T. H. (1854). Review of 'Vestiges of the Natural History of Creation' (1853). *British and Foreign Medico-Chirurgical Review*, **19**, 425–39.

Huxley, T. H. (1870). Paleontology and the doctrine of evolution. In *Collected Essays* (1893), **8**, 340–88. London: Murray.

Huxley, T. H. (1871). Mr. Darwin's critics. *Contemporary Review*, **18**, 443–76.

Huxley, T. H. (1874). On the classification of the animal kingdom. *Nature*, **11**, 101–2.

Huxley, T. H. (1893). *Collected Essays*. London: Murray.

Jardine, N. (1969). The observational and theoretical components of homology. *Biological Journal of the Linnaean Society*, **1**, 327–61.

Laudan, L. (1981). *Science and Hypothesis*. Dordrecht: D. Reidel.

Lewes, G. H. (1852). Goethe as a man of science. *Westminster Review*, **58**, 258–72.

Lurie, E. (1960). *Louis Agassiz: A Life in Science*. Chicago: University of Chicago Press.

Lyell, C. (1830–3). *Principles of Geology*. London: Murray.

Lyell, C. (1889). *Principles of Geology*, 11th edn. London: Murray.

Lyell, K. (ed.). (1881). *Life, Letters and Journals of Sir Charles Lyell, Bart.* London: Murray.

Manier, E. (1978). *The Young Darwin and His Cultural Circle*. Dordrecht: D. Reidel.

Marchant, J. (ed.). (1916). *Alfred Russel Wallace: Letters and Reminiscences*. New York: Harper.

Mayr, E. (1942). *Systematics and the Origin of Species*. New York: Columbia University Press.

Mayr, E. (1969). *Principles of Systematic Zoology*. New York: McGraw-Hill.

Mayr, E. (1982). *The Growth of Biological Thought*. Cambridge, Mass.: Harvard University Press.

Mill, J. S. (1843). *A System of Logic*, London: Longmans.

Mill, J. S. (1874). *Three Essays on Religion* (1969). New York: Greenwood Press.

Nelson, G. & Platnick, N. (1981). *Systematics and Biogeography: Cladistics and Vicariance*. New York: Columbia University Press.

Ospovat, D. (1981). *The Development of Darwin's Theory*. London: Cambridge University Press.

Owen, R. (1846). Report on the archetype and homologies of the vertebrate skeleton. *Report of the Sixteenth Meeting of the British Association for the Advancement of Science*, pp. 164–340. London: Murray.

Owen, R. (1848). *On the Archetype and Homologies of the Vertebrate Skeleton*. London: Van Voorst.

Owen, R. (1849). *On the Nature of Limbs*. London: Van Voorst.

Owen, R. (1851). Review of 'Principles of Geology' by Sir Charles Lyell, *Quarterly Review*, 89, 412–31.

Owen, R. (1860). Darwin on the origin of species. *Edinburgh Review*, 111, 487–532.

Owen, R. (1868). *The Anatomy of Vertebrates*, vol. 3. London: Longmans.

Paradis, J. C. (1978). *T. H. Huxley: Man's Place in Nature*. London: Murray.

Popper, K. R. (1957). *The Poverty of Historicism*. London: Routledge & Kegan Paul.

Popper, K. R. (1974). Darwinism as a metaphysical research program. In *The Philosophy of Karl Popper*, vol. 1, ed. P. A. Schillp, pp. 133–43. LaSalle, Illinois: Open Court Press.

Ridley, M. (1982). Coadaptation and the inadequacy of natural selection. *British Journal for the History of Science*. 15, 45–68.

Rudwick, M. J. S. (1972). *The Meaning of Fossils*. New York: Science History Publications.

Ruse, M. (1971). Is the theory of evolution different? *Scientia*, 106, 765–83, 1069–93.

Ruse, M. (1975). Darwin's debt to philosophy. *Studies in the Philosophy of Science*, 6, 159–81.

Ruse, M. (1977). Karl Popper's philosophy of biology. *Philosophy of Science*, 44, 638–61.

Ruse, M. (1979). *The Darwinian Revolution*. Chicago: University of Chicago Press.

Spencer, H. (1868). The Development hypothesis. In *Essays: Scientific, Political, and Speculative*, 1, 377–83. London: Williams & Norgate.

Whewell, W. (1837). *History of the Inductive Science*. London: Parker.

Whewell, W. (1840). *The Philosophy of the Inductive Sciences, founded upon their History*. London: Parker.

Whewell, W. (1847). *The Philosophy of the Inductive Sciences, founded upon their History*, 2nd edn. London: Parker.

Whewell, W. (1849). *Of Induction, with especial reference to Mr. J. Stuart Mill's System of Logic*. London: Parker.

Wiley, E. O. (1981). *Phylogenetics, the Theory and Practice of Phylogenetic Taxonomy*. New York: Wiley.

Winsor, M. P. (1976). *Starfish, Jellyfish and the Order of Life*. New Haven: Yale University Press.

Yeo, R. (1979). William Whewell, natural theology and the philosophy of science in mid-nineteenth century Britain. *Annals of Science*, **36**, 493–516.

Young, R. M. (1971). Darwin's metaphor: does nature select? *Monist*, **55**, 442–503.

Zangerl, R. (1948). The methods of comparative anatomy and its contributions to the study of evolution. *Evolution*, **4**, 351–74.

5

The several faces of Darwin: materialism in nineteenth and twentieth century evolutionary theory

GARLAND E. ALLEN

Since Darwin first published *The Origin of Species* in 1859, the theory of natural selection has encountered four major periods of resurging controversy: 1859–1880, 1890–1915, 1930–1950, and 1976–present. In the first, third and fourth of these periods religious as well as scientific objections played a highly visible role in the perpetuation, if not the origin, of questions about the validity of Darwinian theory. Among the biological arguments against Darwinism during these four periods, however, certain topics have recurred with persistent regularity.

(1) *Variation in natural populations.* How much variation, and of what kind (large, discontinuous, or small, continuous variation) exists in nature? Is more variation, or less variation adaptive for the individual or population?

(2) *The role of natural selection.* Does natural selection actually play a role in the origin of new species, or does it merely sort out unfit variants from within a heterogeneous population?

(3) *The rate of evolution.* Does evolution proceed gradually, in a step-by-step manner, or does it proceed in leaps and spurts, with periods of rapid change alternating with periods of slow change (or no change at all)?

(4) *The unit of selection.* Does selection act on the individual organism or the whole population of which the individual is merely a part? How can what is adaptive for the individual be adaptive for the population? And conversely, how can what is adaptive for the population also be adaptive for the individual?

(5) *Different genetic mechanisms.* As evolutionary trends develop, especially beyond the divergence of two species, do genetic mechanisms different from those involved in earlier stages of divergence make their

appearance? In other words, are macro-evolutionary developments simply extrapolations of micro-evolutionary genetic events?

In addition, there has lurked in the background of all the periods of controversy the more general question of what the relationship is, or ought to be, between evolutionary theory, and other fields of biology such as genetics, embryology, ecology and biochemistry (including molecular biology). The context in which these questions and problems have arisen is different in each of the periods of controversy and the relative prominence of each problem has varied from one period of time to another. But the remarkable fact is that many of the same problems facing evolutionary theorists today have been around in one form or another since the publication of *The Origin of Species*.

Underlying most if not all of the controversies in twentieth century evolutionary theory is a conflict between two major philosophical schools of thought: what I will refer to as mechanistic materialism and holistic materialism. I will define these terms in the next section, so that their meaning and implications for evolutionary theory are as unambiguous as possible. Suffice it to say here that the history of western philosophy over the past 150 years has shown first the rise of mechanistic materialism in the mid-nineteenth century, and its gradual replacement by slowly-growing concepts of holistic, and later dialectical materialism. Increasing dissatisfaction with the mechanistic materialist outlook in philosophy in general, and in science (including biology) in particular, generated new approaches to understanding complex processes. These new approaches formed only slowly and haltingly out of the superstructure of mechanistic thought. But they gradually merged into a coherent body of philosophy which I will designate as holistic materialism, and its more formal and rigorous extension, dialectical materialism. Although this shift has occurred more rapidly in some fields of thought than others (for example, philosophy, political economy, or physics as contrasted to general history, aesthetics, or biochemistry) it is nonetheless apparent to one degree or another in virtually all aspects of western thought.

In this paper I discuss the changing philosophical scene – that is, the replacement of mechanistic by some forms of holistic and/or dialectical materialism – as reflected in the recurrent controversies in evolutionary theory during the present century. To give the discussion more focus, I will discuss in some detail attempts of workers in the period 1900–55 to understand the problem of variation in natural populations: how extensive it was, the degree to which it was adaptive and non-adaptive, whether it was beneficial for the population but not the individual (or vice versa) and how the study of variation within populations could be related to the study of collateral areas of biology such as individual development. In particular, I will focus in some detail on the shift in philosophical views embodied in the work of R. A. Fisher (1890–1962) in the 1930s, and that of I. Michael Lerner (1910–1977) in the 1950s. In that shift, I will argue, lies a profound change in philosophical outlook in modern evolutionary theory.

The general thesis which I will sketch out will be that: (1) Darwin himself had two philosophical faces, one looking backward to the more mechanistic materialism of the early-to-mid-nineteenth century, and the other looking forward to the more holistic, even dialectical materialism of the mid-to-late-nineteenth and early twentieth centuries; (2) it was the mechanistic side which was most prevalent in the works of many of Darwin's followers as well as in that of his critics; and that this mechanistic view informed the basic structure of the synthesis of Mendelian genetics with Darwinian theory between 1918 and 1930; and finally (3) a slow but steady development of first a holistic, then a more dialectical approach to evolutionary problems can be seen in the work of various investigators during the twentieth century. In tracing this development, I will show how new approaches to old questions have involved at least in part new philosophical views, and how the kinds of questions which investigators ask and the kinds of answers they accept change over time, reflecting changes (sometimes conscious, often not) in philosophical view. More important, I will suggest that a conscious understanding of philosophy in general, and specific philosophical methods in particular can be of great help in the development of actual research work in science in general, and biology in particular.

Mechanistic, holistic and dialectical materialism

I will outline as briefly as possible the nature of the materialist philosophical positions in general, and the distinctions between mechanistic, holistic, and dialectical materialism in particular

All forms of materialist philosophy share several beliefs in common. (1) The belief that a material reality exists outside human consciousness. (2) The belief that reality exists prior to, and independent of our ideas about it. Thus, a corollary of this proposition is that ideas are derived ultimately from experience with the world and not vice versa. (3) The belief that all change in the universe is a result of matter in motion, that is, the action of one material entity on another. Materialism excludes from rational consideration non-material substances such as ether or caloric, or mystical forces such as the *élan vital*. Materialism and idealism are quite distinct (and opposing) philosophical views, especially when applied to the study of the natural world. At its most basic level the clash between science and religion in the past several centuries has been largely a clash between materialistic and idealistic world views, respectively.

Mechanistic materialism includes several distinct but interrelated propositions. (1) The parts of a complex whole are separate and distinct; that is, a chemical reaction, clock, planetary system or living organism, consists of many separate and identifiable parts. A consequence of this view is the notion that (2) the proper study of a complex whole proceeds best by study of the individual parts in isolation. Mechanists strive to characterize each part of a whole in and of itself; while admitting that parts interact, mechanists lay stress on the parts, not the interactions. (3) Consequently, mechanists believe that the whole can be reconstructed as the sum of the individual parts—and no more. If you know the attributes of each part, it is possible to reconstruct the functioning of the whole as the sum of those attributes. (4) Mechanistic materialists believe that change in any system is impressed on the system from the outside, as opposed to the working out of internally-derived forces, or developments from within the system. (5) Finally, mechanistic materialism has long been associated, especially in science, with an atomistic view of nature. From the seventeenth through the twentieth centuries, most mechanists were also atomists. However, merely believing in the particulate nature of matter, without proper demonstration of the existence of atoms themselves, can, as some mechanists have pointed out, come close to pure idealism. In terms of philosophical materialism it is not enough to simply *believe in* the existence of fundamental particles, or atoms. It is also important to seek confirmation that such discrete units of structure actually exist, that is, have material reality. This demonstration becomes one of the crucial obligations of the mechanistic materialist position.

By contrast, holistic materialism argues from several different premises. (1) The parts of a complex whole are interconnected and cannot be studied only by themselves – that is, in isolation. Parts interact with other parts, and those interactions are one of the characteristics of the parts themselves. Thus, by studying parts in isolation a very important characteristic of each is lost. (2) As a consequence, it is necessary to seek new methods of studying the interactions of parts. That is, parts must be studied *in situ*, in their natural interrelationships with one another. (3) The whole of any complex process or structure is more than the mere sum of its parts. It is, in fact the sum of the parts *plus* their interactions. The additional component which makes the whole equal to more than the sum of its parts is not anything mystical or supernatural, but merely the interactions of parts, but which cannot be predicted from a knowledge of only the parts themselves. The term *emergence* has sometimes been used to refer to the new properties of a system deriving from interaction of its parts. 'Emergence' has, however, been used in the past to refer to mystical or vital properties of a system and so I will avoid using it in this discussion. (4) Processes in the world are dynamic and constantly changing. Change in any complex system is partly a result of interactions between that system and factors outside and partly a result of built-in, internal changes within the system itself. Thus, organisms are living, dynamic systems which mature and die not primarily because of accidental changes impressed on them from the environment, but because of internal developments within their chemical and physical organization. Holistic materialism is thus a philosophical view emphasizing the wholeness and integration of complex systems, in contrast to mechanistic materialism which emphasizes parts rather than their relationship to the whole.

While dialectical materialism grows directly out of holistic materialism and shares all of its basic premises, it embodies several further elements. First, and most important, is the notion that the internal change within a system is the result specifically of the interaction of opposing forces, or tendencies, within the system itself. This is one meaning of the term 'dialectic'. Thus, for example, the evolution of adaptations within populations can be seen as the result of a constant dynamic tension between the value for the moment of a high degree of specialization, and the value for the future of retaining as large a level of flexibility as possible. It is an argument of the dialectical approach that it is by the interaction of such opposing forces that any system (in this case population of organisms) is constantly propelled along a path of change and development. Because of dialectical interactions no system remains static. Analysis of the dialectic involved in any case can thus reveal the dynamics of a system's development that otherwise might remain hidden.

A second important difference between dialectical and holistic materialism lies in the notion that quantitative changes lead to qualitative changes. What this means is that the accumulation of small (quantitative) differences within a system can ultimately lead to a large-scale (qualitative), and sometimes rapid transition to a new state. In evolution, a population divided into two geographic isolates undergoes the accumulation of many quantitative changes (genetic variations) which after some point of transition produce a qualitative change – the two populations can no longer effectively interbreed and thus are two separate species. The notion of quantitative changes leading to qualitative changes resolves the old and artificial distinction between the processes of evolution (accumulation of small, quantitative changes) and revolution (the sometimes rapid achievement of a new state of being). Qualitative changes cannot occur without quantitative changes preceding them. Conversely, quantitative changes inevitably produce qualitative changes. The two processes – quantitative and qualitative change – are inextricably linked.

We turn now to a brief look at how these various elements of philosophical methods express themselves, and are indeed intermixed, in the work of Darwin and the twentieth century neo-Darwinians.

The three faces of Darwin: mechanistic, holistic, and dialectical materialism

While Darwin explicitly rejected idealistic explanations for the origin of diversity he incorporated elements of mechanistic, holistic, and dialectical materialism in the formulation of the theory of natural selection. His mechanistic materialism harkens back to the earlier and mid-nineteenth century materialist tradition, seeing change in terms of fundamental units of matter interacting by the laws of blind chance and yet also yielding lawful changes. Darwin's mechanistic outlook involves first and foremost his atomistic view of the living world, which he saw as composed of individual organisms each acting in its own right and for its own survival. The organism was to the laws of evolution what the atom was to the laws of chemical reactions. In Darwin's mechanistic view the individual, atomized units of evolution, that is, individual organisms, interacted with each other through the process of struggle for existence and competition. Which organisms interacted was largely a matter of chance in the same way that it is a matter of chance which atoms or molecules interact during a chemical reaction.

Reference to the laws of chance emerges again in Darwin's treatment of variation. Individual variations occur not because they are needed, nor in any directional sense, but by pure chance. The notion of chance events lying at the basis of lawful natural processes was part and parcel of nineteenth century mechanistic science, deriving as it did from the physical sciences, and implying, as it also did, so much for philosophy. As John Dewey pointed out (Dewey, 1909), perhaps Darwin's greatest impact was his suggestion that even the finely tuned adaptations of living organisms were the result of 'blind chance' and purposelessness. If that evaluation is true, it suggests how closely Darwin's thinking was related, directly or indirectly, to the mechanistic science of his age.

Darwin's formulation of the theory of natural selection is mechanistic in yet another way. While there is no doubt that few naturalists have had a better understanding of the interrelationship between organisms and environment, and the delicate precision of ecological balance and adaptation, Darwin still saw organisms as separate from the total environment in which they existed. He wrote in *The Origin of Species* that a plant on the edge of the desert can as well be considered to be struggling against the environment as against another plant. For Darwin evolution would always take place partly because no adaptations could ever be perfect, and partly because the environment was always undergoing its own changes. Darwin did not emphasize, however, that by their very adaptation organisms alter their environments, thus producing an organism–environment interaction that goes beyond the geological or climatically related changes the environment generates on its own. Thus, Darwin saw the organism as separate from the environment (though obviously affected by it) in the same way that atoms in a chemical reaction are apart from, but affected by, their environment (that is, temperature, pressure etc. in a test tube).

In yet a third way Darwin reflected the mechanistic philosophy of his age. In describing the interactions of organisms Darwin used numerous metaphors from classical nineteenth-century political economy, a field itself dominated by mechanistic philosophy. For example, one of the phrases Darwin uses repeatedly is 'division of labour', a notion that atomizes the totality of the labour process into separate component parts. Darwin uses the term a total of six times in the first edition of the *Origin*, sometimes to refer to individuals within the 'economy of nature', sometimes to refer to anatomical or physiological division of labour within an individual organism (Schweber, 1980, 195–289; Barrett, Weinshank & Gottleber, 1981: 210). Darwin also speaks of the 'admirable division of labour in the community of ants', being as useful to the survival of humans as it is with the social insects (Barrett *et al.*, 1981: 210). As a number of authors have pointed out, the notions of struggle, competition and 'economy of nature' are all drawn from prevailing economic literature of the early nineteenth century.

At the same time, Darwin went beyond the mechanistic view in much of his understanding of interactions among organisms and within ecosystems. A primary aspect of his holistic thinking is his keen awareness of the interconnections among organisms in any natural system, the idea admirably epitomized in Darwin's metaphor of 'the tangled bank' in the final paragraph of the *Origin*. But Darwin was more than holistic – he was specifically dialectical in his view that evolution requires the interaction of two diametrically opposed forces: heredity, or faithful replication, and variation, or unfaithful replication. Without either of these elements evolution as we know it could not occur; and yet with both elements present, evolution becomes inevitable. The process of natural selection is as dialectical a process as one could find in nature, a point which has been commented upon by several previous authors (Sachsse, 1968: 141ff; Hogben, 1927: 107). Darwin's view of evolution is also dialectical in that he saw that through gradual accumulation of imperceptible individual variations it is possible to obtain a qualitatively different outcome – that is, the formation of a new species.

As a result of the interaction of the opposing forces of heredity and variation, evolution in Darwin's view was a constant, developmental process. There was no end to it – it was distinctly *not* teleological. But it was developmental if by that term we mean that the characteristics of each stage grow out of and build upon the stages before – quite different from, say, the changing positions of an atom in a gas. In both cases later stages are causally linked to earlier stages. But in developmental processes, the later stage also incorporates, in either direct or in modified form, aspects of earlier stages. In this sense Darwin's concept of evolution by natural selection is both developmental and dialectical.

Despite Darwin's clear emphasis on a holistic and dialetical approach to evolution, it was the mechanistic side of his thinking that was picked up by his followers throughout the nineteenth and early twentieth centuries – at least up through the 1950s. Only after that time did holistic and dialectical elements of the evolutionary process as described by Darwin begin to gain more attention. To illustrate this point I would like to compare and contrast the work of two twentieth century evolutionary theorists: Ronald A. Fisher, representing the more mechanistic, and I. Michael Lerner, representing the more holistic side of Darwinian theory. Since they worked approximately a generation apart, the shift in views which their work represents provides a basis for characterizing a more general development of evolutionary thinking away from a simplistic and mechanistic approach and towards a more holistic approach during the period 1915–1955.

Mechanistic materialism in population genetics 1900–1935: the case of R. A. Fisher

The development of the mechanistic side (to the neglect of the holistic and dialectical side) of Darwinian theory was not an isolated phenomenon in the late nineteenth and early twentieth centuries. It occurred at a time when biologists were rapidly gaining a new interest in analytical, experimental and quantitative methods for studying living processes. In a word, it was a period which saw the incorporation of attitudes characteristic of mechanistic materialism directly into the philosophy of biological research (Allen, 1978*b*).

In that vein the mechanistic side of Darwin's theory came to have considerable appeal for a generation of biologists concerned with problems of heredity and evolution, yet fed up with only speculative, idealistic methods – such as those of Ernst Haeckel or August Weismann – for solving them. Darwin may not have offered much of a theory of heredity, but he did not pretend to. Moreover, his theory of natural selection was clearly materialistic, anti-mystical, and provided a way of marshalling facts that was clear and forthright. The metaphor of struggle for existence and competition had a ring of reality about it. And Darwin's suggestions (growing stronger with successive editions of the *Origin*) that the environment was the direct material *cause* of specific variations, was materialist to the core. In a period marked by the desire to banish all mysticism and flights of speculation from biology, Darwin's work seemed, for natural history, a prototype of rationality and materialist thinking.

Most important to younger workers, however, was the desire to make the study of evolution experimental and quantitative. Among the most successful, especially in the application of quantitative methods to evolutionary theory, was R. A. Fisher between 1918 and 1939. It has been shown clearly that one of Fisher's major motivations in providing a mathematical approach to evolution was his eugenic leanings and his desire to put eugenic theory on a sound theoretical footing (MacKenzie, 1981: 183ff). Another motivation, however, was his interest in reconciling Darwinian and Mendelian theory – that is, in showing that natural selection could be understood in terms of the discrete genes, or factors, of Mendelian genetics. An architect of what came to be known as the 'evolutionary synthesis', Fisher brought together Mendelian genetics with Darwinian natural selection to provide a quantitative solution to such problems as evolutionary rate, effects of variation and the role of selection within large populations. There is little question that Fisher, among others (including Sewall Wright, J. B. S. Haldane and Lancelot Hogben) made a major contribution to the demonstration of the effectiveness of natural selection as the mechanism of evolutionary change and to the application of mathematics to biology in general, as well as to evolutionary theory in particular.

At the same time, Fisher's approach to this synthesis was highly mechanistic. As he himself put it, one of his major goals was to make evolutionary theory as rigorous and as quantitative as the physical sciences (Fisher, 1922: 321). The manner in which he sought to accomplish this goal emphasized the mechanistic side of Darwinian theory. For example, Fisher abstracted Darwin's individual, atomized organisms down to discrete genes interacting in an abstract space known as a gene pool. The organism thus became an idealized entity, a gene (or, more accurately, a pair of genes), and the population an idealized aggregate of atomized units, the genes. In the gene pool, genes interact at random, but when like gases in an enclosed flask, an effective interaction takes place (that is, fertilization occurs) the genes are taken temporarily out of the gene pool and bound together as an organism. The organism was both the temporary carrier of genes, and the object on which selection directly acted. Fisher's picture of the evolutionary process thus had all the earmarks of mid-nineteenth century billiard ball physics. We will return to this point shortly.

Several additional aspects of Fisher's approach emphasize his mechanistic bias. The first is his highly simplified view of genes governing the production of single traits (one gene = one trait). Although Fisher did suggest that genes interacted with each other, what is called polygenic inheritance, for him the only important components were the simple additive ones, because these could lead to predictions. For purposes of predicting populational developments, then, Fisher viewed the organism as a mosaic of traits, each determined by a single hereditary unit, the gene. The macroscopic mosaic that was the phenotype merely reflected the microscopic mosaic that was the genotype.

Another of Fisher's mechanistic features follows his mosaic treatment of traits. Fisher was able to draw a rigid line between the genotype and the phenotype, effectively ignoring the process of embryonic development and all of the interactions and phenotypic variation to which it gives rise. To separate the genotype from the phenotype meant that Fisher dealt only with selection acting on adult phenotypes, and not on embryonic processes. While such a simplifying assumption had the obvious advantage of easier mathematical manipulation, its lack of biological reality (selection does, of course, act on embryonic phenotypes, and on polygene systems as wholes, as well as on fully-formed adult traits as individual entities) was apparent to some critics from the outset.

A final example of Fisher's mechanistic treatment can be found in his general assumption that the whole is nothing more than the sum of its parts. Fisher maintained that the effects of single-gene variations are additive. Thus, for example, in discussing the origin of dominance, Fisher argued that if a gene for change in the length of a structure of 1 mm has a given adaptive value, a gene for a change in the same structure of 0.1 mm has one-tenth the adaptive value (Fisher, 1930: p. 15–16). In this sense, then, evolution was to Fisher a matter of single-gene substitutions with quantitative, or additive, effects in terms of selective value. Not only was each gene considered separately from every other gene (that is, genes for eye colour separately from genes for hair colour, height, etc.), but also each variation in any single gene was separate from the others in its total phenotypic effect. Again, for reasons of simplification and mathematization, Fisher's model was highly atomistic.

Indeed, Fisher's approach to the genetics of whole populations can be described as the biological counterpart of the mid-nineteenth century kinetic theory of gases. The system was highly deterministic, based on each gene, or variation, having a single measurable effect. If populations were large enough, like collections of gas molecules in a container, any random effects could be ignored, and the system treated as a statistical aggregate. This meant, of course, that the system was deterministic on the macroscopic scale, as a whole, though random on the microscopic (variations, and matings were random, yet in a large enough population the results of selection could still be highly predictable). As Fisher himself wrote, the study of natural selection 'may be compared to the analytical treatment of the Theory of Gases, in which it is possible to make the most varied assumptions as to the accidental circumstances, and even the essential nature of the individual molecules, and yet to develop the general laws as to the behavior of gases, leaving but a few fundamental constants to be determined by experiment.' (Fisher, 1922: 321–2; quoted from Provine, 1971: 149).

It is perhaps worth pointing out that the work of R. A. Fisher was of monumental importance in clarifying a number of issues and in developing the first wholly consistent mathematical fusion of Mendelian and Darwinian theory. Even today, his ideas form the basis of much of population genetics as it is both practised and taught, and his books on statistics and experimental design are still classics in the field. He contributed centrally to the development of population genetics, and must be recognized as one of the leading mathematical population geneticists of the century. To point out his own self-conscious and self-styled mechanistic bias is not to detract from his stature or influence. But it is also true that he was, like all people, a product of his times, with both the strengths and weaknesses which that entails. Mechanistic bias was rampant in biology at the time, and Fisher thought of himself as applying mechanistic thinking to that most refractory area of study at the time: evolutionary theory. (An exception to Fisher's mechanistic concept can be found in his earlier (1915) treatment of sexual selection, where he describes interactive phenomena such as 'runaway processes' in evolution.)

As valuable as Fisher's approach was in providing a quantitative methodology, it ultimately had its limitations, as population geneticists came to realize. The mechanistic approach simply could not encompass or deal with more holistic interactions: of genes with other genes, of feedback and stabilizing processes within individuals or populations, or with varying rates of genetic change (evolution) within populations. It was the work of investigators such as Ivan Schmalhausen, Theodosius Dobzhansky, C. H. Waddington, and most especially I. Michael Lerner, which began to move the genetical theory of natural selection away from its mechanistic past toward a more holistic future.

Towards a more holistic approach: I. Michael Lerner and the concept of genetic homeostasis

While mechanistic trends predominated during the period 1890–1930, some more holistic developments, always part and parcel of evolutionary theory, were growing in importance. The new holistic, and partly dialectical thinking which began to emerge in evolutionary theory in the 1930s grew out of several realizations emerging in standard laboratory genetics in the preceding decade: (1) the growing notion of gene interactions, that is, of polygenic systems, coming out of the *Drosophila* and maize work; (2) an increased emphasis on gene–environment interactions and the importance of genetic diversity and heterosis for evolution; (3) a growing awareness of genetic feedback and control mechanisms functioning not only on the level of the individual organism, but also on the level of whole populations; and (4) finally, an increased interest in relating natural selection to the entire life history of the organism – from reproduction and fertilization to embryonic development and maturity. To see how these newer concerns found some concrete expression we will examine briefly the work on genetic homeostasis of I. Michael Lerner.

Born in Manchuria in 1910 Lerner emigrated to British Columbia in 1927, where he obtained a job caring for chickens at the University of British Columbia's agricultural department in Vancouver (anon, 1978). To keep his job, Lerner had to enroll in the course in animal husbandry, a decision which changed the course of his life. One of the problems most central to students of animal husbandry in the mid-1920s was that of heterosis (the greater vitality and especially fertility of hybrids): how general was it? What is the genetic and/or physiological basis of heterosis, and finally, what is the relationship between outbreeding (which produces heterosis) and inbreeding (which produces increased homozygosity)? The answer to these questions was of great importance for agricultural production, but neither practical tradition nor scientific investigation yielded a clear-cut answer in the 1920s. While in general heterosis seemed to produce beneficial changes, every practical breeder knew that some amount of inbreeding was necessary to fix and stabilize traits in most stocks. (Sewall Wright was also aware of this problem and had already worked on it while employed by the US Department of Agriculture between 1915 and 1925.)

It was during his stay in Vancouver that Lerner met Theodosius Dobzhansky, who came to the University of British Columbia to give a lecture. This meeting not only introduced Lerner to Dobzhansky's energetic and dynamic personality, an attraction which he retained for the remainder of his life, but also to the literature on population genetics which, up to that time he knew only scantily. The combination of conversations with Dobzhansky, and his own reading, soon raised in Lerner's mind a whole host of questions about the relationship between selection, evolution, and, above all, the nature and meaning of heterosis. Bolstered by his own experience in animal husbandry, Lerner focused his interest on two specific questions which guided much of his later work. The first was: How does selection of the individual produce something of value for the population? If selection works ultimately on individuals, and yet if species are integrated populations possessing a common gene pool, selection of individuals and of populations must somehow be interrelated. The second question Lerner asked was quite different, though ultimately related: How can selection for adult traits (as exemplified by the work of neo-Darwinians such as Fisher), be related to selection for embryonic traits such as that described by C. H. Waddington? Since Lerner was becoming convinced from the work of various authors (as will be discussed below) that selection must act not only on individual or adult traits, but on the organism as a whole, he was seeking a way in which selection could be understood as acting effectively, and in a coordinated way, at all phases of the organism's life cycle. Although it was not until the mid-1950s that Lerner was able to put his ideas into a full and comprehensive form, they developed slowly away from the more mechanistic, and along a more holistic line from the late 1920s onward.

One of the most important influences on Lerner's thinking in the years immediately after meeting Dobzhansky was Walter Bradford Cannon's popular physiology book *The Wisdom of the Body* (Cannon, 1932). Cannon, a neurophysiologist at Harvard Medical School, had developed the idea of homeostasis – that is, that living organisms possess a variety of mechanisms by which they maintain, in Claude Bernard's terms, the constancy of their internal environments. Cannon was not the first to note this principle. He was, however, a pioneer in showing some of the mechanisms in the vertebrate body which make such responses possible. The basic principle, Cannon pointed out, was that of negative feedback, operating in a manner similar to a thermostat-furnace system in a house. Homeostatic systems usually involved a number of bodily systems – brain, endocrine and neuromuscular, for example – acting in an integrated way. Homeostasis was of necessity a dynamic and holistic concept, involving the notion of integrated and interacting parts.

An important feature of Cannon's work was that he saw homeostasis operating on a number of different levels: the organ-system, organismic, psychological and social. Along with his Harvard colleague Lawrence J. Henderson, Cannon believed that societies had to develop in-built homeostatic control receptors or else risk collapsing whenever confronted with serious pressures from without. It is not difficult to envision Cannon's concern with the social application of homeostasis when we remember that his book was written, if not conceived, in the early 1930s, in the wake of the great financial crash of 1929. Both Cannon and Henderson drew explicit parallels between the body physiologic and the body politic, and saw an important social lesson to be learned for human social organizers – namely, to build self-regulating processes into the economic, social and political system. (See Allen (1978a) and Russett (1966) for a fuller discussion.)

Significantly, in his larger view of the homeostatic process, Cannon saw its clear evolutionary implications. Homeostatic mechanisms *evolved*, through natural selection, because they were adaptive. Cannon also noted that the vertebrate homeostatic mechanisms developed gradually during ontogenesis, paralleling in a peculiarly Haeckelian way, the phylogenetic path by which homeostasis was acquired in the first place. In a single sentence Cannon connected both ontogeny and phylogeny. Noting that new-born humans lack the physiological stability for regulating such conditions as temperature and blood sugar, Cannon wrote: 'The evidence that homeostasis as seen in mammals is the product of an evolutionary process – that only gradually in the evolution of vertebrates has stability of the fluid matrix of the body been acquired – is interestingly paralleled in the development of the individual.' (Cannon, 1932: 283). This passage could not have helped but make a particular impression on Lerner in a work which he was already impressed with from the start. Here was a way in which Lerner could make the connection between contemporary evolutionary theory and embryology in words that might avoid the gross simplifications of the older school of population genetics, and build a truly integrated concept of genetics, evolution and embryology.

Inspired by Cannon's physiological concept of homeostasis, Lerner drew from several other contemporary works. While the exact contribution of these sources cannot be dealt with here, a brief summary will nevertheless be helpful. The five major ideas on which Lerner drew were C. H. Waddington's view of embryonic canalization, Fisher's and Ford's concept of genetic polymorphism, Ivan Schmalhausen's notion of 'stabilizing selection', Dobzhansky's studies of variation in natural populations and Kenneth Mather's analysis of polygenic inheritance. From Mather (1943a, b) he gained a greater sense of importance of the interaction of genes, as opposed to single gene systems, in determining the development of any embryonic or adult trait. From Dobzhansky (1947) he came to realize empirically that far more variation existed in natural populations than had previously been suspected. (In a survey of *Drosophila pseudoobscura* in the American southwest, a classic description of which was published in 1947, Dobzhansky had shown that natural populations were heterozygous for a large number of chromosome inversions.) From Schmalhausen (1949) Lerner imbibed the idea that heterozygosity is, in fact, more advantageous for a population than homozygosity because it allows for genetic flexibility in the face of environments which vary continually in space and time. From Fisher's theoretical consideration in conjunction with E. B. Ford's (1940) field studies on butterflies, Lerner came to appreciate the significance of balanced polymorphism as one aspect of maintaining genetic diversity within a population. And finally, from Waddington, Lerner came to view the process of embryonic development as a series of downhill channels which buffered the developing embryo against unpredictable external influences. It was the concept of homeostasis that provided a focus and organizing principle around which all of these varying ideas could be gathered, and led to Lerner's notion of 'genetic homeostasis'.

In articulating his theory, Lerner began with the assumption that heterozygosity such as that observed by Dobzhansky or Ford in natural populations must be adaptive (or else it would not exist to such a large extent). The adaptive value lay in the genetic flexibility which hetero-zygosity could provide for the population as a whole, given constantly changing environmental conditions. Fisher and Ford had shown that populations spread over diverse environments (such as a combination of forest and field habitats) often displayed a balanced polymorphism in which heterozygotes, not necessarily best adapted to either habitat in themselves, nevertheless continued to exist as a by-product of the process of selecting for their respective homozygotes. Along the same line, Dobzhansky had shown that the proportions of various inversion heterozygotes in a natural population of *Drosophila* changed with different seasons. It seemed clear to Lerner that heterozygosity acted as a sort of genetic 'buffer' providing flexibility for populations which faced a variety of environmental changes.

At the same time, heterozygosity could also be adaptive for individuals by providing genes for alternative chemical pathways, and hence greater biochemical and physiological flexibility. The greater the diversity of genes, the more alternative biochemical pathways open. Lerner was familiar with the work on genes and biochemical pathways from the 1940s onward, but he wisely left the basic mechanism of such an adaptation obscure. The important point, he felt, was to suggest that heterozygosity could be seen as having an advantage for the individual as well as for the population. At the same time, a genetic and biochemical mechanism of this sort would also explain the longstanding question of the nature of heterosis itself – that is, why hybrid forms are usually more vigorous and fertile than inbred ones. Hybrid individuals had more flexibility and could adjust quickly and effectively to changes in their surroundings (even to such localized changes as ones of diet).

Up to this point, however, Lerner's conception omitted any reference to embryonic development. Using C. H. Waddington's (1942, 1948) concept of embryonic canalization, Lerner was able to incorporate the embryo as well as the adult into his concept of homeostasis. Waddington's concept of canalization was that embryonic cells or tissues can be visualized as if they started development at the top of a hill where all cells are non-specialized. In normal development, the cells can be seen as moving down the hill through a series of branching channels, becoming progressively less flexible in their determination. As a cell enters its first channel at the top of the hill, it is responding to the first internal cue in its determination – it can now follow only where that channel and its ramifications lead. At every branch point fewer and fewer options are open for the future course of the cell's development. At the end of the developmental pathway the cell has been channelled, or as Waddington put it, 'canalized', into a specific developmental pathway by a series of specific and internalized cues.

Canalization was not, in Waddington's view, a totally mechanical process. He visualized that in the early stages of evolution, cells were triggered to go into different developmental 'channels' by environmental agents such as light or temperature. However, as evolution proceeded, internal cues (positions of one cell with respect to other cells, for instance) were substituted for external cues – largely because internal cues were more reliable and hence more adaptive as guiding forces for the complex and step-wise process of embryonic differentiation. Nonetheless, according to Waddington, embryos (especially non-mammalian embryos) were still exposed to some environmental variations, and the problem was how to keep differentiation proceeding along lines which would produce the desired ends (that is, a fully functional organism) in spite of sometimes substantial fluctuations in external conditions. Waddington saw each embryonic cell as able to respond to various cues in direct proportion to its age and cell type. 'Canalization' offered the possibility of some flexibility in development because the channels were often quite general at first, so that cells entering a given one, let us say for producing mesoderm, could subsequently give rise to many kinds of mesoderm. If one portion of mesoderm tissue were injured, it might well be possible for other tissues in the mesoderm channel to take over the lost function, since they themselves were not at that point highly committed to one path of differentiation. This developmental flexibility was enhanced by genetic diversity within the embryo – that is, by heterozygosity. As early as 1895 Hans Driesch had called attention to the embryo's flexibility – its functioning as what he termed a 'harmonious equipotential system'. To Lerner, this functional flexibility was the result of the number of alternative channels for development which was made possible by greater genetic diversity within populations and hence within individuals. Heterosis could be understood by an obvious physical and chemical model that correlated well with much that was known about genetics and development.

Lerner concluded from his studies that at all levels of organization flexibility is an adaptive trait in organisms, selected for by natural selection. Homeostatic mechanisms – at the embryonic, organismic (physiological) and populational levels – work to maintain that flexibility. The point is, however, that selection acts on mechanisms – balancing mechanisms if you will – which generate the capacity for flexibility at all levels or stages of the life cycle. The importance of such flexibility as homeostatic mechanisms generate is, of course, ultimate stability of the living system. Ironically, by a process of constant *change* (homeostatic mechanisms are constantly changing in a dynamic interaction between two or more parts – e.g. thermostat and furnace) the overall system maintains its *constancy*.

Lerner's approach, as expressed in *Genetic Homeostasis* (1954) is just one example of that new more holistic approach to evolution and genetics, emerging in the post-1940s period. Holistic materialism appears in several aspects of Lerner's works.

First, he is explicitly concerned with interrelationships, with showing how the same genetic property (heterozygosity) affects processes at the individual (adult and embryo) and populational levels. Genotype affects all aspects of the organization of living systems.

Second, Lerner's holistic materialism demonstrates clearly that at whatever level one looks – population, organismic, embryonic – the whole of a living system is greater than the sum of its parts. For example, the parts of any homeostatic system are meaningless viewed in isolation from one another. Each part is defined in certain ways by its relationship to the other part (or parts) with which it interacts. The concept of homeostasis *focuses* attention on these interactions. They become part of the very definition of the organic parts of any system. In forcing our attention onto interactions (as opposed to simply paying lip-service to such properties) Lerner's concept of genetic homeostasis is distinctly holistic in outlook.

Third, Lerner emphasized that just as in physiology, so in population genetics homeostasis is a dynamic process. It occurs by the continual interaction of its parts – a thermostat–furnace system is constantly at work. It is not like a balanced see-saw which, once in position, remains the same unless disturbed. Homeostatic mechanisms are ongoing – they never stop unless they break down. Lerner's use of the homeostatic model emphasizes the dynamic nature of such processes.

A fourth feature of Lerner's holistic view is that the mechanisms of genetic homeostasis – at both the individual (adult or embryo) and populational levels – is one internal to the system itself. That is, homeostatic mechanisms, while capable of responding to changes in the external environment, operate primarily through internalized relationships. For example, if heterozygotes within a population are selectively advantageous, a decrease in heterozygotes relative to homozygotes will mean that more of the different homozygotes (proportionally) will find each other as mates, and thus increase the frequency of heterozygotes in the next generation. The response is built into the system itself (in this case a random-breeding population).

Conclusion

In this chapter I have discussed one historical trend within post-Darwinian evolutionary theory, a trend deriving from two philosophical aspects of Darwin's own approach in *The Origin of Species*. On the one hand Darwin's mechanistic materialism gave rise, among most of his followers through the 1930s, to an ultimately oversimplistic view of the evolutionary process (the organism as a mosaic of traits, each trait governed by a single gene, evolution as single-gene substitution). R. A. Fisher was used to illustrate one aspect of this approach. On the other hand, Darwin's far more holistic, even dialectical view of the evolutionary process led to interpretations which were often at odds with the more mechanistic explanations of how natural selection worked. This dissatisfaction was at the heart of many scientific (especially biological) arguments against Darwinian theory, in both the periods 1890–1920, and 1930–1940s. The work of I. Michael Lerner illustrated one attempt – by no means the only one – to break away from and grow beyond mechanistic materialism to a more holistic approach to the evolutionary process. I should emphasize that I chose to focus on Lerner here not because he was the most influential population geneticist who tried to go beyond the work of Fisher and others – in many ways he was far less influential than Dobzhansky or Sewall Wright. I chose Lerner because more than most others, he tried explicitly to bring many problems together in an outwardly holistic vein approach. He is, if you will, a 'favourable organism' for historical study because he represents clearly a general trend, which is visible though perhaps less clearly, in the work of many evolutionists in the 1940s and 1950s. Lerner was *not* a major cause of a shift in thinking among evolutionists, nor was he a major influence – let us say to the degree Fisher was – on evolutionary thinking. Nonetheless, Lerner does illustrate clearly a developing change in philosophical perspective that was becoming increasingly apparent among ultimately more influential investigators such as Dobzhansky, Wright, Haldane, Ernst Mayr, George Gaylord Simpson, G. Ledyard Stebbins and others during the 1940s and 1950s. This is not to say, of course, that Lerner was largely ignored, or bore no influence. A number of references to Lerner's work, especially *Genetic Homeostasis*, can be found in the writings of evolutionists in the 1950s and 1960s. The actual extent of his influence could be the subject of an interesting paper. In the present context, however, Lerner is used more to suggest how holistic materialism was substantively involved in critiques of the older mechanistic view of evolution, than to suggest how Lerner himself influenced the future course of evolutionary theory.

In conclusion, it seems important to point out that the three faces of Darwin – mechanistic, holistic and dialectical materialism – are still with us today in writings on modern evolutionary theory. The debate between the gradualist and punctuated equilibrium schools is a clear example of a controversy between, respectively, the mechanistic and holistic approaches. Gradualists prefer to see all of evolution as the slow accumulation of small, quantitative changes, proceeding not only at a gradual, but at an *even* pace. They have a low tolerance for the holistic or dialectical notion that quantitative changes lead to qualitative changes, and that those qualitative changes can occur rapidly. In other words, evolution leads to revolution, which, among other things, implies a significant change of pace. The notion of punctuated equilibrium, however, as expressed particularly by Gould (1980), shows a clear appreciation of just this point, and builds it into the very process of evolutionary development itself. In punctuated systems, quantitative changes lead to qualitative changes, for example, situations of sudden evolutionary developments which can quickly give rise to speciation events – that is, differentiation of one single species into two or more species (which amounts to, so far as I would define it, a qualitative change).

Mechanistic thought is still prevalent in much of evolutionary theory when, for example, biologists speak of organisms reacting to the environment – as if the environment were not shaped by the organism as well, but rather always stood outside it and apart. A newer approach to this issue, appearing earlier in Ehrlich & Raven (1964) and more recently (and explicitly) in Lewontin & Levins (1978), has come to light and promises to yield a more subtle and refined picture of organism– environment interaction. Explicitly applying a dialectical materialist view, Lewontin & Levins see the organism as altering its environment by the very process of adaptation itself. Hence, even if the environment does not change much through its own processes (geological, climatic, etc.) it will be slowly evolving through organism–environment interactions. While such interactions may seem trivial or unimportant in some cases, the method of considering such interactions is important in all cases precisely because it is a way of looking at the evolutionary process – a frame of mind, so to speak. Whole new areas of investigation may be opened up by such considerations which would be otherwise hidden from view, and possibly ignored.

Finally, let me suggest that one of the lessons we may learn from history is how to develop our own thought processes beyond the obvious and necessary limitations of the prevailing philosophical view. Darwinian theory was born and has developed in an age founded on the principles of temporal gradualism and mechanistic philosophy. This was historically in tune for a theory developing out of mid-nineteenth century industrial society. It need not be a view appropriate to what has been learned in the intervening century since Darwin's death – either in the biological or the social/philosophical realm. In our traditional mind-set, gradualism is comfortable, rapid change – revolution in the political, social or philosophical sense – is uncomfortable. Even on the microscale we seek consistent theories and views. We try to ignore, shove under the rug, or minimize contradictions, or opposing tendencies within what we think are the realities we try to describe. Perhaps another way of viewing ourselves, our thought processes, and our actions, is to *seek out* contradictions, to look for change, even rapid change, and thus to see such change (and this means sometimes rapid, qualitative change) as inevitable in the real world, a component of the reality in which we live. Perhaps in this way we will be able to grasp the dynamics of the evolutionary process – biological and social – in a more holistic, and ultimately more realistic way.

Most of this paper was written while holding a Fellowship at the Charles Warren Center for Studies in American History at Harvard (1981–82). I benefited greatly from the opportunity for research and writing which this Fellowship afforded. The initial ideas and drafts were immeasurably improved from discussions with Ernst Mayr, Richard Lewontin, Steve Gould, Randy Bird, John Beatty, John Niffenegger, and from feedback at the joint meeting of the Society for the Study of Evolution and the American Society of Naturalists, Stony Brook, NY, June 22, 1982. I am particularly indebted to Doug Futyuma for the invitation to present this material at that time.

References

Allen, Garland E. (1978a). *Thomas Hunt Morgan. The Man and His Science.* Princeton: Princeton University Press.

Allen, Garland E. (1978b) *Life Science in the Twentieth Century.* New York: Cambridge University Press.

Anonymous, (1978). Obituary: I. Michael Lerner. *Genetics,* **88,** (Supplement); S 139–40.

Barrett, Paul H., Weinshank, Donald J., & Gottleber, Timothy T. (1981). *A concordance to Darwin's Origin of Species, First Edition.* Ithaca, New York: Cornell University Press.

Cannon, Walter B. (1932). *The Wisdom of the Body*. New York: W. W. Norton.

Dewey, John (1909). The influence of Darwinism on philosophy. In *The Influence of Darwinism on Philosophy and Other Essays*. New York: Henry Holt and Company, Inc.

Dobzhansky, Theodosius (1947). Adaptive changes induced by natural selection in wild populations of *Drosophila*. *Evolution*, 1, 1–16.

Ehrlich, P. R. & Raven, P. H. (1964). Butterflies and plants: a study in coevolution. *Evolution*, 18, 586–608.

Fisher, Ronald A. (1915). The evolution of sexual preference. *Eugenics Review*, 7, 184–92.

Fisher, Ronald A. (1922). On the dominance ratio. *Proceedings of the Royal Society of Edinburgh*, 42, 321–41.

Fisher, Ronald A. (1930). *The Genetical Theory of Natural Selection*. Oxford: Clarendon Press.

Ford, E. B. (1940). Genetic research in the Lepidoptera. *Annals of Eugenics*, 10, 227–52.

Gould, Stephen Jay (1980). Is a new and general theory of evolution emerging? *Paleobiology*, 6, 119–30.

Hogben, Lancelot (1927). *Principles of Evolutionary Biology*. Capetown, South Africa: Juta.

Lerner, I. Michael (1954). *Genetic Homeostasis*. New York: John Wiley.

Lewontin, Richard & Levins, Richard (1978). Evoluzione. in *Enciclopedia*, V, *Divino-Fame*, 995–1051. Torino: Einaudi.

MacKenzie, Donald A. (1981). *Statistics in Britain, 1865–1930*. Edinburgh: Edinburgh University Press.

Mather, Kenneth. (1943a). Polygenic inheritance and natural selection. *Biological Reviews*, 18, 32–64.

Mather, Kenneth (1943b). Polygenics in development. *Nature*, 151, 960.

Provine, William (1971). *The Origins of Theoretical Population Genetics*. Chicago: University of Chicago Press.

Russett, Cynthia Eagle (1966). *The Concept of Equilibrium in American Social Thought*. New Haven: Yale University Press.

Sachsse, Hans (1968). *Die Erkentnis des Lebendigen*. Braunschweig: Friedrich Vieweg & Sohn.

Schmalhausen, Ivan (1949). *Factors of Evolution*. Philadelphia: Blakiston.

Schweber, Sylvan (1980). Darwin and the political economists: divergence of character. *Journal of the History of Biology*, 13, 195–289.

Waddington, C. H. (1942). Canalization of development and the inheritance of acquired characters. *Nature*, 150, 563–5.

Waddington, C. H. (1948). The concept of equilibrium in embryology. *Folia Biotheoretica*, 3, 127–38.

MOLECULAR AND CELLULAR EVOLUTION

6

Self-replication and molecular evolution

MANFRED EIGEN

Selection and evolution governed by natural law

If we, living in the second half of the twentieth century, ask ourselves whose work has furthered our understanding of the phenomenon of 'life' to the greatest extent, then the name Darwin springs at once to mind. 'If you ask whether we shall call this the century of iron, or of steam, or of electricity, then I can answer at once with complete conviction: it will be called the century of the mechanistic understanding of Nature – the century of Darwin', wrote his contemporary, Ludwig Boltzman in 1886 (see Boltzmann, 1905: 28).

It is scarcely possible that the founder of statistical physics, writing these lines, was referring to Darwin the observer, the botanist, the zoologist. It was Darwin's perception to which he was paying tribute. Today, with the improved methods and profound insights of molecular biology, Darwin's gigantic empirical opus has become a matter of historical interest only, an outstanding achievement of the last century – not out-of-date, but out of the front line of modern science. Darwin's true legacy is an insight, which is as relevant in the twentieth century as in the nineteenth, even if today the emphasis is placed rather differently.

What new perception of Nature was encapsulated in the principle of selection? We do not mean here the actual biological process, with all its complicated superstructure – we want to understand the principle itself. What is it? An axiom, an underivable, fundamental law, to which all living matter is subject, which we simply have to observe and accept? Or is it a heuristic device, which simply describes a pattern of observations, but whose seeming regularity disappears when it is rigorously analysed?

It is neither the one nor the other. The principle of selection turns out to be a law which is rigorously derivable from physical premises.

To illustrate this, let us examine a physico-chemical law with which we are all familiar – the Law of Mass Action. It governs the distribution of chemical reactants in equilibrium. In thermodynamics, 'equilibrium' is characterized by an extremum principle, which in this case means that in a system at constant temperature and pressure, which does not exchange matter with its surroundings (a 'closed' system), the Gibbs' free energy tends towards a minimum, and the system is in equilibrium when this minimum is reached. At this point the distribution of reactants and product is determined, exclusively and therefore reproducibly, by the mass-action constant.

The selection principle may be approached in a very similar way. We are interested in the selection of a particular genotype which on the basis of its phenotypic properties may be summed up as 'the fittest'. The genotype, the genetic information of an organism, is represented by an enormous molecular chain, in which each monomeric subunit (nucleotide) corresponds to a 'letter' of the genetic message. Even in the simplest, single-celled organisms, such as the coli bacterium, this message consists of four million nucleotides, corresponding in letters to a 1000-page book. When a genetic message is reproduced, some letters are 'read off' wrongly; we call these 'mutations'. They are the origin of evolutionary progress. Selection thus means two things.

(1) The information representing the fittest phenotype has to be chosen, copied many times over, and made into the starting point for further development.

(2) The information for the fittest phenotype has to be stable enough to remain intact until a better mutant arises. In other words, mistakes must not accumulate but must remain below a critical threshold level.

One might formulate the physical problem which is to be solved by way of selection as follows. What mechanism could guarantee the necessary but spontaneous origin of information?

The complexity of living organisms rules out any solution for this problem such as is offered by the Law of Chemical Mass Action. Even a single gene, which encodes one of many thousand functions, has an inconceivably large number of possible arrangements of its elements, even if the total length and overall composition are known. Accumulation of errors could lead from a given sequence to any other one, and if thermodynamic equilibrium were to be reached all possible sequences would appear, with various probabilities. It is easy to estimate that if each possible sequence were represented by one molecule then the volume of the universe would not be sufficient for more than a tiny fraction of the material necessary.

Only selection can circumvent this dilemma. An essential prerequisite for this is the prevention of equilibrium, and this is only possible by continually supplying free energy to the system. In organisms this requirement is met by metabolism.

Natural selection rests on two further material prerequisites: self-reproduction of molecules in turn presupposes particular structural properties, which are fulfilled, for example, by the nucleic acids. It is these which endow all organisms with the ability to reproduce themselves. Variability is simply a consequence of 'blurred' self-reproduction.

In earlier publications (Eigen, 1971, 1981, 1982; Eigen & Schuster, 1977, 1978a, b; Eigen *et al.*, 1981) we have shown that the principle of selection can be deduced from the premises of a self-replicating system as an extremum principle. It states that inherent linear autocatalysis causes the relative population numbers to take on values which correspond to the highest reproductive efficiency of the system as a whole. The distribution of relative concentrations in the stationary population is, after a short induction period, independent of changes of the system as a whole. The population consists of a uniquely defined wild type (or several equivalent, *i.e.* 'degenerate', variants) and a spectrum of mutants. The wild type, which is the mutant appearing most frequently in the distribution, qualifies as such through its superior selective advantage. Since only a fraction of the wild type is correctly copied in each replication round, the selective advantage relative to the average productivity of all mutants must be sufficiently large to compensate for the loss due to faulty copying. There is a corresponding stability criterion for the selection of the wild type. If this is not fulfilled, then an 'error catastrophe' occurs. The information stored in the wild type gradually gets lost by accumulation of copying errors. Moreover, the appearance of a mutant that has selective advantage over the existing wild type likewise leads to the collapse of the previously stable distribution, whereupon the advantageous mutant establishes itself as the new wild type.

In comparison to the whole set of mutants the wild type can be present in small absolute number. The deficit due to copying errors is compensated by the selective advantage (σ) of the wild type (relative to the average for all mutants). The fraction of correct copyings (Q) decreases with increasing number of symbols in the sequence. If v is the number of symbols in the sequence and $\bar{q}$ the average copying accuracy per symbol, then one has as a good approximation $Q = \bar{q}^\nu$. Maximum utilization of the information capacity then means that ν is maximized and thereby $Q = \bar{q}^\nu$ is minimized (since $\bar{q} < 1$), so that the stability condition $Q\sigma > 1$ is just fulfilled.

Both processes, *selection* as stabilization of a mutant distribution of a wild type that is best adapted with relation to its information content and *evolution* as successive establishment of new distributions with steadily increasing selective value, are inherent characteristics of the system. They are unavoidable – and thus mathematically deducible – consequences of self-reproductive behaviour.

This evolutionary self-organization of replicative forms of matter can be viewed as a mountain climb up a cliff of selective value. There is a difficulty in this analogy, however. The improvement mechanism only permits progress to occur while the climbing route leads uphill. In other words, only when a mutant with a selective advantage actually appears can its selection lead to the system reaching a higher value level. The system can not by itself – without external intervention into the environmental conditions – climb down from a value level that it has once reached; it must be able to leap over 'valleys'. The chances for leaping (multiple mutations) are, to be sure, quite limited. In terms of a landscape that we would encounter on the surface of the earth, the optimization process would soon come to a halt at the peak of some relatively obscure hill. The peak of Mount Everest could hardly be scaled in this way even if the tour were to begin in the highlands of Tibet. From investigations of enzyme reaction mechanisms, however, we know that the level of optimization of proteins is more comparable to Mount Everest, Mont Blanc or the Matterhorn than to a small hill.

The fact that the evolutionary optimization process indeed reaches the 'mountain peaks' and does not get stuck in the 'foothills' lies in the topology of multidimensional mutation space. Consider for example a binary sequence with ν members. We can give each position in the sequence a coordinate axis with two points and thus obtain a ν-dimensional phase space in which each of the 2^ν points represents a mutant. The process of evolution can then be regarded as a route in this space, characterized by a continually rising selection value. The topology of such a multi-dimensional space is not easily imaginable; the 'mountains' are extremely bizarre, for although these are 2^ν points the greatest separation (in terms of mutation steps) is only ν. There are saddle-points of various orders. On them, movement in k directions leads uphill and in $(\nu - k)$ directions downhill. For this reason relatively small mutational jumps suffice to find an uphill route in connected mountain ranges. There will be a range of v-values for which the number of routes is already quite large and the probability of multiple mutations still appreciable, so that at least in a connected region the highest 'peak' can be reached (A. Dress & M. Eigen, unpublished theoretical results).

Let us summarize: selection, evolution and optimization are processes which follow regular physical laws and which can be formulated quantitatively. This of course does not mean that the actual, historical process of evolution can be deduced from theory; the starting-point, the complex boundary conditions and the multitude of superimposed perturbing influences are all more or less unknown. Theory tells us simply what follows when certain premises are set up and certain boundary conditions are imposed. It explains the reproducible, regular phenomenon of Nature in an 'if – then' description. This generalization applies to the theory outlined here, which has helped us to interpret and explain the experimental results described below. It has been confirmed by quantitative measurement under the exactly-defined initial and boundary conditions of the laboratory. For processes occurring in Nature, however, it reveals only trends, minimal requirements, limitations and, perhaps, some consequences. It must also be shown by experimental investigation whether conclusions from the theory have any relevance for naturally-occurring processes. Two such conclusions are especially worthy of mention.

(1) The quantity of information which can be selected in a molecular population depends upon the average error rate and the average selective advantage of the wild type. Crossing the critical error threshold leads to such an accumulation of errors that the information in the wild-type sequence is irretrievably lost.

(2) The capacity of the wild type for adaption is greatest close to the error threshold. The quantity of information compatible with a stable distribution is then in optimal relation to the variety in its spectrum of mutants. Such a system responds very flexibly to changes in its environment. The wild-type is the predominant individual sequence, but it makes up only a small fraction of the complete mutant spectrum.

The predictions of the theory of evolution can be tested on natural systems. Charles Weissmann and his co-workers (Domingo, Flavell & Weissmann, 1976; Batschelet, Domingo & Weissmann, 1976) have studied the mutant distribution of the $Q\beta$ virus wild type. (For characterization of this type of virus see the next section.)

The wild type has a defined sequence, which, however, does not mean that the majority of viruses share exactly the same sequence. It means merely that the superposition of all sequences gives an unambiguous 'majority' or master sequence, namely, that of the wild type.

The cloning of single viruses or single viral RNA molecules followed by their rapid multiplication leads to populations with various sequences. The sequences generally deviate in one or two positions from the wild type, which is itself found in hardly any clone (see Fig. 6.1). The fact that the wild type makes up only a (nearly negligibly) small fraction of the mutant spectrum implies that the information content of the wild type has very closely approached the threshold value of ν.

Deliberately-produced extra-cistronic single mutations (these are non-lethal mutations in portions of the sequence that are not translated) revert to the wild-type. A given nucleotide is changed with a probability of around

Fig. 6.1. In this experiment carried out by Charles Weissmann and his co-workers, single Qβ-RNAs from a wild-type distribution (a) were cloned. After rapid multiplication the clones of individual RNA molecules (b) were analysed and compared by the fingerprint method. Differences were noticed in one or two positions in the sequence. After a further, long period of reproduction (c) the wild-type distribution (d) was found in every clone, *i.e.*, the average sequences had become identical again.

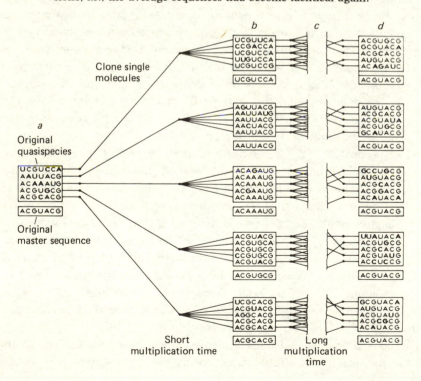

3×10^{-4}. Quantitative analysis of such data allows the determination of the error rate, the selective advantage of the wild type and thus the establishment of the critical error threshold for the transmission of the information in the genome. This value agrees within experimental error with the quantity of information present (4500 nucleotides). The fact that the cloning of individual mutants is possible at all is due to the dominance in the distribution, around the wild type, of mutants whose growth rates are very similar to that of the wild type itself. These are preferentially 'fished out' in the serial dilution steps needed for the cloning of single molecules. Since they multiply nearly as rapidly as the wild type, they begin by producing a spectrum of mutants with an average sequence identical to their own. At some point, the wild type will reappear in this spectrum. However it can only assert itself slowly, the speed with which it does so corresponding to the (small) difference between the growth rates of the wild type and the cloned mutant. Finally each clone is dominated by the wild type (see Fig. 6.1).

Yet again the conclusion can be drawn that all single-stranded RNA viruses are subject to similar restrictions with regard to information content. In Nature there are no (single-stranded) RNA viruses whose replicative unit contains more than the order of 10^4 nucleotides (Table 6.1). All larger viruses possess double-stranded nucleic acids or are composed of several replicative units. These in turn are subject to analogous relationships between the error threshold and the maximum reproducible information content. DNA polymerases in general work more

Table 6.1. *Single-stranded RNA viruses*

Host	Virus	Number of nucleotides	Replicative units
Bacteria	Qβ	4500	1
	R17; MS2	4000	1
	φCb5	~4000	1
	PP7	~4000	1
Plants	Cucumber Mosaic	~3500	1
	Tobacco Mosaic	6800	1
	Como	~9000	1
Animals	Picorna (Polio)	~6000	1
	Toga (German Measles)	12000	Several
	Orthomyxo (Influenza)	12000	8
	Paramyxo (Measles)	22000	1
	Tumour	30000	3

accurately than RNA polymerases. This is due to their additional facility for recognition and correction of errors (Fersht, 1981; Loeb & Kunkel, 1982).

The results of this section lead to the conclusion that selection and evolution are based on self-reproduction in an open (far from equilibrium) system and as such are inevitable consequences of the prerequisites that can be formulated as natural laws.

In the next two sections we shall consider whether bio-molecules can be produced in the laboratory by an evolutionary process on the basis of this natural law, and whether these molecules resemble their natural counterparts in their optimal fulfillment of a prescribed function.

Experiments on molecular evolution

We shall first examine in some detail the mechanism of reproduction of a virus whose genetic information lies encoded in a single-stranded RNA molecule. This virus is able to perform basically four functions, each of which is associated with a different protein unit: a capsid, which encloses the RNA and protects it from degradation by hydrolysis; a penetration enzyme, to inject the genetic material into the host cell; a factor to break down the host cell; finally, a mechanism which re-programmes the complex machinery of the host cell in accordance with instructions from the virus. In the case which we shall discuss, the last function is carried out by a protein molecule that associates with three ribosomal proteins to give an enzyme which recognizes the viral genome and replicates it very rapidly. The apparatus of metabolism and gene translation is thereby turned over to the exclusive purpose of producing new virus particles. This factor thus embodies the essence of viral infection.

The reproduction mechanism of $Q\beta$, an RNA virus capable of infecting bacteria, is shown schematically in Fig. 6.2. During the replication of RNA, the enzyme, which consists of four subunits, moves from the 3' to the 5' end of the template. The newly-formed RNA replica has an internally folded structure and this prevents the formation of a double strand. Spiegelman and his colleagues (Haruna *et al.*, 1963; Haruna & Spiegelman, 1965; Spiegelman *et al.*, 1965; Levisohn & Spiegelman, 1968, 1969), who were the first to isolate the reproduction enzyme of this virus and, using it, to synthesize infectious viral RNA, have shown that annealing destroys the reproductive capacity of the virus. The reproduction does not take place at a steady speed: the enzyme pauses at so-called 'pause sites'. Presumably it has to wait for the next portion of the matrix to melt apart before it can proceed to copy it. The discovery and isolation of a non-infectious RNA component, 220 nucleotides long, are also due to Spiegelman (Mills, Kramer & Spiegelman, 1973; Kramer & Mills, 1978); this RNA molecule, known as 'midivariant', contains the 3'- (and the 5'-) end needed for recognition by Qβ replicase and is therefore replicated by the enzyme – in fact much faster than the real Qβ–RNA. 'Midivariant' is thus a 'scrounger', which cannot infect a cell alone, since it cannot produce the specific reproduction enzyme.

In our laboratory the mechanism of RNA replication by Qβ replicase has been studied quantitatively (Sumper & Luce, 1975; Biebricher, Eigen &

Fig. 6.2. The single-stranded RNA of the bacteriophage Qβ is replicated with the assistance of an enzyme, called Qβ replicase, that consists of four subunits (black dots). The enzyme recognizes the template specifically. It moves from the 3' to the 5' end of the template strand during the elongation process. The replica formed (−) is complementary to the template (+). The 3' and 5' ends are symmetrically related in such a way that both plus- and minus-strands have similar 3' ends; both are recognised by the replicase, and the minus-strand thus acts as a template for the formation of a plus-strand. Internal folding of the two separate strands prevents the formation of a plus-minus double helix.

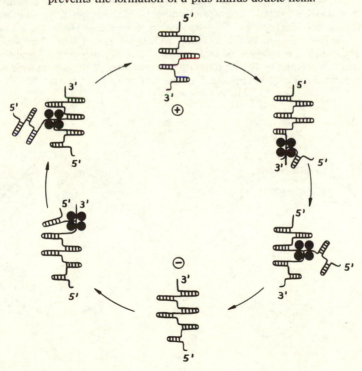

Luce, 1981*a*, *b*; C. Biebricher, M. Eigen & W. Gardiner, unpublished observations). The experiments were begun by Manfred Sumper and Rüdiger
Luce and continued by Christof Biebricher and Rüdiger Luce. Our knowledge of the mechanism of the replication reaction results from a combination of (i) experimental investigation of the rate of replication as a
function of the concentrations of substrate, enzyme and RNA templates,
(ii) the mathematical analysis of a model for the replication (see Fig. 6.3)
and (iii) the computer simulation of this model using realistic values for
parameters, obtained from experimental data. A typical experiment is
sketched out in Fig. 6.4. A ^{32}P-labelled nucleotide is used to measure the
amount of newly-synthesized RNA as a function of time. The concentrations
of the substrates (the four nucleoside triphosphates of A, U, G, and C) and
that of the enzyme are constant. The initial template concentration is varied

Fig. 6.3. A characteristic of the mechanism of RNA replication is the
coupled pair of synthesis cycles for the plus- and minus-strand respectively.
A catalytically active complex consists of the enzyme (replicase) and an
RNA template. Four phases of each cycle can be distinguished: (i)
commencement of replication by the binding of at least two substrate
(nucleoside triphosphate) molecules; (ii) elongation of the replica strand
by successive incorporation of nucleotides; (iii) dissociation of the complete
replica from the replicase; (iv) dissociation of the enzyme from the 5' end
of the template and its reassociation with the 3' end of a new template.
The template is represented by I (information), the enzyme by E and the
reaction product by P. The ultimate reaction product P_n is then used as
a template (I). The substrate S is the triphosphate of one of the four
nucleosides A, U, G and C.

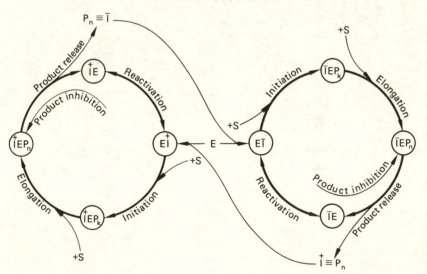

by serial dilution with a constant dilution factor. The increase in RNA concentration in the course of time can be divided into three phases.

(1) An induction period, which increases logarithmically with increasing dilution. Repeated dilution by a constant factor produces a series of constant displacements along the time axis.

(2) A linear increase in RNA concentration, starting at the moment when

Fig. 6.4. If a synthesis mixture (buffer, salts, Qβ replicase, nucleoside triphosphates and RNA templates) containing equal numbers of enzyme molecules and template strands is incubated, the rate of production of RNA molecules is constant in time, since the number of catalytically effective complexes is equal to the (constant) number of enzyme molecules. (The flattening of the curves at high RNA concentration is due to inhibition of the enzyme by binding of excess RNA to the replica site.) If the template concentration is lowered by constant factors, the growth curves are displaced along the time axis by equal intervals. This logarithmic dependence of the induction period indicates that an exponential growth law holds as long as the enzyme is in excess over RNA, during which time the number of catalytically active complexes steadily increases (auto-catalysis). Even if no template is present at the beginning, RNA arises 'de novo' after a long induction period. This serves as template and multiplies rapidly. Such synthesis of RNA 'de novo' is a particular property of Qβ replicase.

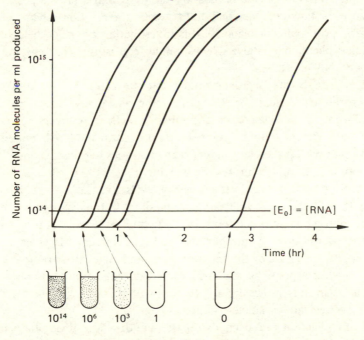

Number of RNA templates initially added

the RNA concentration becomes (approximately) equal to the enzyme concentration.

(3) A plateau, which is not reached until the concentration of the RNA greatly exceeds that of the enzyme.

This behaviour can be explained by the following kinetic considerations. The reaction leading to new RNA templates is catalyzed by a complex of enzyme and template. The affinity between these partners is so high that at the concentrations used every RNA molecule binds to an enzyme. The number of catalytically active complexes then rises exponentially until the RNA concentration becomes equal to the concentration of enzyme. At this point the enzyme is saturated with RNA. From now on the number of catalytically active RNA–enzyme complexes remains constant and the synthesis enters the linear phase, that is, the rate of appearance of new RNA molecules becomes constant. The new RNA molecules, now in excess over the enzyme, bind not only as templates but also, less strongly, at the site of synthesis. This leads to inhibition of synthesis by the excess RNA molecules present in the linear phase, and finally the synthesis comes to a standstill.

By altering the reaction conditions (e.g. substrate concentration, enzyme concentration, initial ratio of plus- to minus-strand) it proved possible to measure the kinetic parameters and to verify the reaction scheme of Fig. 6.3. The principal conclusions of this investigation are given below.

(1) The replication is catalyzed by a complex consisting of one enzyme and one RNA molecule.

(2) The growth rate is proportional to the lower of the two total concentrations (enzyme or RNA). This means for low and high RNA concentrations exponential and linear growth respectively.

(3) If two different mutants compete, then even in the linear region the mutant with the selective advantage continues to grow exponentially, until this mutant saturates the enzyme.

(4) A selective advantage in the linear phase is given solely by differences in the kinetics of enzyme–template binding. In the exponential phase, in contrast, the competition is based upon overall rates of production.

(5) The overall growth rate of plus- and minus-strands is given in the exponential phase by the geometric mean of their rate parameters and in the linear phase by the harmonic mean. The ensemble of plus- and minus-strands grows up in the ratio of the square roots of their respective rate parameters of production.

(6) The replication rate depends on the length of the RNA chain to be synthesized and on the concentrations of the substrates (A, U, G, C).

The latter dependence is weaker than linear as successive substrate molecules are frequently pre-bound to the enzyme.

We now arrive at the actual question which we wished to answer: what are the consequences of the laws of replication? Can new properties be attained by the evolution of a replicating system? Are the characteristics of the replicative system sufficient, or do we need to look for additional necessary properties?

Spiegelman (Mills, Peterson & Spiegelman, 1967) has provided an important stimulus for research in this direction. By serial transfer of RNA templates from one nutrient glass to another (cf. Fig. 6.5) he obtained, at the end of such a series, selected variants of the Qβ genome which were no longer infectious but which showed a higher replication rate (measured per nucleotide). They made use of their higher replication rate to escape the selection pressure of dilution. The significant finding was however that these new variants not only had a higher rate of chain elongation but also possessed about 500 instead of the original 4500 nucleotides, which enabled them to complete a round of replication in a fraction of the original time. Evolution of this sort, resulting in a loss of information, might well

Fig. 6.5. Serial transfer, a method devised by Spiegelman to prolong growth indefinitely, was applied to demonstrate the de novo initiation of RNA synthesis and the evolution of optimal templates. A series of test tubes is prepared with Qβ replicase, necessary growth factors, monomers of A, U, G and C and a few RNA templates. After incubation in the first tube a small portion is transferred to a second tube and incubated again. After many repetitions with decreasing incubation times a single, optimal template is selected. As an indication of the vast multiplication achieved by this method, if the growth in each tube amplifies the product RNA 10 000 times, 10 transfers are equivalent to a growth that would saturate the world's oceans with RNA.

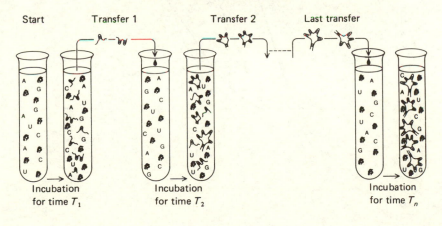

Start   Transfer 1   Transfer 2   Last transfer

Incubation for time T_1 Incubation for time T_2 Incubation for time T_n

Fig. 6.6. A solution of nucleotide triphosphate is incubated in the presence of Qβ replicase for just long enough to assure several replication rounds of any templates which may contaminate the enzyme. The incubation is interrupted before even one template has time to arise 'de novo'. The solution is then divided into portions and the incubation is continued long enough to allow products to arise 'de novo' and to multiply. The RNA formed in each portion is analysed by the fingerprint method; various different reaction products are found. Sometimes the growth curve displays the appearance of a new mutant. While the incubation time of template-instructed synthesis is determined unambiguously, 'de novo' synthesis shows a scatter of induction times. This indicates that the initiation step is a unique molecular process which is then rapidly 'amplified'.

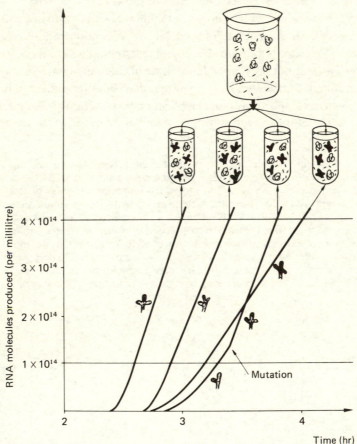

rather be termed 'degeneration'. However, the experiments showed that this replication system is highly capable of adaption – an indispensible pre-requisite for evolutionary behaviour.

These experiments became particularly relevant when, in 1974, a surprising observation was made by Manfred Sumper. In series which were diluted so far that in each sample the probability of finding a template molecule was very small, there arose none the less, reproducibly and homogeneously, a molecule with size and structure similar to the 'midivariant' isolated by Spiegelman. This phenomenon is indicated in Fig. 6.4. In contrast to the template-instructed replication, this synthesis was associated with a disproportionately long induction period which depended critically upon the reaction conditions. Sumper was convinced that he had found a variant which the Qβ enzyme had 'invented' and synthesized 'de novo'. (Other experts in the field believed almost unanimously that he was seeing an impurity carried by the enzyme.) Sumper was able by further experiments to refute the 'impurity' postulate, and since then the existence of synthesis 'de novo' has been amply proven, above all in experiments conducted by Christof Biebricher and Rüdiger Luce. Kinetic measurements have shown that the induction of template-instructed and 'de novo' synthesis are subject to entirely different rate laws. Thus, for example, the template-instructed synthesis proceeds on a single enzyme molecule and the substrates are introduced successively into the growing chain. Synthesis 'de novo', on the other hand, requires a complex of several enzyme molecules, and the rate-determining step is a nucleation by three or even four substrate molecules. However the principal evidence lies in the demonstration that, in the early phase of synthesis, variants appear which lengthen under selection pressure and which lead under different experimental conditions to different end-products. The last observation is also due to Manfred Sumper. He obtained different 'minivariants', which grew normally under conditions – e.g. in the presence of reaction inhibitors – in which the wild-type was no longer capable of existence.

The decisive experiment of Biebricher and Luce is shown in Fig. 6.6. A synthesis medium containing highly purified enzyme and substrates is incubated and maintained at a suitable temperature, for a time adequate to allow the multiplication of any templates present but too short to enable products to arise 'de novo'. Then the solution is divided into portions. Each portion is incubated long enough to allow synthesis 'de novo' and the products are compared by the fingerprint method. If the 'impurity' hypothesis is correct then multiplication of the impurity in the first phase should lead to the same product from each portion of the incubated medium. If the 'de novo' hypothesis is correct then the products should be different, since at the beginning different enzyme molecules were working on different products. Selection – that is, preferential reproduction of *one* rudimentary strand – could not yet take place, since in the first, short incubation period none of the products 'de novo' was complete.

The experiment gave many different products. Only if these were mixed and again incubated did a single, homogeneous, reproducible variant grow up. The earliest products that could be detected directly were about 70 nucleotides long. In the course of evolution there appeared longer chains, e.g., at higher salt concentration the 'midivariant', with about 220 nucleotides.

The important result of these experiments is, however, not just an explanation of this unusual feature of the Qβ system. More important, we now have a flexible replication system at our disposal, with which a series of further interesting studies can be set up. Above all, it can be shown that molecular selection and evolution are inevitable consequences of self-replication, and can as such be investigated quantitatively. For example, the question of rapid optimization under extreme experimental conditions has been answered in detail. The results of the evolution experiments described can be summarized in four principal statements.

(1) The synthesis de novo of RNA by Qβ replicase proceeds by a mechanism fundamentally different from that of template-instructed RNA synthesis. The active reaction complex contains at least two enzyme molecules and requires nucleation by a seed of three or four substrate molecules.

(2) The rate-limiting step is nucleation, while elongation and reproduction follow rapidly. The singular nature of the molecular process which initiates the reaction is reflected in the scatter in induction times for synthesis de novo.

(3) Synthesis de novo produces a broad spectrum of mutants of varying length, containing sequences capable of adaptation to a great variety of environmental conditions.

(4) Initiation of self-reproduction is clearly sufficient to set the process of evolutive optimization in motion.

This finding allows the development of an evolution reactor, in which optimally reproducing RNA sequences can be produced (Eigen *et al.*, 1981).

A Turing machine for biology?

In 1936 Alan M. Turing described a universal automatic calculator that was supposed to compute the values of any given mathematical functions. It used simple algorithms in a finite, but possibly very large, sequence of steps. Turing's computing machine was not, however, a forerunner of the electronic computers that began to be realized about a decade later; indeed, it was not even supposed to be a real machine at all, but an abstract one, a conceptual model. In this respect Turing's machine resembles another well-known ideal machine, the Carnot engine, which also is not supposed to represent a real heat engine, but rather an abstract process that has the effect of converting heat energy into work. With the help of Carnot cycles – operations of the ideal machine which return its state to its starting point – a fundamental question of the theory of heat could be analysed, namely as to the maximum fraction of heat that can theoretically (i.e. using idealized, arbitrarily slow conversion processes) be converted into useful work. Turing's machine addresses a similarly fundamental question of mathematics: are the values of all functions computable? Turing's scheme starts with the notion of an arbitrarily long computing tape, divided into segments. The machine in action is seen to carry out essentially four kinds of operations: the tape moves to the right, to the left, stops, and is written upon. This implies, of course, that the machine is able to read and that it has a memory that contains instructions. It reads signs that appear on the segment of tape that is being worked, changes its mode of operation accordingly in a unique manner, and, if appropriate, rewrites the segment of tape. The change in the mode of operation causes the machine to move the tape to the right or to the left, whereupon the whole procedure repeats itself. Turing's idea had important consequences for numerical analysis and practically introduced the age of electronic data processing. But its real value is its abstract one. The concept of the Turing machine permitted an answer to be found to the basic question it addressed: only those functions can be held to be computable whose values can be obtained in a finite series of operations of the Turing machine. What could we now imagine to be a Turing machine for biology? The idea itself is really not new at all, being based on John von Neumann's model (1966) of the 'self reproducing automaton'. Today, however, as a consequence of our insights into the molecular mechanisms of biological information processing, we can formulate the problem more concretely, indeed in such a manner that the abstract machine can already take on quite concrete form. Our task can be formulated in three central statements.

(1) All forms of life known to us are extremely complex in their form and function; (cf. Schrödinger's (1944) 'aperiodic crystals').
(2) The complexity is organized right at the molecular level.
(3) This organization enables, through adaptation to the external world, an optimization of all characteristics of the organism that are necessary for its survival and development.

If the problem of the Turing machine in mathematics is 'computability', then the problem of the analogous machine for biology can be termed 'optimizability', or more precisely 'functional optimizability of complex structures'. According to the development in the previous sections we already know that the foundation of optimization is *self-reproduction*. In this respect our inquiry does indeed relate to von Neumann's 'self reproducing automata'. We have learned that self-reproduction in open systems leads to selection, that selection under a finite mutation rate leads in turn to evolution, which finally proceeds to the optimal structure. The experiments with RNA molecules already described have shown us that one indeed gets optimal structures on the basis of such a 'Darwinian logic', as long as the error rate limitation on the total information content is not exceeded. The concept of 'optimal' relates here to the reproduction rates that have been measured for viruses *in vivo*. These rates are similar to those found for the RNA strands obtained in evolution experiments.

The scheme of such a Darwinian experiment was sketched in Fig. 6.5. One could view these experiments as the realization of an optimization process based on molecular machinery. In the abstract representation one would only have to divide the chemical processes occurring in the test tube into schematic steps. One would then arrive at the scheme shown in Fig. 6.7, which would be run iteratively.

It is important for understanding the optimization process to know the amount of information (number of nucleotides per RNA strand) stored in each individual mutant. The relative amount of each mutant present (expectation value of a k-error mutant) is determined, for a given total population, by this amount. In addition, the number of reproductive cycles necessary for selection of an optimal mutant is also determined by this information content. In this schematization an abstract mechanical process becomes visible that is relevant to the basic biological question: can a functional system be optimized in a finite number of steps in a genetically determinable manner? In the theoretical treatment one must start from the n-dimensional topology of sequence space.

From the practical point of view the construction of a machine for optimizing the replication of RNA sequences is less interesting. On the one hand the Qβ experiments described above have already confirmed the possibility of a far-reaching optimization of the replication of RNA molecules. On the other hand, these characteristics of the RNA molecules – their adaptation as target structures for the replication enzyme – are only of limited interest. Here the genotype and the phenotype are represented by one and the same structure. With a suitable execution of the experiments the optimal structure simply grows out of the mixture; no mechanical iteration is needed.

Goal-directed optimization by a machine really becomes a challenge when phenotype and genotype are no longer represented by the same molecular structure. Such a genotype–phenotype dichotomy, which we see in all forms of life, comes into play at the molecular level with the translation of the genetic information, i.e. with the transfer of the responsibility for function to protein molecules.

For evolution experiments of the kind described above a dilemma immediately arises: the information is localized in the nucleic acids. Out of their mutant spectrum the suitable information must be selectively amplified. But the selective advantage only comes to expression at the executive level, i.e. in the protein molecules. The advantage recognizable in the function of a protein molecule must somehow be transmitted in some form to the RNA molecule that coded for it; in particular, the very mutant

Fig. 6.7. Schematic abstraction of the evolution experiment shown in Fig. 6.5. The problem behind the 'Turing scheme' is optimizability. The number of generations involved depends on information content, population size and topology of selective values.

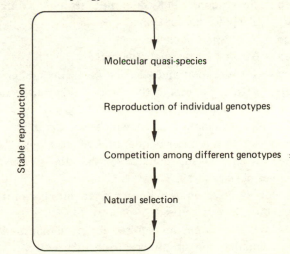

that gave rise to the advantageous protein must be given a favourable reproduction rate in order to be selected. In a living cell a feedback of this kind is provided automatically, for the cell represents a regulatory element that is evaluated as a whole in the selection process. If one wanted to extend the experiments already described to the evolution of proteins, then one would have to provide for this feedback right at the start. One would have to assume the entire complex symbiotic molecular mechanism of the cell, just the thing that is supposed to arise stepwise in such experiments. This mechanism, developed over millions of years in nature, is of such a complexity that any experiments of this kind would be doomed to failure right at the start. But there is a way out of the dilemma. We recall that self-reproduction is required for the natural evolutionary process for two reasons. On the one hand it permits the retention of information, on the other it provides through its autocatalytic nature the mechanical basis for selection. The difficulty that comes into play with the genotype–phenotype dichotomy touches only the question of selection. One therefore has to find a way to separate the two processes from one another. A possible procedure is to replace natural selection by an artificial choice on the basis of a test appropriate for the phenotype. In this way one can still retain the replication of the genotype and its translation to phenotype as a means of conserving information. (Otherwise one would have to synthesize the phenotype, including all of its modifications, continuously anew). In the stepwise mechanical realization this process would work as follows: (cf. also Fig. 6.8).

Step 1: Production of a mutant spectrum of self-reproducing templates.

Step 2: Separation of individual mutants and cloning.

Step 3: Amplification of clones.

Step 4: Expression of clones.

Step 5: Testing for optimal phenotype.

Step 6: Identification and selection of the optimal genotype.

Step 7: Return the 'best' of the clones – after mutagenous multiplication – to Step 1, and start again.

Behind this schematic process, which remains an abstract one for the time being, is hidden a series of quantitative questions. The essential one is of course whether a mentionable optimization can be attained in a finite number of steps at all. Thus:

(1) How large an information content (number of nucleotides per RNA sequence) can be used?

(2) How many clones are required, for given information content, in order to populate the mutant spectrum sufficiently?

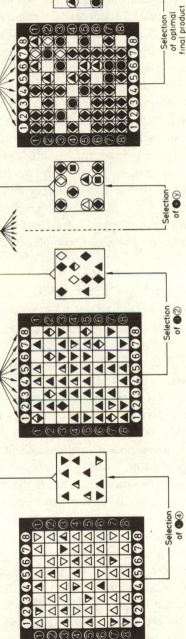

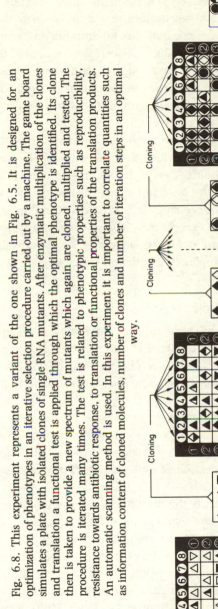

Fig. 6.8. This experiment represents a variant of the one shown in Fig. 6.5. It is designed for an optimization of phenotypes in an iterative selection procedure carried out by a machine. The game board simulates a plate with isolated clones of single RNA mutants. After enzymatic multiplication of the clones and translation a functional test is applied through which the optimal phenotype is identified. Its clone then is taken to provide a new spectrum of mutants which again are cloned, multiplied and tested. The procedure is iterated many times. The test is related to phenotypic properties such as reproducibility, resistance towards antibiotic response, to translation or functional properties of the translation products. An automatic scanning method is used. In this experiment it is important to correlate quantities such as information content of cloned molecules, number of clones and number of iteration steps in an optimal way.

(3) How many iterations are required to reach an optimum?

The three questions are obviously closely coupled to one another. Thus for example an RNA sequence of 300 nucleotides in a population of 10^{15} molecules would have sufficient representation of all mutants up to those with six errors. A plate containing one million clones, however, has at best only all the 2-error mutants. The question about the continuous climb to the optimum depends decisively on populating the mutant spectrum. If the route to the optimum exists, then for a gene length of 300 nucleotides fewer than 1000 iterations are required to traverse it.

Many additional questions present themselves if one faces the practical realization of such a machine. It is obvious that – in view of the large numbers of clones and the multiplicity of iterations – an automatic execution of the process is required. The clones have to be addressable, the analysis must utilize parallel processing (e.g. autoradiography) and automatic sampling. With such experiments biology some day might become 'Big Science'. This ought not to be surprizing, as the physical problem of biology is indeed 'complexity and its reproducibility'.

From principles to reality

Physical principles indeed lie behind real processes, but reality, overspread by a myriad of extraneous influences, cannot be 'reduced' to a single principle. Inattention to the difference between Principle and Reality is the sole cause of the fruitless wrangling between the so-called Holists and Reductionists.

It is clear that selection and evolution on a molecular scale, based on the self-reproduction of nucleic acids, is a necessary condition for the genesis of life. Does this further provide a sufficient explanation for the molecular organization of the living cell, of which we now know so much?

The answer is: no.

The Darwinian principle contains the definition of its own limitation, which appears in the error threshold for information transfer. The reproduction of the original sequences was limited in its accuracy by the molecular forces of base-pairing. Even in the case of the stable G–C pair the length of a reproducible sequence does not exceed the magnitude of 100 nucleotides. This has been confirmed by measurements on model systems (Pörschke, 1977; Lohrmann, Bridson & Orgel, 1980; Orgel, personal communication). Such an 'information crisis' could only be overcome by the evolution of a mechanism of enzymic reproduction. Evolution requires the storage of information in reproducible sequences.

To make the information available, a process of decoding is necessary. However, the simplest decoding equipment requires much more information than could ever have been accommodated in one of the primordial sequences (Eigen & Schuster, 1977, 1978a, b). Thus a further necessary condition was the functional integration of various replicative units. The various carriers of information could indeed all have emerged from a single, selected distribution of mutants, but they then had to adapt to their various functions. Darwin's principle is not sufficient to account for this, for the following conditions must all have been fulfilled simultaneously.

(1) Each replicative unit within the functionally integrated system must remain in continuous competition with its own mutants, and therefore work so well that no information is lost from one generation to another.

(2) Competition between the replicative units belonging to the functional system must be abolished and replaced by a mutual regulation of concentration.

(3) The functional unit must as a whole be able to compete successfully with alternative units.

Theory (Eigen & Schuster, 1977, 1978a, b) tells us that a circular, closed network of coupled reactions – we call it a hypercycle – is the only system capable of fulfilling these three conditions simultaneously and thus of overcoming the information crisis.

A further problem arises in connection with the evolution of a translation apparatus which we have referred to already as the Genotype–Phenotype Dichotomy. The information, which is later selected, resides in the genotype. However, its selection is based on the phenotypic properties displayed by the products of its translation. These must interact selectively with their 'own' genotype, or otherwise the evolution of phenotypic properties will not be possible.

As an example, let us consider what happens when a bacterial cell is infected by RNA phage. In order to assure preferential reproduction of its genome, the virus codes for a protein which combines with proteins of the host cell to produce an RNA replicase; this recognizes and reproduces the viral genome but not any of the RNA sequences of the host cell. Here again we encounter a hyperbolic growth law as typical for a hypercycle. The rate of reproduction is on the one hand proportional to the number of templates (copies of the viral genome) and on the other to the number of functioning replicase units, which is itself proportional to the number of copies of the viral genome, since the replicase is a decoding product of the genome. The resulting quadratic, self-catalytic rate law gives rise to hyperbolic growth.

The origin of the translation apparatus poses problems of this kind, soluble only by postulating both hypercyclic organization and compartmentation of the functionally integrated system. The growth laws resulting from these necessitated a once-and-for-all decision on the organization of the translation apparatus and the structure of the genetic code. The result of this decision has been binding on all living beings ever since.

If on the one hand Darwin's principle is insufficient to account for the origin of life, it is on the other hand an essential premise for consequences whose importance reaches far beyond the borders of biology. Wherever there is natural selection, the result is self-organization into sharply-differentiated, discrete states instead of the broad Gaussian distribution to which we are used in stochastic systems at thermodynamic equilibrium. The principle of natural selection could thus explain why sharply-defined states occur at all, instead of continua: why there is 'something' and not 'everything'. It would be interesting to look for such applications in elementary-particle physics (H. J. Queisser, personal communication), neurobiology (Malsburg, 1979; Cowan, 1980; Edelman, 1981) or sociobiology; this means looking for the prerequisites of such behaviour in each case.

Although Darwin cannot have foreseen such detailed extensions of his concept of selection and evolution, there is no doubt that he recognised its general implications. Thus he writes to Nathaniel Wallich in 1881: 'You expressed quite correctly my views where you said that I had intentionally left the question of the Origin of Life uncanvassed as being altogether ultra vires in the present state of our knowledge, and that I dealt only with the manner of succession. I have met with no evidence that seems in the least trustworthy, in favour of so-called Spontaneous Generation. I believe that I have somewhere said (but cannot find the passage) that the principle of life will hereafter be shown to be a part, or consequence, of some general law.'

This chapter has been prepared using two earlier publications (Eigen, 1981, 1982). I am indebted to Ruthild Winkler-Oswatitsch for helpful comments and assistance in preparing the manuscript and drawings, and to Paul Woolley and William C. Gardiner for their excellent work in translating into English.

Further reading

Eigen, M., Gardiner, W., Schuster, P. & Winkler-Oswatitsch, R. (1981). The origin of genetic information. *Scientific American* **244** (4), 78–94.

Eigen, M. & Schuster, P. (1977). The hypercycle. A principle of natural self-organization. Part A: emergence of the hypercycle. *Naturwissenschaften*, **64**, 541–65.

Eigen, M. & Schuster, P. (1978). The hypercycle. A principle of natural self-organization. Part B: the abstract hypercycle. *Naturwissenschaften*, **65**, 7–41.

Eigen, M. & Schuster, P. (1978). The hypercycle. A principle of natural self-organization. Part C: the realistic hypercycle. *Naturwissenschaften*, **65**, 341–69.

Eigen, M. & Winkler-Oswatitsch, R. (1982). *Laws of the game. How the principles of Nature govern chance*. Translated by R. & R. Kimber. London: Penguin Books.

References

Batschelet, E., Domingo, E. & Weissmann, C. (1976). The proportion of revertant and mutant phage in a growing population, as a function of mutation and growth rate. *Gene*, **1**, 27–32.

Biebricher, C. K., Eigen, M. & Luce, R. (1981a). Product analysis of RNA generated *de novo* by Qβ replicase. *Journal of Molecular Biology*, **148**, 369–90.

Biebricher, C. K., Eigen, M. & Luce, R. (1981b). Kinetic analysis of template-instructed and *de novo* RNA synthesis by Qβ replicase. *Journal of Molecular Biology*, **148**, 391–410.

Boltzmann, L. (1905). *Der zweite Hauptsatz der Mechanischen Wärmetheorie*. Leipzig: J. A. Barth.

Cowan, J. D. (1980). Symmetry-braking in embryology and in neurobiology. In *Symmetry in sciences*, ed. B. Gruber & B. R. S. Millman. New York: Plenum Press.

Domingo, E., Flavell, R. A. & Weissmann, C. (1976). In vitro site-directed mutagenesis: generation and properties of an infectious extracistronic mutant of bacteriophage Qβ. *Gene*, **1**, 3–25.

Edelman, G. M. (1981). Group selection as the basis for higher brain function. In *The Organization of the Cerebral Cortex*, ed. F. O. Schmitt, F. G. Worden, G. Adelman & S. G. Dennis, pp. 535–63. Cambridge, Mass.: MIT Press.

Eigen, M. (1971). Selforganization of matter and the evolution of biological macromolecules. *Naturwissenschaften*, **58**, 465–523.

Eigen, M. (1981). Darwin and molecular biology. *Angewandte Chemie International Edition*, **20**, 233–41.

Eigen, M. (1982). In *Proceedings of the Second International Kyoto Conference on New Aspects of Organic Chemistry*, and *Aharon Katchalsky Lecture*.

Eigen, M., Gardiner, W., Schuster, P. & Winkler-Oswatitsch, R. (1981). The origin of genetic information. *Scientific American*, **244** (4), 78–94.

Eigen, M. & Schuster, P. (1977). The hypercycle. A principle of natural self-organization. Part A: emergence of the hypercycle. *Naturwissenschaften*, **64**, 541–65.

Eigen, M. & Schuster, P. (1978a). The hypercycle. A principle of natural self-organization. Part B: the abstract hypercycle. *Naturwissenschaften*, **65**, 7–41.

Eigen, M. & Schuster, P. (1978b). The hypercycle. A principle of natural self-organization. Part C: the realistic hypercycle. *Naturwissenschaften*, **65**, 341–69.

Fersht, A. R. (1981). Enzymic editing mechanisms and the genetic code. *Proceedings of the Royal Society of London Series B*, **212**, 351–79.

Haruna, I., Nozu, K., Ohtaka, Y. and Spiegelman, S. (1963). An RNA 'replicase' induced by and selective for a viral RNA: isolation and properties. *Proceedings of the National Academy of Sciences, USA*, **50**, 905–11.

Haruna, I. & Spiegelman, S. (1965). Specific template requirements of RNA replicases. *Proceedings of the National Academy of Sciences, USA*, **54**, 579–87.

Kramer, F. R. & Mills, D. R. (1978). RNA sequencing with radioactive chain-terminating ribonucleotides. *Proceedings of the National Academy of Sciences, USA*, **75**, 5334–8.

Levisohn, R. & Spiegelman, S. (1968). The cloning of a self-replicating RNA molecule. *Proceedings of the National Academy of Sciences, USA*, **60**, 866–72.

Levisohn, R. & Spiegelman, S. (1969). Further extracellular Darwinian experiments with replicating RNA molecules: diverse variants isolated under different selective conditions. *Proceedings of the National Academy of Sciences, USA*, **63**, 805–11.

Loeb, L. A. & Kunkel, T. A. (1982). Fidelity of DNA synthesis. *Annual Review of Biochemistry*, **51**, 429–57.

Lohrmann, R., Bridson, P. K. & Orgel, L. E. (1980). Efficient metal-ion catalyzed template-directed oligonucleotide synthesis. *Science*, **208**, 1464–5.

Malsburg, C. von der (1979). *Biological Cybernetics*, **32**, 49.

Mills, D. R., Kramer, F. R. & Spiegelman, S. (1973). Complete nucleotide sequence of a replicating RNA molecule. *Science*, **180**, 916–27.

Mills, D. R., Peterson, R. L. & Spiegelman, S. (1967). An extracellular Darwinian experiment with a self-duplicating nucleic acid molecule. *Proceedings of the National Academy of Sciences, USA*, **58**, 217–24.

Neumann, J. von (1966). *Theory of self-reproducing automata*. Urbana: University of Illinois Press.

Pörschke, D. (1977). *Molecular Biology, Biochemistry and Biophysics*, **24**, 191.

Schrödinger, E. (1944). *What is life?* Cambridge: Cambridge University Press.

Spiegelman, S., Haruna, I., Holland, I. B., Beaudreau, G. & Mills, D. (1965). The synthesis of a self-propagating and infectious nucleic acid with a purified enzyme. *Proceedings of the National Academy of Sciences, USA*, **54**, 919–27.

Sumper, M. & Luce, R. (1975). Evidence for *de novo* production of self-replicating and environmentally adapted RNA structures by bacteriophage Qβ replicase. *Proceedings of the National Academy of Sciences, USA*, **72**, 162–6.

Turing, A. M. (1936). On computable numbers with an application to the 'Entscheidungsproblem'. *Proceedings of the London Mathematical Society Series 2*, **42**, 230–65.

7

Molecular tinkering in evolution

FRANÇOIS JACOB

If a man were to make a machine for some special purpose but were to use old wheels, springs and pulleys, only slightly altered, the whole machine, with all its parts, might be said to be especially contrived for that purpose. Thus throughout nature almost every part of each living being has probably served, in a lightly modified condition, for diverse purposes, and has acted in the living machinery of many ancient and distinct specific forms.

Charles Darwin, 1886

If an engineer were asked, starting from scratch, to manufacture a frog, it seems unlikely that he would first design such a swimming precursor as a tadpole and transform it later into a land animal. If he were asked to build a human baby, there is little chance that he would go through an embryonic stage with the gill slits and the complicated pattern of circulation that persists up to the adult stage of fishes.

In contrast to the engineer, evolution does not produce innovations from scratch. It works on what already exists, either transforming a system to give it a new function or combining several systems to produce a more complex one. The way evolution proceeds has no analogy with any aspect of human behaviour. If one wanted to use a comparison, however, one would have to say that this process resembles, not engineering, but tinkering, *bricolage*, as it is called in French. While the engineer's work relies on his having the raw materials and the tools that exactly fit his project, the tinkerer manages with odds and ends. Often without even knowing what he is going to produce, he uses whatever he finds around him, old cardboard, pieces of strings, fragments of wood or metal, to make some kind of workable object. What the tinkerer ultimately produces is often related to no special project. It merely results from a series of contingent events, from all the opportunities he has had to enrich his stocks with leftovers. When evolution turns a leg into a wing or a part of jaw into a piece of ear, it behaves somewhat like a tinkerer who makes a fan out of an old car wheel or a book case out of an old table. In a way, evolution proceeds like a tinkerer who, during millions of years, has slowly modified his products, retouching, cutting, lengthening, using all opportunities to transform and create (Jacob, 1977).

Proteins and novelties

It is at the molecular level that the tinkering aspect of evolution is the most apparent. Evidence concerning molecular evolution has long come mainly from comparing sequences of isofunctional proteins in different organisms or in different tissues of the same organism. When performed on proteins from species whose time of evolutionary divergence is known, this work has allowed an estimate of the rates at which mutations, especially amino-acid substitutions, have occurred in proteins. Most surprisingly, proteins were found to change at nearly constant rates (Zuckerkandl & Pauling, 1962), although this rate is not necessarily the same for proteins with different functions. In other words, the sequences in genes and proteins evolve in a clockwise manner (see discussion by Wilson, Carlson & White, 1977). By means of this evolutionary clock it has become possible to derive phylogenetic trees from sequence data of protein families. Such trees represent the relationship and branching orders between lineages leading to present-day molecular structures from a common ancestor.

Some examples of molecular tinkering can be found in these lineages; for instance in the derivation, from a common precursor, of two proteins with some similarity of sequence but different functions: lysozyme and lactalbumin (see Dayhoff, 1978). Lysozyme has kept what appears to have been the function of the ancestor protein. It is an hydrolytic enzyme that destroys the mucopolysaccharide component of bacterial cell walls. This activity is found in most tissues of vertebrates. On the other hand, a similar sequence, with only minor modifications, is used in a completely different system: lactalbumin, one of the main components of milk, constitutes also in humans the B chain of lactose synthetase, a dimeric enzyme present in the lactating mammary gland. The other polypeptide chain of this enzyme, the A chain, exhibits an N-acetyllactosamine synthetase activity. While lactalbumin itself has no known enzymatic activity, its binding to the A chain increases enormously the affinity of the latter to the substrates and, under certain conditions, catalyzes the synthesis of lactose. With only small modifications, similar polypeptide chains can thus be used in rather different ways. Small alterations are probably sufficient to change the possibilities of interaction of a protein. And it is presumably the appearance of new capacities of interaction that represents the main part of functional novelties at the molecular level.

Primordial genes

In recent years, new insights on molecular evolution have been provided by recombinant DNA and DNA sequencing techniques. An example is given by the information gained on collagen, which has a long evolutionary history and is present both in vertebrates and invertebrates, including primitive organisms such as sponges and sea anemones (see Solomon & Cheah, 1981).

In vertebrates, the structure of most tissues depends on a network of bundles of collagen fibrils. A single fibril contains more than 10^6 molecules per cm which are staggered from each other along the axis of the fibril by 234 amino acids, a distance called a D unit (see Trelstad, 1982). In all tissues, collagen molecules appear to be very similar: they are semi-rigid, rod-shaped structures built of three polypeptide chains in helical conformation. In recent years, however, it turned out that from different tissues, and even within the same tissue, collagen molecules can differ in both their primary structure and post-translational reactions. The individual polypeptides, called α-chains, contain some 1000 amino acids and have a simple triplet gly–X–Y as a basic structure in which the X and Y positions can be occupied by a variety of amino acids, proline being frequently the X and hydroxyproline the Y. The tripeptide is thus repeated several hundred times in the α-chains which in addition contain, at both termini, small non-helical regions of about 15–20 amino acids. Each α-chain is synthesized in a precursor form, or procollagen, which has large peptide extensions at both termini of the polypeptide chain. Pro-α-chains are converted into α-chains by limited cleavage with proteases (see Grant & Jackson, 1976; Bornstein & Sage, 1980).

In recent years, the number of collagen types was found to be more important than previously thought. In vertebrates, there must exist at least nine different genes to code for the different known α-chains. Furthermore, other proteins, such as the C1q component of the complement or the enzyme acetylcholinesterase have been found to contain short, triple helical collagenous sequences (Reid & Porter, 1976; Rosenberry & Richardson, 1977). The number of genes in the collagen family is thus probably at least 15 or 16.

The evolutionary relationship between the collagen chains has long been based on amino-acid sequence data. Because of the general similarity of the collagen chains, they have been assumed to be evolutionarily related and to descend from a common ancestor gene. An evolutionary tree can be sketched in which successive gene duplications lead to separate genes coding for the different chains. As for the origin of the ancestor gene, the repeat of the gly–X–Y tripeptide led to the assumption that the gene was formed by successive duplications of a fundamental genetic unit coding for the triplet. There was also the possibility of an initial expansion of this basic genetic unit to a small primordial gene, which in turn could expand by duplication to form the gene coding for the long chain. The model of an intermediate gene predicts the existence of internal homologies in the collagen polypeptides. Such repeats have been observed at a distance, D, of 234 amino acids, or 78 triplets, a distance which corresponds to the stagger between adjacent triple helical molecules in the collagen fibril (McLachlan, 1976). Besides the D unit, other periodicities such as D/3 (78 residues), D/6 (39 residues) and more particularly D/13 (18 residues) can also be observed (Hofman, Fietzek & Kühn, 1980).

Further evidence in favour of a small primordial collagen gene has been obtained by the use of recombinant DNA technology. Although still incomplete these studies have already brought about two results. On the one hand, the coding information for each of the α-chains studied is interrupted by a large number – up to 50 for the gene coding for the $\alpha2(I)$ procollagen chain of chicken – of intervening sequences of various sizes from 100 to 3000 base pairs. Small fragments of coding sequences amounting to 5000 bases are thus buried in more than 40000 bases, making this gene one of the largest genes identified so far (Yamada *et al.*, 1980; Boyd *et al.*, 1980; Ohkubo *et al.*, 1980; Wozney *et al.*, 1981*a*; Wozney, Hanahan, Talé & Boedtker, 1981*b*).

On the other hand, a number of fragments coding for the helical part of the same $\alpha2(I)$ chain from chicken have been sequenced. All these fragments turned out to contain multiples of the nine base pairs coding for the gly–X–Y triplet. More than half contain 54 base pairs. The others have 45 (54−9), or 99 (54+45) or 108 (2 × 54) base pairs. These data can best be accommodated in a model that assumes the existence of a primordial collagen gene of 54 base pairs, a gene which was duplicated during evolution probably by recombinational events, some of which could have resulted in insertions or deletions of multiples of nine base pairs. Additional recombinational events with other, non-collagenous genetic units would have provided the non-helical parts of the procollagen gene in its C- and N-terminal regions (Yamada *et al.*, 1980; Wozney *et al.*, 1981*a, b*). Traces of such a primordial gene, however, were not found in the structure of a collagen in *Drosophila* (Monson, Natzle, Friedman & McCarthy, 1982).

The collagen family also brings some information on two other aspects of molecular tinkering. One is found in the C1q subcomponent of complement. This molecule is a complex composed of several subunits made of three polypeptide chains A and B and C, of mol. wt 23–24000, in equimolar amounts. These chains contain, close to their N-terminus, collagen-like sequences of 78 amino acids which are in a triple helix configuration (Reid & Porter, 1976). It is likely that the ancestor of the three C1q genes was formed by recombination between a collagen genetic unit and a non-collagen one.

In the other protein containing also a collagenous element, acetyl-cholinesterase, the situation is somewhat different. There are two known forms of acetylcholinesterase molecules in the neuromuscular junctions; globular and asymmetric forms. Globular forms contain an hydrophobic region and are membrane-bound. Asymmetric forms have no hydrophobic region, contain collagen-like elements and are bound, not to cellular membranes but to basal lamina. In this case, the collagenous part of the molecules is not covalently linked to the globular part endowed with cholinesterase activity. It is a subunit made of three polypeptide chains of mol. wt 40000 (instead of 100000 in α-chains of collagen) in a triple-helix configuration. This collagenous subunit gives the molecule its affinity towards basal lamina (Rosenberry & Richardson, 1977; Bon, Vigny & Massoulié, 1979). In acetylcholinesterase, the evolutionary tinkering has been performed at the level, not of the gene, but of the protein.

Another beautiful example of a primordial gene was recently provided by a comparative study of mammalian serum albumin (SA) and α-fetoprotein (AFP) (see Eiferman, Young, Scott & Tilghman, 1981). The similarity between the two proteins indicates that their genes probably arose through duplication of an ancestral gene some three to five hundred million years ago. Each of the proteins is made of a thrice repeated domain suggesting that the ancestral gene was made by a thrice tandem repeat of a primordial gene coding for about 200 amino acids. Nucleotide sequence analysis shows that each repeat is made of four coding blocks interrupted by three intervening sequences. And sequence homologies among the four coding blocks that constitute a single domain – i.e. probably the primordial gene – also suggest that these blocks were derived, at least in part, from a common sequence which underwent successive amplification and divergence.

Despite the changes in sequence which have occurred throughout subsequent evolution, collagen, SA and AFP molecules still keep traces of the process by which they have been formed. There is also some indication that the ancestral genes for immunoglobulin variable regions of heavy and light chains might have arisen from about 12 tandem repeats of a 48 base-long primordial sequence (Ohno & Matsunaga, 1982). In the case of the troponin C family (see Dayhoff, 1978), a number of genes found to-day appear to derive from an ancestral gene which probably coded for a calcium-binding polypeptide of about 39 amino acids. Long before eukaryotic cells appeared, two successive internal duplications produced a gene four times as long as this ancestral gene. Further duplications and divergence led to a large family of proteins, either contractile proteins such as the light chains of myosin, or calcium-binding proteins such as troponins C or parvalbumins. Remarkably enough, another protein which a priori seems to be completely unrelated to the troponin family, the lysozyme of bacteriophage T4, contains a sequence of about 30 amino acids closely resembling a region of parvalbumins.

All these cases suggest that new and longer genes can arise either through internal duplications of a sequence or through some mechanism of joining at random preexisting sequences. Such a phylogenetic model of gene formation by combination and permutation of smaller DNA sequences is supported by the ontogenetic formation of immunoglobulin genes: every gene coding for a heavy or a light chain of immunoglobulins is prepared during embryonic development by the joining of several DNA stretches, each one sampled from a pool of similar, but not identical sequences (see Leder, 1982).

As to the origin of present-day genes, it is likely that it all began with small stretches of 30 to 50 nucleotides produced by chemical evolution, each able to code for some 10 to 15 amino acids. Only secondarily could such stretches be joined at random by some ligating process to form longer polypeptide chains, some of which turned out to be useful and were then selected and refined. If this is correct, one should expect to find many fragments of amino-acid sequences that are common to what look like unrelated proteins. Such a partial homology is observed, for instance, between the human hormone gastrin and the middle T protein of polyoma virus. While middle T protein has a molecular weight of about 50000, gastrin exists in several forms, mainly a chain of 34 amino acids, produced by proteolysis of a much larger prohormone. Despite these differences, statistical comparison of the two sequences provides unequivocal evidence of an evolutionary relationship between the C-terminal 17 amino acids of the two proteins (Baldwin, 1982). As the analysis of DNA and protein sequences proceeds, not only should new families and domains appear, but also subfamilies and sub-subfamilies. Once again, it is difficult to see how molecular evolution could have proceeded, if not by turning old into new by knotting pieces of DNA together – that is by tinkering.

Genome plasticity

Until recently, the mechanism of genetic tinkering was difficult to visualize. The chromosomes of an organism were viewed as compact and almost intangible structures, containing exactly the amount of genetic information required for the production and functioning of that organism. This order could only be altered by rare gene mutations and chromosome rearrangements. To climb up the ladder of life, evolution had simply to refine constantly the old protein structures and to produce new ones by adding progressively more DNA sequences.

In the last two decades, however, this picture has been destroyed. The first blow came from the finding that multicellular organisms contain an amount of nuclear DNA in vast excess of what appears to be the gene-coding requirement of the organism. Worse, the ladder of DNA content per haploid genome does not parallel the ladder of life since the amount of DNA is some 20 times higher in certain organisms, such as salamanders, than in humans. Molecular biologists found it hard to believe that 20 times more genes are required to build a salamander than to build a human being! This situation appeared therefore as a paradox, the so-called C-value paradox (see Orgel & Crick, 1980).

The second surprise came with the nature of the excess DNA. On the one hand, the structural genes of higher organisms often contain too much DNA for the protein they code for; in many cases, the coding sequences are interrupted by a variable number of intervening sequences that are transcribed into RNA but spliced out from RNA before translation (see Cold Spring Harbor Symposia on Quantitative Biology, 1980). There also exist 'pseudogenes' in eukaryotic genomes, i.e. sequences that are homologous to structural genes but contain mutations precluding the formation of a functional product (see, for instance, Proudfoot & Maniatis, 1980; Nishioka, Leder & Leder, 1980). On the other hand, a large proportion of the eukaryotic DNA is composed of families of non-coding repetitive sequences in which the length of repeats is highly variable, from two to 6000, as well as the size of the families, from a few copies to several millions. In each species, the genome can contain several hundreds of different families that are distributed in many different ways relative to the chromosomes and relative to each other. Some families are localized close to centromeres and telomeres. Other families occur in complex patterns of interspersion with each other and with single copy DNA (see reviews in Dover & Flavell, 1982).

Equally surprising is the fact that a large fraction of the repeat DNA is subject to a diversity of sequence rearrangements. While structural genes of related species remain very similar and appear to change at the slow pace of the evolutionary clock, there are frequent and massive changes in the multiple-copy families. Instead of a structure in a frozen state, the eukaryotic genome must be viewed as composed of islands of rather stable coding sequences immersed within a genomic medium in a constant dynamic flux. There has long been genetic evidence for the existence, in the eukaryotic genome, of mobile or transposable elements (McClintock, 1956). Yet it is only recently that the importance of these units has been fully realized. These mobile elements can be inserted in, or excised from, many sites in the host DNA where they can produce mutations, deletions, inversions, duplications and transpositions of neighbouring sequences (see Cold Spring Harbor Symposia on Quantitative Biology, 1980). Additional tricks can be performed by viruses, especially retroviruses with their terminal repeats and reverse transcriptase: for instance the formation of cDNA from host mRNA and the insertion of such cDNA into the host genome to produce a pseudogene (Lueders, Leder, Leder & Kuff, 1982) or the activation of host genes, as a result of the adjacent insertion of a virus (Hayward, Neel & Astrin, 1981; Payne, Bishop & Varmus, 1982).

It is clear that these features – a large excess DNA, the fragmentation of structural genes, the presence of mobile elements in large number – all provide exactly the tools required for genetic tinkering. Fragments of DNA sequences can be cut or lengthened, can move back and forth from one genomic environment to another, can become linked to, or separated from other sequences, etc. (see Gilbert, 1978). The critical question about the excess DNA concerns mainly its function, its usefulness for the host if any. The catalogue of proposed functions for the various repetitive families includes all the unsolved problems of molecular and evolutionary biology, but none has been clearly demonstrated. Some of the repeated sequences are transcribed into RNA while others are not, but transcription does not prove function and one can imagine functions which do not involve transcription.

Although no known function does not mean no function, it seems likely that a part at least of the excess DNA has no phenotypic effect on the host and just spreads by a 'selfish' process of natural selection within genomes (Dawkins, 1976; Doolittle & Sapienza, 1980; Orgel & Crick, 1980). This non-specific DNA has a great virtue: for a large part it may be considered as 'junk', and the ladder of content in useful DNA then parallels the ladder of life. Yet this idea of junk DNA is not accepted by everyone, especially by those who consider that every structure of an organism is under constant surveillance by natural selection, any junk being rapidly eliminated. Actually, non-specific DNA should be characterized by a much greater freedom of drifting.

Of course, no function does not mean no consequence on the host phenotype and no importance in the evolutionary process. Although there is not as yet an adequate description of the number and distribution of the multiple-copy families in any species, the polymorphic array of interspersed highly repeated sequences as well as the specificity of the so-called satellite DNA are sufficiently distinctive to allow a diagnosis of species, at least among the rodents and primates that have been studied. Massive and 'concerted' changes of repeat sequences forming satellite DNA have been assumed to play a role in the process of sympatric speciation (see Dover *et al.*, 1982). Similarly, growth rate of the host organism may be influenced by the amount of DNA (Cavalier-Smith, 1978).

Actually the presence of useless DNA in the genome is no more disturbing than Darwin's 'bits of useless anatomy'. The idea of selfish DNA fits in quite nicely with that of genomic tinkering. When the opportunity occurs, a tinkerer gives some useless piece of his stock an unexpected function. It seems likely that occasionally the host organism might also find some role for a useless sequence that happens to be located at a particular site.

Obviously, one of the possible uses for wandering sequences concerns control systems (see Britten & Davidson, 1969; Davidson & Britten, 1979). Sequences that become scattered at random through the genome are not likely to find themselves in suitable places to have any usefulness. On some occasions, however, a family of non-coding sequences might influence allometric growth rates of an organism. Or, in very rare cases, a sequence might even, by mere chance, become located at a site that enables it to help turning on or off some neighbouring structural sequence. If useful, this new control system would be rapidly refined by natural selection. It might also be scattered at random through the genome and, in rare instances, subject different and dispersed coding sequences to a coordinated regulation. In the evolution of complex organisms, there are many situations in which such systems are clearly required, for instance when, after duplication and dispersion, structural genes of the same family, coding for isoforms of the same protein, become expressed in different tissues at different developmental stages and must, therefore, depend on different control systems; or conversely, when different structural genes, coding for different proteins, become part of the same developmental programme expressed in a given tissue at a certain stage and must therefore be coordinately regulated.

Occasionally, genetic tinkering can even transgress the species barrier. For instance, in the plant tumour called crown gall, the abnormal growth pattern of the plant cells has been shown to be the direct consequence of the presence and expression of specific 'tumour' genes. These genes have been transferred into the plant nuclei by bacteria which contain natural gene vectors in the form of large plasmids. It is therefore a mechanism of gene transfer from bacteria to plants that has thus evolved (Schell *et al.*, 1979). Another remarkable example is provided by the enzyme superoxide dismutase of the bioluminescent bacterium *Photobacter leiognathi*. Three major groups of superoxide dismutase have been described. The first contains copper and zinc, the second contains manganese and the third contains iron. Comparison of amino-acid sequences indicates that the iron- and manganese-enzymes are derived from a common ancestral protein, while the copper–zinc enzyme represents an independent line of descent. The copper–zinc enzyme is characteristic of eukaryotic cells, the iron one of prokaryotes while the manganese enzyme has been found in prokaryotes and mitochondria. Remarkably enough, the superoxide dismutase of *P. leiognathi* turned out to be a copper–zinc enzyme which by its structure and properties closely resembles that of teleost fishes, especially of ponyfish. Now *P. leiognathi* is a symbiont of the ponyfish where it is found in a specific gland. It seems likely that the gene specifying the copper–zinc enzyme has been transferred from the fish to the bacterium during their coevolutionary history (see Martin & Fridovich, 1981).

Developmental constraints

If many of the guesses about unknown functions deal with regulatory systems, it is because, in this domain, our ignorance matches their importance. It is now clear that novelties in coding sequences, i.e. in protein structures, can have hardly been a main driving force in the diversification of multicellular organisms. What distinguishes a butterfly from a lion, a hen from a fly, or a worm from a whale is much less a difference in chemical constituents than in the organization and distribution of these constituents. The few really big steps in evolution clearly required the acquisition of new information. But specialization and diversification took place by using differently the same structural information. In vertebrates, chemistry is the same. As has been often emphasized (see King & Wilson, 1975), differences between vertebrates are a matter of regulation rather than of structure. Minor changes in regulatory circuits during the development of the embryo can deeply affect the final result, i.e. the adult animal, by changing the growth rate of different tissues, or the time of synthesis of certain proteins, speeding up here, slowing down there. Tinkering then operates at the cell level.

Evolution is described in phylogenetic terms, that is, as differences between adult organisms. Yet differences between adult organisms merely reflect differences in the developmental processes that produce them. To really know how evolution proceeds, it is necessary to understand embryonic development and its limitations. Unfortunately, to this day, very little is known about regulatory circuits and embryonic development. The only logic that biologists really master is one-dimensional. If molecular biology was able to develop rapidly, it was largely because information in biology happens to be determined by linear sequences of building blocks. Thus the genetic message, the relations between the primary structures, the logic of heredity, everything turned out to be one-dimensional.

Yet, during the development of an embryo, the world is no longer merely linear. The one-dimensional sequence of bases in the genes determines in some way the production of two-dimensional cell layers that fold in a precise way to produce the three-dimensional tissues and organs that give the organism its shape and its properties. How all this occurs is still a complete mystery. While the molecular anatomy of a human hand is known in some detail, almost nothing is known about how the organism instructs itself to build that hand, what algorithm it uses to design a finger, what means it finds to sculpt a nail, how many genes are involved and how these genes interact. Even the principles of the regulatory circuits involved in embryonic development are not known. Nor is it known whether the arrangement of genes along the chromosomes has any importance for development. As long as the opportunities offered, and the constraints imposed by embryonic development are not clarified, it will remain difficult to really understand the rules of evolutionary tinkering.

References

Baldwin, G. S. (1982). Gastrin and the transforming protein of polyoma virus have evolved from a common ancestor. *FEBS Letters*, **137**, 1–5.

Bon, S., Vigny, M. & Massoulié, J. (1979). Asymmetric and globular forms of acetylcholinesterase in mammals and birds. *Proceedings of the National Academy of Sciences of the United States of America*, **76**, 2546–50.

Bornstein, P. & Sage, H. (1980). Structurally distinct collagen Types. *Annual Review of Biochemistry*, **49**, 957–1003.

Boyd, C. D., Tolstoshev, P., Schafer, M. P., Trapnell, B. C., Coon, H. C., Kretschmer, P. J., Nienhuis, A. W. & Crystal, R. G. (1980). Isolation and characterization of a 15-kilo-base genomic sequence coding for part of the pro α2-chain of sheep type I collagen. *Journal of Biological Chemistry*, **255**, 3212–20.

Britten, R. J. & Davidson, E. H. (1969). Gene regulation for higher cells. A theory. *Science*, **165**, 349–58.

Cavalier-Smith, T. (1978). Nuclear volume control by nuclear skeletal DNA, selection for cell volume and cell growth rate, and the solution of the DNA C-value paradox. *Journal of Cell Science*, **34**, 274–8.

Cold Spring Harbor Symposia on Quantitative Biology. (1980). Volume 45. *Movable genetic elements.*

Darwin, C. (1886). *On The Various Contrivances by which British and Foreign Orchids are fertilized by Insects and on the Good Effects of Intercrossing*. London: John Murray.

Davidson, E. H. & Britten, R. J. (1979). Regulation of gene expression: possible role of repetitive sequences. *Science*, **204**, 1052–9.

Dawkins, R. (1976). *The Selfish Gene*. Oxford: Oxford University Press.

Dayhoff, M. O. (1978). Atlas of protein sequences and structures. *National Biomedical Research Foundation, Washington, D.C.*, Vol. 5, Suppl. 5, p. 273.

Doolittle, W. F. & Sapienza, C. (1980). Selfish genes, the phenotype paradigm and genome evolution. *Nature*, **284**, 601–3.

Dover, G., Brown, S., Coen, E., Dallas, J., Strachan, T. & Trick, M. (1982). The dynamics of genome evolution and species differentiation. In *Genome Evolution*, ed. G. A. Dover & R. B. Flavell. New York: Academic Press.

Dover, G. A. & Flavell, R. B. (ed.) (1982). *Genome Evolution*. New York: Academic Press.

Eiferman, F. A., Young, P. R., Scott, R. W. & Tilghman, S. M. (1981). Intragenic amplification and divergence in the mouse α-fetoprotein gene. *Nature*, **294**, 713–18.

Gilbert, W. (1978). Why genes in pieces? *Nature*, **271**, 501.

Grant, M. E. & Jackson, D. S. (1976). The biosynthesis of procollagen. *Essays in Biochemistry*, **12**, 77–113.

Hayward, W. S., Neel, B. G. & Astrin, S. M. (1981). Activation of a cellular *onc* gene by promoter insertion in ALV-induced lymphoid leukosis. *Nature*, **290**, 475–80.

Hofmann, H., Fietzek, P. P. & Kühn, K. (1980). Comparative analysis of the sequences of the three collagen chains α1(I), α2 and α1(III). Functional and genetic aspects. *Journal of Molecular Biology*, **141**, 293–314.

Jacob, F. (1977). Evolution and tinkering. *Science*, **196**, 1161–6.

King, M. C. & Wilson, A. C. (1975). Evolution at two levels in humans and chimpanzees. *Science*, **188**, 107–16.

Leder, P. (1982). The genetics of antibody diversity. *Scientific American*, **246**, 5, 72–83.

Lueders, K., Leder, A., Leder, P. & Kuff, E. (1982). Association between a transposed α-globin pseudogene and retrovirus-like elements in the BALB/c mouse genome. *Nature*, **295**, 426–8.

Martin, J. P. & Fridovich, I. (1981). Evidence for a natural gene transfer from the ponyfish to its bioluminescent bacterial symbiont *Photobacter leiognathi*. *Journal of Biological Chemistry*, **256**, 6080–9.

McClintock, B. (1956). Controlling elements and the gene. *Cold Spring Harbor Symposia on Quantitative Biology*, **21**, 197–216.

McLachlan, A. D. (1976). Evidence for gene duplication in collagen. *Journal of Molecular Biology*, **107**, 159–74.

Monson, J. M., Natzle, J., Friedman, J. & McCarthy, B. J. (1982). Expression and novel structure of a collagen gene in Drosophila. *Proceedings of the National Academy of Sciences, USA*, **79**, 1761–5.

Nishioka, Y., Leder, A. & Leder, P. (1980). An unusual alpha globin-like gene that has cleanly lost both globin intervening sequences. *Proceedings of the National Academy of Sciences, USA*, **77**, 2806–9.

Ohkubo, H., Vogeli, G., Mudryi, M., Avvedimento, V. E., Sullivan, M., Pastan, I. & Combrugghe, B. de (1980). Isolation and characterization of overlapping genomic clones covering the chicken α2 (Type I) collagen gene. *Proceedings of the National Academy of Sciences, USA*, **77**, 7059–63.

Ohno, S. & Matsunaga, T. (1982). The 48-base long primordial building block of immunoglobulin light chain variable regions is complementary to the primordial building block of heavy chain variable regions. *Proceedings of the National Academy of Sciences, USA*, **79**, 8338–41.

Orgel, L. E. & Crick, F. H. C. (1980). Selfish DNA: the ultimate parasite. *Nature*, **284**, 604–7.

Payne, G. S., Bishop, J. M. & Varmus, H. E. (1982). Multiple arrangements of viral DNA and an activated host oncogene in bursal lymphomas. *Nature*, **295**, 209–14.

Proudfoot, J. H. & Maniatis, T. (1980). The structure of a human α-globin pseudogene and its relationship to α-globin gene duplication. *Cell*, **21**, 537–44.

Reid, K. B. M. & Porter, R. R. (1976). Subunit composition and structure of subcomponent C1q of the first component of human complement. *Biochemical Journal*, **155**, 19–23.

Rosenberry, T. L. & Richardson, J. M. (1977). Structure of 18 S and 14 S acetylcholinesterase. Identification of collagen-like subunits that are linked by disulfide bonds to catalytic subunits. *Biochemistry*, **16**, 3550–8.

Schell, J., van Montagu, M., De Beuckeler, M., De Block, M., Depicker, A., De Wilde, M., Engler, G., Genetello, C., Hernalsteens, J. P., Holsters, M., Seurinck, J., Silva, B., van Vliet, F. & Villaroel, R. (1979). Interactions and DNA transfer between *Agrobacterium Tumefaciens*, the Ti-plasmid and the plant host. *Proceedings of the Royal Society of London, B.*, **204**, 251–66.

Solomon, E. & Cheah, K. S. E. (1981). Collagen evolution. *Nature*, **291**, 450–1.

Trelstad, R. L. (1982). Multistep assembly of type I collagen fibrils. *Cell*, **28**, 197–8.

Wilson, A. C., Carlson, S. C. & White, T. J. (1977). Biochemical evolution. *Annual Review of Biochemistry*, **46**, 573–639.

Wozney, J., Hanahan, D., Morimoto, R., Boedtker, H. & Doty, P. (1981a). Fine structural analysis of the chicken pro α2 collagen gene. *Proceedings of the National Academy of Sciences, USA*, **78**, 712–16.

Wozney, J., Hanahan, D., Talé, V., Boedtker, H. (1981*b*). Structure of the pro α2(I) collagen gene. *Nature*, **294**, 129–35.

Yamada, Y., Avvedimento, V. E., Mudryi, M., Ohkubo, H., Vogeli, G., Irani, M., Pastan, I. & Combrugghe, B. de (1980). The collagen gene: evidence for its evolutionary assembly by amplification of a DNA segment containing an exon of 54 bp. *Cell*, **22**, 887–92.

Zuckerkandl, E. & Pauling, L. (1962). Molecular disease, evolution and genic heterogeneity. In *Biochemistry*, ed. M. Kasha & B. Pullman, pp. 189–225. New York: Academic Press.

8

Intimations of evolution from the three-dimensional structures of proteins

D. C. PHILLIPS, M. J. E. STERNBERG AND B. J. SUTTON

Protein structures

Protein molecules are, of course, the primary products of gene expression and their structures have therefore long been studied for evidence of evolutionary mechanisms and relationships. Until 1960 this growing effort was concentrated on analyses of the chemical structures of protein (their amino-acid sequences) but in that year publication of the X-ray crystallographic analysis of myoglobin (Kendrew *et al.*, 1960) inaugurated a new age in which the three-dimensional structures of protein molecules became available for study. By the present time the structures of more than one hundred globular proteins have been determined in atomic detail and relationships between them are emerging which provide tantalising intimations of evolutionary mechanisms.

Principles of protein folding

Chemically, proteins are long, chain-like molecules in which the individual links in the chain are amino-acid residues. Each of these residues contributes six atoms (with one minor exception) to the main chain which therefore has the repeating structure

$$-NH.C^{\alpha}H(R_1).CO.[NH.C^{\alpha}H(R_2).CO.]NH.C^{\alpha}H(R_3).CO-.$$

In a typical protein there may be some 200 amino-acid residues linked in this way and they are of twenty different kinds defined by twenty different side chains, $R_1, R_2, R_3 \ldots$etc. Each protein molecule has a characteristic amino-acid sequence in which the different side chains occur in a seemingly random order. These sequences are defined by the structures of the genes and are related to them by the genetic code.

In globular proteins the protein chains are folded in complex ways to form compact, roughly-spherical molecules and these are the structures examined crystallographically. Happily it has been established that many such proteins display their characteristic activity in the crystal form and this observation, together with other experimental evidence, strongly suggests that in general the crystallographic structure represents that of the functional protein. Experimental evidence from renaturation experiments shows that for many proteins the functional, folded structure in an appropriate environment is a conformation of minimum free energy. Other studies suggest that the protein searches for the native structure by folding according to a programme or pathway rather than by sampling every possible conformation. The information specifying both the final structure and the pathway of folding thus appears to be encoded in the amino-acid sequence of a protein, at least when the order in which the chain is synthesized and the environment are taken into account. Fig. 8.1 illustrates the path of the polypeptide chain between the C^α atoms of each residue of hen egg-white lysozyme (Blake *et al.*, 1965), a relatively small protein

Fig. 8.1. Perspective drawing of the main-chain conformation of hen egg-white lysozyme. Only the positions of the C^α atoms are shown.

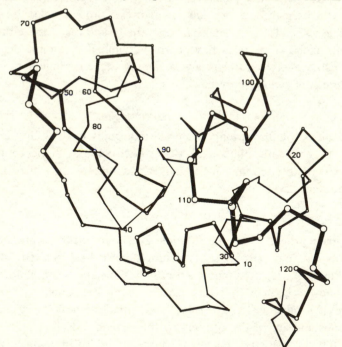

with 129 residues. In the complete structure the atoms are packed closely together effectively filling the space occupied by the molecule.

Many proteins exist in an aqueous environment and a basic feature of their structures can, on a crude analogy, be described as similar to that of soap in water. Each soap molecule has one polar and one nonpolar end. In water the soap molecules form micelles in which their nonpolar tails aggregate and their polar ends interact favourably with the surrounding water. The surface of a protein consists mainly of polar atoms, both main chain and side chain, that interact favourably with the solvent. Although some nonpolar atoms are found on the surface, the majority are buried in the interior of the protein. A major influence on protein folding clearly is the so-called hydrophobic effect that causes nonpolar atoms to avoid contact with the solvent. However, the main-chain component of each amino-acid residue includes polar atoms that carry partial electric charges – the NH and CO groups – and it is thermodynamically unfavourable for either of these groups to be buried in a nonpolar medium without the compensating presence of a group of the opposite polarity. This problem is overcome by the formation of main chain N—H...O=C hydrogen bonds and in all but a few of the smallest proteins, a regular pattern of these bonds leads to the formation of one or both types of repeating, or secondary

Fig. 8.2. (*a*) The right-handed α-helix. All of the atoms in the main chain are shown but side chains are represented by a single 'atom' labelled R. Hydrogen bond contacts between polar atoms are shown by broken lines. (*b*) The parallel β-pleated sheet.

(a)

(b)

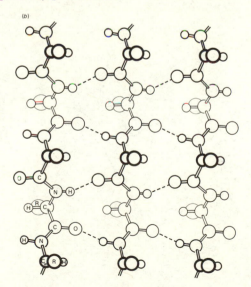

structures. These are the right-handed α-helix and the β-pleated sheet (Fig. 8.2), whose presence in proteins was predicted by Pauling & Corey (1951).

In the three-dimensional structures of globular proteins, α-helices and β-strands pack together to bury nonpolar residues on their surfaces. Levitt & Chothia (1976) have identified four classes of protein molecules based upon different combinations of these α- and β-structures: these are all-α, all-β, α/β and α+β (see Fig. 8.3).

Fig. 8.3. Schematic drawings by Jane Richardson of the main chain conformation in (*a*) Haemoglobin β-subunit – an α protein; (*b*) Cu/Zn superoxide dismutase – a β protein; (*c*) flavodoxin – an α/β protein; (*d*) One subunit of triose phosphate isomerase – an α/β protein (barrel); (*e*) Hen egg-white lysozyme – an α+β protein.

(*a*)   (*b*)   (*c*)

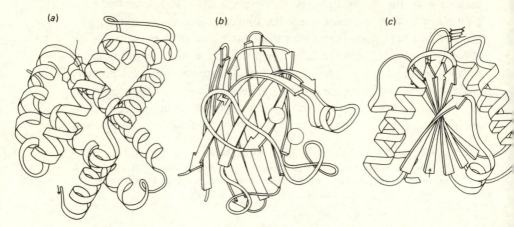

(*d*)   (*e*)

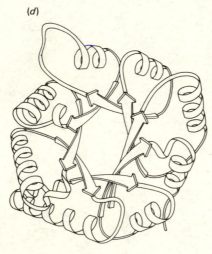
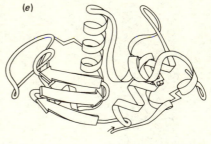

In the all-α class, much of the polypeptide chain adopts an α-helical conformation and there is little or no β-sheet structure. Typical of this class is the first protein whose structure was determined – sperm whale myoglobin, the oxygen-storage protein of muscle tissue. Similar three-dimensional structures are adopted by the α- and the β-chains of haemoglobin (Fig. 8.3a; note that α- and β- denote different but related amino-acid sequences and do *not* indicate the conformation of the polypeptide chains). This protein consists of four polypeptide chains, two α- and two β-chains (Fermi & Perutz, 1981).

The all-β proteins have extensive β-sheet structures but few or no α-helical regions. A common packing motif, illustrated in Fig. 8.3b by Cu/Zn superoxide dismutase (Richardson *et al.*, 1976), is the formation of two β-sheets that stack together. The polypeptide chain meanders between the two sheets, each of which tends to have antiparallel strands.

In the α/β class of proteins, the regular secondary structure tends to alternate between α-helix and β-strand as one progresses along the polypeptide chain. A common folding motif is the packing of the α-helices against both sides of a multi-stranded β-sheet that has exclusively or predominantly parallel strands. This motif is illustrated by flavodoxin (Burnett *et al.*, 1974) in Fig. 8.3c. Another motif is found in chicken triose

Fig. 8.4. (*a*) The association of β-strands and α-helices in βαβ units: The right-handed arrangement is almost always found. (*b*) Diagrams of βαβ units and related conformations in which the α-helix is replaced by a β-strand(s). α-Helices are denoted by circles and β-strands by triangles (upright triangles viewed from amino terminus: inverted triangles viewed from carboxyl terminus).

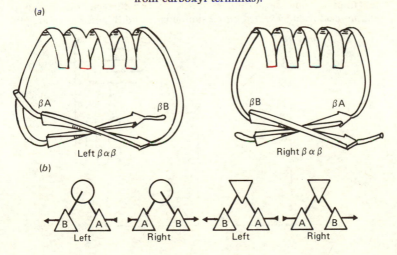

(a)

βA   βB   βB   βA

Left βαβ   Right βαβ

(b)

B   A   A   B   B   A   A   B
Left   Right   Left   Right

phosphate isomerase (Banner *et al.*, 1975) in which a pure parallel β-sheet curls up to produce a β-sheet in which each β-strand forms hydrogen bonds with neighbours on both sides. α-Helices pack against the outside surface of the β-sheet (Fig. 8.3*d*).

The α + β class is rather less well defined but the α-helices and β-strands tend to segregate into distinct regions along the polypeptide chain. Hen egg-white lysozyme is representative of this class (Fig. 8.3*e*; see also Fig. 8.1).

Comparative analyses of the known structures have revealed geometrical and topological rules governing the packing of α-helices and β-strands in protein molecules (for a review see Richardson, 1981). For example the geometry of α/β proteins is such that α-helices tend to lie roughly parallel to and about 10 Å above the β-sheet (Janin & Chothia, 1980; Cohen, Sternberg & Taylor, 1982). A major topological restriction on α/β proteins (Fig. 8.4) is that two parallel β-strands with a helical or other connections between them can yield structures with different chiralities depending on the relative positions of the strands. It has been shown that 99% of these units are right-handed (Sternberg & Thornton, 1976; Richardson, 1976).

Domains and sectors

This description of protein structure has so far taken into account only the folding of the polypeptide chain into a single, compact, globular unit. However Wetlaufer (1973) noted that several proteins fold into two or more spatially-separate, compact units which he referred to as domains.

Fig. 8.5. Diagonal plot showing distribution of contacts between residues less than a certain distance marked X, for a hypothetical protein with *n* amino-acid residues folded into 2 domains A and B joined at residue *i*.

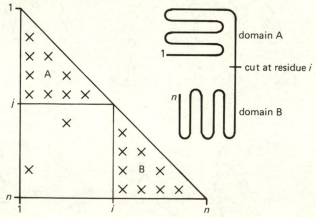

Since then this term has been used to refer to a variety of subdivisions of a protein structure into recognizably separate units. In this paper we use the term domain to refer to part of the tertiary structure of a protein that forms a spatially distinct separate folding unit whose structure is reminiscent of a small, compact protein (c.f. Richardson, 1981). This definition of a domain means that several proteins such as hen egg-white lysozyme and sperm whale myoglobin should be regarded as single-domain proteins. There is no strict size definition for a domain; the number of residues varies from 80 to 400 and tends to be different for the four structural classes (Richardson, 1981).

One approach to identifying domains from the atomic coordinates of a protein involves the use of distance diagrams (Phillips, 1970; Rossmann & Liljas, 1974). In such a diagram the sequence numbers for both the horizontal and vertical axes and the distance between the C$^\alpha$ atom of any

Fig. 8.6. Gō plot: diagonal plot showing distribution of contacts between residues greater than a certain distance for a hypothetical protein with a single domain. The division into sectors is shown by hatching in the structure diagram and by horizontal and vertical lines in the Gō plot.

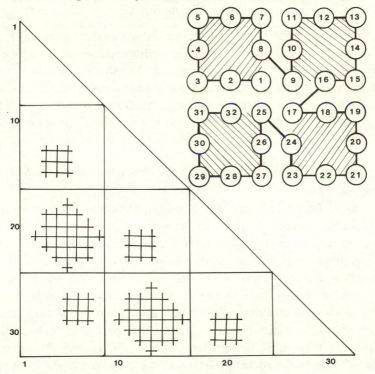

two residues i and j is denoted at position i, j. Fig. 8.5 illustrates the identification of the two domains in a hypothetical protein by the indication of C^α–C^α separations closer than a certain minimum distance (e.g. 6 Å). A suitable division into domains leads to few close contacts (i.e. few crosses in Fig. 8.5) between the domains.

Recently Gō (1981) has proposed another use for distance diagrams that involves the identification of distant rather than close C^α–C^α separations. Fig. 8.6 shows such a plot for a hypothetical protein and illustrates how the diagram can be divided into regions by lines drawn where there are no crosses (no distant C^α—C^α separations). Each region is part of the polypeptide chain that is most distant from other parts of the molecule. The region starts near the centre of the protein, goes towards the surface, and returns to the centre. We propose that, by analogy with solid geometry, such regions should be called sectors. Sectors are sub-divisions of domains and, as we shall see later, Gō (1981) has suggested that they may be related to subdivisions of the genes.

Gene duplication and structural relationships between proteins

A common mechanism that is believed to lead to the evolution of new proteins is gene duplication. According to this mechanism, a single copy of a gene duplicates to form two or more copies and the new copies evolve independently by point mutation and minor amino-acid sequence insertions and deletions to yield proteins with new functions. A variation on this theme is that after duplication (and perhaps mutation) the genes fuse to yield a larger protein. This process will be called internal gene duplication.

Similarities of amino-acid sequence and three-dimensional structure

Protein families

The globin family of proteins is a well-documented instance of gene duplication (see also Jeffreys, this volume). The amino-acid sequences of myoglobin and the α- or the β-chain of haemoglobin can be aligned to yield around 20 % identity of amino-acid residues between equivalenced sequence positions and a large number of positions where the equivalenced residues are of similar chemical type. The myoglobin and haemoglobin chains adopt very similar three-dimensional structures and, qualitatively, they convey the impression that three-dimensional structures change more slowly than amino-acid sequences during the evolution that results from gene duplication. It must be noted, however, that a major constraint on the fold of the globins is the preservation of their common function of binding oxygen to a haem group.

The same degree of similarity observed in the globin family also exists between the amino-acid sequences and the three-dimensional structures of egg-white lysozyme and mammalian α-lactalbumin (Brew, Vanaman & Hill, 1967; Smith, 1982). However these proteins have very different functions; lysozyme cleaves the polysaccharide that is a component of some bacterial cell walls while α-lactalbumin is a component of lactose synthetase that modifies the specificity of a galactosyl transferase.

Internal gene duplication

Evidence for internal gene duplication can be obtained from amino-acid sequence and structural repeats within a protein. In *Peptococcus aerogenes* ferredoxin residues 1 to 26 can be aligned with 27 to 53 so that there are 12 identities (Tsunoda, Yasunobu & Whiteley, 1968). The protein folds into a single domain and has a striking structural repeat, so that a two-fold symmetry axis can be placed through the centre of the molecule relating the first half of the protein chain to the second (Adman, Sieker & Jensen, 1973). A rotation of about 180° leads to 26 C^α atoms from the first half superimposing onto equivalent atoms in the second half with a root mean square deviation (r.m.s.d.) between equivalenced atoms of 1.25 Å (Rossmann & Argos, 1976). A possible evolutionary mechanism is that the primitive gene coded for a protein of about 26 residues that dimerized to form the functional protein. After gene duplication and fusion the role of the dimer was replaced by a monomer with roughly double the number of residues.

Evidence primarily from similarities between three-dimensional structures

A structural repeat similar to that in ferredoxin has been observed in bovine liver rhodanese (Phloegman *et al.*, 1978). This protein, with 293 residues, folds into two spatially separate domains formed from roughly residues 1 to 160 and 161 to 293. Each domain comprises a five-stranded parallel β-sheet flanked by α-helices and, in a superposition of the two domains, 117 pairs of C^α atoms can be equivalenced with a r.m.s. deviation of 1.95 Å. This superposition is the result of a rotation of one domain by 179° and a translation along the rotation axis of less than 1 Å. However of these 117 pairs of equivalenced residues only 15 positions have identical amino-acids. A standard approach to the comparison of amino-acid sequences when there are few identities is to evaluate the minimum number of base changes per codon required to change one amino-acid sequence into the other. For the 117 equivalent residues the average value is 1.27 bases/codon. This can be compared to the value of 0.81 bases/codon for the similarity of the two parts of ferredoxin (Rossmann & Argos, 1976).

Convergence or divergence?

The comparisons for rhodanese raise the general question of what sequence and structural similarities might be expected to arise by convergent evolution because of stereochemical and/or functional requirements. If sequence and/or structural similarities are closer than expected from convergence, then divergent evolution following gene duplication has to be considered as an explanation for them. But the power of convergence must not be underestimated. It is demonstrated, in an unimpeachable example, by the very similar geometric arrangements of the residues that form the active sites of trypsin and subtilisin (Kraut *et al.*, 1972). As these enzymes have quite different sequences and three-dimensional structures, divergent evolution is a highly improbable explanation for the similarities of their active sites.

Because rhodanese is an α/β protein, this class is emphasized in the following discussion, but the conclusions are likely to be generally applicable. The first question is what stereochemical constraints are there on the topology of an α/β domain that could lead to two identical topologies in the domains of rhodanese. In theory, there should be α/β proteins formed from β-sheets with different numbers of β-strands, within which there are different strand orders, and both right- and left-handed connections. However α/β proteins appear to be severely constrained: their β-sheets tend to have between 4 and 9 β-strands; there are restrictions on the strand order; and nearly all the connections are right-handed (Cohen *et al.*, 1982). Nevertheless, we estimate that there are between 10^2 and 10^4 possible α/β topologies. Thus the fact that two proteins or two domains have an identical topology is suggestive of divergence.

Given two identical topologies, geometric restrictions on α/β packing will lead to similar three-dimensional structures. Folding simulation (Cohen *et al.*, 1982) suggest that these geometric restrictions can yield by convergence a r.m.s.d. of around 5.0 Å for 100 residues, and so structural similarities closer than this are likely to be an indication of divergence.

Finally, the sequences of two similar α/β domains may never be completely unrelated as stereochemical requirements will place the same or chemically similar residue types in equivalent positions in the two structures. The requirements can, by convergence, lead to a minimum base change of between 1.2 and 1.3 bases per codon between any two sequences that adopt a similar structure. Thus lower base changes are required to suggest divergence.

In addition to these stereochemical constraints, functional requirements may dictate a structural and/or sequence similarity between two domains. Hol, van Duijnen & Berendsen (1978) have proposed that the binding of negatively charged groups, for example the phosphate groups of coenzymes such as nicotinamide adenine dinucleotide (NAD), may be favourable at the amino (N) end of helices because of the presence of a region of positive charge arising from the aligned peptide dipoles. Thus similar α/β structures could have arisen because of convergence to a fold favourable for the binding of certain moieties.

For rhodanese, no functional argument has been proposed to explain why the two domains should have similar structures. By our criteria, the structural similarities seem significant and therefore suggest internal gene duplication, but the sequences of the two halves have diverged and now have little or no similarity. Several other proteins have evidence for internal gene duplication from structural repeats, but without the corresponding sequence repeats. For example a two-fold repeat has been observed in L-arabinose binding protein (Guilliland & Quiocho, 1981) and in glutathione reductase (Schulz *et al.*, 1978), and a four-fold repeat in γ-crystallin (Blundell *et al.*, 1981). These observations strengthen the impression that three-dimensional structures change more slowly than amino-acid sequences during evolutionary changes following gene duplication.

Domains: the Rossmann fold

Further intimations of evolution come from the observation that certain domain topologies are found in identical or closely related forms in several proteins. One of these topologies, often referred to as a Rossmann fold (following the pioneer work of Rossmann, Moras & Olsen (1974) in this field), consists of an α/β structure with 6 parallel β-strands in the order FEDABC with right-handed connections that generally include an α-helix between sequential strands (Fig. 8.7a). This fold has been found in the domain that is involved in binding NAD in dogfish lactate dehydrogenase (Holbrook *et al.*, 1975), in horse liver alcohol dehydrogenase (Eklund *et al.*, 1976), in the ATP-binding domain of phosphoglycerate kinase (e.g. Banks *et al.*, 1979), in a domain in phosphorylase that binds NADH (e.g. Johnson *et al.*, 1980) and in the CTP-binding domain of aspartate carbamoyl-transferase (Monaco, Crawford & Lipscomb, 1978). Furthermore the NAD-binding domain of glyceraldehyde 3-phosphate dehydrogenase (Moras *et al.*, 1975) consists of a 9-stranded β-sheet flanked by α-helices, and a core of 6-strands with their connecting α-helices adopts the Rossmann fold.

Structural comparisons of the different Rossmann folds yield a r.m.s.d. of around 3 Å for between 50 and 110 residues (Ohlsson, Norstrom & Brändén, 1974; Jenkins *et al.*, 1981) compared to the upper limit of a random r.m.s.d. of 5.0 Å for 100 residues. At the sequence level, between alcohol and glyceraldehyde-3-phosphate dehydrogenases the minimum base change for about 70 equivalenced residues is ∼ 1.1 base/codon (Rossmann *et al.*, 1974) which is lower than the random value of between 1.2 and 1.3 bases.

One explanation for these sequence and structural similarities is that there was a gene coding for a domain with a Rossmann fold that had nucleotide binding activity. Gene duplication led to several copies of this gene, and each fused with one or several other genes to form proteins with different activities that share a common architecture for their nucleotide binding region. However, each of the Rossmann folds is involved in binding a negatively charged group such as NAD at the N-termini of α-helices and this must be taken into account in assessing the probability that the similarity of α/β fold could be the result of convergence. On balance, in our view, the structural similarities are sufficiently close for divergence to

Fig. 8.7. (*a*) Schematic diagrams showing arrangement of main chains in a 'Rossmann Fold' in lactate dehydrogenase (LDH), phosphoglycerate kinase (PGK), phosphorylase (PHO) and glyceraldehyde 3-phosphate dehydrogenase (GPD). (*b*) Arrangement of main chains as α/β barrels in triose phosphate isomerase and pyruvate kinase (cf. Fig. 8.3*d*).

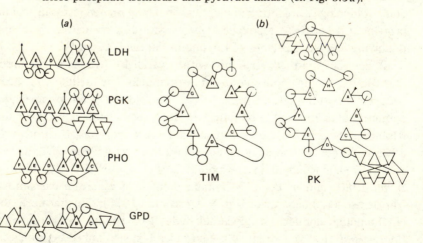

be the more probable explanation for the common occurrence of the Rossmann fold.

Domains: the α/β barrel

Another domain fold that has been found in several proteins is that of the 8-stranded α/β barrel of the glycolytic enzyme triose phosphate isomerase (Fig. 8.3*d*). This motif (Fig. 8.7*b*) occurs in one of the three domains of another glycolytic enzyme, cat pyruvate kinase (Stuart *et al.*, 1979), in 2-keto-3-deoxy-6-phosphogluconate (KDPG) aldolase (Lebioda *et al.*, 1982), in the main domain of Taka-amylase (Kusunoki *et al.*, 1981) and possibly also in the larger domain of spinach glycolate oxidase (Lindqvist & Brändén, 1980). The atomic coordinates of the barrel in triose phosphate isomerase (TIM), pyruvate kinase and KDPG aldolase have been super-imposed and some 150 C^{α} atoms can be equivalenced between any two enzymes with a r.m.s.d. of around 3.2 Å (Stuart *et al.*, 1979; Lebioda *et al.*, 1983). The only sequence information available is for TIM and KDPG aldolase where there is an average minimum base change of 1.43 bases/codon between 150 equivalenced residues (Lebioda *et al.*, 1983) which is effectively a random value. KDPG aldolase is in the hexose monophosphate pathway and is considered to be functionally very similar, and possibly structurally related, to the glycolytic enzyme fructose 1–6 diphosphate aldolase that operates on hexose sugars. TIM and pyruvate kinase are glycolytic enzymes involved in reactions on triose sugars. In all three enzymes the substrate binding site is at the carboxyl (C) end of the β-strands and the three functions are related in that they all activate a C—H bond adjacent to a carbonyl group (Rose, 1981). Furthermore in glycolate oxidase and in Taka-amylase the substrate binding is also at the C-end of the β-strands. As with the Rossmann fold, divergence or conver-gence are possible explanations for these structural similarities. At present the evidence is sparse and more amino-acid sequences and refined crystal structures are needed. But it is a tenable hypothesis that, here again, a common domain structure is being observed that has been propagated by gene duplication and divergent evolution, and has become associated with a variety of other domains in the generation of diverse enzymes.

Gene structure

So far we have seen how similarities between the three-dimensional structures of different proteins, and between parts of the same or different proteins, suggest their evolutionary history. These intimations are sometimes supported by similarities between the corresponding amino-acid sequences, but this evidence may be lacking since it appears that sequences may change more rapidly than three dimensional structures. In search of further evidence we turn now to the genes themselves, because recent discoveries have suggested that eukaryotic genes contain more information about their evolutionary history than had been expected.

In 1977, when the new techniques for rapid nucleic acid sequencing were first applied to eukaryotic DNA, it was found that the contiguous amino-acid sequence of the protein was represented in the DNA molecule by fragments of the expected sequence of bases, interrupted by long and apparently untranslated sequences. These intervening sequences were first discovered in the chicken ovalbumin gene (Breathnach, Mandel & Chambon, 1977; Doel *et al.*, 1977), and the β-globin gene of rabbit (Jeffreys & Flavell, 1977) and of mouse (Konkel, Tilghman & Leder, 1978). Fig. 8.8 illustrates the correspondence between the DNA sequence and that of the mRNA for mouse β-globin. Gilbert (1978) has suggested the term 'exon' for a sequence which appears in the mature mRNA and is expressed, and 'intron' for an intervening sequence which is lost. Rabbit, mouse and human β-globin genes all contain three exons, and the introns occur at exactly the same points in the sequence (see also Jeffreys, this volume).

In view of our observations concerning protein structures, the first question to consider is whether the exon sequences correspond to either secondary or tertiary structural units, or to functional units of the protein: 'Do Genes-in-pieces imply Proteins-in-pieces?' (Blake, 1978).

Single-domain proteins

The globins

We begin with an example of a small, single-domain protein, the β-globin sub-unit of haemoglobin, for which both the gene structure and three-dimensional protein structure have been studied extensively. The three

Fig. 8.8. Mosaic structure of the mouse β-globin gene, with three exons separated by two introns, and corresponding mRNA coding for protein.

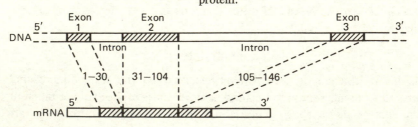

exons correspond to residues 1–30, 31–104 and 105–146, and as Fig. 8.3*a* shows, there is clearly no correlation with secondary structure since residues 31 and 105 are each within α-helical segments of the protein.

However, closer analysis reveals a striking observation. Gō (1981) has constructed a diagonal plot (as described above in Fig. 8.6) of all the distances between the C^{α} atoms of β-globin (Fig. 8.9). Vertical and horizontal lines are drawn to divide the sequence between residues 30 and

Fig. 8.9. Gō plot with minimum distance 27Å for the β-subunit of haemoglobin. Exon/intron boundaries at residues 30 and 104 are shown by full lines and the boundary subsequently found in leg-haemoglobin at residue 68 is shown by broken lines.

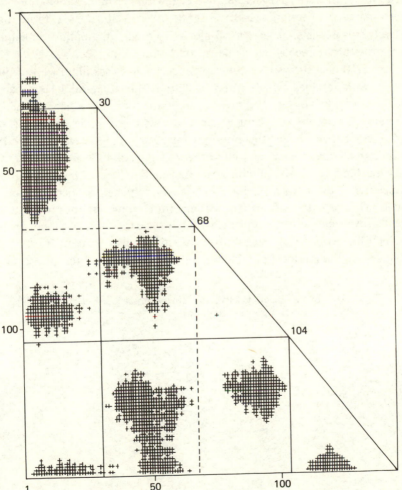

31, and 104 and 105, which are the exon/intron boundaries. Remarkably, these lines (shown solid in Fig. 8.9) appear to divide the shaded regions also, and therefore delineate sectors of the protein. The three sectors are shown in Fig. 8.10, from which it can be seen that only the central sector has a relatively compact structure.

However, if a correlation really exists between sectors and exons, then Fig. 8.9 in fact predicts that there should be four rather than three exons for β-globin, since the central sector can clearly be divided at about residue 66 (Gō, 1981) as indicated by the broken lines in Fig. 8.9. Perhaps two pre-existing exons 'fused' to form the central exon. This prediction was almost immediately confirmed when the leg-haemoglobin gene from soybean was sequenced (Jensen *et al.*, 1981) and found to consist of four exons corresponding to residues 1–32, 32–68, 69–103 and 104–C-terminus. Clearly it is possible, at least for β-globin, to predict the pattern of exons from the three-dimensional structure of the protein.

Fig. 8.10 also shows that the central sector includes almost all of the amino-acid residues which interact with the haem, and it has been suggested by Blake (1979) that this represents a primitive haem-binding protein. The results of recent experiments with a peptide consisting of residues 31–104 support this notion (Craik, Buchman & Beychok, 1980), although all three sectors as well as the α-globin sub-unit are required for oxygen binding (Craik, Buchman & Beychok, 1981). This and other functions of the tetrameric haemoglobin molecule may therefore have evolved by the juxtaposition of the first and third exons with the pre-existing central exon, followed by gene duplication to yield the α- and β-globin genes (which have identical exon/intron junctions after alignment of the sequences), the products of which could form dimers and tetramers. The

Fig. 8.10. Exon products in the three-dimensional structure of β-globin.

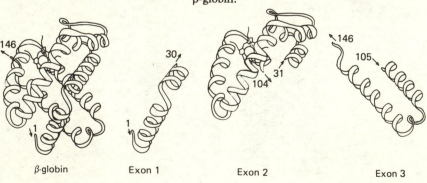

β-globin   Exon 1   Exon 2   Exon 3

fact that the three exons of β-haemoglobin correspond to identifiable though not compactly folded sectors of the tertiary structure may well reflect this process of stepwise addition of stretches of polypeptide chain in the evolution of the protein.

Evidence that such a mechanism might also lead to the generation of proteins with new functions, comes from the observation that the central sector of the β-globin molecule can be structurally superimposed upon the cytochrome c_{551} molecule, and the central haem-binding part of the cytochrome b_5 molecule (Argos & Rossmann, 1979). The prediction of a single exon for the former, and three for the latter molecule (Argos & Rossmann, 1980) has yet to be tested, but these molecules may have a common ancestral gene fragment.

Lysozyme

Further evidence which suggests the involvement of exons in protein evolution comes from the comparison of another single domain protein lysozyme, from two different sources, avian and bacteriophage.

The hen egg-white lysozyme gene contains 4 exons (Jung *et al.*, 1980). There is no correlation with secondary structural units or with the bilobal shape of the enzyme, but as for β-globin, a Gō plot (not shown) delineates sectors of the molecule, with boundaries at residues 28, 81 and 108, which correlate remarkably well with the exon/intron boundaries, although the plot would predict that the second exon was once divided by an intron at about residue 56.

Hen egg-white lysozyme and T4 'phage lysozyme have different three-dimensional structures, but approximately the first 80 residues of T4 lysozyme do have a three-dimensional structure resembling residues 27–109 of hen egg-white lysozyme (Rossmann & Argos, 1976; Remington & Matthews, 1978). This structural similarity also includes the two catalytic residues and the substrate binding site (Matthews *et al.*, 1981). Is this an example of convergence to a similar structure, or do these two proteins have a common evolutionary origin? The striking observation is that exons 2 and 3 of hen egg-white lysozyme together code for residues 28–108, which is almost exactly the part of this molecule that closely resembles the functionally equivalent part of T4 'phage lysozyme (Artymiuk, Blake & Sippel, 1981). If these two proteins do indeed share a common ancestral gene fragment, then it would appear that different proteins can be generated by the juxtaposition of different combinations of exons, as first proposed by Gilbert (1978).

So far we have considered only small, single-domain proteins. In the larger multi-domain proteins the correlation between exons and sectors might be expected within domains, but perhaps domain boundaries will also coincide with exon/intron boundaries. Unfortunately the gene sequences of the proteins which contain the common domains such as the eight-stranded β-barrel or the Rossmann fold are still unknown, but the genetics of one family of multi-domain proteins, the immunoglobulins, is known in considerable detail.

A multi-domain protein: the immunoglobulin molecule

Immunoglobulin molecules (also known as antibodies) are multifunctional proteins which bind foreign antigens, and interact with serum proteins (such as complement) and cell-surface receptor molecules, to initiate clearance of the antibody–antigen complex. Most remarkable is the fact that while the range of potential antigens is virtually unlimited, each antibody is highly specific, so that an enormous diversity of antibodies with different antigen binding sites must be generated by the immune system. The mechanisms by which this diversity is generated may shed some light on evolution processes.

Protein structure

The structure of an immunoglobulin molecule is shown schematically in Fig. 8.11. The amino-acid sequence of each of the two identical heavy and light chains consists of a stretch of approximately 110 residues (including an internal disulphide bridge) which is repeated with closely similar sequences four times, and twice, respectively. The sequence of the N-terminal region of each chain is different for each immunoglobulin molecule, whereas the other regions are constant for a given type of light chain or class of heavy chain. Fig. 8.11 explains the nomenclature for the variable (V) and constant (C) regions, and shows also the short 'hinge' region which breaks the pattern between C_H1 and C_H2. Each region of conserved sequence corresponds to a compactly folded domain of all-β structure which is very similar to that of Cu/Zn superoxide dismutase (compare Figs. 8.3 b and 8.11). These domains associate in pairs more or less closely in different parts of the molecule.

Fig. 8.11 also shows that the antigen binding part of the molecule consists of the two variable domains, whereas all other functions are performed by the constant domains. Furthermore, within each variable region are three loop regions of even greater sequence variability, and it is these residues which, when brought together by the $V_L:V_H$ association, constitute the antigen binding site itself and determine the specificity of the immunoglobulin molecule.

Fig. 8.11. Correlation of the mosaic gene structure of an immunoglobulin molecule (class G) with the amino-acid sequence of the polypeptide chain (which is segmented accordingly) and its three dimensional structure. The structure of the molecule is divided into domains and the fold of the polypeptide chain within each of the four domains (V_L, C_L, V_H and C_H1) is shown. Of the two β-sheets in each domain, the one involved in domain pairing ($V_L:V_H$ and $C_L:C_H1$) is represented by clear ribbons, the other by cross-hatched ribbons. The dashed segments of V_L and V_H correspond to the hypervariable regions of the chain (designated L1, H1 etc. and are indicated on the schematic 4-chain diagram). These segments together form the antigen-binding site in the three-dimensional structure. Interactions with complement and cell surface molecules occur in the C_H2 and C_H3 domains, each of which has a similar fold to C_H1. The solid lines connecting the heavy (H) and light (L) chains, and within domains, of the 4-chain diagram represent disulphide bridges. The C exons correspond to whole domains; the J exons correspond to a single strand of β-sheet (designated by close hatching); the heavy chain D exon corresponds to part of the third hypervariable loop of V_H (designated by the thick broken line). The latter, together with the other hypervariable loop regions (designated by thin broken lines), contribute to the antigen-binding site.

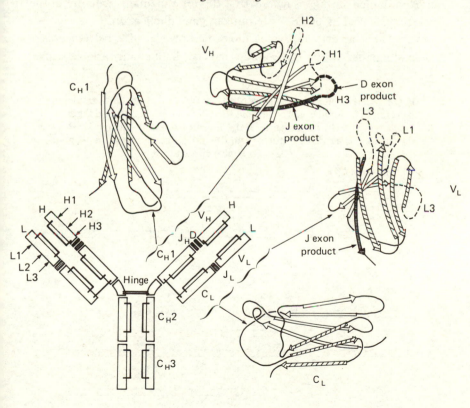

Gene structure of immunoglobins

As early as 1965, the existence of light chains with different V_L but identical C_L sequences prompted the suggestion that the many V_L and very few C_L sequences might be encoded by separate genes which could be combined in different ways (Dreyer & Bennett, 1965). In the last few years this proposition has been confirmed, and an uninterrupted stretch of DNA coding for the C_L region has been found well separated from the V_L genes, which are clustered together elsewhere on the same chromosome, as shown in Fig. 8.12a (for review see Tonegawa, 1981). This same arrangement has been found for the heavy chain genes, although instead of a single C gene, the constant region genes for each of the five classes of heavy chain are found together (see Fig. 8.12b). There is another difference however: the heavy chain C genes contain introns, and remarkably, the exons correspond precisely to the domains and to the hinge region, as Fig. 8.12b also shows (Sakano *et al.*, 1979).

In the constant part of the molecule therefore, each domain is coded by an exon, which strongly suggests that the multi-domain structure of the molecule evolved by processes of tandem gene duplication.

In the variable domains of the molecule however, this pattern breaks down altogether. Of the 108 residues in every light chain variable region,

Fig. 8.12. The relative arrangement in the DNA of (*a*) the immunoglobin light chain V, J and C genes, and (*b*) the heavy chain V, D, J and C genes. One C gene is shown for each of the immunoglobulin classes, M, D, G, E and A (subclasses are omitted) and the mosaic structure of the C_γ gene (class G) is also shown.

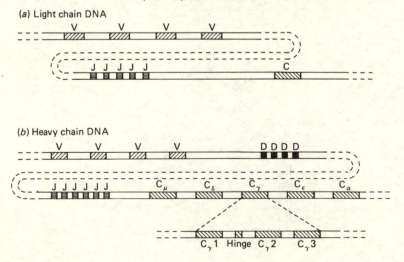

approximately the first 95 are encoded by one exon (called the V exon), but the rest are encoded by another (designated J) which lies in a small cluster of related sequences much nearer to the single C_L gene copy than to the cluster of V exons (see Fig. 8.12a). In the heavy chain variable region there is a third exon (D) which lies in a cluster of similar sequences between the V and J exons (Fig. 8.12b). (For a recent review, see Leder, 1982.) Rearrangements both at the DNA level (during lymphocyte differentiation) and RNA level (splicing) serve to bring any combination of exons together, as summarized in Fig. 8.13.

The correspondence between the pattern of exons and the structure of the immunoglobulin molecule is shown in Fig. 8.11. The V, C and J exons clearly do not code for domains as the C exons do, but exon D does correspond to part of the third hypervariable loop of V_H, and part of exon J to the very end of this loop. The potential antigen binding site is therefore formed from amino-acid residues coded for by a total of five exons, and the number of possible permutations of exon recombinations is more than sufficient to account for the observed diversity of immunoglobin molecules.

Evolutionary implications

In the constant region of the molecule where interaction with other invariant structures must be preserved, exons code for protein domains. The molecule probably evolved by tandem gene duplication with the appearance of new functions. For example, the C_H3 domains are not

Fig. 8.13. Schematic representation of the rearrangements of immuno-globulin light chain V, J and C genes, between the embryonic DNA and mature lymphocyte DNA, and between the primary RNA and the mature messenger RNA transcript which is translated into the protein polypeptide chain.

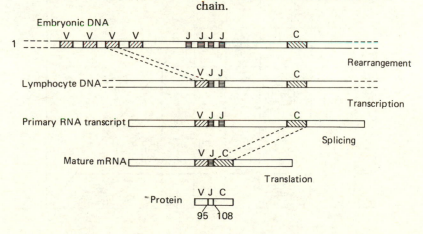

required for complement activation, but both C_H2 and C_H3 domains are required for most cell surface interactions, and immunoglobulins of class M, which have an extra domain, C_H4, form pentamers of the four-chain molecule of Fig. 8.11 and activate complement even more effectively.

In the variable region however, where variations upon a basic antigen combining site structure are required, some shorter exons are found which code only for extended stretches of polypeptide chain (see Fig. 8.11) and the antigen combining site apparently lies at the conjunction of five exon products. Here, in the mechanism by which immunoglobulin diversity is generated, is evidence of exon recombination of the kind suggested by the

Fig. 8.14. Gō plot (minimum distance 27Å) for the V_L domain of a human immunoglobulin G. Nine sectors are identified, but none of the junctions between them corresponds to the V/J join at residues 90/91. (The residue numbering for this protein differs from that of Fig. 8.13.)

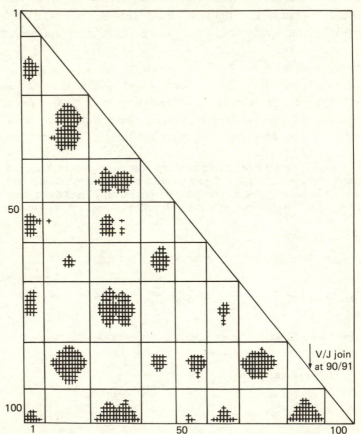

comparisons of proteins with common domain structures that were discussed above.

In the search for evidence of evolutionary mechanisms in the extant machinery of immunoglobulin synthesis it is also instructive to consider whether the exons which code for the V domains could have been predicted from a Gō diagram. Fig. 8.14 shows such a plot for a V_L domain. Clearly the structure can be divided into many sectors but this is not at all surprising since any excursion of the polypeptide chain from the centre of a domain and back (i.e. each loop between β-strands) constitutes a sector and will generate a cluster of large C^α–C^α distances on the Gō plot. The question is to what extent the sectors correlate with exons. The predicted exon/intron boundaries occur at the mid-point of each β-strand, but in fact the V/J join occurs at the end of the third hypervariable loop (at residues 90/91 in the numbering of Fig. 8.14) and the D exon corresponds only to a part of the corresponding loop of V_H. The Gō plot therefore identifies sectors of the V domains (as it would also for C domains), but there is no correlation between these sectors and the exons of immunoglobulin domains. Clearly these plots must be interpreted with caution, but one intriguing question is raised. Since the V_L domain is coded for by two exons, and V_H three, were there ever any more to correspond to the other hypervariable regions? Certainly of all the different combining sites produced by V, D, J recombinations, a disproportionate number will differ from each other in only the third hypervariable regions of V_L and V_H, which is only a part of the potential antigen binding site. Unfortunately the question of the evolution of the exon/intron structure of eukaryotic genes remains a matter for speculation.

Conclusions

The three-dimensional structures of more than one hundred globular proteins are now known and a complex pattern of relationships between them is beginning to emerge.

The starting point for assessment of this evidence is the long-accepted view that close correspondence between the amino-acid sequences of proteins is indicative of evolutionary relationships. Thus, the sequences of cytochrome c molecules from various plant and animal species have long been used in the derivation of evolutionary trees (e.g. Dickerson, 1980). Similar evidence has also been taken to establish family and evolutionary relationships between proteins of different function (e.g. the globins) even when their functions are apparently unrelated (e.g. lysozyme and α-lactalbumin). Sequence comparisons have also established the importance of gene duplication for creating larger and more complex proteins (such as ferredoxin and the immunoglobulins), in addition to the generation of extra copies of gene sequences which can then diverge to produce variant proteins.

In general, the analysis of three-dimensional structures has lent weight to these conclusions in that molecules with closely related sequences have been found also to have closely similar structures. But structural analyses have also shown that recognized members of the same protein family which are distantly related enough to have very different amino-acid sequences still have closely similar three-dimensional structures. This evidence shows that, during evolutionary change, three-dimensional structure is conserved more strongly than amino-acid sequence.

A difficulty arises when closely similar three-dimensional structures are found in molecules with apparently unrelated amino-acid sequences. In these instances, when the foregoing evidence that three-dimensional structure is strongly conserved is taken into account, it seems that the three-dimensional structures may be providing evidence of divergent evolution that has been lost from the sequences. But the alternative model of convergent evolution has also to be considered and the degree of similarity must be compared with that which might be expected to arise independently from the operation of topological constraints on polypeptide chain folding and the consequent stereochemical constraints upon the structure. In addition, for proteins with different but related functions, the resolution of this ambiguity probably requires a deeper understanding than we have at present of how similar two structures are bound to be in order to perform similar functions.

A similar difficulty arises when the detailed structures of the genes are taken into account. If the exon/intron patterns of the genes coding for two structurally similar proteins are identical it is tempting to suppose that they have a common evolutionary origin. But, given our present uncertainties about the genesis of the mosaic structure of genes and the evidence from the analysis of pseudogenes that introns can be lost (Vanin *et al.*, 1980), and that the exon/intron patterns of related nucleotide sequences can be quite different (Leicht *et al.*, 1982), it is obviously necessary to be cautious. Identical mosaic structures appear to provide strong evidence for divergent evolution but differences in mosaic structure can be explained away all too easily.

Nevertheless, correlation of the gene structures of proteins such as the globins and lysozymes with their three-dimensional structures has suggested a plausible mechanism for the evolution of complex proteins by the stepwise accretion of simpler exon products. Furthermore, this kind of analysis has also revealed, or confirmed, structural relationships between parts of protein molecules that are not in themselves compelling but which may have evolutionary significance. It is clear that there is no unique correlation between exons and one particular type of protein fragment (domain, sector, strand or functional unit) but common structures of various kinds are being recognized in a number of proteins and a persuasive model of protein evolution appears to be emerging that has parallels in the processes that still exist for the generation of diversity among the immunoglobulins.

Clearly more evidence is needed from studies of gene sequences to compare with the three-dimensional structures of their protein products, and this evidence is now accumulating rapidly. However, since it appears that the tertiary structure of a protein may out-last changes both in its chemical structure and the chemical and mosaic structure of its gene, the only surviving evidence for divergent evolution sometimes may lie in the three-dimensional structure of the protein. Furthermore, it is only through the study and comparison of the tertiary structures of proteins unrelated by amino-acid or gene sequence, that convergent evolutionary processes can be revealed. The three-dimensional structures of proteins may provide the best intimations of evolution that are accessible to us.

Further reading

Leder, P. (1982). The genetics of antibody diversity. *Scientific American*, **246**, 5, 72–83.
Richardson, J. S. (1981). The anatomy and taxonomy of protein structure. *Advances in Protein Chemistry*, **34**, 167–339.
Rossmann, M. G. & Argos, P. (1981). Protein folding. *Annual Review of Biochemistry*, **50**, 497–532.

References

Adman, E. T., Sieker, L. C. & Jensen, L. H. (1973). The structure of a bacterial ferredoxin. *Journal of Biological Chemistry*, **248**, 3987–96.
Argos, P. & Rossmann, M. G. (1979). Structural comparisons of heme binding proteins. *Biochemistry*, **18**, 4951–60.
Argos, P. & Rossmann, M. G. (1980). The relationships between coding sequences and function in some heme binding proteins. *Journal of Molecular Evolution*, **16**, 149–50.
Artymiuk, P. J., Blake, C. C. F. & Sippel, A. E. (1981). Genes pieced together – exons delineate homologous structures of diverged lysozymes. *Nature*, **290**, 287–8.

Banks, R. D., Blake, C. C. F., Evans, P. R., Haser, R., Rice, D. W., Hardy, G. W., Merrett, M. & Phillips, A. W. (1979). Sequence, structure and activity of phosphoglycerate kinase: a possible hinge-bending enzyme. *Nature*, **279**, 773–7.

Banner, D. W., Bloomer, A. C., Petsko, G. A., Phillips, D. C., Pogson, C. I., Wilson, I. A. (1975). Structure of chicken muscle triose phosphate isomerase determined crystallographically at 2.5 Å resolution using amino acid sequence data. *Nature*, **255**, 609–14.

Blake, C. C. F. (1978). Do genes-in-pieces imply proteins-in-pieces? *Nature*, **272**, 267.

Blake, C. C. F. (1979). Exons encode protein functional units. *Nature*, **277**, 598.

Blake, C. C. F., Koenig, D. F., Mair, G. A., North, A. C. T., Phillips, D. C. & Sarma, V. R. (1965). Structure of hen egg-white lysozyme – a three-dimensional Fourier synthesis at 2 Å resolution. *Nature*, **206**, 757–63.

Blundell, T. L., Lindley, P., Miller, L., Moss, D., Slingsby, C., Tickle, I. J., Burnell, B. & Wistow, G. (1981). The molecular structure and stability of the eye lens: X-ray analysis of γ-crystallin II. *Nature*, **289**, 771–7.

Breathnach, R., Mandel, J. L. & Chambon, P. (1977). Ovalbumin gene is split in chicken DNA. *Nature*, **270**, 314–9.

Brew, K., Vanaman, T. C. & Hill, R. L. (1967). Comparison of the amino acid sequence of bovine α-lactalbumin and hens egg white lysozyme. *Journal of Biological Chemistry*, **242**, 3747–9.

Burnett, R. M., Darling, G. D., Kendall, D. S., LeQuesne, M. E., Mayhew, S. G., Smith, W. W. & Ludwig, M. L. (1974). The structure of the oxidized form of Clostridial flavodoxin at 1.9 Å resolution. Description of the flavin mononucleotide binding site. *Journal of Biological Chemistry*, **249**, 4383–92.

Cohen, F. E., Sternberg, M. J. E. & Taylor, W. R. (1982). Analysis and prediction of the packing of α-helices against a β-sheet in the tertiary structure of globular proteins. *Journal of Molecular Biology*, **156**, 821–62.

Craik, C. S., Buchman, S. R. & Beychok, S. (1980). Characterization of globin domains: heme binding to the central exon product. *Proceedings of the National Academy of Sciences, USA*, **77**, 1384–8.

Craik, C. S., Buchman, S. R. & Beychok, S. (1981). O_2 binding properties of the product of the central exon of β-globin gene. *Nature*, **291**, 87–90.

Dickerson, R. E. (1980). Cytochrome C and the evolution of energy metabolism. *Scientific American*, **242**, 3, 136–53.

Doel, M. T., Houghton, M., Cook, E. A. & Carey, N. H. (1977). The presence of ovalbumin mRNA coding sequences in multiple restriction fragments of chicken DNA. *Nucleic Acids Research*, **4**, 3701–13.

Dreyer, W. J. & Bennett, J. C. (1965). The molecular basis of antibody formation: a paradox. *Proceedings of the National Academy of Sciences, USA*, **54**, 864–8.

Eklund, H., Nordstrom, B., Zeppezauer, E., Sonderlund, G., Ohlsson, I., Hoiwe, T., Soderberg, B-O., Tapia, O., Branden, C-I. & Akeson, A. (1976). Three-dimensional structure of horse liver alcohol dehydrogenase at 2.4 Å resolution. *Journal of Molecular Biology*, **102**, 27–59.

Fermi, G. & Perutz, M. F. (1981). *Atlas of Molecular Structures in Biology 2 Haemoglobin & Myoglobin.* Oxford: Clarendon Press.

Gilbert, W. (1978). Why genes in pieces? *Nature*, **271**, 501.

Gō, M. (1981). Correlation of DNA exonic regions with protein structural units in haemoglobin. *Nature*, **291**, 90–2.

Guilliland, G. L. & Quiocho, F. A. (1981). Structure of the L-arabinose-binding protein from Escherichia coli at 2.4 Å resolution. *Journal of Molecular Biology*, **146**, 341–62.

Hol, W. G. J., van Duijnen, P. T. & Berendsen, H. J. C. (1978). The α-helix dipole and the properties of proteins. *Nature*, **273**, 443–6.

Holbrook, J. J., Liljas, A., Steindel, J. & Rossmann, M. G. (1975). Lactate dehydrogenase. In *The Enzymes*, ed. P. D. Boyer, 3rd edn. Vol. 11, pp. 191–292. New York: Academic Press.

Janin, J. & Chothia, C. (1980). Packing of α-helices onto β-pleated sheets and the anatomy of α/β proteins. *Journal of Molecular Biology*, **143**, 95–128.

Jeffreys, A. J. & Flavell, R. A. (1977). The rabbit β-globin gene contains a large insert in the coding sequence. *Cell*, **12**, 1097–108.

Jenkins, J. A., Johnson, L. N., Stuart, D. I., Stura, E. A., Wilson, K. S. & Zanotti, G. (1981). Phosphorylase: control and activity. *Philosophical Transactions of the Royal Society, London B*, **293**, 23–41.

Jensen, E. O., Paludan, K., Hyldig-Nielsen, J. J., Jorgensen, P. & Marcker, K. A. (1981). The structure of a chromosomal leghaemoglobin gene from soybean. *Nature*, **291**, 677–9.

Johnson, L. N., Jenkins, J. A., Wilson, K. S., Stura, E. A. & Zanotti, G. (1980). Proposals for the catalytic mechanism of glycogen phosphorylase b prompted by crystallographic studies on glucose-1-phosphate binding. *Journal of Molecular Biology*, **140**, 565–80.

Jung, A., Sippel, A. E., Grez, M. & Schitz, G. (1980). Exons encode functional and structural units of chicken lysozyme. *Proceedings of the National Academy of Sciences, USA*, **77**, 5759–63.

Kendrew, J. C., Dickerson, R. E., Strandberg, B. E., Hart, R. G., Davies, D. R., Phillips, D. C. & Shore, V. C. (1960). Structure of myoglobin – a three-dimensional Fourier synthesis at 2Å resolution. *Nature*, **185**, 422–7.

Konkel, D. A., Tilghman, S. M. & Leder, P. (1978). The sequence of the chromosomal mouse β-globin major gene: homologies in capping, splicing and poly(A) sites. *Cell*, **15**, 1125–32.

Kraut, J., Robertus, J. D., Birktoft, J. T., Alden, R. A., Wilcox, P. E. & Poers, J. C. (1972). The aromatic substrate binding site in subtilisin BPN and its resemblance to chymotrypsin. *Cold Spring Harbor Symposium on Quantitative Biology*, **36**, 117–23.

Kusunoki, M., Harada, W., Tanaka, N. & Kakudo, M. (1981). The structure of taka-amylase A at 3.0 Å resolution and the D-Fourier studies of substrate binding sites. *Acta Crystallographica A37, Abstracts for the Twelfth International Congress*. 02.1-32.

Lebioda, L., Hatada, M. H., Tulinsky, A. & Mavridis, I. (1982). Comparison of the folding of 2-keto-3-deoxy-6-phosphogluconate adolase, triose phosphate isomerase and pyruvate kinase and its implications in molecular evolution. *Journal of Molecular Biology*, **162**, 445–58.

Leder, P. (1982). The genetics of antibody diversity. *Scientific American*, **246**, 5, 72–83.

Leicht, M., Long, G. L., Chandra, T., Kurachi, K., Kidd, V. J., Mace, M., Davie, E. W. & Woo, S. L. C. (1982). Sequence homology and structural comparisons between chromosomal human α_1-antitrypsin and chicken ovalbumin genes. *Nature*, **297**, 655–9.

Levitt, M. & Chothia, C., (1976). Structural patterns in globular proteins. *Nature*, **261**, 552–8.

Lindquist, Y. & Brändén, C-I., (1980). Structure of glycolate oxidase from spinach at a resolution of 5.5 Å. *Journal of Molecular Biology*, **143**, 201–11.

Matthews, B. W., Grutter, M. G., Anderson, W. F. & Remington, S. J. (1981). Common precursor of lysozymes of hen egg-white and bacteriophage T4. *Nature*, **290**, 334–5.

Monaco, H. L., Crawford, J. L. & Lipscomb, W. N. (1978). Three-dimensional structures of aspartate carbamoyltransferase from Escherichia coli and of its complex with cytidine triphosphate. *Proceedings of the National Academy of Sciences, USA*, **75**, 5276–80.

Moras, D., Olsen, K. W., Sabesan, M. N., Buehner, M., Ford, G. C. & Rossmann, M. G. (1975). Studies of asymmetry in the three-dimensional structure of lobster D-glyceraldehyde-3-phosphate dehydrogenase. *Journal of Biological Chemistry*, **250**, 9137–62.

Ohlsson, I., Norstrom, B. & Branden, C-I. (1974). Structural and functional similarities within the coenzyme binding domains of dehydrogenases. *Journal of Molecular Biology*, **89**, 339–54.

Pauling, L. & Corey, R. B. (1951). The pleated sheet, a new layer configuration of polypeptide chains. *Proceedings of the National Academy of Sciences, USA*, **37**, 251–6.

Phillips, D. C. (1970). The development of crystallographic enzymology. In *British Biochemistry Past and Present*, ed. T. W. Goodwin, pp. 11–28. London: Academic Press.

Ploegman, J. H., Drent, G., Kalk, K. H. & Hol, W. G. J. (1978). Structure of bovine liver rhodanese. I- structure determination of 2.5 Å resolution and a comparison of the conformation and sequence of its two domains. *Journal of Molecular Biology*, **123**, 557–94.

Remington, S. J. & Matthews, B. W. (1978). A general method to assess similarity of protein structures, with applications to T4 bacteriophage lysozyme. *Proceedings of the National Academy of Sciences, USA*, **75**, 2180–4.

Richardson, J. S. (1976). Handedness of crossover connections in β-sheets. *Proceedings of the National Academy of Sciences, USA*, **73**, 2619–22.

Richardson, J. S. (1981). The anatomy and taxonomy of protein structure. *Advances in Protein Chemistry*, **34**, 167–339.

Richardson, J. S., Richardson, D. C., Thomas, K. A., Silverton, E. W. & Davies, D. R. (1976). Similarity of three-dimensional structure between the immunoglobulin domain and the copper, zinc superoxide dismutase subunit. *Journal of Molecular Biology*, **102**, 221–35.

Rose, I. A. (1981). Chemistry of proton abstraction by glycolytic enzymes (aldolase, isomerases and pyruvate kinase). *Philosophical Transactions of the Royal Society London B*, **293**, 131–43.

Rossmann, M. G. & Argos, P. (1976). Exploring structural homology of proteins. *Journal of Molecular Biology*, **105**, 75–95.

Rossmann, M. G. & Liljas, A. (1974). Recognition of structural domains in globular proteins. *Journal of Molecular Biology*, **85**, 177–81.

Rossmann, M. G., Moras, D. & Olsen, K. W. (1974). Chemical and biological evolution of a nucleotide-binding protein. *Nature*, **250**, 194–9.

Sakano, H., Rogers, J. H., Hüppi, D., Brack, C., Traunecker, A., Maki, R., Wall, R. & Tonegawa, S. (1979). Domains and the hinge region of an immunoglobulin heavy chain are encoded in separate DNA segments. *Nature*, **277**, 627–33.

Schulz, G. E., Schirmer, R. H., Sachsenheimer, W. R., Pai, E. F. (1978). The structure of the flavoenzyme glutathione reductase. *Nature*, **273**, 120–4.

Smith, S. G. (1982). Structural studies of the lactose synthetase system. *D. Phil. Thesis, University of Oxford*.

Sternberg, M. J. E. & Thornton, J. M. (1976). On the conformation of proteins: the handedness of the β-strand—α-helix—β-strand unit. *Journal of Molecular Biology*, **105**, 367–82.

Stuart, D. I., Levine, M., Muirhead, H. & Stammers, D. K. (1979). Crystal structure of cat muscle pyruvate kinase at a resolution of 2.6 Å. *Journal of Molecular Biology*, **134**, 109–42.

Tonegawa, S. (1981). Somatic recombination and mosaic structure of immuno-globulin genes. In *The Harvey Lectures*, Series 75, pp. 61–83. London & New York: Academic Press.

Tsunoda, J. N., Yasunobu, K. T. & Whiteley, H. R. (1968). Non-heme iron proteins. IX. The amino acid sequence of ferredoxin from *Micrococcus aerogenes*. *Journal of Biological Chemistry*, **243**, 6262–72.

Vanin, E. F., Goldberg, G. I., Tucker, P. W. & Smithies, O. (1980). A mouse α-globin-related pseudogene lacking intervening sequences. *Nature*, **286**, 222–6.

Wetlaufer, D. B. (1973). Nucleation, rapid folding and globular intrachain regions in proteins. *Proceedings of the National Academy of Sciences, USA*, **70**, 697–70.

9

Evolution of gene families: the globin genes

ALEC J. JEFFREYS, STEPHEN HARRIS, PAUL A. BARRIE, DAVID WOOD,
ALAIN BLANCHETOT AND SUSAN M. ADAMS

One of the best studied multigene families is the set of genes that code for vertebrate haemoglobins. A wide variety of globin amino-acid sequences from many different organisms has been accumulated over the past two decades, and more recently recombinant DNA technology has enabled molecular biologists to study the structure and chromosomal organization of the globin genes themselves. Although much of this research has been directed towards an understanding of the molecular mechanisms that underlie the developmental regulation of these gene families, these studies have revealed a wide array of molecular phenomena which throw considerable light on the evolution of these genes.

In this paper, we discuss generally some of the more recent ideas in molecular evolution that have emerged from studies on multigene families, in particular the globin genes. It should be stressed that the evolutionary behaviour of globin gene families is probably fairly typical of most eukaryotic gene families, and that many of these ideas have been strongly influenced by research into numerous other systems including repetitive and satellite DNAs as well as protein coding genes (see Jeffreys, 1981; Dover & Flavell, 1982).

Globin phylogeny

In addition to adult haemoglobins, mammals produce embryonic and in some cases foetal specific globins. All of these haemoglobins are tetramers of two α-like and two β-like globin polypeptides, and amino-acid sequence analysis has shown that all of these globins share significant sequence homology and are therefore most likely the products of globin gene duplications that have occurred during evolution. The sequence divergence between globins can be summarized in a dichotomizing dendrogram (Dayhoff *et al.*, 1972; Efstratiadis *et al.*, 1980; Czelusniak *et al.*, 1982), as shown in Fig. 9.1 for the human embryonic (ε, ζ), foetal ($^G\gamma$, $^A\gamma$) and adult (α1, α2, β, δ) globins. This diagram clearly implies a hierarchy of gene duplications starting with an initial α – β duplication followed by expansion and diversification of both the α- and β-globin gene subfamilies.

Such dendrograms can also be constructed for homologous globin sequences determined from different species (see Dayhoff *et al.*, 1972; Czelusniak *et al.*, 1982). The molecular phylogenies obtained generally accord well with classical phylogenetic data, and by comparing protein divergence with the time of common ancestry of two species, as estimated from the palaeontological record, the rate of amino-acid substitution during evolution can be deduced. Surprisingly, this rate appears to be remarkably constant for a given protein, irrespective of the evolutionary lineage being studied (see Wilson, Carlson & White, 1977). This molecular clock provides a powerful method for estimating the divergence time between species and between members of a multigene family within one species, as shown for the human globins in Fig. 9.1. According to this diagram, the initial duplication to give distinct α- and β-related globin genes is ancient and occurred some five hundred million years ago, early in the evolution of the vertebrates. This is supported by the existence of distinct α- and β-globin gene families in mammals, birds, reptiles, amphibians and bony fish but not in primitive vertebrates (lamprey, hagfish) which only

Fig. 9.1. An evolutionary tree of human globins. This tree was deduced from amino-acid and DNA sequence data and shows the estimated times of gene divergence in millions of years. Adapted from Dayhoff, Hunt, McLaughlin & Jones (1972), Efstratiadis *et al.* (1980) and Czelusniak *et al.* (1982).

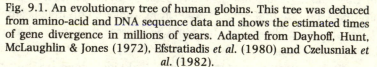
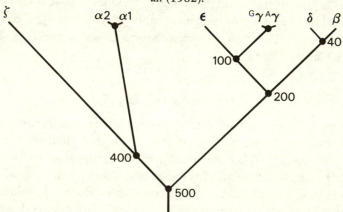

possess a monomeric haemoglobin specified by a globin gene which has not undergone an α-β duplication (see Dayhoff *et al.*, 1972).

The gene duplication story in Fig. 9.1 can readily be interpreted as a series of molecular adaptations during evolution to various vertebrate modes of life. The initial α–β globin gene duplication gave rise to diverged gene products capable of associating into multimeric haemoglobins suitable for cooperative oxygen binding within large and metabolically active organisms. There is some indication that many of the critical differences in the α- and β-polypeptides were accumulated rapidly, in a non-clock-like fashion, soon after the initial gene duplication (Goodman, Moore & Matsuda, 1975; Czelusniak *et al.*, 1982; however, see Kimura, 1981*a*). If so, this would be an example, at the molecular level, of rapid adaptive Darwinian evolution. The later embryonic and foetal globin gene duplications can also be interpreted as molecular adaptations to altered oxygen demands in vertebrate embryos and later in large placental mammals with a distinct foetal stage of development.

Throughout this process, functionally important amino-acid residues, such as those involved in binding haem, have remained relatively invariant in evolution. This type of correlation between function and evolutionary conservation has repeatedly led to the general idea that regions containing slowly evolving sequences (in DNA, RNA or protein) will correspond to functional segments maintained by conservative selection (see Kimura & Ohta, 1974).

What is the evolutionary significance of relatively recent amino-acid changes in globins, and why should these changes have accumulated in a clock-like fashion? Diametrically opposed explanations are given by selectionists and neutralists. The former argue that most or all changes represent adaptive 'fine-tuning', despite a lineage-independent constant rate of substitution that is apparently not related to widely varying tempos of environmental and morphological change during evolution. The neutral theory (see Kimura, 1979) maintains that most or all changes are non-adaptive (neutral) and have become fixed by genetic drift; to explain the molecular clock, this theory requires that the mutation rate itself must be lineage-independent and not influenced by factors such as generation time.

To summarize, protein data now complemented by DNA sequence information have made it possible to construct molecular phylogenies and to deduce the gene duplication histories of multigene families as shown in Fig. 9.1. Before going any further, it might be useful to emphasize a few points that are often considered to be implicit in such gene family diagrams.

All members of the family are assumed to have evolved at the same rate. While this is approximately true for adult α-and β-globins, insufficient data exist for other proteins such as the ζ- and ε-globin chains.

Nodes in the tree are presumed to represent gene duplication events; after duplication, genes are assumed to diverge independently. As we shall see below, these two assumptions are often false.

The tempo of gene duplication is considered to be rather slow; thus in Fig. 9.1, the human globin gene family has apparently fixed only six duplications over the last five hundred million years. However, it is possible that many more duplications have occurred and that genes other than the eight shown in Fig. 9.1 have subsequently been lost. As discussed below, direct molecular evidence for supernumerary duplicates now exists.

Structure and organization of globin genes

All of the human globin genes shown in Fig. 9.1 have been isolated by recombinant DNA methods and characterized in great detail (see Efstratiadis *et al.*, 1980). Their general structure and organization in human chromosomal DNA is typical of animal genes and multigene families, and show many unusual features that are of considerable evolutionary interest.

Each human globin gene contains two intervening sequences (introns) interrupting the coding sequence (see Jeffreys, 1981). These introns are transcribed together with the regions of coding sequence (exons) to produce a large precursor RNA, which is rapidly processed by RNA splicing in the nucleus to produce functional globin mRNA plus excised intron RNA whose fate within the nucleus is still largely unclear. Every vertebrate α- and β-globin gene so far examined in species ranging from amphibians to mammals contains these two introns at precisely homologous intragenic locations. This suggests that the number and positions of introns within globin genes are stable in evolution and were established before the α–β duplication some five hundred million years ago. Despite this stability, no functions of these globin introns have yet been discovered, and detailed DNA sequence comparisons of homologous introns in globin genes isolated from different species have shown that they diverge much more rapidly in evolution than protein coding sequences (Van Den Berg *et al.*, 1978). It therefore seems possible that these rapidly evolving sequences have no role in gene expression but instead are relics of ancient events in the evolution of globin genes.

Two basic models of gene evolution have been proposed to account for the split structure of globin genes and numerous other eukaryotic (but not prokaryotic) genes. The first hypothesis elaborated by Gilbert (1978) and Blake (1979) is that each exon of a gene specifies a discrete structural domain within a protein and evolved as a discrete functional element, a 'minigene'. Movement of these exon elements around the genome during evolution would create novel combinations of exons, and their fusion, by inclusion within a single transcriptional unit coupled with RNA splicing, would produce novel proteins containing already stable domains. This hypothesis implies an important new mechanism for generating new biochemical functions that is quite different from the more traditional models of gene duplication and divergence, and is supported by numerous correlations between protein domains and exons in various gene systems, including globin (Blake, 1979; Craik, Buchman & Beychok, 1980; Gō, 1981). So far though, no newly evolved protein created by exon shuffling has been discovered. The second hypothesis is that globin genes and other eukaryotic split genes were originally continuous, and were invaded by transposable DNA elements (Darnell, 1978; Crick, 1979). Such an invasion could be tolerated provided that the elements carry information to direct their own removal by RNA splicing. This model of 'selfish' transposable DNA (Doolittle & Sapienza, 1980) as generators of introns neatly explains the precise colinearity of exons and protein even in very highly split genes. However, no definite example of intron gain by a gene during recent evolution has yet been found, although the highly variable structure of actin genes suggests that insertions might occur (Davidson, Thomas, Scheller & Britten, 1982).

The eight functional human globin genes (Fig. 9.1), each comprised of three exons and two introns, are organised into two large unlinked gene clusters, one of which contains the members of the α subfamily arranged in the order ζ–$\alpha 2$–$\alpha 1$ and the other the β subfamily organized ε–$^G\gamma$–$^A\gamma$–δ–β (see Efstratiadis *et al.*, 1980). This organization reflects the two basic modes of gene family organization in eukaryotes: gene clustering, and dispersal of related genes around the genome. Similar patterns of organization are also seen in repetitive DNA families: clustered repeats of, for example, ribosomal RNA and histone genes are taken as evidence of sequence amplification by unequal crossing over, whereas dispersed repetitive DNA families such as the AluI family in human DNA suggests that these elements may be mobile (transposable) (Jagadeeswaran *et al.*, 1981).

In both the α- and β-globin gene clusters, genes are arranged in the same orientation and in developmental sequence, with embryonic genes preceding adult genes. Surprisingly little of the DNA in these clusters specifies globin; 8% of the DNA constitutes the globin exons with an additional 8% being found in globin introns. The remaining 84% of cluster DNA is found between genes and comprises a complex mixture of unique sequence and repetitive DNAs, including elements such as AluI family repeats that are also found at many other locations elsewhere in the genome (Fritsch, Lawn & Maniatis, 1980; Jagadeeswaran *et al.*, 1981). The function of these extensive regions of intergenic DNA is unknown. Evidence from human mutants who carry deletions of various regions of the cluster suggests that at least some of the intergenic DNA does not contain essential non-globin genes (Flavell *et al.*, 1978), but might contain regulatory elements involved in globin gene switching during development (Fritsch, Lawn & Maniatis, 1979).

It has been suggested that most of the intergenic DNA in gene clusters is not functional at all, but instead represents junk DNA that has accumulated by duplication of selfish DNA sequences in these regions or by repeated insertion of selfish transposable elements (Ohno, 1980; Orgel & Crick, 1980). The first possibility is supported by the discovery of pseudogenes within several gene clusters (see Proudfoot, 1980). Pseudogenes were originally detected as extra gene-like sequences by cross-hybridization with functional gene DNA. DNA sequence analysis of cloned pseudogenes has shown that they have accumulated a wide variety of mutations, including alteration of normally invariant codons as well as deletions, insertions and frameshifts within the coding sequence, which would prevent them from coding a functional protein. In addition, it seems likely that many pseudogenes are not transcribed, although a transcribed interferon pseudogene has been described (Goeddel *et al.*, 1981). The human globin gene clusters contain at least three pseudogenes. The α cluster contains a $\psi\zeta$ and $\psi\alpha$ globin pseudogene arranged between functional genes in the order $\zeta-\psi\zeta-\psi\alpha-\alpha2-\alpha1$. Similarly, a globin pseudogene has been found in the β-globin gene cluster, in the intergenic region between the $^A\gamma$- and δ-globin genes (see Little, 1982). As implied in the nomenclature, the $\psi\zeta$ and ζ genes are closely homologous, as are the $\psi\alpha$ and α genes, suggesting that these pseudogenes were the products of excess ζ- and α-globin gene duplications respectively. Homologies of the pseudogene in the β-globin gene cluster with ε-, γ- and β-globin genes have not yet been reported.

The question now is whether pseudogenes, once silenced, become totally functionless, or whether instead they still retain some features of active genes that might be used to modulate the expression of genes and gene clusters – perhaps by supplying regulatory functions once associated with active genes or by mimicking functional genes in maintaining the local structure of chromatin domains. It might be possible to test these alternatives by analysing the evolutionary behaviour of pseudogenes to see whether they diverge rapidly or slowly in evolution and whether they consistently occupy specific regions of gene clusters.

Initial comparisons of pseudogene sequences with their active relatives suggest that at least some pseudogenes were initially functional genes, and that after silencing, they did diverge rapidly in sequence (Lacy & Maniatis, 1980; Miyata & Yasunaga, 1981). Indeed, pseudogenes are already being used as models for the neutral evolution of functionless DNA (Li, Gojobori & Nei, 1981). However, these rate arguments presuppose the existence of a neutral DNA clock (not yet proven) and the genetic isolation of members of a multigene family (not necessarily correct, since DNA sequences can be exchanged by elements of a family; see below). As yet, no really reliable divergence rates of pseudogenes are available, and their 'neutral' behaviour is still a speculation.

The size of the pseudogene component of a multigene family is also unclear. There might be a strictly limited number of (functional?) pseudogenes; alternatively, those pseudogenes so far detected might represent only the tip of an iceberg of globin pseudogenes ranging from recently silenced genes to pseudogenes so far diverged as to be completely undetectable by nucleic-acid hybridization.

Recent evolutionary history of the β-globin gene cluster

Many of the evolutionary questions connected with gene clusters, including the timing of gene duplications, appearance and divergence of pseudogenes and the general evolutionary behaviour of gene clusters and intergenic DNA in relation to the junk DNA hypothesis, could be resolved by a detailed phylogenetic analysis of mammalian gene clusters.

The β-globin gene cluster has been characterized in the rabbit (Lacy, Hardison, Quon & Maniatis, 1979), mouse (Jahn *et al.*, 1980) and goat (Cleary, Schon & Lingrel, 1981) as well as in man (see Fig. 9.2). In evolutionary terms, these four mammals are approximately equally related and share common ancestors at the time of the mammalian radiation round about eighty-five million years ago. These clusters all have certain features in common, including at least one embryonic gene at the head of the cluster, adult genes at the opposite end, and, interestingly, a pseudogene consistently present on the 5' side of the adult gene(s). However, the divergence between species has been sufficient for the accumulation of major shifts in arrangement of these clusters. There have been substantial changes in cluster lengths and in the numbers of genes, implying that gene duplication and contraction are not infrequent events in evolution. In the goat, a duplication of an entire ψβ–β gene pair seems

Fig. 9.2. The arrangement of various mammalian β-globin gene clusters. The human cluster was determined by Fritsch, Lawn & Maniatis (1980), the rabbit cluster by Lacy, Hardison, Quon & Maniatis (1979), the mouse cluster by Jahn *et al.* (1980) and the goat family by Haynes *et al.* (1980). In the goat, linkage of all members of the family has not yet been shown. Each gene is transcribed from left to right and contains two intervening sequences. The developmental stage of expression, where known, is indicated by asterisks: *, embryonic; **, foetal; ***, adult. Genes homologous to human β are shown by filled boxes, γ-like genes by hatched boxes and ε-like genes by open boxes; these homologies were determined primarily from sequence data (see Czelusniak *et al.*, 1982). Additional pseudogenes are shown by crosses, although the pseudogene status of the human ψβ2 and mouse βh2 genes have not yet been confirmed by direct sequencing. The scale bar is in kb (base pairs × 10⁻³).

Man

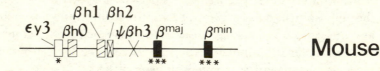

Rabbit

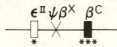

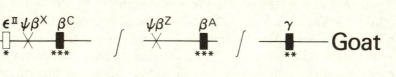

Mouse

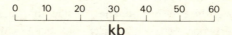

Goat

to have arisen recently in evolution; although the linkage arrangement between the two ψβ–β pairs has not yet been established (Cleary *et al.*, 1981). Also, there are some remarkable instances where genes have apparently switched their developmental expression during evolution. For example, the β3 gene in the rabbit is orthologous to the human foetal ($^G\gamma$ and $^A\gamma$) globin genes (Czelusniak *et al.*, 1982) yet is expressed along with the β4 gene in the embryo (Hardison *et al.*, 1979). Similarly, the goat foetal specific globin gene is β-like, not γ- like, in sequence (Czelusniak *et al.*, 1982). However, despite these complications, it is still evident that distinct ε-, γ- and β-like genes had appeared before the mammalian radiation, in accordance with the gene duplication times estimated for the human gene family (Fig. 9.1).

Because the human, mouse, rabbit and goat β-globin gene clusters are too far diverged for any detailed analysis of cluster evolution, we have characterized the organization of this cluster in a representative group of primates ranging from prosimians to man, with divergence times that neatly divide the interval leading from the mammalian radiation to contemporary primates (Barrie, Jeffreys & Scott, 1981; see Fig. 9.3). What emerges from this analysis is that the β-globin gene cluster appears to change during evolution in a rather discontinuous fashion, with long periods of stable organization. Thus the β-globin gene cluster is indistinguishable in organization in man, gorilla, chimpanzee and baboon, suggesting that over the last twenty five million years these primate clusters have not been subjected to a continuous rearrangement of selfish DNA sequences. In addition, the DNA sequences between the genes appear to have been evolving very slowly in these species, on average at about 10–20% of the rate of evolution of functionless DNA (as deduced from estimates of rates of silent base substitutions within functional gene sequences: these rates are generally taken to be a minimal estimate of the rate of neutral evolution, see Efstratiadis *et al.*, 1980; Kimura, 1981*b*; Miyata, Yasunaga & Nishida, 1980). Both the organizational stability and the apparent evolutionary conservation of intergenic DNA sequences seem to be incompatible with the notion that the β-globin gene cluster is simply a collection of genes embedded in a matrix of junk intergenic DNA. Instead, it seems more likely that the cluster in its entirety, including the intergenic DNA, has some functional significance and might represent a single coadapted supergene.

Clearly, there are many questions that still need to be answered. Do globin gene clusters in other lineages show similar periods of stasis and conservation? What is the detailed pattern of sequence conservation within clusters? Are pseudogenes and introns really exempt from conservation? If a cluster is a functional entity, where are its boundaries outside the functional genes and where are the next genes or gene clusters along the chromosome? What were the molecular events that resulted in the major shifts in organization seen when comparing, for example, Old World and New World monkeys (Fig. 9.3)? Do these shifts have any adaptive significance or are they merely rare neutral events that are still compatible with cluster function? Can functional assays of cloned gene clusters be devised, to test for adaptively relevant sequences? With current DNA technology, it should be possible to begin to answer many of these questions.

Fig. 9.3. Evolution of the β-globin gene cluster in man and the primates. The arrangement of the human cluster is taken from Fritsch *et al.* (1980) and the primate clusters from Barrie, Jeffreys & Scott (1981) and, for the chimpanzee, from P. A. Barrie & A. J. Jeffreys (unpublished data). β-, γ- and ε-like genes, as determined by sequencing and hybridization analysis, are represented by filled, hatched and open boxes respectively. Human globin pseudogenes are indicated by crosses; corresponding pseudogenes probably exist in the β-globin gene clusters of the gorilla, chimpanzee and baboon. The linkage between owl monkey ε- and γ-globin genes is provisional, and the ε–γ to δ–β linkage has not yet been established. Divergence times, estimated primarily from molecular data, are taken from Sarich & Cronin (1977) and from Romero-Herrera, Lehmann, Joysey & Friday (1973). The divergence time between the brown and ruffed lemurs is uncertain (Jeffreys *et al.*, 1982).

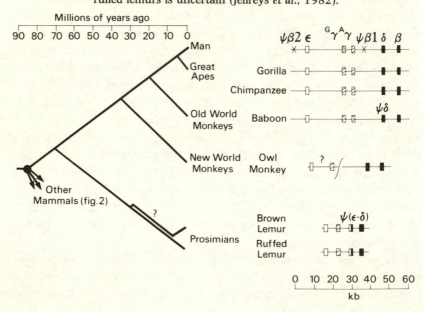

Histories of globin gene duplications

Globin gene cluster phylogenies such as that shown in Fig. 9.3 provide a new means for timing globin gene duplications. How well do these phylogenetic dates compare with dates deduced from amino-acid sequence comparisons of human globins (Fig. 9.1)?

Consider the γ-globin genes. A duplicated γ locus is present in man, apes and an Old World monkey, but not in rabbit, prosimian or New World monkey. This suggests that the γ-globin gene duplication is ancient, and arose about twenty to forty million years ago (Fig. 9.3). In contrast, the human γ-globin genes appear to have diverged very recently (Fig. 9.1), and are closely homologous, particularly over a well-defined 1500 base pair region extending from the 5' side of the gene and terminating within the second intron (Shen, Slightom & Smithies, 1981). This major discrepancy suggests that the duplicated γ-globin genes were not free to diverge independently after duplication, and instead, it seems that some form of sequence interchange has recently occurred between the γ loci, leading to the 'concerted' evolution of the ^{G}γ- and ^{A}γ-globin genes (Slightom, Blechl & Smithies, 1980; Shen *et al.*, 1981). A number of mechanisms can be envisaged which would lead to this interchange, all of which involve the unequal pairing of ^{G}γ- and ^{A}γ-globin genes. Recombination could then lead to three-γ and one-γ chromosomes, or to a localized reciprocal interchange of segments of ^{G}γ- and ^{A}γ-globin genes, or to a non-reciprocal gene conversion in which both partners in the recombination event undergo heteroduplex repair to generate non-parental DNA segments on each chromosome. In each case, homogenization of sequence differences between the duplicated γ loci could ensue. Interestingly, the homology block shared by the human ^{G}γ- and ^{A}γ-globin genes terminates at the simple sequence $(TG)_n$ (Slightom *et al.*, 1980). This sequence can, in principle, adopt a left-handed helical configuration (Wang *et al.*, 1979) and it seems possible that such an unusual structure in DNA might somehow have initiated the recombination event that recently homogenized the 1500 base pair sequence block shared by the duplicated γ-globin genes.

Another example illustrating the complexities underlying gene duplication and divergence is given by the δβ-globin gene duplication in man. The 'classic' duplication time is about forty million years ago (Fig. 9.1). At first sight, this time accords well with the specific presence of a δ-globin polypeptide in New World monkeys, great apes and man but not in prosimians, suggesting a duplication time of thirty-eight to seventy-five million years ago. However, DNA analysis has shown that the lemur, a prosimian, does in fact contain a δ-globin gene located as usual between the γ- and β-globin genes (Jeffreys *et al.*, 1982; Fig. 9.3). This gene is closely homologous to the human δ-, but not β-, globin gene over the second intron, third exon and 3′ flanking region. In contrast, the beginning of the gene is abnormal, rendering the whole gene a pseudogene, and is comprised of sequences related to an ε- or possibly γ-globin gene. Furthermore, initial DNA sequencing of the lemur β-globin gene is indicating that the human and lemur β-globin genes are closely related yet show little or no similarity in their non-coding regions to the δ-globin gene. The picture that emerges from the analysis, particularly of non-coding DNA, is that the δβ-globin gene duplication, far from having arisen forty million years ago, is extremely ancient, and that most of the sequence differences between δ and β had accumulated before the divergence of prosimians and simians about seventy million years ago (Fig. 9.4). Subsequently the lemur δ-globin gene must have been silenced and must have acquired ε-

Fig. 9.4. Evolutionary history of the primate δ-globin gene. The approximate times of major events so far discovered in the evolution of this gene are indicated by broken lines. The age of the δβ-globin gene duplication is completely unknown, although it presumably predates the divergence of the Lemuridae and Anthropoidea. (Reproduced from Jeffreys *et al.*, 1982, with permission of Academic Press.)

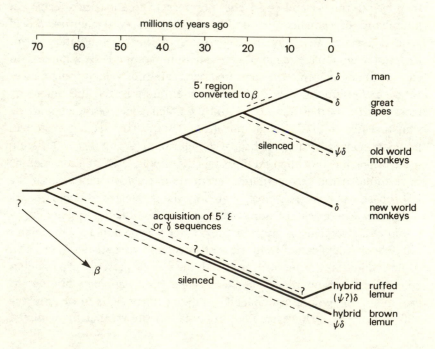

or γ-globin sequences by some form of recombination or gene conversion with a non-adult globin gene. In contrast, the human δ- and β-globin genes are remarkably similar over the 5′ region of the gene, suggesting that a correction of this region of δ by β occurred recently in human evolution (Fig. 9.4), analogous to the $^G\gamma$–$^A\gamma$ homogenization discussed above. This analysis also demonstrates that a functional gene in one organism can be a pseudogene in another and opens up the possibility that phylogenetic analysis could help determine the timing of gene silencing and hence the evolutionary behaviour of pseudogenes after the silencing event. Curiously, the δ-globin gene seems also to have been silenced in Old World monkeys (Martin, Zimmer, Kan & Wilson, 1980), raising problems concerning the role of the δ-globin gene, which has been maintained in an active or silent state over the entire history of primate evolution yet whose protein product is apparently inessential.

These complex sequence interchanges that can occur between members of a multigene family after gene duplication seem to be a common phenomenon in many gene systems (see Baltimore, 1981; Dover, 1982), and make the interpretation of gene family phylogenies (Fig. 9.1) very difficult. Each node on the phylogenetic dendrogram represents an estimate not only of the original duplication time but also of the times of succeeding localized homogenization events. Clearly, such trees will have to be replaced by more complex networks, and it is not yet clear how complex these will have to be. Some genes seem to be relatively immune from these processes, for example, the embryonic and adult globins which have remained quite discrete in sequence in a wide variety of mammals. This immunity might reflect some feature of the gene that blocks homogenization; alternatively interchanges might freely occur but generate products that are in some way dysfunctional and are eliminated by selection.

To understand the dynamics of sequence interchange amongst members of a multigene family, it will be important to investigate a newly arisen family with a reasonably clearly defined time of amplification, rather than families such as globin that have no obvious time of origin. We have recently investigated the fibroblast interferon gene and have found that a single gene exists in almost all vertebrates examined, including a wide variety of mammals. The exceptions are the cow and blackbuck (Family Bovidae) both of which contain a complex multigene family that must have been amplified recently during the evolution of the artiodactyls (P. Boseley & A. Jeffreys, unpublished). This new interferon family should be ideal for examining the rate of sequence interchange and homogenization in gene families, to see whether there are preferred donors and recipients of sequence blocks, and also whether the process involves reciprocal recombination events or asymmetric gene conversions.

To date, there is no definite evidence for true gene conversion occurring in any animal multigene family. If conversion does take place, then Dover (1982) has argued that biases will arise, with certain sequence types preferentially correcting other variants of the sequence. This 'molecular drive', in addition to processes of drift, could lead to the rapid homogenization of gene families and repeated DNA sequences both within a genome and within a population. Obviously, gene families such as globin are not, and must not be, driven to homogenization and one can only wonder at the genetic load imposed on a species by undesirable sequence interchanges within functional gene families.

From the analysis of primate β-globin gene clusters (Fig. 9.3) it can be seen that only one functional gene duplication, the γ-globin duplication, has arisen in this cluster in recent primate evolution. What molecular mechanisms were responsible for this duplication? Shen *et al.* (1981) have sequenced over 11000 base pairs of DNA around the human foetal globin genes, and have shown that this region consists of a large (5000 base pair) duplication encompassing each γ-globin gene and flanked by a short direct repeated DNA sequence (r) to give an organization $r-^G\gamma-r-^A\gamma-r$. The simplest explanation is that the γ-globin locus duplicated by unequal crossing over between r elements flanking an ancestral γ-globin gene (Fig.

Fig. 9.5. Consequences of recombination between dispersed repeated DNA elements near structural genes. Short repeated DNA sequences (solid boxes) are shown near a gene (open box); in C, the repeats are orientated in opposite directions. (Adapted from Jeffreys & Harris, 1982, with permission of MacMillan Journals.)

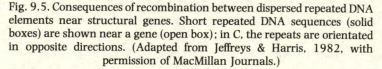
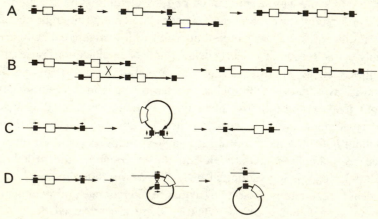

9.5). In this way, the length of the duplication is determined by the location, fortuitous or otherwise, of dispersed repeated DNA elements known to be scattered throughout these gene clusters. Depending on their position, more than one gene could be included within the duplication, and this could be the mechanism responsible for the ψβ–β block duplication seen in the goat β-globin gene cluster (Cleary *et al.*, 1981; Fig. 9.2). Indeed, these dispersed DNA elements would tend to have a general destabilizing influence on gene clusters; as well as generating tandem duplicates (Fig. 9.5A) which in turn would provide excellent targets for rapid gene amplification (Fig. 9.5B) and the related process of genetic homogenization, these elements could also promote localized inversions (Fig. 9.5C) and complete excision of small DNA circles containing genes (Fig. 9.5D). These circles could in turn insert themselves elsewhere in the genome via recombination with another dispersed repeated DNA element, leading to the progressive dispersal of gene clusters.

Dispersal of globin gene clusters

Many gene families are to a greater or lesser extent dispersed around the genome. While the functional significance of gene dispersal is far from obvious, the processes of dispersal are becoming more clear.

All mammals and birds so far examined contain two unlinked (dispersed) clusters of α- and β-globin genes. In contrast, the amphibian *Xenopus tropicalis* has a single cluster containing a closely-linked adult α- and β-globin gene (Jeffreys *et al.*, 1980). This strongly suggests that the α- and β-globin genes initially arose by a tandem gene duplication, perhaps similar to the foetal globin gene duplication in man. If so, then in the amphibians the α–β duplicates have since remained closely associated, whereas this linkage must have been disrupted somewhere along the lineage leading to birds and mammals. Several mechanisms could have led to this dispersal. Perhaps one or other gene was excised and relocated by the mechanism shown in Fig. 9.5D, a form of non-replicative transposition. Alternatively, a full chromosome translocation with a break point between the α- and β-globin genes could have been responsible. Another set of possibilities involves chromosome duplication or polyploidization, to generate two complete unlinked α–β clusters which could evolve towards contemporary mammalian α- and β-clusters by silencing of one or other gene to give a ψα–β and α–ψβ cluster, accompanied by duplicative expansion of the non-silenced gene. An analogous event which has generated globin diversity is found in *Xenopus laevis*, which has undergone tetraploidization in the recent past and now possesses two unlinked α–β clusters, one of which specifies the major adult α- and β-globins and the other some minor adult haemoglobins (Jeffreys *et al.*, 1980; Patient *et al.*, 1980).

Some more bizarre processes of gene dispersal have recently been discovered, particularly in an analysis of the α-globin gene family of the mouse. Mouse α-globins are specified by a single gene cluster containing one embryonic and two adult genes (Leder *et al.*, 1981). In addition, there are several diverged α-like genes in mouse DNA, none of which is linked to the α-globin gene cluster. One of these genes, ψα4, looks like a 'conventional' pseudogene and might represent a pseudogene generated within the parent cluster and moved out for example by the mechanism in Fig. 9.5D (Leder *et al.*, 1981). A second dispersed pseudogene, ψα3, (Leder *et al.*, 1981) is not so conventional; it has had both intervening sequences precisely removed (Nishioka, Leder & Leder, 1980; Vanin, Goldberg, Tucker & Smithies, 1980) and is flanked by the relics of retroviral sequences (Leuders, Leder, Leder & Kuff, 1982). The simplest explanation for this is that an α-globin gene was captured by a proretrovirus and transposed via spliced retroviral RNA. However, the recent discovery of processed intronless immunoglobulin and tubulin pseudogenes, both of which possess an oligo (dA) tract at the 3' end corresponding to the poly (A) tail of mature mRNA, strongly suggests that reverse transcription and genomic integration of DNA copies of RNA transcripts including mRNA itself can occur (Hollis *et al.*, 1982; Wilde *et al.*, 1982). These integration events must occur in the germ line, but we see no compelling reason to presuppose that the mRNA itself must necessarily have been transcribed in germ cells. Instead, it seems at least theoretically feasible that some mRNA copies might be acquired from the soma, opening the possibility that the doctrine of Weissmann need not be absolute and that a feedback of sequence data from soma to germ line might occasionally occur (Steel, 1979). However, it is far from clear whether such a process would have any evolutionary significance, or whether instead these mRNA copies are simply yet more examples of selfish DNA.

Ancient events in globin gene evolution

All of the preceding discussion has been devoted to relatively recent events in the evolution of globin gene families. There are compelling reasons for trying to probe further back into the past, in particular to see whether any shifts from the three exon–two intron structure of vertebrate globin genes have occurred, and whether these would help resolve between the basic hypotheses of exon shuffling and insertion elements as generators of split genes.

Myoglobin is a relatively distant member of the vertebrate haemoglobin family, and has diverged from haemoglobin according to various estimates between five hundred million years ago (Czelusniak *et al.*, 1982) and eight hundred million years ago (Hunt, Hurst-Calderone & Dayhoff, 1978), before the α–β globin gene duplication. We have recently used mRNA from seal muscle rich in myoglobin to isolate the seal myoglobin gene and are currently investigating the internal structure of this gene. In addition to vertebrate globins, a wide variety of haemoglobins appear sporadically in invertebrates, and there is one report of a small haemoglobin-like protein in the protozoan *Paramecium* (see Terwilliger, 1980). At least some of these haemoglobins show homology to animal globins (Hunt *et al.*, 1978) and form part of the globin family. In addition, some very large invertebrate globins exist which may well have evolved by internal reduplication of a conventional globin gene. So far, the entire invertebrate kingdom of haemoglobins is completely virgin ground for DNA analysis. Finally, there are sporadic examples of monomeric plant haemoglobins, the leghaemoglobins, produced in the root nodules of nitrogen-fixing plants. DNA analysis has shown that leghaemoglobin genes bear an uncanny similarity to animal genes. They exist in soybean DNA as a multigene family of genes and pseudogenes, show gene clustering and, most extraordinarily, share a common split internal organization that is identical to animal haemoglobin genes except for an additional intron between two haem binding domains that are fused together in the central exon of the animal genes (Gō, 1981; Jensen *et al.*, 1981; Hyldig-Nielsen *et al.*, 1982). There are various possible explanations for this surprising similarity of plant and animal genes: they are the products of convergent evolution (with the convergence including protein length, polypeptide folding, haem binding plus the detailed internal organization of the genes themselves, even though haemoglobin and leghaemoglobin have very different physiological roles); they are directly related via the common ancestors of plants and animals (possible, though this would predict that haemoglobins are common to all plants and animals; this appears not to be the case); and finally, the leghaemoglobin gene or gene family was acquired from an animal recently in evolution by horizontal gene transmission (Hyldig-Nielson *et al.*, 1982; Jeffreys, 1982).

The last possibility of horizontal transfer is bizarre but mechanistically feasible, and could be tested by a detailed phylogenetic analysis of leghaemoglobin genes and by comparing these genes with a more representative range of animal globin genes including myoglobins and invertebrate globins. A precedent has already been set for horizontal transfer by the recent discovery of a sea-urchin histone gene family which appears to have undergone a recent horizontal transmission between two distantly related sea urchin species (Busslinger, Rusconi & Birnstiel, 1982). Horizontal transfer of genes is unlikely to occur on a wide scale, otherwise molecular phylogenies would be nonsensical. Nevertheless, any horizontal transfer is potentially immensely important, since in principle it opens up the possibility of adaptive evolution by the simple expedient of acquiring the necessary genes from another already-adapted species, followed by fixation of these new genes by natural selection. Hopefully, further DNA analysis will reveal the extent of horizontal transmission that might occur in evolution, and if it is proven to be a real phenomenon, then we shall await with interest to see what impact it has on neo-Darwinian theory.

Further Reading

Jeffreys, A. J. (1981). Recent studies of gene evolution using recombinant DNA. In *Genetic Engineering*, **2**, ed. R. Williamson, pp. 1–48. London: Academic Press.

References

Baltimore, D. (1981). Gene conversion: some implications for immunoglobulin genes. *Cell*, **24**, 592–4.

Barrie, P. A., Jeffreys, A. J. & Scott, A. F. (1981). Evolution of the β-globin gene cluster in man and the primates. *Journal of Molecular Biology*, **149**, 319–36.

Blake, C. C. F. (1979). Exons encode protein functional units. *Nature*, **277**, 598.

Busslinger, M., Rusconi, S. & Birnstiel, M. L. (1982). An unusual evolutionary behaviour of a sea urchin histone gene cluster. *EMBO Journal*, **1**, 27–33.

Cleary, M. L., Schon, E. A. & Lingrel, J. B. (1981). The related pseudogenes are the result of a gene duplication in the goat β-globin locus. *Cell*, **26**, 181–90.

Craik, C. S., Buchman, S. R. & Beychok, S. (1980). Characterization of globin domains: heme binding to the central exon product. *Proceedings of the National Academy of Sciences, USA*, **77**, 1384–6.

Crick, F. (1979). Split genes and RNA splicing. *Science*, **204**, 264–71.

Czelusniak, J., Goodman, M., Hewett-Emmett, D., Weiss, M. L., Venta, P. J. & Tashian, R. E. (1982). Phylogenetic origins and adaptive evolution of avian and mammalian haemoglobin genes which are expressed differentially during ontogeny. *Nature*, **298**, 297–300.

Darnell, J. E. (1978). Implications of RNA splicing in evolution of eukaryotic cells. *Science*, **202**, 1257–60.

Davidson, E. H., Thomas, T. L., Scheller, R. H. & Britten, R. J. (1982). The sea urchin actin genes, and a speculation on the evolutionary significance of small gene families. In *Genome Evolution*, ed. G. A. Dover & R. B. Flavell, pp. 177–91. London: Academic Press.

Dayhoff, M. O., Hunt, L. T., McLaughlin, P. J. & Jones, D. D. (1972). Gene duplications in evolution: the globins. In *Atlas of Protein Sequence and Structure 1972*, ed. M. O. Dayhoff, pp. 17–30. Washington: National Biomedical Research Foundation.

Doolittle, W. F. & Sapienza, C. (1980). Selfish genes, the phenotype paradigm and genome evolution. *Nature*, **284**, 601–3.

Dover, G. (1982). Molecular drive: a cohesive mode of species evolution. *Nature*, **299**, 111–17.

Dover, G. A. & Flavell, R. B. (eds.) (1982). *Genome Evolution*. London: Academic Press.

Efstratiadis, A., Posakony, J. W., Maniatis, T., Lawn, R. M., O'Connell, C., Spritz, R. A., DeRiel, J. K., Forget, B. G., Weissman, S. M., Slightom, J. L., Blechl, A. E., Smithies, O., Baralle, F. E., Shoulders, C. C. & Proudfoot, N. J. (1980). The structure and evolution of the human β-globin gene family. *Cell*, **21**, 653–68.

Flavell, R. A., Kooter, J. M., De Boer, E., Little, P. F. R. & Williamson, R. (1978). Analysis of the β-δ-globin gene loci in normal and Hb Lepore DNA: direct determination of gene linkage and intergene distance. *Cell*, **15**, 25–41.

Fritsch, E. F., Lawn, R. M. & Maniatis, T. (1979). Characterisation of deletions which affect the expression of fetal globin genes in man. *Nature*, **279**, 598–603.

Fritsch, E. F., Lawn, R. M. & Maniatis, T. (1980). Molecular cloning and characterization of the human β-like globin gene cluster. *Cell*, **19**, 959–72.

Gilbert, W. (1978). Why genes in pieces? *Nature*, **271**, 501.

Gō, M. (1981). Correlation of DNA exonic regions with protein structural units in haemoglobin. *Nature*, **291**, 90–2.

Goeddel, D. V., Leung, D. W., Dull, T. J., Gross, M., Lawn, R. M., McCandliss, R., Seeburg, P. H. Ullrich, A., Yelverton, E. & Gray, P. W. (1981). The structure of eight distinct cloned human leukocyte interferon cDNAs. *Nature*, **290**, 20–6.

Goodman, M., Moore, G. W. & Matsuda, G. (1975). Darwinian evolution in the genealogy of haemoglobin. *Nature*, **253**, 603–8.

Hardison, R. C., Butler, E. T., Lacy, E., Maniatis, T., Rosenthal, N. & Efstratiadis, A. (1979). The structure and transcription of four linked rabbit β-globin genes. *Cell*, **18**, 1285–97.

Haynes, J. R., Rosteck, P., Schon, E. A., Gallagher, P. M., Burks, D. J., Smith, K. & Lingrel, J. B. (1980). The isolation of the β^A, β^C and γ globin genes and a presumptive embryonic globin gene from a goat DNA recombinant library. *Journal of Biological Chemistry*, **255**, 6355–67.

Hollis, G. F., Heiter, P. A., McBride, O. W., Swan, D. & Leder, P. (1982). Processed genes: a dispersed human immunoglobulin gene bearing evidence of RNA-type processing. *Nature*, **296**, 321–5.

Hunt, T. L., Hurst-Calderone, S. & Dayhoff, M. O. (1978). Globins. In *Atlas of protein sequence and structure*, ed. M. O. Dayhoff, pp. 229–51. Washington: National Biomedical Research Foundation.

Hyldig-Nielsen, J. J., Jensen, E. O., Paludan, K., Wiborg, O., Garrett, R., Jørgensen, P. & Marcker, K. A. (1982). The primary structure of two leghaemoglobin genes from soybean. *Nucleic Acids Research*, **10**, 689–701.

Jagadeeswaran, P., Biro, P. A., Tuan, D., Pan, J., Forget, B. G. & Weissman, S. M. (1981). The interspersed repetitive DNA sequences of the human genome: are they transposons? In *Proceedings of the 6th International Congress of Human Genetics*.

Jahn, C. L., Hutchison, C. A., Phillips, S. J., Weaver, S., Haigwood, N. L., Voliva, C. F. & Edgell, M. H. (1980). DNA sequence organization of the β-globin complex in the BALB/c mouse. *Cell*, **21**, 159–68.

Jeffreys, A. J. (1981). Recent studies of gene evolution using recombinant DNA. In *Genetic Engineering* Vol. 2, ed. R. Williamson, pp. 1–48. London, New York: Academic Press.

Jeffreys, A. J. (1982). Evolution of globin genes. In *Genome Evolution*, ed. G. A. Dover & R. B. Flavell, pp. 157–76. London: Academic Press.

Jeffreys, A. J., Barrie, P. A., Harris, S., Fawcett, D. H., Nugent, Z. J. & Boyd, A. C. (1982). Isolation and sequence analysis of a hybrid δ-globin pseudogene from the brown lemur. *Journal of Molecular Biology*, **156**, 487–503.

Jeffreys, A. J., & Harris, S. (1982). Processes of gene duplication. *Nature*, **296**, 9–10.

Jeffreys, A. J., Wilson, V., Wood, D., Simons, J. P., Kay, R. M. & Williams, J. G. (1980). Linkage of adult α- and β-globin genes in *X. laevis* and gene duplication by tetraploidization. *Cell*, **21**, 555–64.

Jensen, E. O., Paludan, K., Hyldig-Nielsen, J. J., Jørgensen, P. & Marcker, K. A. (1981). The structure of a chromosomal leghaemoglobin gene from soybean. *Nature*, **291**, 677–9.

Kimura, M. (1979). The neutral theory of molecular evolution. *Scientific American*, **241**, 98–126.

Kimura, M. (1981a). Was globin evolution very rapid in its early stages? A dubious case against the rate-constancy hypothesis. *Journal of Molecular Evolution*, **17**, 110–13.

Kimura, M. (1981b). Estimation of evolutionary distances between homologous nucleotide sequences. *Proceedings of the National Academy of Sciences, USA*, **78**, 454–8.

Kimura, M. & Ohta, T. (1974). On some principles governing molecular evolution. *Proceedings of the National Academy of Sciences, USA*, **71**, 2848–52.

Lacy, E., Hardison, R. C., Quon, D. & Maniatis, T. (1979). The linkage arrangement of four rabbit β-like globin genes. *Cell*, **18**, 1273–83.

Lacy, E. & Maniatis, T. (1980). The nucleotide sequence of a rabbit β-globin pseudogene. *Cell*, **21**, 545–53.

Leder, A., Swan, D., Ruddle, F., D'Eustachio, P. & Leder, P. (1981). Dispersion of α-like globin genes of the mouse to three different chromosomes. *Nature*, **293**, 196–200.

Leuders, K., Leder, A., Leder, P. & Kuff, E. (1982). Association between a transposed α-globin pseudogene and retrovirus-like elements in the BALB/c mouse genome. *Nature*, **295**, 426–8.

Li, W.-H., Gojobori, T. & Nei, M. (1981). Pseudogenes as a paradigm of neutral evolution. *Nature*, **292**, 237–9.

Little, P. F. R. (1982). Globin pseudogenes. *Cell*, **28**, 683–4.

Martin, S. L., Zimmer, E. A., Kan, Y. W. & Wilson, A. C. (1980). Silent δ-globin gene in Old World monkeys. *Proceedings of the National Academy of Sciences, USA* **77**, 3563–6.

Miyata, T. & Yasunaga, T. (1981). Rapidly evolving mouse α-globin-related pseudo gene and its evolutionary history. *Proceedings of the National Academy of Sciences, USA* **78**, 450–3.

Miyata, T., Yasunaga, T. & Nishida, T. (1980). Nucleotide sequence divergence and functional constraint in mRNA evolution. *Proceedings of the National Academy of Sciences, USA* **77**, 7328–32.

Nishioka, Y., Leder, A. & Leder, P. (1980). Unusual α-globin-like gene that has cleanly lost both globin intervening sequences. *Proceedings of the National Academy of Sciences, USA,* **77,** 2806–9.

Ohno, S. (1980). Gene duplication, junk DNA, intervening sequences and the universal signal for their removal. *Rev. Brazil Genet III,* **2,** 99–114.

Orgel, L. E. & Crick, F. H. C. (1980). Selfish DNA: the ultimate parasite. *Nature,* **284,** 604–7.

Patient, R. K., Elkington, J. A., Kay, R. M. & Williams, J. G. (1980). Internal organization of the major adult α- and β-globin genes of *X. laevis. Cell,* **21,** 565–73.

Proudfoot, N. J. (1980). Pseudogenes. *Nature,* **286,** 840–1.

Romero-Herrera, A. E., Lehmann, H., Joysey, K. A. & Friday, A. E. (1973). Molecular evolution of myoglobin and the fossil record: a phylogenetic synthesis. *Nature,* **246,** 389–95.

Sarich, V. M. & Cronin, J. E. (1977). Generation length and rates of hominoid molecular evolution. *Nature,* **269,** 354.

Shen, S., Slightom, J. L. & Smithies, O. (1981). A history of the human fetal globin gene duplication. *Cell,* **26,** 191–203.

Slightom, J. L., Blechl, A. E. & Smithies, O. (1980). Human fetal $^G\gamma$- and $^A\gamma$-globin genes: complete nucleotide sequences suggest that DNA can be exchanged between these duplicated genes. *Cell,* **21,** 627–38.

Steel, E. J. (1979). *Somatic selection and adaptive evolution.* Toronto and London: Williams-Wallace International and Croom Helm.

Terwilliger, R. C. (1980). Structure of invertebrate haemoglobins. *American Zoologist,* **20,** 53–67.

Van Den Berg, J., Van Oöyen, A., Mantei, N., Schamböck, A., Grosveld, G., Flavell, R. A. & Weissmann, C. (1978). Comparison of cloned rabbit and mouse β-globin genes showing strong evolutionary divergence of two homologous pairs of introns. *Nature,* **276,** 37–44.

Vanin, E. F., Goldberg, G. I., Tucker, P. W. & Smithies, O. (1980). A mouse α-globin-related pseudogene lacking intervening sequences. *Nature,* **286,** 222–6.

Wang, A. H. J., Quigley, G. J., Kolpak, F. J., Crawford, J. L., van Boom, J. H., van der Marel, G. & Rich, A. (1979). Molecular structure of a left-handed double helical DNA fragment at atomic resolution. *Nature,* **282,** 680–6.

Wilde, C. D., Crowther, C. E., Cripe, T. P., Gwo-Shu Lee, M. & Cowan, N. J. (1982). Evidence that a human β-tubulin pseudogene is derived from its corresponding mRNA. *Nature,* **297,** 83–4.

Wilson, A. C., Carlson, S. S. & White, T. J. (1977). Biochemical evolution. *Annual Review of Biochemistry,* **46,** 573–639.

10

Gene clusters and genome evolution

W. F. BODMER

There are three major aspects to the understanding of evolution. The first, which is the classical approach to evolutionary studies, involves taxonomy, palaeontology and the natural history of organisms, and is represented by Darwin's own work. The second is the study of the genetics, physiology and development of organisms. This forms the basis of understanding function at the molecular level, and so understanding the differences on which natural selection acts, and the nature of that action. This second strand of evolutionary study is represented by the work of the geneticists and molecular biologists, Mendel, Morgan and the *Drosophila* school, and, of course, Watson and Crick and all that follows in molecular biology. The third basic aspect of evolutionary study is population genetics, as developed originally by Fisher, Haldane and Sewall Wright, which is the quantitative theory of evolution, and which forms the essential background against which all evolutionary ideas must be tested. All these three aspects of the study of evolution need to be brought together for its complete under-standing. Unfortunately, all too often, discussions on evolution lack this comprehensiveness through focussing mainly on one or two of these major aspects. This centenary volume brings all three together.

Darwin formulated his ideas without a knowledge of the mechanisms of inheritance, and so without a proper quantitative theory of evolution. It is therefore remarkable that his genius and intuition forced him to believe in a theory which in many respects was incompatible with the prevailing views on the mechanisms of inheritance. Fisher, Haldane and Sewall Wright, the originators of the quantitative theory of evolution, were really the first to be in a position to combine all three aspects of the understanding of evolution, at least in terms of the knowledge available to them at the time. Palaeontological studies provide the essential time scale for an assessment of the validity of quantitative predictions given by population genetics, while genetic mechanisms tell us what changes determine the evolutionary process, and whose rate of accumulation must be measured. As Sidney Brenner has pointed out, many of the current apparent problems in understanding evolutionary processes will disappear as we eventually learn to understand the complexity of living organisms at the physiological and molecular levels.

The concept of the gene has proceeded from the phenomenological level of Mendel's 'elemente', to the view that came from the *Drosophila* school of geneticists that the 'elemente' must be physical entities strung together in a defined sequence on the chromosome, to the simple and elegant notion that came from Garrod's inborn errors and Beadle and Tatum's one gene–one enzyme concept that there must be a one-to-one relationship between DNA and amino-acid sequences. Then in the late 1970s, as discussed in a number of previous contributions, came the discovery that genes occur in pieces that may be put together and shuffled in various ways (see Jacob, this volume). The hierarchy of genetic organization now proceeds from the nucleotide, through the codon to the exon, which is the minimal functional unit from which genes and their clusters are built. The gene, in its simplest form, is a series of exons separated by introns, which together determine a polypeptide, and the gene cluster is a collection of adjacent, evolutionarily, and functionally related genes. Beyond this level of organization lies the chromosome arm, the chromosome and eventually the karyotype, or genome as a whole.

In this brief review, I shall emphasize the notion that the gene cluster is the basic genetic functional unit and discuss some aspects of its evolution, having first considered the broader question of the evolution of the karyotype as a whole.

Karyotype evolution

The haploid DNA content of mammals varies comparatively little from one species to another while, on the other hand, the number of chromosomes varies over at least a five-fold range (see e.g. Ohno, 1970). This fact alone suggests that there is not any strong evolutionary pressure for the conservation of a particular karyotype or chromosome organization. Nevertheless, karyotype changes, when they do occur in evolution, must be subject to the same forces of natural selection and random genetic drift as are other genetic variations. Ohno (1967) emphasized the extensive homology between different mammalian X chromosomes and suggested that the X had been relatively conserved over long evolutionary periods. Subsequently, Ohno (1973) suggested that this conservation should be extended to the autosomes, and that homologous linkages would be preserved between species, essentially by chance, because of the very slow rate at which chromosome rearrangements would be fixed during evolution. Recent data, especially from somatic cell genetics, have shown a surprising degree of conservation of mammalian linkage groups (see, e.g. *Human Gene Mapping 5*, 1979) though not perhaps to the extent that Ohno had predicted. A previous estimate (Bodmer, 1975) of the rate of karyotype evolution in mammals, based on the then available data on chromosome banding patterns, especially in the higher primates, gave a value of approximately six break points substituted per chromosome arm during mammalian evolution. This estimate, however, allowed for a considerable margin of error in the detection of break points using chromosome banding techniques. As a result it clearly turns out, based on presently available data, to be an over-estimate, possibly by as much as a factor of three. The number of break points substituted during evolution is, thus, far too small to explain the majority of changes in the control of important genetic functions that might underlie basic differences between mammalian species, as has been suggested, for example, by Wilson, Sarich & Maxson (1974).

Population genetics theory applied to the increase of chromosome mutations clearly indicates that a translocation or an inversion will only increase in frequency if there is some positive selection to counterbalance the intrinsic initial disadvantage of such chromosome mutations, resulting from their disruption of meiosis in heterozygotes (Bengtsson & Bodmer, 1976). The simplest form of selection in favour of such chromosome mutations occurs when two particular alleles at different genetic loci which interact favourably, are brought together by an inversion or translocation. The combination of the pair of alleles may then have a much increased selective advantage because of the reduced recombination frequency in inversion or translocation heterozygotes, as first proposed by Fisher (1930) in his brief but classical theoretical analysis of the interaction between selection and linkage. On this view, the substitution of chromosome break points in evolution is an adventitious phenomenon that depends on the particular combination of genes occurring in the particular inversion or translocation being substituted. Karyotype evolution does not, therefore, seem to reflect any grand overall design in the organization of genes on chromosomes. Indeed the major overall feature of organisms influencing rates of karyotype evolution, at least in mammals, is likely to be their pattern of reproductive behaviour. Thus, it is at a much finer level of genetic organization, namely that of the gene cluster, that one must seek an understanding of the major features of genome evolution.

Gene clusters: the basic genetic functional units

The significance of gene duplication for the evolution of new genetic functions has been discussed since Bridge's original description of the phenomenon. The simple theory is that, once a gene has been duplicated, copies can diverge from the original to produce new functions without jeopardizing the function of the original gene. Classical examples of gene clusters before the advent of molecular biology include the plant incompatibility systems, such as that in the common primrose, and blood group systems such as the Rhesus blood groups, (see e.g. Bodmer & Parsons, 1962; Bodmer, 1979). Fisher was probably the first to appreciate the significance of such clusters and, indeed, drew on the example of the primula incompatibility system for his classical analysis of the Rhesus blood groups as a system of closely linked genes (personal communication and unpublished correspondence). These early notions led to the idea of a supergene, which is essentially a set of closely linked genes with related functions, generally held together by mutually interdependent selective advantages. This mechanism of selection, by which, under appropriate circumstances, interacting genes are favoured if they are closer together on a chromosome, was first suggested by Fisher in 1930 and underlies nearly all subsequent discussions of the way that natural selection might influence the relative position of genes on chromosomes. Fisher's model was formulated without postulating a mechanism by which such genes interact but, of course, subsequent developments at the molecular level have provided many possible explanations.

The first molecular model for a gene cluster came from studies of the haemoglobins and their constituent polypeptides, as discussed by Jeffreys (this volume). With increasing knowledge of protein structure, as discussed by Phillips (this volume), it became apparent that proteins could often be divided into more-or-less independently acting domains. This clearly foreshadowed the idea that genes might occur in pieces, as later demonstrated so dramatically at the DNA level.

The more genes in higher organisms are now studied at the molecular level, the more evidence there is for the existence of gene clusters. At the simplest level, enzymes such as those of the glycolytic pathway may not be clustered. However, many other genetic functions, ranging in complexity from the comparatively simple haemoglobin system to the much more complex immunoglobulin and major histocompatibility systems and including the genes for major structural proteins such as actin and collagen, are controlled by genes in clusters.

If gene clusters are indeed the basic functional units, then overall genetic complexity should be considered in terms of the number of gene clusters, rather than in terms of the numbers of any other genetic unit. The mammalian genome contains approximately 3×10^9 nucleotide pairs. The average human chromosome, for example, may therefore contain about 150×10^6 nucleotide pairs, while the smallest chromosome fragment visible under the light microscope may contain some 10 to 15×10^6 nucleotide pairs. In terms of DNA clones, this means that the human genome would be covered by some 60000 clones with the present maximum insert size of about 50000 nucleotide pairs, or 3×10^6 clones of length 1000. This gives some indication of the scale of the problem of clonal analysis of the human genome using presently available techniques. If one took the classical view of a gene, based on a simple one-to-one relationship between DNA and protein sequences, and assuming all the DNA was functional, then the estimate of the total number of genes would be about three million. Data at the molecular level now clearly indicate, however, that a radically different view of genetic complexity is appropriate and the question is how to obtain a reasonable estimate of the total number of gene clusters.

Present indications are that, perhaps, no more than 50% of the total of 3×10^9 nucleotide pairs is informational, namely is not 'selfish' (Doolittle & Sapienza, 1980; Orgel & Crick, 1980). Assuming the number of genes within a cluster is on average 15, which is a reasonable compromise between the size of the haemoglobin and HLA, or major histocompatibility gene, clusters, assuming an average size for protein products corresponding to about 1000 nucleotide pairs and assuming that, on average, the coding ratio (the proportion of the nucleotide sequence within a cluster that codes for protein) is about 1/30 as in the case of haemoglobin, then one can estimate the total number of gene clusters to be $(1.5 \times 10^9/30) \times 15 \times 1000$, or approximately 3300 (Bodmer, 1981*a*).

Clearly these figures are subject to a considerable margin of error. But it is, perhaps, unlikely that the upper limit for the number of gene clusters is more than, say, between 10000 and 15000. Measured in terms of the number of possible proteins, this represents say from 50000 to 150000 proteins, of which only a relatively small proportion are likely to be expressed in any given cell. Considering, further, that gene clusters occur in related families such as, for example, the two sets of haemoglobin chains or the different chains of the immunoglobulin molecule, the overall level of genetic complexity, in terms of numbers of basic functions as represented by gene clusters, is certainly much, much less than was at one time supposed. Functional complexity in higher organisms is not, therefore, on the whole likely to be based on an increase in the number of different genetic functions, but more on a clever use of combinations of functions, and their control.

One obvious important implication of this limit to the number of basic genetic functions is that we may be much nearer to knowing something about a significant proportion of these functions than has previously been supposed. Another implication is that there may be a greater possibility of finding the genetic basis for any particular complex phenotypic difference than might at first seem apparent, simply by chance associations of genetic functions with the phenotype together with a little intuition as to the types of genetic functions that may be relevant. There should also be a greater overlap between phenotypes at the genetic level, since a given basic genetic function may be associated with a wider variety of effects than would be supposed if the number of basic genetic functions were much more numerous and each, therefore, more specific. We can, for example, no longer think of actin as just a muscle protein.

Another important practical application of this view of the number of basic genetic functions is that appropriate genetic and phenotypic sub-divisions may lead to an easier identification of unique genetic functions. For example, the number of genetic functions involving a particular class of products coded for on any given chromosome could in some cases be quite small. The small human chromosome 21 might, for example, contain no more than about 100 gene clusters and, among these, the number that code for surface glycoprotein receptors could well be at most two or three. Thus, if one has a technique for cloning such genes from chromosome 21, and these are now available, the chance that an arbitrary clone may correspond to a particular known function on this chromosome could be as high as 30 to 50%.

Evolution of gene clusters

The evolution of a gene cluster may be quite complex and must now be assumed to involve not just the classical processes of duplication, followed by divergence mediated by simple mutations, and adaptive selection or random genetic drift. Mutational events themselves may involve changes in exons as a whole. For example, exons may be duplicated individually within a gene sub-cluster or between clusters located in different parts of the genome. Exons may be deleted, inverted or split by the insertion, by transposition, of an alien intron sequence from elsewhere. DNA sequences can be read in six different ways, so that even the same exon under different circumstances could code for different proteins as is the case, for example, for some of the DNA tumour viruses. Differential splicing of the primary RNA transcripts may then lead to different functional messages, coding for different protein products. An interesting example of this is the synthesis of alternative forms of immunoglobulin which are either secreted or membrane associated depending on which exon sequence is incorporated into the functional message. (For more detailed discussion of some of these ideas, and for further references see Bodmer, 1979, 1981*a, b.*)

In considering the complex evolution of gene clusters and the novel types of events at the DNA level that might be involved, such as transposition or gene conversion, it is important to distinguish the mechanisms for production of a variant from the processes that lead to their propagation in the population. The rich variety of mechanisms which can now be seen to underlie the production of genetic variants, or mutations in their broadest sense, can provide an explanation for even the most complex change. These mechanisms, however, still occur at a comparatively low frequency, and so cannot in general explain why a particular new, perhaps complex, variant increases in frequency in a population. Such an increase still needs to be explained in terms of the fundamental ideas of population genetics.

There are basically three types of causes for the propagation of a new variant in a population, namely (1) random genetic drift, if the variant is adaptively neutral; (2) the classical Darwinian process of natural selection; and (3) non-Mendelian processes which lead to preferential segregation of one or other allele in a heterozygote. A classical example of the last is the phenomenon of 'meiotic drive'. As pointed out by Gutz & Leslie (1976), based on an earlier suggestion by Chovnik, and as much discussed recently by, for example, Baltimore (1981), Bodmer (1981*a*) and others, gene conversion may be a particularly important mechanism for non-Mendelian propagation of new variants in a population. In particular, non-homologous gene conversion involving adjacent duplicate genes may be responsible for spreading a variant along a gene cluster (Baltimore, 1981). Asymmetric gene conversion between homologous allelic variants, on the other hand, may be responsible for increasing the frequency of complex new mutations (Bodmer, 1981*a*). The former possibility is suggested by the existence of regions with anomolous homologies in different segments of a gene cluster (see Jeffreys, this volume), or more poetically, of homogeneous islands in a background sea of genetic diversity. The latter possibility is suggested by rates of evolution that are higher than those expected based on the neutral gene theory, as expounded particularly by Kimura (1981), using for example the rate of evolution of synonymous silent mutations as a guide for the expected rate of evolution for an adaptively neutral mutation. Hollis *et al.* (1982) and others have postulated a cycle of events, by which a pseudogene lacking introns can be formed by making a DNA copy of an RNA message (which might have been propagated for some time in an RNA virus) and reincorporating this back into a novel site in the genome. So long as these insertions occur in a region of the genome where they have no obvious functional effect, they are likely to be adaptively neutral. The only mechanism then which could propagate such a variant into the population at a rate higher than that expected from the neutral gene theory is that of asymmetrical gene conversion. The classical notion of recurrent mutation pushing a gene into a population is obviously quite inappropriate for such a complex situation. Although gene conversion, as measured classically in fungi, is a comparatively rare event, it does seem possible that large differences between chromosomes in homologous positions, involving for example either insertions or deletions of substantial amounts of genetic material, could be recognized by DNA repair systems rather readily. This could, perhaps, lead to relatively high rates of asymmetric gene conversion, with a polarity which favours, for example, an inserted segment.

Fisher (1930) pointed out that very closely linked genes in sexual organisms behave as though they were a part of an asexual organism. One can perhaps, therefore, think of the functional gene cluster as a mini-organism in its own right, subject to quite complex patterns of evolution.

This discussion of the evolution of gene clusters has not encompassed the complex phenomena associated with highly repetitive DNA sequences and their dispersion throughout the genome by a process sometimes called 'concerted evolution' (see e.g. Dover *et al.*, 1982). These sequences, which are the canonical selfish DNA sequences, may be subject to quite different forms of intra-cellular or intra-organismal selection than are other functional gene sequences. They may also be propagated rapidly through-out the genome by a form of infective replication and dispersal analogous to that associated with the transforming RNA viruses, or transposons (see Shapiro, this volume). Their evolution is, therefore, clearly in most respects distinct from that of the evolution of functional gene clusters.

The HLA system: a model complex gene cluster

The HLA system, one of the most complex gene clusters so far known and studied, illustrates many of the features of the structure and evolution of gene clusters already discussed. The HLA system is the major histocompatibility system of man, and has its counterparts in other species such as the H-2 system in the mouse, and more generally comparable systems in at least the mammals, birds and reptiles. In mammals it seems to encompass about one thousandth of the genome, or about three million nucleotide pairs which code for at least four apparently quite different sets of gene products. The first are the HLA-ABC determinants found on most nucleated cells, which are 43 000 molecular weight glycoproteins associated with another protein, β_2-microglobulin, that is coded for on a different chromosome. Second are the HLA-DR and related sets of products, which are composed of two glycoprotein chains both coded for in the HLA region. These two sets of products play a fundamental role in interactions between cells of the immune system and, through that, in the control of the immune response. The remaining two sets of products in the HLA system form part of the cascade of complement components which also have important effector functions in the immune system. (For a brief review and further references see e.g. Bodmer, 1981*a*.)

Analysis of the primary structure of the HLA-ABC and similar products clearly indicates that these are divided into three protein domains outside the cell, followed by transmembrane and cytoplasmic regions. The first two domains, starting from outside the cell surface, vary substantially between individuals, while the third domain, closest to the membrane, is the least variable, even between different products. This domain is probably responsible for the association with β_2-microglobulin at the cell surface. Analysis of these gene sequences at the genomic DNA level shows, as is the case for many other eukaryotic proteins, that the exons correspond remarkably closely to the protein domains. It is, furthermore, clear that there may be between thirty and fifty related sets of genes in this class, or sub-cluster of genes, within the HLA genetic region.

A similar analysis of the HLA-DR related set of molecules also indicates a close relationship at the genomic level between exons and domains, and reveals another subcluster, but of somewhat lesser complexity, involving perhaps half a dozen or so each of the two types of chains which together form an HLA-DR like product. When sequences are compared between these different sets of products, it becomes clear that each has a domain close to the membrane of the cell which shares substantial homology with the others while other parts of the molecules may show little or no detectable homology. Thus, the average homology between the corresponding domains of the HLA-ABC and related products, the HLA-DR α and β chains and β_2-microglobulin is between 30 and 35%. These homologies clearly indicate a common evolutionary origin with a separation comparable to that of the haemoglobin α and β chains, namely of some five hundred million years and so long predating the evolutionary divergence of the mammals. These data, therefore, suggest that the HLA region and its homologues in different species have a very long evolutionary history during which many of the complex events that underlie the evolution of a gene cluster, as discussed above, may have taken place. (For a brief review of the molecular data, see e.g. Bodmer *et al.*, 1983.)

During this evolution, changes within a species involving any one gene product must be compatible with the needs for functionally effective interactions with other gene products, such as for example, between HLA-ABC and β_2-microglobulin or between the HLA-DR α and β chains. Thus, within the species genetic changes must be selected for that are compatible with each other. Between species, however, there is no such pressure for co-adaptation of genetic changes and therefore, for example, the corresponding products, such as HLA-ABC from man and β_2-microglobulin from the mouse, may associate much less efficiently. This is, of course, just one example of how reproductive isolation between species leads inevitably to a degree of physiological or molecular incompatibility. The initial stages of reproductive isolation may often depend on only a few genes. It is only after two populations have become reproductively isolated that their separate evolutionary pathways inevitably diverge, leading to physiological and other incompatibilities that give rise to species hybrid infertility and inviability. Here then is an example of how a well-known complex evolutionary phenomenon can readily be explained at the molecular level.

Conclusion

In this brief overview of genome evolution it has been emphasized that gene clusters are the basic functional genetic units and that it is their detailed evolution that dominates the pattern of genome evolution. Overall karyotypic evolution, while it may be selectively mediated, most probably does not play a major functional role in the genetic strategy of organisms.

The principle of common descent implies that one or other of the present patterns of genetic organization at the molecular level, as seen in prokaryotes or eukaryotes, must have come first. It seems most likely that it is the prokaryotic arrangement, without split genes, that was the original one and that it was the evolution of the eukaryotic nuclear organization, coupled with the evolution of RNA processing and RNA transfer from nucleus to cytoplasm, that gave rise to the possibility of split genes. This, in turn, dramatically enhanced the opportunities for the evolution of new genes by the process of exon shuffling, allowing the ready construction of new genetic combinations from old. As suggested by Leslie Orgel (personal communication) it may well be those genes whose functions, and so evolutionary origins, are common to the prokaryotes and eukaryotes, and so in this sense are more primitive, which do not show a good correspondence between exons and protein domains. This correspondence may have been a product of the development of the eukaryotic system of gene organization and so may involve predominantly those genes which evolved after the divergence of the eukaryotic from the prokaryotic line of descent. Hence the striking correspondence between exons and domains as seen, for example, in the decidedly eukaryotic immunoglobulin and HLA systems. Similarly, perhaps, it is these comparatively newer genes that are found in complex gene clusters, reflecting again the wider opportunities for evolution of gene functions provided by the eukaryotic system. It will be interesting to see to what extent consistent differences in the pattern of gene organisation emerge as more of the basic functions common to prokaryotes and eukaryotes, such as glycolysis, are analysed at the DNA level.

There is no doubt that the understanding of genetic functions, in all their complexity, must eventually underlie the understanding of the evolutionary process as a whole. If Charles Darwin were alive now, he would surely be delighted by present-day levels of understanding at the molecular level, and would revel in the interpretations of evolution made possible by genetic and molecular advances. He would undoubtedly not be a nineteenth-century Darwinian, but a modern student of evolution who combines Darwinian ideas with molecular genetics and with the quantitative evolutionary theory of population genetics.

References

Baltimore, D. (1981). Gene conversion: some implications for immunoglobulin genes. *Cell*, **24**, 592–4.

Bengtsson, B. O. & Bodmer, W. F. (1976). On the increase of chromosome mutations under random mating. *Theoretical Population Biology*, **9**, 260–81.

Bodmer, W. F. (1975). Linkage analysis by somatic cell hybridization and its conservation by evolution. In *Chromosome Variation in Human Evolution* (*Symp. Study Hum. Biol. 14*) pp. 53–61. New York: Halsted Press.

Bodmer, W. F. (1979). Gene clusters and the HLA system. In *Human Genetics: possibilities and realities*. CIBA Foundation, Series 66, pp. 205–29. Holland: Elsevier.

Bodmer, W. F. (1981a). The William Allan Memorial Award Address: Gene Clusters, Genome Organization, and Complex Phenotypes. When the sequence is known, what will it mean? *American Journal of Human Genetics*, **33**, 664–82.

Bodmer, W. F. (1981b). Introductory remarks to the fourth session: gene organization and evolution. *Philosophical Transactions of the Royal Society of London, B*, **293**, 173–6.

Bodmer, W. F. & Parsons, P. A. (1962). Linkage and recombination in evolution. *Advances in Genetics*, **11**, 1–100.

Bodmer, W. F., Bodmer, J. G., Crumpton, M. J., Lee, J. & Trowsdale, J. (1983). The structure of HLA. *Transplantation Proceedings*, (In Press).

Doolittle, W. F. & Sapienza, C. (1980). Selfish genes, the phenotype paradigm and genome evolution. *Nature*, **284**, 601–3.

Dover, G., Brown, S., Coen, E., Dallas, J., Strachan, T. & Trick, M. (1982). The dynamics of genome evolution and species differentiation. In *Genome Evolution*, ed. G. A. Dover & R. B. Flavell, pp. 343–72. London: Academic Press.

Fisher, R. A. (1930). *The genetical theory of natural selection*. Oxford: Clarendon Press.

Gutz, H. & Leslie, J. F. (1976). Gene conversion: a hitherto overlooked parameter in population genetics. *Genetics*, **83**, 861–6.

Hollis, G. F., Hieter, P. A., McBride, O. W., Swan, D. & Leder, P. (1982). Processed genes: a dispersed human immunoglobulin gene bearing evidence of RNA-type processing. *Nature*, **296**, 321–5.

Human Gene Mapping 5 (1979): Fifth International Workshop on Human Gene Mapping. Birth Defects: Original Article Series XV, 11, 1979, The National Foundation, New York: Basel: S. Karger AG.

Kimura, M. (1981). Estimation of evolutionary distances between homologous nucleotide sequences. *Proceedings of the National Academy of Sciences*, **78**, 454–8.

Ohno, S. (1967). Sex chromosomes and sex linked genes. In *Monographs on endocrinology*, ed. A. Labhart, T. Mann & L. T. Samuels. Heidelberg: Springer-Verlag.

Ohno, S. (1970). *Evolution by Gene Duplication*. Heidelberg: Springer-Verlag.

Ohno, S. (1973). Ancient linkage groups and frozen accidents, *Nature*, **244**, 259–62.

Orgel, L. E. & Crick, F. H. C. (1980). Selfish DNA: the ultimate parasite. *Nature*, **284**, 604–7.

Wilson, A. C., Sarich, V. M. & Maxson, L. R. (1974). The Importance of Gene Rearrangement in Evolution: evidence from studies on rates of chromosomal protein, and anatomical evolution. *Proceedings of the National Academy of Sciences, USA*, **71**, 3028–30.

11

The primary lines of descent and the universal ancestor

CARL R. WOESE

In Darwin's time, it was impossible to know or even to guess at the enormous world represented by bacteria and their evolution. Indeed it would be another century before this could begin to be appreciated and before the problem of the origin of the first cells and the nature of the Universal Ancestor would emerge from the complacent conceptual confines of Darwin's warm little pond.

Before the 1950s, macrofossils determined our perception of the fossil record. These cover a mere six hundred million years of evolutionary time (with fossil eukaryotic algae extending back about 10^9 years) (Schopf, 1968). A central question then was why it took life so long to arise on this planet. No burning question this, for the origin of life was viewed as a series of high unlikely happenings (a scientific transliteration of the religious view that it was indeed miraculous), and the Oparin Ocean scenario seemed to give us Darwin's warm little pond on a rather grand scale.

Over the past quarter century, the unfolding story of microfossils has forced a drastic revision in our estimate of when life first arose on Earth and a rethinking of the pernicious notion that the origin of life represents a series of highly unlikely events. Fossil structures of bacterial origin – both microscopic fossils of bacteria themselves and the fossils of structures produced by certain photosynthetic bacteria, called stromatolites – are, if not abundant, readily seen in sedimentary rocks dating back 3.5, perhaps 3.8×10^9 years (Knoll & Barghoorn, 1978; Lowe, 1980; Schidlowski, 1980; Pflug, 1982). These are the oldest known sedimentary rocks. Since at this early stage (photosynthetic) *sublines* of eubacteria seem already to have arisen (i.e., cyanobacteria and/or *Chloroflexus* species), it seems likely that life arose on this planet well before this time, within five hundred million years of the Earth's formation. Accounting for a long delay in life's origin no longer becomes a problem. Instead, we must now rationalize why life arose so rapidly. (The pushing of life's origin near to the time of the earth's formation brings into question the entire Oparin Ocean scenario – that simmering, unchanging ocean from which life ultimately would spring. Are perhaps the early, obviously drastic transitions in the physical state of the earth – due to core formation, to changes in rate of bombardment, to changes in the amount of radiogenic heating, etc. – somehow the driving force, the conditions that define the early events in the origin of life? These considerations, however, go beyond our present topic.)

Thus we come to see the enormous scope of bacterial evolution, the important role bacteria have played in the physical evolution of the planet, and the central position they will have in defining for us the evolution of the cell. There existed an approximately three thousand million year interval – properly called the Age of Microorganisms – during which the first cells emerged, the Earth's atmosphere became oxidizing (brought about by cyanobacteria), the bulk of the iron ores, the banded irons, were formed (again perhaps through bacterial intervention), and the eukaryotic cell was fashioned. Bacteria were the main if not the sole participants in this grand drama.

Another main reason for our failure to appreciate the enormity of bacterial evolution was that until recently, we had no idea of bacterial phylogeny. This does not reflect failure to recognize the problem. On the contrary; the problem had been evident for a century; the techniques available to attack it have, until recently, been sorely inadequate. (Little in the way of reliable phylogenetic classification can result when one is confined to spherical, rod, or spiral shapes, simple biochemistries, etc.)

Past prejudices and perceptions

To appreciate properly what an understanding of bacterial evolution will mean to biology, we must understand our present concept of early evolution, for most of the current notions regarding early evolutionary events, cherished dogmas, will turn out to be misplaced emphases at best, misleadingly wrong at worst.

In Darwin's time, the living world was divided into plants and animals. So strong was the belief in this basic, archaic division of living forms, that the biologist automatically divided the incredibly diverse world of microorganisms likewise – the motile ones were animals, the non-motile, sometimes green ones, plants. This latter division would not bear close scrutiny, however, and the biologist ultimately realized that major categories other than animal and plant must exist. The so-called Five Kingdom scheme (Whittaker, 1977) generally taught today – animals, plants, fungi, protists, and monera – represents little improvement in phylogenetic classification, however.

One of the most significant developments in phylogeny occurred with the realization by Chatton in the 1930s (Chatton, 1937), that a division of living systems, more fundamental than the customary one (based upon whole organism characteristics) was possible. On the basis of *cellular organization* all animals and plants (as well as protists and fungi) fall into one class 'eukaryotes', while all bacteria (inclusive of the cyanobacteria), fall into a class 'prokaryotes'. Eukaryotic cells are large, possess a nucleus defined by a nuclear membrane, and have certain other intracellular structures (mitochondria, chloroplasts, etc.). 'Prokaryotes' have a smaller type of cell with no well-defined nucleus (no nuclear membrane) and none of the other intracellular structures seen in eukaryotic cells (Stanier & van Niel, 1941; Margulis, 1970; Stanier, 1970).

While the prokaryote–eukaryote distinction was a most significant conceptual development, it embodied two prejudices that strongly shaped our subsequent understanding of phylogeny and cellular evolution. (1) That the prokaryote–eukaryote distinction is a phylogenetic dichotomy: every (self-replicating) living system on earth belongs to either the prokaryotic or the eukaryotic line of descent. The best proof of this assertion is the fact that in making the many molecular distinctions between prokaryotes and eukaryotes, the biologist seldom opted to study other than the 'typical prokaryote', *Escherichia coli*. (As we now know, this prejudice is false.) (2) That the prokaryote was the evolutionary precursor of the eukaryote – which is implied by the name '*prokaryote*'. Most speculations concerning the origin of the eukaryotic cell start with this or that prokaryote, which evolved into this or that eukaryotic structure – 'host cell', mitochondrion, chloroplast, cilium, etc. (Margulis, 1970; Stanier, 1970). (The best that can be said for this notion is that it misses the essence of the eukaryotic cell.) However, that there are *at least* two major phylogenetic units, one that can properly be called 'eukaryotes', as distinct from the 'prokaryotes', can no longer be doubted. Differences between the two at the molecular level are too many and too profound to conclude otherwise. (To avoid a certain confusion, I choose to call the latter unit the 'eubacteria', rather than 'prokaryotes'; see below.)

The biologist's accounts of the transition from an abiotic primitive Earth to a living one can be little more than *Just So Stories*. Chief among these is the Oparin–Haldane scenario for the origin of life (Oparin, 1924, 1957;

Haldane, 1929, 1954). The Oparin thesis gives rise to the generally accepted notion that the first cells were anaerobic, fermentative, entirely heterotrophic, non-photosynthetic entities. This led to the idea, proposed by Horowitz (1945), that enzymatic pathways have been evolved backwards, one step at a time. In other words, as the primitive oceanic store of, say, an amino acid was exhausted by the first organisms, there then arose a type of organism that could produce that amino acid from some immediate available chemical precursor thereof – which development was repeated when the oceanic supply of the precursor became exhausted, and so on. When all stores of energy-rich compounds became exhausted, then the time to evolve photosynthesis arrived (Oparin, 1957)!

The eukaryote–prokaryote dichotomy and the Oparin Ocean scenario (together with their corollary notions) have formed the basis for our understanding of early evolutionary events for almost the last half century.

Measurement of evolutionary relationships

Zuckerkandl and Pauling in the 1950s were the first to comprehend fully the possibility of using macromolecular sequences to measure evolutionary relationships (Zuckerkandl & Pauling, 1965). The molecular phylogenies determined over the subsequent 25 years bear witness to the power of their insight.

The virtue of comparative sequencing lies in its being a *genotypic*, not a phenotypic, measure of relatedness. Genetic sequence space is truly enormous – to the point that extensive sequence homology can only be interpreted to mean common ancestry – i.e., it never reflects convergent evolution. Moreover, a significant fraction of genetic changes appears to be selectively neutral. To the extent this is true, the rate of fixation of mutations is uniform (provided the mutation rate is likewise), and as a result, genes can be used as molecular chronometers (Wilson, Carlson & White, 1977).

The metazoan phylogenetic tree constructed on the basis of cytochrome *c* sequence comparisons is well known to us (Fitch & Margoliash, 1967). It confirms quite precisely the phylogenetic tree constructed on the basis of classical, phenotypic criteria. It also incorporates an evolutionary distance (time) measure, which agrees well with divergence times estimated from the fossil record. And it detects deeper branchings than can the classical approaches. It is obvious that this or similar genotypic measures should be capable of settling the otherwise intractable problems of bacterial phylogeny.

There persists a notion even to this day that bacterial phylogenies, for any of several reasons, cannot be determined. (This would seem to have developed out of the chronic failure of microbiologists to determine them (Stanier, 1970).) It is often claimed that lateral, interspecific transfer of genes among bacteria is rapid and extensive enough that the relationships determined through comparative sequence analysis reflect only the histories of the individual genes studied (Ambler *et al.*, 1979; Ambler, Meyer & Kamen, 1979). In other words, evolutionarily, bacteria are total chimeras.

Fortunately, this is not the case; although some bacterial genes are subject to extensive lateral transfer, others appear not to be. Sequence comparisons using both 16S ribosomal RNA and cytochrome *c* from the same group of (purple photosynthetic) bacteria give very nearly the same phylogenetic tree – which is highly unlikely were either of the systems subject to appreciable lateral transfer (Dickerson, 1980; Woese, Gibson & Fox, 1980). Thus, it *is* possible to determine bacterial phylogenies if the correct genotypic criteria are used.

The choice of proper molecular chronometers for such studies is something I wish to discuss only briefly. The appropriate molecule should: (1) be universally distributed (easy to isolate and so on), (2) be functionally constant (to minimize *selected* changes in sequence), and (3) have a sequence that changes slowly enough to retain traces of the universal ancestral version. The 16S-like ribosomal RNAs approximate these requirements to a high degree, in addition to which the molecule is relatively large, being 1500 residues in length or thereabouts (Woese, 1982). (For reasons I will not go into, larger molecules are better, smoother molecular chronometers than smaller ones.)

The actual technique used to measure relatedness among bacteria will also not be described in detail (Sanger, Brownlee & Barrell, 1965; Uchida *et al.*, 1974; Woese *et al.*, 1976): the 16S-like ribosomal RNA from a given species is digested with T_1 ribonuclease to produce a set of oligonucleotides, ranging in size from one to 15–20 bases. The oligonucleotides are electrophoretically separated from one another and sequenced. Oligonucleotides of lengths greater than five bases tend to occur but once in the sequence and so be position specific. The set of such oligonucleotides (i.e., those with more than 5 bases) constitutes an oligonucleotide catalogue that is characteristic of the rRNA and so of the organism in question. Oligonucleotide catalogues can be compared in a variety of ways. Usually this is done through a binary association coefficient, or S_{AB}, which measures the fraction of bases in any two catalogues found in

Fig. 11.1. Dendrogram of relationships among the eubacteria. The clostridia and relatives includes the genera *Clostridium, Bacillus, Lactobacillus, Streptococcus, Staphylococcus, Peptococcus, Sarcina, Ruminococcus, Mycoplasma, Spiroplasma,* and *Acholeplasma.* The actinomycetes and relatives includes the genera *Cellulomonas, Nocardia, Oerskovia, Micrococcus, Arthrobacter, Corynebacterium, Actinomyces, Streptomyces, Chainia, Kitasotoa, Streptosporangium, Ampulariella, Dactylosporangium, Micromonospora, Propionibacterium,* and *Bifidobacterium.* The purple non-sulphur group I includes members of the genera *Rhodospseudomonas, Rhodospirillum, Rhodomicrobium, Paracoccus, Rhizobium, Agrobacterium,* and *Pseudomonas* (*diminuta*). Purple non-sulphur II includes the genera *Rhodopseudomonas* (*gelatinosa*), *Rhodospirillum* (*tenue*), *Pseudomonas, Chromobacterium, Alcaligenes, Aquaspirillum, Spirillum, Sphaerotilus,* and *Thiobacillus.* The purple sulphur group includes the genera *Chromatium, Thiocapsa, Escherichia, Yersinia, Proteus, Serratia, Aeromonas, Pasteurella, Benekea, Photobacterium, Oceanospirillum,* and *Pseudomonas* ('fluorescent' group). The cyanobacterial group includes *Aphanocapsa, Nostoc, Fischerella, Synechococcus, Oscillatoria,*

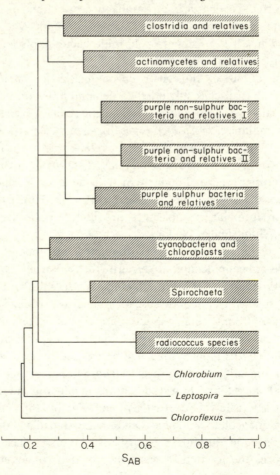

oligonucleotides common to the two catalogues (Fox, Pechman & Woese, 1977). From a table of S_{AB} values covering a set of organisms, a phylogenetic tree can be constructed using any of several treeing procedures.

The phylogeny of bacteria and the universal phylogenetic tree

At the time the work on bacterial phylogeny using 16S rRNA cataloguing began, the microbiologist generally assumed that there existed two main phylogenetic groups of bacteria, the Gram-positive and the Gram-negative. Doubt was sometimes expressed as to whether cyanobacteria (which had formerly been considered plants) fell into either of the categories; many workers also felt that the mycoplasmas were more primitive phylo-genetically than bacteria and so demanded separate status (Razin, 1978). The microbiologist also generally accepted, in keeping with Oparin, that the first bacteria to arise were non-photosynthetic anaerobic (extreme) heterotrophs – which meant that photosynthetic or autotrophic bacteria then each arose in one or a few sublines derived from the common fermentative ancestor (van Niel, 1946).

As work with 16S rRNA catalogues progressed, the phylogenetic outline of what would become known as the true bacteria (or eubacteria) began to emerge (Woese & Fox, 1977a).

The eubacteria (see Fig. 11.1) fall into a number of major subgroupings (Fox *et al.*, 1980). The typical Gram-positive eubacteria constitute one of these subgroups (Stackebrandt, Lewis & Woese, 1980; Woese, Maniloff & Zablen, 1980; Stackebrandt & Woese, 1981a; Tanner, Stackebrandt, Fox & Woese, 1981). However, the Gram-negative eubacteria rather than being a single, coherent group, comprise a number of phylogenetic groupings, each as remote from the others as it is from the Gram-positive group (Gibson *et al.*, 1979; Fox *et al.*, 1980; Stackebrandt & Woese, 1981b; Woese, Blanz, Hespell & Hahn, 1982). (The majority of recognized

Caption to Fig. 11.1 (*cont.*)

Prochloron, and the chloroplasts of *Lemna*, *Zea mays*, *Euglena* and *Porphy-ridium*. The spirochaetes include the genera *Spirochaeta* and *Treponema*. The 'Radiococcus' group are the unusual micrococci that are radio-resistant. The Chloroflexus group includes the genera *Chloroflexus* and *Herpetosiphon*. (Bonen & Doolittle, 1975; Doolittle *et al.*, 1975; Zablen *et al.*, 1975; Gibson *et al.*, 1979; Brooks *et al.*, 1980; Fox *et al.*, 1980; Stackebrandt & Woese, 1981a, b; Seewaldt & Stackebrandt, 1982.)

Gram-negative species, however, fall into the so-called purple bacteria group, so there may be virtue in restricting the definition of a Gram-negative bacterium to encompass this general bacterial phenotype only – and thereby retain the name 'Gram negative' as a phylogenetic unit.) Note also that neither photosynthetic bacteria nor autotrophic bacteria are restricted to small, exclusive subgroups of the eubacteria, as the Oparin thesis was taken to imply. Four of the major eubacterial subgroups contain photosynthetic bacteria. Rather than photosynthetic (and autotrophic) eubacteria having arisen from a fermentative, heterotrophic, non-photosynthetic ancestor, the data seem more consistent with the opposite notion. Finally, note that all of the phenotypes that correspond to deep (ancient) phylogenetic units are anaerobic. A number of aerobic phenotypes occur throughout the eubacterial tree, suggesting independent evolutions of aerobic metabolism, but each is surrounded phylogenetically by related anaerobic phenotypes. This result constitutes strong phylogenetic support for the notion that an oxygen atmosphere arose relatively late in the earth's history.

As the rRNA cataloging data revealed the coherence of 'prokaryotes' as a phylogenetic unit, they also made more evident the depth of the division between 'prokaryotes' and the other known major life form, the eukaryotes. The question of why so much difference existed between the 'prokaryotic' and eukaryotic versions of the 16S-like rRNA sequence became a critical one. Was it possible that the eukaryotic version had begun as a typically 'prokaryotic' one and then changed drastically during the common line of eukaryotic descent? If so, what kinds of selective pressures would bring about these changes in the molecule? Alternatively, could the eukaryotic and 'prokaryotic' versions of the ribosomal RNAs have diverged from one another long before their individual ancestral versions had become established? Questions like these, concerning the eukaryote–prokaryote relationship, were not new (Margulis, 1970; Stanier, 1970). However, at that stage in our understanding they could be neither answered convincingly nor refined. The break in the problem was to come in a totally unexpected way.

When (in collaboration with Ralph Wolfe's laboratory) we encountered methanogen rRNAs, it was immediately apparent that something was unusual. These RNAs were not typically 'prokaryotic'; they had very few oligonucleotides in common with 'prokaryotic' rRNAs. Simple interpretation of the finding suggested that methanogens represented an ancient divergence in the bacterial line of descent, far deeper than any bacterial divergence previously encountered (Balch *et al.*, 1977). Were these rRNAs as unlike typical bacterial rRNAs as were the eukaryotic examples?

The answer was 'yes', and what is more, to the extent they resembled anything, the methanogen rRNA catalogues appeared as close to those of the eukaryotes as they did to those of 'prokaryotes' (Fox *et al.*, 1977; Woese & Fox, 1977*a*). Thus it could not be that the methanogen rRNAs were merely a rapidly evolving variant of the 'prokaryotic' line of rRNA descent – unless, of course, both methanogens and eukaryotes were so, and were *specifically* related to one another (Woese, 1982)!

This unexpected state of affairs was not long in defining itself (Magrum, Luehrsen & Woese, 1978; Woese, Magrum & Fox, 1978; Fox *et al.*, 1980; Woese, Maniloff & Zablen, 1980). Certain other bacteria were found to be specifically related to the methanogens – the extreme halophiles and certain thermoacidophiles – and the grouping so defined (by rRNA cataloging) came to be called the *archaebacteria*, to distinguish these organisms from the true bacteria (or *eubacteria*) to which they were not specifically related, and to denote antiquity of ancestry; see Fig. 11.2. Apparently, there existed two separate groups of prokaryotes, unrelated to one another.

Fig. 11.2. Dendrogram of relationships among the archaebacteria. The *Methanococcaceae* include *M. vaniellii* and *M. voltae*, the *Methanomicrobiaceae* include *Methanomicrobium mobile*, *Methanogenium cariaci*, *M. marisnigri*, and *Methanospirillum hungatei*. The *Methanobacteriaceae* include *Methanobacterium formicicum*, *M. bryantii*, *M. thermoautotrophicum*, *Methanobrevibacter arboriphilus*, *M. ruminantium*, and *M. smithii*. The *Halobacteriaceae* include *Halobacterium halobium*, *H. volcanii*, *H. sodomense*, *H. saccharovorum*, and *Halococcus morrhuae*. (Fox *et al.*, 1980; Stackebrandt & Woese, 1981*b*).

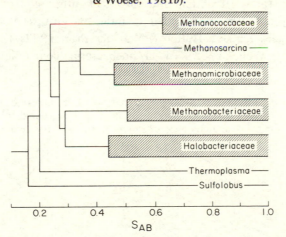

So grand an evolutionary conclusion could not rest on a single criterion, one that, moreover, was merely quantitative, devoid of phenotypic quality. (And indeed initially many biologists would not accept that a 'third form of life' could exist.) The great differences seen in the purely quantitative, genotypic measure had to be supported by comparably striking differences at the phenotypic level (and the alternative interpretations of the data had to be dealt with) if the conclusion were to be believed.

The problem in interpretation is essentially that of rooting the phylogenetic tree. If they all evolve at the same rate, the problem is simple, and then the three lines of descent all stem from the same common ancestor. However, if one of the lines evolves more slowly than do the other two, then the two faster evolving lines must bear a specific relationship to one another in order to accommodate the data. The alternatives can be seen in Fig. 11.3.

In a strict sense neither of the two rooted trees can be excluded. The decision between them, if there is to be one, comes from a consideration of the molecular phenotypes of the three groups, in effect saying that situations *b* and *c* in Fig. 11.3 cannot reasonably lead to the same kind of phenotypic relationships among the three groups.

Archaebacteria

Relatively little is known about archaebacteria at present. However, there is no doubt that they are prokaryotes by the cytological definition thereof. But otherwise they are definitely unique. The methanogens, for example, both possess several major co-factors that occur among eubacteria or eukaryotes rarely, if at all, and also lack other co-factors that are universally or widely distributed in the latter two groups (Wolfe & Higgins, 1979; Vogels, Keltjens, Hutten & Van der Drift, 1982; Thauer, 1982; Romesser & Wolfe, 1982). All archaebacteria have unique cell walls

Fig. 11.3. Alternative phylogenetic trees. (*a*) Unrooted tree; (*b*) rooted isochronic tree – all lines evolve at the same rate; (*c*) rooted anisochronic tree – one line evolves slower than the other two.

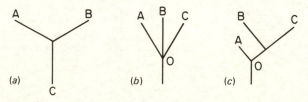

Table 11.1. *A comparison of various molecular characteristics among archaebacteria, eubacteria and eukaryotes*

	Archaebacteria	Eubacteria	Eukaryotes	References
RNA polymerase	Unique subunit structure	Unique subunit structure	Unique subunit structure	Zillig et al., 1982
Cell wall	Variety of types; none incorporates muramic acid	Variety within one type; all incorporate muramic acid	No cell wall in animal cells; variety of types in other phyla	Kandler, 1979, 1982
Membrane lipids	Ether-linked branched aliphatic chains	Ester-linked straight aliphatic chains	Ester-linked straight aliphatic chains	Langworthy et al., 1982
Transfer RNAs				
Ribothymidine in 'common' arm	Absent; replaced by 1-Me-pseudouridine or pseudouridine	Present in most transfer RNAs of most species	Present in most transfer RNAs of all species	Sprinzl & Gauss, 1982; Pang et al., 1982
Dihydrouracil	Absent in all but one genus	Present in most transfer RNAs of all species	Present in most transfer RNAs of all species	Gupta & Woese, 1980
Amino acid carried by initiator transfer RNA	Methionine	Formylmethionine	Methionine	—
Ribosomes				
Subunit sizes	30S, 50S	30S, 50S,	40S, 60S	Woese, 1982
Approximate size of 16S (18S) RNA	1400–1500 nucleotides	1500–1600 nucleotides	1800 nucleotides	Woese, 1982
Approximate size of 23S (25–28S) RNA	2900 nucleotides	2900 nucleotides	3500 nucleotides or more	Woese, 1982
Translation–elongation factor	Reacts with diphtheria toxin	Does not react with diphtheria toxin	Reacts with diphtheria toxin	Kessel & Klink, 1980
Sensitivity to chloramphenicol	Insensitive	Sensitive	Insensitive	—
Sensitivity to anisomycin	Sensitive	Insensitive	Sensitive	—
Sensitivity to kanamycin	Insensitive	Sensitive	Insensitive	—
Messenger-RNA binding site AUCACCUCC at 3' end of 16S (18S) RNA	Present	Present	Absent	Steitz, 1978

(Kandler, 1979, 1982). Their lipids are of a common and unique type (Kates, 1978; Langworthy, Tornebene & Holzer, 1982). In their transfer and ribosomal RNAs, novel base modifications occur, while some otherwise 'universal' modifications are missing (Gupta & Woese, 1980; Pang et al., 1982). Archaebacterial enzymes show unique subunit structure (Zillig, Stetter & Janekovic, 1978; Zillig, Stetter & Janekovic, 1979; Zillig et al., 1982). And repeat sequence DNAs occur in their genomes (Sapienza & Doolittle, 1982a, b; Pfeifer, Ebert, Weidinger & Goebel, 1982). The general conclusion one draws from all this is that the archaebacteria seem as different from the other two primary kingdoms in molecular phenotype as the other two are known to be from one another (Woese, 1982).

This is not to say that archaebacteria bear no *specific* resemblances to either of the other two types. Indeed they do. And at present, it would seem that archaebacteria have more specific points in common with eukaryotes than they do with eubacteria. Table 11.1 summarizes the molecular phenotypic similarities among the three primary kingdoms. We still know too little, however, to draw conclusions about specific relationships between archaebacteria and eukaryotes, and so on. Nor is there a need to do so. The issue, at least in part, turns on the nature of the (chimeric) eukaryotic cell. And simplistic pronouncements as to its relationship to archaebacteria, and so on, would serve only to impede our understanding the problem.

The universal ancestor state: progenotes

All extant life has arisen from a common ancestor. While the evidence for this is completely convincing, our understanding of it is naive. We take this circumstance for granted, see nothing special in it. Yet the so-called Universal Ancestor hides what could be one of the more interesting biological problems. I maintain that the Universal Ancestor is a rudimentary type of entity called a *Progenote*, and that its universality is not explained in any conventional way. Rather 'Universal Ancestor' is a name that covers a diverse class of primitive entities and an unusual evolutionary era about which we know almost nothing.

What is a progenote?

It can be argued that because the translation apparatus is so complex (of the order of 100 components) and has no basis in a single physical interaction (as nucleic acid replication does in base pairing), the process began with a simple, imprecise mechanism that evolved through a series of stages, in which its size (number of components) and accuracy greatly increased (Woese, 1970, 1973; Woese & Fox, 1977*b*). An entity with a rudimentary translation apparatus would still be evolving the link between its genotype and its phenotype – i.e., in terms of its precision. Such an entity is properly called a *Progenote*.

The consequences of, and limitations imposed by, an imprecise translation mechanism are global. Proteins of normal length (by modern standards) could not be produced free of error, so the organism would be fashioned from smaller proteins and/or error-ridden proteins (Woese, 1965, 1970). The result would be an organism with fewer, more general functions. The most severe consequence of imprecise translation is that nucleic-acid replication would be a less accurate process – which consequently would drastically limit the number of genes the organism could carry. Biochemical pathways, too, would be different, less specific – they would produce families of related compounds, not single, well-defined products (Yčas, 1974). The progenote would lack all those refinements we associate with the high biological specificity of organisms today.

Why consider the universal ancestor a progenote?

The differences in molecular phenotype among eubacteria, archaebacteria, and eukaryotes are of a more striking quality than the differences in molecular phenotype that develop within each of the three primary kingdoms. Yet the former, more drastic differences evolved over a shorter time span than did the latter. (The eubacteria and archaebacteria seem each to have been in existence for over 3×10^9 years.) Thus both the mode and the tempo of evolution at the time of the Universal Ancestor appear different, strongly implying that the Ancestor itself was a different type of entity.

Then too, the fact that the molecular phenotypes of eubacteria, archaebacteria, and eukaryotes differ from one another in significant details of just about every system in the cell – enzyme-subunit structure, lipid composition, cell-wall structure, tRNA and rRNA modification pattern, antibiotic sensitivity, etc. – is difficult to account for by any conventional evolutionary course. (Intuitively, it would seem that once basic cell functions have evolved and become integrated into the fabric of the cell, it would not be possible to alter them significantly in any wholesale way.)

I would argue that the differences among the translation mechanisms of the three primary kingdoms involve refinements of the mechanism. For example, almost all tRNAs of almost all eubacterial and eukaryotic species contain ribothymine (T) at position 54, in the so-called 'common arm'

(Sprinzl & Gauss, 1982). None of the archaebacterial tRNAs contain this nucleoside (Gupta & Woese, 1980). Rather, the homologous position is occupied by either pseudouridine or 1-Me-pseudouridine ($^1_m\Psi$) (Pang *et al.*, 1982). T and $^1_m\Psi$ would base pair the same and have the same 'molecular profile', as Fig. 11. 4. shows. Yet they must be introduced into tRNA by different biochemical mechanisms. This would seem to be an example of convergence in molecular function. Other examples of this sort are found in ribosomal RNAs (Woese, 1982).

It is reasonable that base modifications and the like, in general, represent refinements, fine tuning of the translation apparatus (Woese, 1970, 1973; Woese & Fox, 1977*b*). It would seem then that a number of refinements in the translation function were evolved independently in the individual primary lines of descent, which in turn implies that the universal ancestor mechanism did not possess these refinements – i.e., it was a rudimentary mechanism.

If the universal ancestor had been a prokaryote, it would have been difficult to explain how it could have changed in so many, and so profound ways in evolving to become the eubacteria and the archaebacteria. However, it is not so difficult to picture an ancestral progenote doing so. The molecular details, the refinements in which the three primary kingdoms differ from one another, would not have been possessed by the ancestral progenote. These details would have been acquired independently in each primary line of descent, as it passed from the progenote to the more advanced prokaryote or eukaryote state (Woese, 1970; Woese & Fox, 1977*b*). Thus the transition becomes a matter of evolutionary refinement, not of replacement of one functional detail by another.

Fig. 11.4. Comparison of the structure of ribothymine and 1-methylpseudouridine.

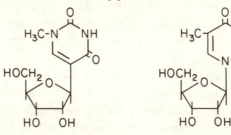

l-methylpseudouridine   thymidine

The genomic and overall organization of the progenote

Inaccuracy of information transfer would manifest itself in two ways that strongly shape the nature of the progenote. It would severely limit the number of different genes the progenote could carry, and the generally lower biological specificity would then mean that overall, fine-tuned, feed-back mechanisms that make the cell today the smooth functioning whole that it is, would be missing. Consequently, the progenote would not be a highly integrated, well-defined whole, but rather would probably resemble more an ill-defined, loosely and imprecisely coupled collection of subcellular elements, which have more autonomy than do subcellular entities today. Such a concept would help to explain two of the progenote's presumed characteristics: (1) how, in spite of inaccurate replication, it could carry a reasonable number of genes, and (2) why the progenote ancestor appears to be universal.

The key to progenote organization is that selection at this stage operates *directly* on individual genes and subcellular elements (not just indirectly on them, through action on the organism as a whole). The biologist would perceive such mechanisms (at least many of them) as error detection–elimination processes. And sophisticated versions of these are very much in evidence today. (Thus error detection mechanisms are ancient (selection) mechanisms, not mechanisms evolved *after* the accuracy of macro-molecular synthesis *per se* had been pushed to its limit.) Indeed many cellular processes today (e.g., various 'RNA processing' interactions) may reflect ancient mechanisms that selected against faulty macromolecules. (However, I will not deal with subcellular selection in its entirety; it extends even to self-recognition and Lamarckism. The topic will be confined to detection of aberration.)

The basis of error detection lies in being able to recognize general properties of molecules. The chief such property is, of course, double helical structure, but general properties of proteins, protein–nucleic acid complexes, and other general properties of nucleic acids could all conceivably be recognized as well.

Let me then suggest a model for the progenote as follows. Its genes are RNA (Jeon & Danielli, 1971). (This may be unavoidable, but in any case, saves one information transfer step, transcription, in which error can be introduced.) Its genome is physically disaggregated, comprising a collection of gene-sized pieces. These would in a sense be competing with one another in replication, and for this and perhaps other reasons, each of its genes would exist in highly multiple copies in the progenote. In that proteins are in primitive stages of evolution at this time, the functional units of the progenote are taken not to be enzymes and the like, but single stranded RNAs and RNA–protein complexes. (tRNAs, ribosomes, and the class of small nuclear ribonucleoproteins (Lerner & Steitz, 1981) are seen as surviving members of a larger class of primitive functional RNA particles.) At the progenote stage, the RNA gene then has both a genotypic (double stranded) and a phenotypic form (often in combination with protein).

Gene replication for the progenote is modelled after viral RNA replication, e.g., the Qβ virus (Pace, Bishop & Spiegelman, 1968; Kramer & Mills, 1982). In other words, the replicate strand does not form a double helical structure with the parental template strand; two complementary strands are not necessarily present in equal amounts. Levels of the various RNA strands are, to some extent, determined by unspecified feed-back mechanisms based on functioning of the whole.

Error elimination could occur, to some extent, at the functional level (as occurs in cells today, for example, with certain proteases (Goldberg, 1972; Goldberg & St John, 1976). However, by requiring complementary strands to reassociate (in relation to replication), most errors incurred in replication and function of RNA could, in principle, be eliminated: reassociation would only rarely involve two strands that had participated in the same preceding replicating event, and neither would have had the same subsequent single stranded history. Therefore, each strand would reveal errors in the other. By postulating a simple function that then detects mispairing and triggers a mechanism for destruction (not repair) of a double stranded gene detected as faulty, it is possible to eliminate a high fraction of errors from the progenote genome. (This model of the progenote genome bears certain similarities to the macronucleus of some ciliates, in which genes exist as separate units in high multiplicity, and cell division entails a random segregation of the genes in the macronucleus into two approximately equal groups.)

Universality

The emerging picture of the progenote is one of diversity, non-uniformity. It is one in which individual genetic units have more autonomy, are less coordinated, than in cells today; the higher level unit, the cell, may not yet be a well-defined entity. How could such variety lead to universality of the ancestor? Not only can universality emerge from the picture of the progenote developed here, but it is an automatic consequence of it. The progenote state is a state of ready interchange of genetic units, subcellular entities, 'viruses', and the like. Such an exchange could be rationalized in a number of ways, e.g., lack of cell walls in progenotes, the common types of evolutionary problems all progenotes face, difficulty in defining and distinguishing self and non-self. However, the underlying reason for all would be that at the progenote stage, evolution proceeds on the subcellular level to a larger extent than it would later. Genetic exchange of this sort would cause progenotes to behave, to appear as a single species, but *not* because they are uniform.

As translation became a more precise process, the progenote type of organization would ultimately give way to more complex, well-defined cells, and separate lines of cellular descent would break out from the 'universal' progenote state. This point in evolutionary history could be called the origin of species (of speciation among cells). At this stage – i.e., when the cell was able to propagate more genetic information – genetic information would shift (in each line of descent) to a DNA-in-linear-organization arrangement. If this be true, it follows that genome organizations in the three primary kingdoms would differ greatly from one another. This is certainly the case for eubacteria versus eukaryotes. We await the answer for archaebacteria.

Origin of the eukaryotic cell

We perceive the origin of the eukaryotic cell (as opposed to that of the prokaryotic cell) as an interesting problem primarily because of the complicated internal structure it alone contains. Although there has been little solid evidence to go on, our concept of eukaryotic cellular origins has both refined and narrowed over the past 20 years, a reflection of the growing molecular understanding of the cell. All is about to change, however. The recent and unexpected awareness that there are three, not two, primary lines of descent, plus the revolution in nucleic acid sequencing, have given us unanticipated perspective and power in attacking cellular evolution. While this is not a time to put forth a definitive model for eukaryotic cellular evolution – for we have yet to reap the sequencing harvest – it is a time to begin realizing what the new facts are suggesting, to relate eukaryotic and prokaryotic cellular evolution, and above all, to rid ourselves of mistaken or misleading preconceptions of eukaryote evolution that would inhibit development of a proper understanding.

Today, accounting for eukaryotic origins is tantamount to identifying the prokaryotes that were the endosymbiotic ancestors of various intracellular structures or that wall-less prokaryote that hosted the endosymbionts (Margulis, 1970). The trouble with this view is not that it is incorrect, but that it is too restrictive. To frame all of eukaryotic cellular evolution in terms of the prokaryote endosymbionts model is to overlook its essence.

Two other mistaken prejudices that need to be discarded are the notions that the eukaryotic cell has arisen from prokaryotes and that it is more advanced than are prokaryotes. The basic eukaryotic cell (sometimes called the urkaryote (Woese & Fox, 1977b)) evolved in parallel with the two prokaryotic lines, not from either of them (as the name prokaryote suggests). The organization of the eukaryotic cell is then different from that of the prokaryotes (and theirs from each other), not more advanced.

The origin of the eukaryotic cell actually constitutes two problems, the second of which is generally obscured by the first. The obvious one concerns the origin of organelles through endosymbiosis. This is in a sense the more superficial problem; organelles and the like seem relatively recent additions to the basic urkaryotic cell and are relatively loosely coupled thereto. There can be no doubt that the origins of the chloroplast and the mitochondrion were by processes of endosymbiosis, and the frequent reviews of these two topics make extensive discussion of them here unnecessary.

The chloroplast clearly came from a cyanobacteria-like ancestor (Bonen & Doolittle, 1975; Zablen, Kissel, Woese & Buetow, 1975). The ancestry of the mitochondrion is still a matter of debate. The reason for this seems to be that the mitochondrial genome (especially in animals and fungi) and those mitochondrial functions it encodes have undergone a rapid and degenerative evolution, making it difficult to trace mitochondrial ancestry through them (Mahler, 1980; Brown, Prager, Wang & Wilson, 1982). However, mitochondrial cytochrome c, the gene for which resides in the nucleus, is specifically related in its sequence to the cytochromes c of a particular subgroup of the so-called purple (eu)bacteria (Almassy & Dickerson, 1978; Gibson et al., 1979). Such an origin for the mitochondrion is also suggested by the plant mitochondrial ribosome RNA (whose gene is in the mitochondrial genome), but the measurement in this instance does not permit a very precise localization of the ancestral line (Cunningham, Bonen, Doolittle & Gray, 1976; Woese & Fox, unpublished calculations). Since the purple group of eubacteria is basically photosynthetic (but contains a number of independent examples of apparent evolution of an aerobic phenotype from a photosynthetic one) it is intriguing to consider that the endosymbiotic ancestor of the mitochondrion was an anaerobic photosynthetic bacterium (Woese, 1977). In other words, the original mitochondrion was a photosynthetic organelle – analogous, but unrelated, to the chloroplast – that evolved intracellularly to become aerobic, when the atmosphere became oxidizing. (I stress this point because conventional wisdom never questions the mitochondrion's having arisen from an aerobic bacterial ancestor.)

The second of the two problems in eukaryotic cellular evolution involves the basic molecular organization of the cell. It can now be seen as the general problem of how and why molecular organization differs among eubacteria, archaebacteria, and eukaryotes. Its essence lies in the nature of the universal ancestor and the transitions therefrom to the three basic cell types, I cannot overstress the importance of three-way comparisons of the cell's organization in unravelling this problem.

The three major cell types seem to differ from one another to the same degree on the molecular level. (Of course, much more needs be known about archaebacteria, the least studied in this regard.) However, the basic eukaryotic cell, the urkaryote, is unlike the others in possessing a nucleus. Thus, the origin of the nucleus – more properly, the origin of the nuclear–cytoplasmic relationship – would seem to be the critical issue in eukaryotic evolution. Some time ago, Hartmann (1975) made the useful observation that the eukaryotic cell is unique in the extent to which it uses and manipulates a variety of RNAs. If the universal ancestor possessed an RNA genome, then the eukaryote, of the three cell types, would seem to have retained more of the aboriginal organization than have its prokaryotic counterparts. The nucleus then can be looked at in a number of ways. It could be the residue of the genomes of prokaryotic endosymbionts in a more primitive, RNA-based, cell (Jeon & Danielli, 1971). Such endo-symbiosis would be consistent with the nucleus having a double membrane. Possession of a nucleus could represent the aboriginal, the ancestral state, which would mean it was then lost in the two prokaryotic lines. Or, it could be that the nucleus itself resembles the ancestral condition, the cytoplasm then arising as a highly organized environment around the original cell. The complexity of the RNA processing reactions that occur within the nucleus are suggestive of an entity that may have existed in its own right at some point.

Perhaps the critical characteristic of the nucleus is that it separates transcription from translation, complete gene expression from gene replication. This is a looser coupling than occurs in prokaryotes, again perhaps suggesting a primitive system. This could be a general mechanism for preventing expression of genes that are not integrated into the system, for controlling the expression of foreign genetic elements and so on. Again a primitive type of mechanism is suggested.

The eukaryotic nuclear genome is chimeric. It contains genes (e.g., for cytochrome *c*) that are of *specific* eubacterial origin, some at least of which have come from endosymbionts. It contains genes (e.g., for some 50S ribosomal proteins (Matheson & Yaguchi, 1982)) that are specifically related to their archaebacterial counterparts. It contains genes (e.g., for 16S-like ribosomal RNA) that do not appear specifically related to either eubacterial or archaebacterial counterparts (Woese & Fox, 1977a). A growing feeling among some biologists would have the urkaryote specifically related to, *coming from*, archaebacterial ancestry. This reflects the growing number of specific molecular similarities between the two. However, the area of cellular evolution is only beginning to open up both experimentally and conceptually, and I see no point in focusing the problem so narrowly at this time. The eukaryotic cell is a genetic, evolutionary chimera. We do not know in what ways and for what reasons. The eukaryotic cell is structurally unique and perhaps primitive. The eukaryotic cell is a fascinating evolutionary problem. That is all that needs be said for now.

Conclusion

The advances in sequencing technology over the past fifteen years have led to the unravelling of the problem of bacterial phylogeny, which in turn has extended our evolutionary perspective fivefold in time. We are now poised to understand the evolution that accompanied major changes in the state of the Earth, such as the advent of an oxygen atmosphere or the formation of major mineral deposits. More importantly, we are in a position to probe the important origin questions – the origin of the major eukaryotic and prokaryotic lines of descent, the origin of eukaryotic organelles and the like. Most importantly, we can – given that there are three, not two, primary lines of descent – begin a meaningful attack on the fundamental evolutionary problem, the nature of the Universal Ancestor and its radiation into the primary lines. For the first time, a bridge between what has been called prebiotic evolution (and studied by non-historical, non-biological approaches) and true biological evolution can at least be imagined.

Further reading

Fox, G. E., Stackebrandt, E., Hespell, R. B., Gibson, J., Maniloff, J., Dyer, T. A., Wolfe, R. S., Balch, W. E., Tanner, R. S., Magrum, L. J., Zablen, L. B., Blakemore, R., Gupta, R., Bonen, L., Lewis, B. J., Stahl, D. A., Luehrsen, K. R., Chen, K. N. & Woese, C. R. (1980). The phylogeny of prokaryotes. *Science*, **209**, 457–63.
Woese, C. R. (1981). Archaebacteria. *Scientific American*, **244**, 98–122.

References

Almassy, R. J. & Dickerson, R. E. (1978). Pseudomonas cytochrome C_{551} at 2.0 A resolution: enlargement of the cytochrome c family. *Proceedings of the National Academy of Sciences, USA*, **75**, 2674–8.

Ambler, R. P., Daniel, M., Hermoso, J., Meyer, T. E., Bartsch, R. G. & Kamen, M. D. (1979). Cytochrome c_2 sequence variation among the recognized species of purple nonsulphur photosynthetic bacteria. *Nature*, **278**, 659–60.

Ambler, R. P., Meyer, T. E. & Kamen, M. D. (1979). Anomalies in amino acid sequences of small cytochromes c and cytochrome c' from two species of purple photosynthetic bacteria. *Nature*, **278**, 661–2.

Balch, W. E.. Magrum, L. J., Fox, G. E., Wolfe, R. S. & Woese, C. R. (1977). An ancient divergence among the bacteria. *Journal of Molecular Evolution*, **9**, 305–11.

Bonen, L. & Doolittle, W. F. (1975). On the prokaryotic nature of the red algal chloroplast. *Proceedings of the National Academy of Sciences, USA*, **72**, 2310–14.

Brooks, B. W., Murray, R. G. E., Johnson, J. L., Stackebrandt, E., Woese, C. R. & Fox, G. E. (1980). Red pigmented micrococci: a basis for taxonomy. *International Journal of Systematic Bacteriology*, **30**, 627–46.

Brown, W. M., Prager, E. M., Wang, A. & Wilson, A. C. (1982). Mitochondrial DNA sequences of primates: tempo and mode of evolution. *Journal of Molecular Evolution*, **18**, 225–39.

Chatton, E. (1937). *Titres et travaux scientifique*. Sottano Sete.

Cunningham, R. S., Bonen, L., Doolittle, W. F. & Gray, M. W. (1976). Unique species of 5S, 18S, and 26S ribosomal RNA in wheat mitochondria. *FEBS Letters*, **69**, 116–22.

Dickerson, R. E. (1980). Evolution and gene transfer in purple non-sulfur bacteria. *Nature*, **283**, 210–12.

Doolittle, W. F., Woese, C. R., Sogin, M. L., Bonen, L. & Stahl, D. (1975). Sequence studies on 16S ribosomal RNA from a blue-green alga. *Journal of Molecular Evolution*, **4**, 307–15.

Fitch, W. M. & Margoliash, E. (1967). Construction of phylogenetic trees. *Science*, **155**, 279–84.

Fox, G. E., Magrum, L. J., Balch, W. E., Wolfe, R. S. & Woese, C. R. (1977). Classification of methanogenic bacteria by 16S ribosomal RNA characterization. *Proceedings of the National Academy of Sciences, USA*, **74**, 4537–41.

Fox, G. E., Pechman, K. R. & Woese, C. R. (1977). Comparative cataloging of 16S rRNA: Molecular approach to prokaryotic systematics. *International Journal of Systematic Bacteriology*, **27**, 44–57.

Fox, G. E., Stackebrandt, E., Hespell, R. B., Gibson, J., Maniloff, J., Dyer, T. A., Wolfe, R. S., Balch, W. E., Tanner, R. S., Magrum, L. J., Zablen, L. B., Blakemore, R., Gupta, R., Bonen, L., Lewis, B. J., Stahl, D. A., Luehrsen, K. R., Chen, K. N. & Woese, C. R. (1980). The phylogeny of prokaryotes. *Science*, **209**, 457–63.

Gibson, J., Stackebrandt, E., Zablen, L. B., Gupta, R. & Woese, C. R. (1979). A phylogenetic analysis of the purple photosynthetic bacteria. *Current Microbiology*, **3**, 59–64.

Goldberg, A. L. (1972). Degradation of abnormal proteins in *Escherichia coli*. *Proceedings of the National Academy of Sciences, USA*, **69**, 422–6.

Goldberg, A. L. & St John, A. C. (1976). Intracellular protein degradation in mammalian and bacterial cells. *Annual Review of Biochemistry*, **45**, 747–803.

Gupta, R. & Woese, C. R. (1980). Unusual modification patterns in the transfer RNAs of archaebacteria. *Current Microbiology*, **4**, 245–9.

Haldane, J. B. S. (1929). The origin of life. *The Rationalist Annual*.

Haldane, J. B. S. (1954). The origins of life. *New Biology*, **16**, 12–27.

Hartmann, H. (1975). The centriole and the cell. *Journal of Theoretical Biology*, **51**, 501–9.

Horowitz, N. H. (1945). On the evolution of biochemical syntheses. *Proceedings of the National Academy of Sciences, USA*, **31**, 153–7.

Jeon, K. W. & Danielli, J. F. (1971). Micrurgical studies with large free-living amebas. In *International Review of Cytology*, vol. 30, ed. G. H. Bourne & J. F. Danielli, pp. 49–90. New York: Academic Press.

Kandler, O. (1979). Zellwandstrukturen bei Methanbakterien, Zur Evolution der Prokaryoten. *Naturwissenschaften*, **66**, 95–105.

Kandler, O. (1982). Cell wall structures and their phylogenetic implications. *Zentralblatt fur Bacteriologie, Mikrobiologie und Hygiene*, Originale C3, 149–60.

Kates, M. (1978). The phytanyl ether-linked polar lipids and isoprenoid neutral lipids of extremely halophilic bacteria. *Progress in Chemistry, Fats and Other Lipids*, **15**, 301–42.

Kessel, M. & Klink, F. (1980). Archaebacterial elongation factor is ADP-ribosylated by diphtheria toxin. *Nature*, **287**, 250–1.

Knoll, A. H. & Barghoorn, E. S. (1978). Archaen microfossils showing cell division from the Swaziland system of South Africa. *Science*, **198**, 396–8.

Kramer, F. R. & Mills, D. R. (1982). Secondary structure formation during RNA synthesis. *Nucleic Acids Research*, **9**, 5109–24.

Lerner, M. R. & Steitz, J. A. (1981). Snurps and scyrps. *Cell*, **25**, 298–300.

Langworthy, T. A., Tornebene, T. G. & Holzer, G. (1982). Lipids of archaebacteria. *Zentralblatt fur Bacteriologie, Mikrobiologie und Hygiene*, Originale C3, 228–44.

Lowe, D. R. (1980). Stromatolites 3,400-M yr old from the Archaen of western Australia. *Nature*, **284**, 441–3.

Magrum, L. J., Luehrsen, K. R. & Woese, C. R. (1978). Are extreme halophiles actually 'bacteria?' *Journal of Molecular Evolution*, **11**, 1–8.

Mahler, H. R. (1980). Recent observations bearing on the evolution and possible origin of mitochondria. In *The Origins of Life and Evolution*, ed. H. O. Halvorson & K. E. van Holde, pp. 103–23. New York: Alan R. Liss.

Margulis, L. (1970). *Origin of Eucaryotic Cells*. New Haven, Conn.: Yale University Press.

Matheson, A. T. & Yaguchi, M. (1982). The evolution of the archaebacterial ribosome. *Zentralblatt fur Bacteriologie, Mikrobiologie und Hygiene*, Originale C3, 192–9.

Oparin, A. I. (1924). *Proiskhozhdenie Zhizny*. Izd. Moskovski Rabochii: Moscow (translated by Ann Synge).

Oparin, A. I. (1957). *The Origin of Life on the Earth*, 3rd edn. Edinburgh and London: Oliver and Boyd.

Pace, N. R., Bishop, D. H. L. & Spiegelman, S. (1968). The immediate precursor of viral RNA in the Qβ-replicase reaction. *Proceedings of the National Academy of Sciences, USA*, **59**, 139–44.

Pang, H., Ihara, M., Kuchino, Y., Nishimura, S., Gupta, R., Woese, C. R. & McCloskey, J. A. (1982). Structure of a modified nucleoside in archaebacterial tRNA which replaces ribosylthymine. *The Journal of Biological Chemistry*, **257**, 3589–92.

Pfeifer, F., Ebert, K., Weidinger, G. & Goebel, W. (1982). Structure and function of chromosomal and extra-chromosomal DNA in halobacteria. *Zentralblatt fur Bakteriologie, Mikrobiologie und Hygiene*, Originale C3, 110–19.

Pflug, H. D. (1982). Early diversification of life in the Archaean. *Zentralblatt fur Bakteriologie, Mikrobiologie und Hygiene*, Originale C3, 53–64.

Razin, S. (1978). The mycoplasmas. *Microbiological Reviews*, **42**, 414–70.

Romesser, J. A. & Wolfe, R. S. (1982). CDR factor, a new coenzyme required for carbon dioxide reduction to methane by extracts of *Methanobacterium thermoautotrophicum*. *Zentralblatt fur Bakteriologie, Mikrobiologie und Hygiene*, Originale C3, 271–6.

Sanger, F., Brownlee, G. G. & Barrell, B. G. (1965). A two-dimensional fractionation procedure for radioactive nucleotides. *Journal of Molecular Biology*, **13**, 373–98.

Sapienza, C. & Doolittle, W. F. (1982a). Unusual physical organization of the halobacterial genome. *Nature*, in press.

Sapienza, C. & Doolittle, W. F. (1982b). Repeated sequences in the genomes of halobacteria. *Zentralblatt fur Bakteriologie, Mikrobiologie und Hygiene*, Originale C3, 120–27.

Schidlowski, M. (1980). Antiquity of photosynthesis: possible constraints from Archaean carbon isotope record. In *Biogeochemistry of Ancient and Modern Environments*, ed. P. A. Trudinger & M. R. Walter, pp. 47–54. Berlin: Springer.

Schopf, J. W. (1968). Microflora of the Bitter Springs formation, Late Precambrian, central Australia. *Journal of Paleontology*, **42**, 651–88.

Seewaldt, E. & Stackebrandt, E. (1982). Partial sequence of 16S ribosomal RNA and the phylogeny of Prochloron. *Nature*, **295**, 618–20.

Sprinzl, M. & Gauss, D. H. (1982). Compilation of tRNA sequences. *Nucleic Acids Research*, **10**, r1–r55.

Stackebrandt, E., Lewis, B. J. & Woese, C. R. (1980). The phylogenetic structure of the coryneform group of bacteria. *Zentralblatt fur Bakteriologie, Mikrobiologie und Hygiene*, Originale C1, 137–49.

Stackebrandt, E. & Woese, C. R. (1981a). Towards a phylogeny of the Actinomycetes and related organisms. *Current Microbiology*, **5**, 197–202.

Stackebrandt, E. & Woese, C. R. (1981b). The evolution of prokaryotes. In *Molecular and Cellular Aspects of Microbial Evolution*, Society for General Microbiology Ltd., Symposium 32, ed. M. J. Carlile, J. F. Collins & B. E. B. Moseley, pp. 1–31. Cambridge: Cambridge University Press.

Stanier, R. Y. (1970). Some aspects of the biology of cells and their possible evolutionary significance. In *Organization and Control in Prokaryotic and Eukaryotic Cells*, 20th Symposium Society for General Microbiology, ed. H. P. Charles & B. C. J. G. Knight, pp. 1–38. Cambridge: Cambridge University Press.

Stanier, R. Y. & van Niel, C. B. (1941). The main outlines of bacterial classification. *Journal of Bacteriology*, **42**, 437–66.

Steitz, J. A. (1978). Methanogenic bacteria. *Nature*, **273**, 10.

Tanner, R. S., Stackebrandt, E., Fox, G. E. & Woese, C. R. (1981). A phylogenetic analysis of *Acetobacterium woodii*, *Clostridium barkeri*, *Clostridium butyricum*, *Clostridium lituseberense*, *Eubacterium limosum* and *Eubacterium tenue*. *Current Microbiology*, **5**, 35–8.

Thauer, R. K. (1982). Nickel tetrapyrroles in methanogenic bacteria: structure function and biosynthesis. *Zentralblatt fur Bakteriologie, Mikrobiologie und Hygiene*, Originale C3, 265–70.

Uchida, T., Bonen, L., Schaup, H. W., Lewis, B. J., Zablen, L. & Woese, C. (1974). The use of ribonuclease U_2 in RNA sequence determination. *Journal of Molecular Evolution*, **3**, 63–77.

van Niel, C. B. (1946). The classification and natural relationships of bacteria. *Cold Spring Harbor Symposium of Quantitative Biology*, **11**, 285–301.

Vogels, G. D., Keltjens, J. T., Hutten, T. J. & Van der Drift, C. (1982). Coenzymes of methanogenic bacteria. *Zentralblatt fur Bakteriologie, Mikrobiologie und Hygiene*, Originale C3, 258–64.

Whittaker, R. H. (1977). Broad classification: the kingdoms and the protozoans. In *Parasitic Protozoa*, vol. 1, ed. J. P. Kreier, pp. 1–34. New York: Academic Press.

Wilson, A. C., Carlson, S. S. & White, T. J. (1977). Biochemical evolution. *Annual Review of Biochemistry*, **46**, 573–639.

Woese, C. R. (1965). On the evolution of the genetic code. *Proceedings of the National Academy of Sciences, USA*, **54**, 1546–52.

Woese, C. R. (1970). The genetic code in prokaryotes and eukaryotes. In *Organization and Control in Prokaryotic and Eukaryotic Cells*, 20th Symposium Society for General Microbiology, ed. H. P. Charles & B. C. J. G. Knight, pp. 39–54. Cambridge: Cambridge University Press.

Woese, C. R. (1973). Evolution of the genetic code. *Naturwissenschaften*, **60**, 447–59.

Woese, C. R. (1977). Endosymbionts and mitochondrial origins. *Journal of Molecular Evolution*, **10**, 93–6.

Woese, C. R. (1982). Archaebacteria and cellular origins: an overview. *Zentralblatt fur Bakteriologie, Mikrobiologie und Hygiene*, Originale C3, 1–17.

Woese, C. R., Blanz, P., Hespell, R. B. & Hahn, C. M. (1982). Phylogenetic relationships among various helical bacteria. *Current Microbiology*, **7**, 119–24.

Woese, C. R. & Fox, G. E. (1977a). Phylogenetic structure of the prokaryotic domain: the primary kingdoms. *Proceedings of the National Academy of Sciences, USA*, **74**, 5088–90.

Woese, C. R. & Fox, G. E. (1977b). The concept of cellular evolution. *Journal of Molecular Evolution*, **10**, 1–6.

Woese, C. R., Gibson, J. & Fox, G. E. (1980). Do genealogical patterns in purple photosynthetic bacteria reflect interspecific gene transfer? *Nature*, **283**, 212–14.

Woese, C. R., Magrum, L. J. & Fox, G. E. (1978). Archaebacteria. *Journal of Molecular Evolution*, **11**, 245–52.

Woese, C. R., Maniloff, J. & Zablen, L. B. (1980). Phylogenetic analysis of the mycoplasmas. *Proceedings of the National Academy of Sciences, USA*, **77**, 494–8.

Woese, C., Sogin, M., Stahl, D., Lewis, B. J. & Bonen, L. (1976). A comparison of the 16S ribosomal RNAs from mesophilic and thermophilic bacilli: some modifications in the Sanger method for RNA sequencing. *Journal of Molecular Evolution*, **7**, 197–213.

Wolfe, R. S. & Higgins, I. J. (1979). Microbial biochemistry of methane – a study in contrasts. In *Microbial Biochemistry*, vol. 21, ed. J. R. Quayle, pp. 267–300. Baltimore: University Park Press.

Ycas, M. (1974). On earlier states of the biochemical system. *Journal of Theoretial Biology*, **44**, 145–60.

Zablen, L. B., Kissel, M. S., Woese, C. R. & Buetow, D. E. (1975). Phylogenetic origin of the chloroplast and prokaryotic nature of its ribosomal RNA. *Proceedings of the National Academy of Sciences, USA,* **72,** 2418–22.

Zillig, W., Stetter, K. O. & Janekovic, D. (1978). DNA-dependent RNA polymerase from *Halobacterium halobium. European Journal of Biochemistry,* **91,** 193–9.

Zillig, W., Stetter, K. O. & Janekovic, D. (1979). DNA-dependent RNA polymerase from the archaebacterium *Sulfolobus acidocaldarius. European Journal of Biochemistry,* **96,** 597–604.

Zillig, W., Stetter, K. O., Schnabel, R., Madon, J. & Gierl, A. (1982). Transcription in archaebacteria. *Zentralblatt fur Bakteriologie, Mikrobiologie und Hygiene,* Originale C3, 218–27.

Zuckerkandl, E. & Pauling, L. (1965). Molecules as documents of evolutionary history. *Journal of Theoretical Biology,* **8,** 357–66.

12

Experimental evolution

PATRICIA H. CLARKE

Darwin and microorganisms

The living world appeared to Darwin to present a series of naturally occurring experiments in evolution. He saw the immense variety of different species as representing the end results of different ways of successful adaptation to life in the seas and on the land. He was of course considering mainly the higher plants and animals. Birds, fishes, insects and crustaceans came under his scrutiny but, although there was very little to say about them at that time, the microbial inhabitants of the same world were not completely forgotten. I am indebted to Mr Richard Freeman for the information that, in a letter of 10 December 1881 to G. J. Romanes, Darwin recalled a meeting at University College London at which bacteria were discussed (Freeman, 1982). In *The Origin of Species* Darwin (1859) went so far as to say that 'all the organic beings which have ever lived on this earth may be descended from some one primordial form'. The lower plants and fungi were very familiar to nineteenth century biologists and the existence of living organisms invisible to the naked eye had by that time been known for two centuries. Antonie van Leeuwenhoek (1677) reported in a letter to the Royal Society that he had observed 'exceedingly small animalcules' with his microscopes and it is generally accepted that some of these were bacteria. His correspondence extended over many years and he reported the conditions under which his animalcules could be observed. A particularly significant observation was that he did not find any animalcules in freshly collected rain water but that they appeared after a few days or weeks when samples of pepper, ginger, clove or nutmeg were added to the vessels. This was the first direct observation that micro-organisms appeared spontaneously in aqueous environments containing organic nutrients. This colonization happened so readily that it is easy to understand how it was possible to entertain the idea of spontaneous generation of microorganisms although Leeuwenhoek himself thought that they were already present in some form in the surrounding air. The work of Pasteur and his followers enabled that particular problem to be pushed back into an earlier and more primitive stage of evolutionary history. Pasteur's famous experiments in resolving this problem were carried out at about the same time as the publication of the *Origin*. He demonstrated that microorganisms did not grow if the potential nutrient material was first rendered sterile by heating and then sealed from contact with the world outside. Pasteur was concerned to show that life came from life; that yeasts were descended from yeasts and bacteria from bacteria. Once this was established the way was clear for devising techniques for studying the morphology, physiology, biochemistry and eventually the genetics of bacteria and other microorganisms and to attempt to understand how they might have evolved into species. (It is convenient for micro-biologists to use the term 'species' to designate a cluster of bacteria with similar characteristics but for any particular group the number of recognized species is related more to the number of bacteriologists who have studied that group than to the occurrence of those bacteria in nature. Among reviews discussing this aspect of bacterial classification and taxonomy are Cowan (1962) and Sneath (1962).)

What kinds of bacteria?

In the world of today microorganisms can be found almost everywhere. Some of the ecological niches that have been colonized are commonly thought to be very hostile environments and inimical to life. The majority of microorganisms live in 'moderate' environments, at approximately neutral pH, at temperatures ranging from 20 to 40 °C and under normal atmospheric pressure. However, bacteria are known that grow at tempera-tures approaching the boiling point of water (Brock, 1978) and extreme psychrophiles are a problem to the food industry (Roberts, Hobbs, Christian & Skovgaard, 1981). High salt concentrations inhibit the growth of most bacteria but the extreme halophiles can grow only in media containing 20% or more of sodium chloride. Acid environments are tolerated, or even obligatory, for some groups of bacteria while others flourish under alkaline conditions, with optimal growth at pH 10–11 (Kushner, 1978; Horikoshi & Akiba, 1982). Some of the extreme halophiles and thermoacidophiles have been found to differ significantly in several respects from the more commonly studied bacteria and have now been assigned to a separate group – the archaebacteria – to distinguish them from more widely known eubacteria (Fox *et al.*, 1980; Woese, this volume). However, some of the bacteria that grow in these extreme environmental conditions are otherwise very similar in structure and metabolism to one or other of the main groups of eubacteria.

Adaptation of bacteria to a particular ecological niche can be related to other physiological properties. Some autotrophic bacteria obtain energy from photosynthesis and others from the oxidation of inorganic compounds. While some of these are obligate autotrophs, others are also capable of heterotrophic growth based on the oxidation or fermentation of organic compounds. The heterotrophic bacteria differ among themselves in the number and different kinds of organic compounds that can support growth. For example, the obligate methylotrophs are restricted to growth on one-carbon compounds whereas species of *Pseudomonas* are known to be able to utilize for growth any of a large number of different carbon compounds (Den Dooren de Jong, 1926; Stanier, Palleroni & Doudoroff, 1966). The diversity of biochemical activities among bacteria led Marjory Stephenson, in the first edition of *Bacterial Metabolism* (1930), to speculate. 'Perhaps bacteria may tentatively be regarded as biochemical experimenters; owing to their relatively small size and rapid growth variations must arise more frequently than in more differentiated forms of life and they can, in addition, afford to occupy more precarious positions in natural economy than larger organisms with more exacting requirements.' The implication of this line of thought was that investigations might show that new variants were constantly arising in bacterial populations that might enable them to take over ecological niches not available to the parent population. Two decades of rapid progress in microbial biochemistry reinforced her views. In 1949 she offers this challenge, 'but with bacteria, constant evolutionary changes occur under our eyes and can be controlled and imitated in the laboratory' (Stephenson, 1949).

Bacterial variation and new enzymes

It was not difficult, even at that time, to obtain bacterial variants that differed in properties from the parental cultures from which they had been derived but it was not until major advances had been made in microbial genetics that these predictions of Marjory Stephenson could be fully realized. From time to time attempts have been made to adapt mesophilic bacteria to life under more extreme physicochemical conditions (Kogut, 1980) but studies on experimental evolution have been mainly directed towards the acquisition of new catabolic activities. Most of these investigations have been particularly concerned with the possibility of altering the structure of an enzyme so that it could fulfil a new catabolic function.

Comparative studies of protein, or nucleic acid, sequences represent an extension to the molecular level of the comparative morphology of traditional biology. Comparisons of the morphological structures to be found in living organisms, and in fossil remains, provided the raw material from which phylogenetic trees could be constructed. In a similar manner the amino-acid sequences of proteins could be compared and phylogenetic trees could be constructed on the hypothesis that successive mutations had resulted in sequence divergence of proteins with similar functions (Dayhoff, 1972). At the same time it was discovered that there are close similarities in sequence, and even more in tertiary structure, of enzymes that carry out similar catalytic reactions but on different substrates. This made it reasonable to assume that families of enzymes with related functions, such as the serine proteases, had been evolved from common ancestral proteins (Hartley, 1974). The answer to the question 'Where do new enzymes come from?' could be, 'They come from old enzymes'. The challenge to the experimental evolutionists is to find a system in which an enzyme with new properties appearing 'under controlled conditions in the laboratory' can be shown to have been derived from a pre-existing enzyme.

One of the main advantages of working with bacteria is that it is possible to grow large numbers of them; a culture growing in a good liquid medium will contain more than 10^9 individual cells in one millilitre. Another advantage is that a single bacterial cell can give rise to a clone very rapidly since the mean generation time may be only 30 min. It is therefore theoretically possible to detect very rare mutational events if a selection can be devised to allow the novel mutant to grow on a medium on which the parental cells are unable to do so. This is most convenient using growth of colonies on solid agar medium in a petri dish. It is not always possible to achieve an all-or-nothing distinction between mutant and parent and an alternative approach, which has been successful in some cases, is to set up a system in which the desired mutation confers a growth advantage so that in time the mutant is expected to outgrow the parent (Clarke, 1974; Hartley, 1974). If either of these approaches is to be used it is essential that the enzyme concerned should be rate-limiting for growth, and this has to be related to the overall metabolism of the chosen organism. Microbial growth depends on many enzymes working together in complex metabolic pathways. If the initial enzyme for the attack on a particular compound is missing, then that compound will not support growth even if all the other enzymes of the pathway are present. An extension of this reasoning is that if the first enzyme of the pathway can be changed in structure so that it can attack a novel compound then, providing that the product can be fed into an existing catabolic pathway, the novel compound will support growth.

The changes we expect to occur with a mutation of this sort have to do with the fit between the enzyme protein and its substrate molecule. The long chain that makes up the enzyme protein folds up in a shape, determined by the amino-acid sequence, to give a cleft or pocket into which the substrate fits. The enzyme specificity depends on the exactness of the fit between the substrate and the binding site of the enzyme. The optimal substrate for an enzyme is the one that fits the binding site in such a way that the catalytic reaction can proceed at the maximal rate for that enzyme. However, many enzymes that have high affinities for their optimal substrates may also be able to bind related compounds to some extent. The starting point for most of the experiments in enzyme evolution is the belief that substitutions in one or more amino acids, as a result of point mutations, might result in changes in the conformation of the enzyme protein so that other compounds, related in structure to the original substrate of the enzyme, could be bound at the active site and undergo catalytic change. We now have a number of cases in which such evolutionary events have been observed. Newly evolved enzymes with novel substrate specificities include phosphatase and alcohol de-hydrogenase of the yeast *Saccharomyces cerevisiae* as well as the bacterial enzymes, ribitol dehydrogenase of *Klebsiella* (*Aerobacter*) *aerogenes*, a second β-galactosidase of *Escherichia coli* and amidase of *Pseudomonas aeruginosa*.

Divergent evolution of *P. aeruginosa* amidase

The amidase of *P. aeruginosa* hydrolyses aliphatic amides containing two or three carbon atoms to produce ammonia and the corresponding carboxylic acid.

$$RCONH_2 + H_2O \rightarrow RCOOH + NH_3$$

Ammonia can be used as the nitrogen source for growth and the carboxylic acid is a potential carbon source. Aliphatic acids containing up to ten carbon atoms in the side chain are readily metabolised but amides containing between four and ten carbon atoms cannot be hydrolysed by the amidase produced by the wild-type strain (amidase A). It was thought that growth on higher amides might therefore be achieved by mutations that altered the substrate specificity of this enzyme. Selection for mutants able to utilize amides containing four or more carbon atoms as growth substrates led to the isolation of a family of mutants producing enzymes with altered substrate specificities. The first novel enzyme identified was amidase B which enables mutant B6 to grow on the 4-carbon amide, butyramide. Further mutational steps led in succession to novel enzymes that hydrolysed the 5-carbon and 6-carbon amides, valeramide and hexanoamide, and to enzymes that hydrolysed phenylacetamide (Figs 12.1, 12.2).

The amino-acid sequence of these enzymes has been investigated and

Fig. 12.1. Experimental evolution of amidases of *Pseudomonas aeruginosa*. The wild-type strain PAC1 grows on acetamide and mutants have been obtained that grow on butyramide, valeramide, phenylacetamide and acetanilide. (Clarke, 1978.)

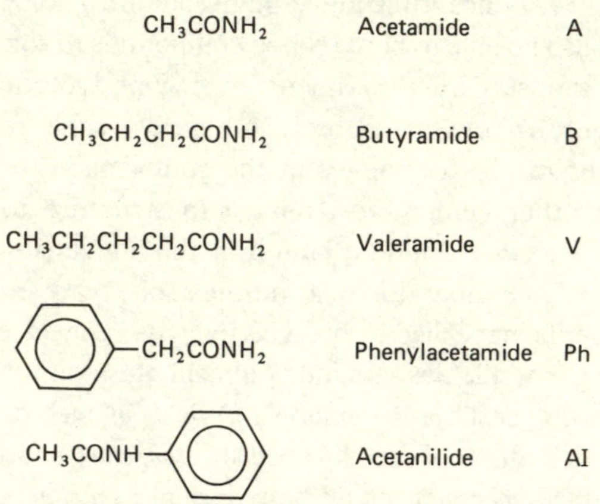

CH_3CONH_2	Acetamide	A
$CH_3CH_2CH_2CONH_2$	Butyramide	B
$CH_3CH_2CH_2CH_2CONH_2$	Valeramide	V
⬡—CH_2CONH_2	Phenylacetamide	Ph
CH_3CONH—⬡	Acetanilide	AI

Fig. 12.2. Amidase evolution in *Pseudomonas aeruginosa*. Strains C11, CB4 and L10 are constitutive for amidase synthesis and carry mutations in the *amiR* gene. Strains in boxes have mutations in the gene *amiE* and produce enzymes with altered substrate specificities. (Clarke, 1978.)

Pseudomonas aeruginosa

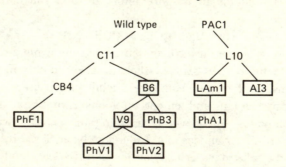

some of the amino-acid substitutions are now known. The difference in the activities of the wild-type amidase A and amidase B, produced by mutant B6, is due to a change in the DNA sequence that results in the substitution of a phenylalanine residue for a serine at position seven from the N-terminus (Paterson & Clarke, 1979). This minor change in the primary sequence is sufficient to allow butyramide to be an effective substrate for growth. Within the family tree of amidase mutants is the subfamily derived from mutant B6. The amidase of mutant V9, derived from B6 by a single mutational step, allows growth on valeramide while those produced by mutants PhB3, PhV1 and PhV2 allow growth on phenylacetamide. Analysis of the amidases produced by these strains showed that the original B6 mutation has been retained during the subsequent evolutionary steps. Thus we can see that successive mutations in the structural gene can introduce progressively greater alterations in the substrate specificities of an enzyme.

The physicochemical and kinetic properties of the novel phenyl-acetamides were not identical. As well as the mutants derived from strain B6 were others obtained from other parental strains (CB4 and L10). Fig. 12.2 shows that phenylacetamidases could be derived from the wild-type amidase by at least five different lines of descent. The enzymes produced by strains PhB3 and PhV1 are relatively thermostable and although they hydrolyse phenylacetamide, valeramide and butyramide they have negligible activity towards acetamide. The enzymes produced by strains PhV2, PhF1 and PhA1 have retained weak activity for acetamide as well as being active on the higher amides, but are much more thermolabile (Betz & Clarke, 1972). In terms of amino-acid sequence and enzyme properties these phenylacetamidases provide an example of divergent evolution. On the other hand in terms of advantage to the organism, since they all confer the ability to utilize phenylacetamide for growth, they can also be regarded as examples of convergent evolution at the molecular level. Most of the novel amide substrates for the strains shown in Fig. 12.2 were altered in the structure of the aliphatic side-chain but it was also possible to obtain amidases that hydrolysed the N-substituted amide, acetanilide. The enzyme produced by strain AI3 hydrolyses acetanilide to give acetate and aniline. This was a relatively thermostable enzyme and had good activity towards acetamide but none towards butyramide and the higher aliphatic amides. In this case the difference in substrate specificities is again due to a single amino-acid substitution in which a threonine residue of the wild-type enzyme has been replaced by isoleucine (Brown & Clarke, 1972). One reason for choosing *P. aeruginosa* for this work was the known biochemical versatility of this genus and it was thought that it was reasonable to expect it to be capable of further evolution of its biochemical repertoire.

Evolved β-Galactosidase (ebg)

One of the best-known of all enzymes is the β-galactosidase of *Escherichia coli*, coded by the *lacZ* gene. Attempts to change the substrate specificity of this enzyme have been uniformly unsuccessful (B. Müller-Hill, personal communication). However, strains in which the *lacZ* gene had been deleted gave rise to lactose-utilizing mutants producing a novel enzyme known as evolved β-galactosidase (Ebg) (Campbell, Lengyel & Langridge, 1973). This apparently orphan enzyme is induced by lactose in wild-type *E. coli* K12 but its activity is so low that it had previously been overlooked (Fig. 12.3). Hall and colleagues have been very successful in using this enzyme system as a model for their studies on acquisitive evolution. They showed that spontaneous point mutations in the enzyme structural gene, *ebgA*, resulted in alterations in the enzyme protein that allowed lactose to be hydrolysed at a rate sufficient for growth to occur (Hall, 1977). This is an example of an enzyme that is virtually non-functional in the wild-type

Fig. 12.3. Evolution of a second β-galactosidase in *Escherichia coli* K12. Wild-type strains carrying a deletion of the *lacZ* gene are unable to grow on lactose. Mutations in *ebgA* result in increased activity, and in *ebgR* in increased amount of a second β-galactosidase. (Campbell *et al.*, 1973; Hall, 1977, 1981.)

$$\text{Lactose} \xrightarrow[ebgR^+ \; ebgA^0]{lacI^+ \; lacZ^+Y^+} \text{Galactose + Glucose}$$

Fig. 12.4. Evolution of β-galactosidases in *Escherichia coli* K12 in strains carrying deletion of the *lacZ* gene. Numbers in parenthesis indicate the number of mutations in the *ebgA* gene. (From Hall, 1981.)

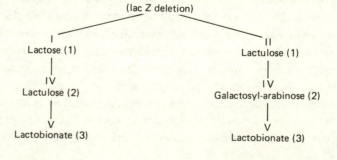

Escherichia coli K12

(lac Z deletion)

I	II
Lactose (1)	Lactulose (1)
IV	IV
Lactulose (2)	Galactosyl-arabinose (2)
V	V
Lactobionate (3)	Lactobionate (3)

strain that can be brought into action if needed. The *ebg* genes of the wild-type are quiescent rather than completely silent but have no significant role when the *lac* operon is present and functional. As I shall discuss later, the altered enzyme mutants have been derived in strains that are constitutive for enzyme synthesis. The earliest altered enzyme mutants were those that enabled growth to occur on lactose but mutants were obtained later that produced enzymes with activity on lactulose, galactosyl-arabinose and lactobionate (Hall, 1981). Several classes of mutants could be distinguished and activity on some of these novel substrates could be achieved by combining two mutations in different regions of the structural gene. Class I mutants have single-site mutations resulting in enzymes that allow growth on lactose, while Class II mutants produce enzymes that are more active on lactulose than lactose. The Class I and Class II mutations are located in regions I and II of the *ebgA* gene and are about one kilobase apart. A strain carrying both Class I and Class II mutations may grow rapidly on lactose, moderately on lactulose and slowly on galactosyl-arabinose. These are the Class IV mutants that can be obtained by recombination between Class I and Class II mutants or by successive mutational steps. Selection on lactobionate from a Class IV mutant gives the Class V mutants carrying three mutations in the structural gene (Fig. 12.4). Detailed studies have been made on the enzymes produced by these different classes of mutants and significant changes in K_m and V_{max} values have been found (Hall, 1981). These results show very clearly how a dormant gene has the potential to evolve a variety of novel metabolic activities.

Gene duplication

We have seen that a single-site mutation can result in an alteration in the substrate specificity of an enzyme. When this occurs the original enzyme may be sufficiently altered for a novel carbon compound to be used as a growth substrate. The experiments with amidase and Ebg β-galactosidase have shown that a new enzyme can evolve from an old enzyme. We would like to know if *two* new enzymes could evolve from *one* old enzyme. Gene duplication, followed by mutational divergence, would be one way in which this could happen and some authors have suggested that gene duplication might be accompanied by inactivation of one copy which would evolve in silence and subsequently revive with a new function (Koch, 1972; Hartley, 1974). The *ebgA* gene is almost silent and it might be suspected that the *ebgA* and the *lacZ* β-galactosidase were related in this way but the two enzymes appear to be dissimilar (Campbell *et al.*, 1973). Gene duplication occurs very readily when selection is made for growth on very low concentrations of poor growth substrates. Novick & Horiuchi (1961) observed duplication of the *lac* genes of *E. coli* cultures grown in a chemostat with limiting lactose. Duplication of the ribitol dehydrogenase genes of *Klebsiella* (*Aerobacter*) *aerogenes* was found to occur in chemostat cultures grown with xylitol as the carbon source. Ribitol dehydrogenase has low activity for xylitol and the gene duplication means that the culture produces more enzyme so that xylitol can be utilized at a faster rate. Instead of producing a better enzyme these bacteria produce high levels of a low activity enzyme (Hartley, 1974). In other experiments the response to this severe selection pressure was the production of mutants with altered ribitol dehydrogenases that had higher affinities for xylitol. We see here that under conditions when ribitol dehydrogenase is rate-limiting for growth an alternative to a single-site mutation to give a more active enzyme is a duplication of the existing gene to give more of the original enzyme. In most cases the gene duplication is unstable and the extra copies are lost when the 'hyper-strain' is removed from the selection pressure of chemostat growth. So far there have been no examples of gene duplication followed by mutation of one of the gene copies but even if this occurred it would be unlikely to confer a significant growth advantage under these experimental conditions. It is possible that mutational divergence following gene duplication might be detected if the chemostat culture that had already undergone gene duplication were to be offered the challenge of two novel substrates at the same time, but this has not yet been attempted.

There are some interesting features of the pentitol pathway genes of *K. aerogenes* that may have some bearing on this. Of the four pentitols only ribitol and D-arabitol are of common occurrence and these two are widely used as growth substrates. On the other hand although the two 'unnatural' pentitols, xylitol and L-arabitol, are utilized by few bacterial species it is not difficult to obtain mutants that do so. In *K. aerogenes* the genes for the ribitol and D-arabitol pathways are closely linked and the similarities of the enzymes of the two pathways suggested that the two operons might have arisen by duplication. This has implications for the duplications that occur in response to the selection for mutants growing faster on xylitol. Inderlied & Mortlock (1977) found that the gene duplication that occurred in chemostat selection with xylitol extended only to the ribitol genes and not to the closely linked D-arabitol genes. We may be seeing here a transitional stage in the evolution of yet a third pentitol operon.

Regulation of gene expression

An aspect of acquisitive evolution that has not been discussed so far is whether or not a novel enzyme can be brought under the regulatory control of its new substrate. Many catabolic enzymes are made only when they are required for growth. In the absence of the substrate the cell contains very low levels of enzyme but when the substrate is present in the growth medium the amount of enzyme may increase more than a thousandfold. In this way the bacteria gain the advantage of high enzyme activities when they are needed for growth but do not waste energy on unnecessary protein synthesis when other growth substrates are available. The classical negative control of the synthesis of a catabolic enzyme, as first described for the *lac* operon of *E. coli*, involves a regulator gene that codes for a repressor protein. In the absence of an inducer (this may be the enzyme substrate or a related molecule) the repressor binds to the operator site adjacent to the structural gene for the enzyme and prevents gene expression. When the inducer is present the repressor is released from the operator and enzyme synthesis takes place. Since it is of little use for an organism to gain a new enzyme activity unless the synthesis of the new enzyme can be switched on at the appropriate time, we might expect that to get new catabolic activities it would be necessary to get mutations in regulatory genes as well as the structural genes for the enzymes concerned.

If there is sufficient ambiguity with respect to substrate specificity a regulatory mutation alone may allow growth on the novel compound. It has been suggested that regulatory mutations may be an essential first step in the evolution of new enzyme activities (Lin, Hacking & Aguilar, 1976). In the case of ribitol dehydrogenase the enzyme has low activity for xylitol but is not induced by it so that the wild type cannot grow on xylitol. A constitutive mutation is one that results in enzyme synthesis in the absence of the normal inducer and a constitutive mutant synthesizes that enzyme even when it is not needed for growth. Strains that are constitutive for ribitol dehydrogenase are able to grow slowly with xylitol as carbon source since the constraints of inducer specificity no longer apply (Wu, Lin & Tanaka, 1968). Ribitol dehydrogenase is rate-limiting for growth of the constitutive strain because the affinity for xylitol is very low. This constitutive strain was the starting point for the experiments on experimental evolution of ribitol dehydrogenase described previously. The mutants that were isolated in response to selection for faster growth on xylitol were, as we saw earlier, either carrying gene duplications or producing ribitol dehydrogenases with higher affinity for xylitol (Hartley,

Altosaar, Dothie & Neuberger, 1976). All of these had retained the constitutive mutation and synthesised their ribitol dehydrogenases in an unregulated manner.

In other systems, in which an enzyme has some activity on a substrate that is not a normal inducer, the regulatory mutation that allows growth to occur has been found to alter the specificity of induction. D-Arabinose can be acted on by the enzyme L-fucose isomerase induced by L-fucose in both *E. coli* and in *K. aerogenes*. Some of the mutants selected for growth on D-arabinose were found to be constitutive for L-fucose isomerase but others had acquired mutations that enabled the enzyme to be induced by both L-fucose and D-arabinose, thus extending the range of possible inducers (Leblanc & Mortlock, 1971; St Martin & Mortlock, 1977). This is a more satisfactory solution from the point of view of cell economy since it avoids unnecessary protein synthesis.

The Ebg enzyme in wild-type *E. coli* K12 (unlike the *lacZ* β-galactosidase) is induced by its substrate, lactose. There are two reasons why the *E. coli* strains carrying the *lacZ* deletion are unable to use the *ebg* β-galactoside for growth. First, the affinity of the enzyme for lactose is very low and secondly the amount of enzyme induced by lactose is also very low. All the lactose-utilizing mutants carry mutations in the *ebgR* regulator gene as well as in the structural gene for the enzyme, *ebgA*. Some of the regulatory mutations result in constitutive enzyme synthesis but others confer increased inducibility. Thus, in this system a double mutational event may

Fig. 12.5. Evolution of an amidase in *Pseudomonas aeruginosa* inducible by its novel substrate, butyramide. (Turberville & Clarke, 1981.)

Wild type		A-Inducible	Amidase A
	Mutation		
C11	↓	Constitutive	Amidase A
	Mutation		
B6	↓	Constitutive	Amidase B
	Mutation		
PhB3	↓	Constitutive	Amidase PhB3
	Recombination		
IB10	↓	A-Inducible	Amidase B
	Mutation		
BB1		B-Inducible	Amidase B

give strains that can be induced to produce adequate amounts of an enzyme with improved activity for lactose. This is an example of the acquisition of a new catabolic function together with its regulatory control. There remained one other requirement for lactose utilization and that was a system for lactose uptake. The normal *lac* operon includes genes *lacZ* for β-galactosidase and *lacY* for the β-galactoside uptake system. The natural inducer for the *lac* operon is allolactose which is produced in vivo from lactose by the action of β-galactosidase always present at low levels in strains carrying the *lacZ* gene. During the experiments on the evolution of the novel β-galactosidases it was necessary to find ways of inducing the *lacY* β-galactoside uptake system. Recently Hall (1982) has obtained a strain with a novel Ebg enzyme (a Class IV mutant) that can convert lactose to allolactose so that in lactose medium both the *ebgA* β-galactosidase and the *lacY* permease can be induced.

The amidase of *P. aeruginosa* is an active enzyme that is induced to high levels by its substrate amides. Constitutive strains can be readily isolated and all the experiments to obtain altered enzyme mutants were carried out with constitutive strains to avoid the problems of regulatory controls. It was of interest to see whether altered enzyme activity could be combined with altered inducibility. So far most of the genetic changes that I have described for the experiments in enzyme evolution have been related to mutation and selection. There seemed no reason why recombination should not be allowed to play a part and this turned out to be a useful step for amidase evolution. Strain PhB3 is constitutive and carries two mutations in the structural gene, one of which is the original B6 mutation. Recombination with an amidase-negative mutant produced a strain carrying a wild-type regulator gene and a structural gene coding for the butyramide-hydrolysing enzyme. This strain could not grow on butyramide since the enzyme was not induced but although most of the butyramide-utilizing mutants isolated from it were constitutive for amidase synthesis there was one which was butyramide-inducible. From the wild-type strain, inducible by acetamide and producing amidase A, we have obtained by mutation, recombination and further mutation a strain that is inducible by butyramide and produces amidase B (Fig. 12.5).

Gene capture

About ten years ago it began to be reported that the genes for certain catabolic enzymes were located not on the bacterial chromosome but on transmissible plasmids (Rheinwald, Chakrabarty & Gunsalus, 1973). Some of these extrachromosomal genetic elements have a wide host range and are capable of being transmitted to many different bacterial hosts. This lateral transmission of genetic information offers further possibilities for experimental evolution. It can be asked whether a catabolic gene can be captured from the chromosome of one species and carried by a transmissible plasmid to another bacterial species. It was known that in *E. coli* K12 it was possible to obtain F′ plasmids with a segment of the chromosome inserted into the *E. coli* sex factor, F. The F′ plasmids can be transferred to other *E. coli* K12 strains. Some drug resistance plasmids have restricted host range but others, including those of the Inc-P group, can be maintained in *E. coli*, *P. aeruginosa* and many other bacterial species. Several authors have reported that certain of these drug resistance plasmids can pick up chromosomal genes to form R′ plasmids carrying both drug resistances and a number of chromosomal genes. Insertion of chromosomal genes into a plasmid is a relatively rare event and to detect its occurrence it is essential to have a good selection system. It is also important to minimize the chances of recombination with homologous regions of the bacterial chromosome.

Holloway and colleagues have used the drug resistance plasmid R68.45 as a sex factor for *P. aeruginosa* and have also used it to obtain R′ plasmids carrying segments of the chromosome of strain PAO. Most of these R′ plasmids have carried genes for biosynthetic enzymes that were able to complement auxotrophic markers (Holloway, 1978; Morgan, 1982). The amidase genes of *P. aeruginosa* are located near two genes for arginine biosynthesis. A strain carrying functional amidase genes and wild-type genes for arginine biosynthesis can be used as a donor, with a *P. putida argF* auxotroph as a recipient and selection made for arginine independence. The *P. aeruginosa* genes do not normally integrate into the chromosome of *P. putida* and the recombinants of such a cross are found to carry the *P. aeruginosa argF* gene on a R′ plasmid. Among the recombinants will be some which also carry amidase genes on the plasmid. *P. putida* does not produce an aliphatic amidase so that the presence of the amidase genes can be detected both by growth phenotype and by enzyme assay. Here we have an example of the acquisition of new catabolic activity by the capture of genes from the chromosome of a different species. There is no need to restrict such selection to donor strains carrying wild-type amidase genes and if mutant amidase genes are used the recombinant will acquire the corresponding mutant growth phenotype. We can see here the way in which the evolution of new catabolic activities can evolve in one species and the new character be transferred by the agency of transmissible plasmids to another species. Further, R′ plasmids carrying amidase genes can also be transferred to *E. coli*, another species lacking amidase activity. The *P. aeruginosa* genes are less well expressed in *E. coli* than in their original host but selection can be made for increased gene expression thus continuing the evolutionary process in a species belonging to a different genus (Clarke, unpublished).

The transposable genetic elements discussed by Jim Shapiro (this volume) provide another means of gene capture by translocating DNA sequences to other positions on chromosomes and plasmids. Such translocations may have important consequences for assembling new catabolic pathways.

Bacteria as experimentalists

The cautious biochemist prefers to deal with the evolution of one enzyme at a time but the bacteria are much more enterprising than that. The chemical industry provides a vast array of new organic compounds to be used as herbicides, insecticides, detergents and drugs. It is less surprising that some of these persist for long periods in the soil, and in biological material, than that so many are rapidly broken down and may be used as sole growth substrates. In many cases it is far from clear how the catabolic pathways for these man-made chemicals have evolved. There are indications from the work of Hans Knackmuss with strains growing on chlorinated benzoates that there may have been duplication and mutational divergence of whole blocks of structural and regulatory genes for pathways for breaking down benzoate and related aromatic compounds (Dorn & Knackmuss, 1978; Schmidt & Knackmuss, 1980). In another instance gene capture, together with inactivation of certain genes, resulted in a new catabolic pathway being assembled from elements of two independent pathways (Reinecke & Knackmuss, 1980). Aromatic compounds are widely distributed in nature as a result of the synthetic activities of plants. There is a well-known aphorism which states that 'if an organic compound can be made biologically then some bacteria will evolve enzymes for breaking it down'. With the evolution of strains able to degrade the synthetic products of the chemical industry we see the extension of this principle to many man-made organic compounds.

If Charles Darwin had been with us now he could have been asked whether it is possible to observe evolution in action. I suggest that he might have replied in the manner of Christopher Wren: 'Si evolutionis gestionem requiris – Circumspice'.

Further reading

Clarke, P. H. (1978). Experiments in microbial evolution. In *The Bacteria*, vol. 6, ed. L. N. Ornston & J. R. Sokatch, pp. 137–218. New York: Academic Press.
Hegeman, G. D. & Rosenberg, S. L. (1970). The evolution of bacterial enzyme systems. *Annual Review of Microbiology*, **24**, 429–62.
Hütter, R. & Leisinger, T. (eds) (1981). *Microbial Degradation of Xenobiotics and Related Compounds*. London: Academic Press.

References

Betz, J. L. & Clarke, P. H. (1972). Selective evolution of phenylacetamide-utilizing strains of *Pseudomonas aeruginosa. Journal of General Microbiology*, **73**, 161–74.
Brock, T. D. (1978). *Thermophilic Organisms and Life at High Temperatures*. New York: Springer-Verlag.
Brown, P. R. & Clarke, P. H. (1972). Amino acid substitution in an amidase produced by an acetanilide-utilizing mutant of *Pseudomonas aeruginosa. Journal of General Microbiology*, **70**, 287–98.
Campbell, J. H., Lengyel, J. A. & Langridge, J. (1973). Evolution of a second gene for β-galactosidase in *Escherichia coli. Proceedings of the National Academy of Sciences, USA*, **70**, 1841–5.
Clarke, P. H. (1974). The evolution of enzymes for the utilization of novel substrates. In *Evolution in the Microbial World. 24th Symposium of the Society for General Microbiology*, ed. M. J. Carlile & J. J. Skehel, pp. 183–217. Cambridge: Cambridge University Press.
Clarke, P. H. (1978). Experiments in microbial evolution. In *The Bacteria*, vol. 6, ed. L. N. Ornston & J. R. Sokatch, pp. 137–218. New York: Academic Press.
Cowan, S. T. (1962). The microbial species – a macromyth? In *Microbial Classification, 12th Symposium of the Society for General Microbiology*, ed. G. C. Ainsworth & P. H. A. Sneath, pp. 433–55. Cambridge: Cambridge University Press.
Darwin, C. (1859). *On the Origin of Species*. London: John Murray.
Dayhoff, M. O. (1972). *Atlas of Protein Sequence and Structure*, vol. 4, Washington: National Biomedical Research Foundation.
Den Dooren de Jong, L. E. (1926). *Bijdrage tot de kennis van het mineralisatie-proces*. Rotterdam: Nigh and Van Ditmar.
Dorn, E. & Knackmuss, H-J. (1978). Chemical structures and biodegradability of halogenated aromatic compounds. Two catechol 1,2-dioxygenases from a 3-chlorobenzoate-grown pseudomonad. *Biochemical Journal*, **174**, 73–84.
Fox, G. E., Stackebrandt, E., Hespell, R. B., Gibson, J., Maniloff, J. Dyer, T. A., Wolfe, R. S., Balch, W. E., Tanner, R. S., Magrum, L. J., Zablen, L. B., Blakemore, R., Gupta, R., Bonen, L., Lewis, B. J., Stahl, D. A., Luehrsen, K. R., Chen, K. N. & Woese, C. R. (1980). The phylogeny of prokaryotes. *Science*, **209**, 457–63.
Freeman, R. B. (1982). *Darwin and Gower Street. Exhibition 19 April 1982 Catalogue item 14*, London: University College, London.
Hall, B. G. (1977). Number of mutations required to evolve a new lactase function in *Escherichia coli. Journal of Bacteriology*, **129**, 540–3.

Hall, B. G. (1981). Changes in the substrate specificities of an enzyme during directed evolution of new functions. *Biochemistry*, **20**, 4042–9.

Hall, B. G. (1982). Transgalactosylation activity of ebg β-galactosidase synthesizes allolactose from lactose. *Journal of Bacteriology*, **150**, 132–40.

Hartley, B. S. (1974). Enzyme families. In *Evolution in the Microbial World. 24th Symposium of the Society for General Microbiology*, ed. M. J. Carlile & J. J. Skehel, pp. 151–82. Cambridge: Cambridge University Press.

Hartley, B. S., Altosaar, I., Dothie, J. M. & Neuberger, M. S. (1976). Experimental evolution of xylitol dehydrogenase. In *Structure-function relationships of proteins*. ed. R. Markham & R. W. Horne. Amsterdam: North Holland.

Holloway, B. W. (1978). Isolation and characterization of an R′ plasmid in *Pseudomonas aeruginosa*. *Journal of Bacteriology*, **133**, 1078–82.

Horikoshi, K. & Akiba, T. (1982). *Alkalophilic Microorganisms*. Tokyo: Japan Scientific Societies Press.

Inderlied, C. B. & Mortlock, R. P. (1977). Growth of *Klebsiella aerogenes* on xylitol; implication for bacterial evolution. *Journal of Molecular Evolution*, **9**, 181–90.

Koch, A. L. (1972). The importance of untranslatable intermediates. *Genetics*, **72**, 297–316.

Kogut, M. (1980). Are there strategies of microbial adaptation to extreme environments? *Trends in Biochemical Sciences*, **5**, 15–8.

Kushner, D. J. (ed.) (1978). *Microbial Life in Extreme Environments*. New York: Academic Press.

Leblanc, D. J. & Mortlock, R. P. (1971). Metabolism of D-arabinose: a new pathway in *Escherichia coli*. *Journal of Bacteriology*, **106**, 90–6.

Leeuwenhoek, A. van (1677–8). Observations communicated to the publisher in a Dutch letter of 9th of October 1676. Concerning little animals by him observed in rain, well, sea and snow water; as also in water wherein pepper had lain infused. *Philosophical Transactions of the Royal Society of London*, **12**, 821–31.

Lin, E. C. C., Hacking, A. J. & Aguilar, J. (1976). Experimental models of acquisitive evolution. *Bioscience*, **26**, 548–55.

Morgan, A. D. (1982). Isolation and characterization of *Pseudomonas aeruginosa* R′ plasmids constructed by means of interspecific mating. *Journal of Bacteriology*, **149**, 654–61.

Novick, A. & Horiuchi, T. (1961). Hyperproduction of β-galactosidase by *Escherichia coli*. *Cold Spring Harbor Symposium on Quantitative Biology*, **26**, 239–45.

Paterson, A. & Clarke, P. H. (1979). Molecular basis of altered enzyme specificities in a family of mutant amidases from *Pseudomonas aeruginosa*. *Journal of General Microbiology*, **114**, 75–85.

Reinecke, W. & Knackmuss, H-J. (1980). Hybrid pathway for chlorobenzoate metabolism in *Pseudomonas* sp. B13, derivatives. *Journal of Bacteriology*, **142**, 467–73.

Rheinwald, J. G., Chakrabarty, A. M. & Gunsalus, I. C. (1973). A transmissible plasmid controlling camphor oxidation in *Pseudomonas putida*. *Proceedings of the National Academy of Sciences, USA*, **70**, 885–9.

Roberts, T. A., Hobbs, G., Christian, J. H. B. & Skovgaard, N. (eds.) (1981). *Psychrotrophic Microorganisms in Spoilage and Pathogenicity*. London: Academic Press.

Schmidt, E. & Knackmuss, H-J. (1980). Chemical structure and biodegradability of halogenated aromatic compounds. Conversion of chlorinated muconic acids into maleoylacetic acid. *Biochemical Journal*, **192**, 339–47.

Sneath, P. H. A. (1962). The construction of taxonomic groups. In *Microbial Classification, 12th Symposium of the Society for General Microbiology*, ed. G. C. Ainsworth & P. H. A. Sneath, pp. 289–332. Cambridge: Cambridge University Press.

St Martin, E. J. & Mortlock, R. P. (1977). A comparison of alternate metabolic strategies for the utilization of D-arabinose. *Journal of Molecular Evolution*, **10**, 111–22.

Stanier, R. Y., Palleroni, N. J. & Doudoroff, M. (1966). The aerobic pseudomonads: a taxonomic study. *Journal of General Microbiology*, **43**, 159–271.

Stephenson, M. (1930). *Bacterial Metabolism*. 1st Edition, Monographs on Biochemistry. London: Longmans, Green & Co. Ltd.

Stephenson, M. (1949). *Bacterial Metabolism*. 3rd Edition. London: Longmans, Green & Co. Ltd.

Turberville, C., Clarke, P. H. (1981). A mutant of *Pseudomonas aeruginosa* PAC with an altered amidase inducible by the novel substrate. *FEMS Microbiology Letters*, **10**, 87–90.

Wu, T. T., Lin, E. C. C. & Tanaka, S. (1968). Mutants of *Aerobacter aerogenes* capable of utilizing xylitol as a novel carbon source. *Journal of Bacteriology*, **96**, 477–56.

13

Variation as a genetic engineering process

JAMES A. SHAPIRO

> *Perdita*: For I have heard it said
> There is an art which in their piedness shares
> With great creating nature.
> *Polixenes*: Say there be;
> Yet nature is made better by no mean
> But nature makes that mean: so over that art,
> Which you say adds to nature, is an art
> That nature makes.
>
> *The Winter's Tale*

One of the main insights we are celebrating here is the realization that hereditary variation within natural populations is the source of evolutionary change. The term 'variation' has multiple meanings, generally referring to the many processes that create new genomic configurations in cell lineages and thereby alter the characteristics of organisms. These processes include the reassortment of Mendelian factors, the generation of recombinant chromosomes, the creation of new allelic forms at specific loci, changes in chromosome structure and number, alterations in the structure and distribution of repeated elements, and the introduction of new hereditary determinants. Each of these different kinds of genomic change relies on a complex series of cellular events and can often be carried out by more than one alternative biochemical pathway. Given their importance in the formation of organic diversity, the richness of variational mechanisms should not be surprising. Certainly it would be an error to pretend to understand the process of evolution without accounting for the factors which are responsible for the nature and timing of variational events.

From the perspective of a bacterial geneticist, I view variational events as those cellular differentiations which can be transmitted through sufficient cell divisions to form a distinct characterizable cell population. (In my experiments this population is generally a colony on a petri dish or a liquid culture derived from a single colony.) No change is final because new variations can always arise, and the stability of various cell types can cover a wide range depending on genomic and environmental conditions. A major effort in bacterial genetics has been directed to identifying and characterizing the agents that mediate these cellular differentiations, and we currently possess a great deal of information about the ones that alter DNA organization. Although most evolutionary theories concentrate on changes *within* genomes, it is important to note that a major component of bacterial DNA variation results from infectious processes – the acquisition of plasmids, prophages and genomic fragments by adsorption of bacterio-phage particles and cell-to-cell transfer (conjugation) – in addition to purely intracellular changes (Jacob & Wollman, 1961; Hayes, 1968). These infectious processes have their parallels in higher organisms which have viral, cellular and multicellular parasites and which generally utilize cell fusion at some stages in their reproductive cycles.

Rather than the cellular viewpoint I have outlined above, we have learned from molecular biology to think of variation in chemical terms – specifically as the formation of new glycosidic and phosphodiester linkages in DNA molecules. It is certainly the case that a variant cell is chemically distinct from its ancestors, and the primary difference can often be identified as a change in nucleotide sequence organization. Nonetheless, a detailed investigation of the variational process reveals the coordinated action of biochemical systems whose specificities and regulation are beyond simple chemical explanations. To illustrate my meaning, I will discuss three different experimental situations that are clearly relevant to the formulation of evolutionary theories: ultraviolet radiation-induced mutation in *E. coli*, spontaneous mutation in *E. coli*, and spontaneous mutation in *Drosophila melanogaster*. These examples are by no means exceptional, and similar conclusions could be reached by consideration of other well-studied systems, such as spontaneous and radiation-induced mutation in yeast or spontaneous mutation in corn.

Ultraviolet mutagenesis in *E. coli*: the SOS system

When a population of *Escherichia coli* cells isolated from nature is subjected to ultraviolet irradiation, there are multiple consequences. These include cell death (i.e. loss of ability to form progeny cells), temporary inhibition of cell division, derepression of proviral genomes leading to bacteriophage formation, and the appearance of altered cell types (mutants) among the survivors. The action spectrum of u.v. light for these various effects shows a peak at 257 nm, indicating that photochemical damage to DNA is the primary effect of irradiation. When mutant cells are examined, they have altered DNA nucleotide sequences in the genomic regions that determine the altered characteristics (generally single or double base-pair substitutions or frameshifts). These facts suggest an obvious hypothesis to explain u.v. mutagenesis – namely, that damaged DNA is a poor template for replication so that errors accumulate upon the duplication or repair of u.v. irradiated genomes. Over two decades of intensive study have provided a thorough test of this hypothesis. The results (summarized chiefly by Radman, 1974, and Witkin, 1976) reveal a different, more complex and more interesting picture.

(1) Post-irradiation treatment of the bacterial cells affects their response. Exposure to near-u.v. visible light activates an enzyme that cleaves pyrimidine dimers, the major photoproduct, and so reverses u.v. damage. Incubation in the dark under conditions where there is no increase in cell mass enhances both survival and mutagenesis. Inhibition of protein synthesis in this 'dark repair' period reduces survival and mutagenesis, indicating the existence of inducible systems involved in both repair of lethal damage and the induction of mutations. The ability of cells to re-pair lethal damage and create mutations in irradiated phage genomes is likewise induced by exposure to u.v. light.

(2) Although irradiated DNA is more prone to u.v.-induced mutagenesis than non-irradiated DNA, the latter is also indirectly subject to the mutagenic effect of u.v. light. This can be observed in two experimental situations: (i) where non-irradiated DNA (such as a plasmid or phage genome) is introduced into an irradiated cell, or (ii) where irradiated DNA is transferred to a non-irradiated cell and leads to increased mutagenesis of the cellular genome, as well as to other consequences that mimic the effect of radiation (this phenomenon is called 'indirect induction', Devoret & George, 1967; Monk, 1967). Thus, direct photochemical damage to a DNA sequence is not an essential component of u.v.-induced mutagenesis of that sequence. (In yeast, there is a particularly interesting parallel phenomenon where the presence of one irradiated nucleus in a heterokaryon induces variation in the other non-irradiated nucleus (Fabre & Roman, 1977).) There are also treatments other than irradiation, such as thymine starvation or addition of nalidixic acid, which induce a mutagenic response and the other sequelae of u.v. treatment.

(3) There are *E. coli* strains with specific hereditary differences that affect one or more aspects of the response to u.v. irradiation. Analysis of these strains has made it possible to identify a remarkable 'SOS system' of coordinated response to radiation damage and other treatments that inhibit DNA replication. We know that this inducible system is responsible for the mutagenic effect of u.v. light because lesions in specific SOS loci either prevent u.v. mutagenesis (without necessarily blocking other responses to radiation) or, conversely, facilitate SOS mutagenesis in the absence of irradiation. The *recA* and *lexA* loci exert overall control of the SOS system: *lexA* encodes a repressor that blocks expression of SOS functions (e.g. inhibitors of cell division and repair enzymes), and *recA* encodes an ATP-dependent protease that is activated by gapped DNA to cleave the *lexA* repressor (as well as prophage repressors; Roberts, Roberts & Craig, 1978; Little *et al.*, 1980). Cells lacking the *recA* protease cannot derepress SOS functions and show no increase in mutations among survivors following irradiation, while *tif* cells with an altered *recA* protease that is activated at high temperature in the absence of gapped DNA show the full SOS response (including mutagenesis) after a simple temperature shift. Strains with a non-cleavable *lexA* repressor are similarly not u.v. mutable. At least three SOS functions are specifically involved in the variational process. The *recA* protein catalyses the reannealing of complementary DNA strands and is required for homologous recombination of chromosomal DNA (Weinstock, McEntee & Lehman, 1979). The *umuC* product is required for mutagenesis but not for filamentation or for the bulk of the repair process (Kato & Shinoura, 1977). The *himA* locus encodes a protein that is required for site-specific recombination in bacteriophage lambda insertion and excision and for expression of essential functions by the transposable mutator phage Mu (Miller, Kirk & Echols, 1981). Functions analogous to the *umuC* product are encoded by several plasmids (e.g. Mortelmans & Stocker, 1976; Walker & Dobson, 1979), and it is interesting to note that the Ames test for mutagens/carcinogens depends mainly on detection of SOS induction rather than of direct damage to the genetic material (McCann *et al.*, 1975).

(4) Just as certain *E. coli* variants are no longer mutable by u.v. irradiation, other species of bacteria have repair systems that do not show mutagenesis.

These observations show that mutagenesis following u.v. irradiation results from a cellular biochemical response to DNA damage, not from the damage itself. In other words, *E. coli* cells do not leave it to the physics of photochemistry to change their hereditary information but instead have special biochemical mechanisms to do this. Naturally, it is an important question why some bacteria and plasmids encode mutator functions like *umuC* while other bacteria do not have them.

Spontaneous mutation in *E. coli*

The discovery of chemical mutagenesis and the elucidation of the structure of DNA led to the formulation of hypotheses about spontaneous hereditary changes as the result of chemical fluctuations in the constitutents of DNA molecules, such as keto–enol tautomerism in the purine and pyrimidine bases during replication (Watson & Crick, 1953). As with u.v. mutagenesis, detailed investigation of spontaneous mutation has revealed instead multiple unanticipated mechanisms for reorganizing DNA molecules and controlling the nature of the changes that occur.

One experimental approach to spontaneous variation in bacteria is to look at the frequency with which a specific genetic change takes place, often the 'reversion' of a particular lesion rendering a strain unable to form colonies on certain media. In each case, this approach examines a limited range of genetic events, such as transition or transversion base substitutions or frameshifts, and parallel assays can define the relative frequencies of different mutational changes at the various tester loci. These assays can also be used to isolate variant strains where the level of one or more mutational events has been altered. These are the so-called 'mutator' strains reviewed by Cox (1976). The results of this kind of study have shown that many loci control the rate of appearance of hereditary changes as well as the relative proportions of different types of changes that occur in the specific tester systems used (Table 13.1). The integrated effects of different alleles of these loci over the entire bacterial genome are not known, but some strains (such as *mutD*) do serve as efficient mutator hosts for a wide variety of target sequences. In some cases the specific biochemical alteration is known, and there are reasonable hypotheses to explain the change in mutability. Examples include mutant DNA polymerases (Hall & Brammar, 1973; Konrad, 1978), strains defective in the removal of uracil residues formed by spontaneous deamination of cytosine (Duncan, Rockstroh & Warner, 1978) and cells with altered methylation activities (increased *or* decreased) that apparently cannot distinguish between the parental and newly synthesized DNA strands to carry out efficient post-replication mismatch correction (Glickman & Radman, 1980). Not all mutator phenotypes result from loss of proofreading functions. Several alleles of *mutD* are dominant and are stimulated in their mutagenic activity

by alterations of the growth medium, such as the addition of thymidine
(Cox, 1976; Cox & Horner, 1982). From these observations we see that
the pattern of base substitution and frameshift mutability is a function of
the biochemical state of the bacterial cell rather than a consequence of the
chemical properties of DNA. Indeed, we can understand mechanisms like
mismatch repair and uracil excision as a protection against the vagaries
of random chemical processes.

A different experimental approach to spontaneous variation is to isolate
large numbers of spontaneous changes in a particular target, such as an
operon or cistron where loss of function can readily be detected. Sixteen

Table 13.1. *Some* E. coli *loci involved in controlling spontaneous mutation*
(*Bachmann & Low, 1980*)

Locus	Function or Mnemonic	Mutational consequences of novel alleles
dam	DNA adenine methylation	$\sim$ 60-fold increased mutagenesis (lack of post-replication mismatch repair)
dcm	DNA cytosine methylation	(In *E. coli* B) Disappearance of hotspots where methyl cytosine deaminates to thymine rather than to uracil
dnaE (= *polC*)	DNA biosynthesis (encodes DNA polymerase III)	Increased mutation in some *ts* mutants
dnaQ	DNA biosynthesis	Increased mutation (= *mutD?*)
himA	host integration mutant; affects site-specific recombination	Reduced excision of IS insertions
mutD	mutator	> 10^4-fold increased rate of transitions, transversions, frameshifts; activated by thymidine
mutH	mutator	> 10^3-fold increased rate of transitions, transversions, frameshifts
mutL	mutator	> 10^3-fold increased rate of AT $\rightleftharpoons$ GC transitions
mutS	mutator	$\sim$ 70-fold increased rate of AT $\rightleftharpoons$ GC transitions
mutT	mutator	> 10^3-fold increase of AT $\rightarrow$ GC transitions only
ung	uracil-DNA-glycosidase	Increased GC $\rightarrow$ AT transitions; eliminates hot-spot at methylated cytosine residues
uvrD (= *mutU*)	u.v. repair	Increased mutation

years ago I followed a suggestion by Sydney Brenner and Roger Freedman
and isolated a collection of spontaneous mutants lacking galactokinase
activity (Shapiro, 1967; Adhya & Shapiro, 1969). Of the 77 strains
examined in detail, 17 had pleiotropic defects and lacked other galactose
metabolism enzymes. This high proportion of pleiotropic mutants was very
different from the distribution of u.v.-induced *gal* mutations, where only
about 1% had significant effects on more than one enzyme. This indicated
a fundamental difference in the spontaneous and u.v. (i.e. SOS) mutagenic
processes. Further examination of the spontaneous pleiotropic lesions in
the *gal* operon showed many of them to result from the insertion of specific
DNA segments (Table 13.2; Shapiro, 1969; Fiandt, Szybalski & Malamy,
1972) which we now call insertion sequences or IS elements (Starlinger
& Saedler, 1972; Bukhari, Shapiro & Adhya, 1977).

Study of IS elements has revealed three important features critical to
their role in genomic reorganization.

(1) The same IS element can insert in (or transpose to) many different
loci, and there are a limited number of IS elements that appear repeatedly
at different genomic locations (Szybalski, 1977; Calos & Miller, 1980;
Kleckner, 1981; Iida, Meyer & Arber, 1983). In all systems studied, a
significant proportion and often more than half of the spontaneous
variants not limited to particular sequence changes result from IS insertions.
Thus, IS elements are specific important components of the mechanisms
for spontaneous variation in *E. coli*.

(2) Each IS element has a complex internal organization, including
polypeptide coding sequences, transcription signals and recombination
signals. The presence of an IS element in a locus affects its expression in
characteristic ways, frequently altering transcription patterns (Kleckner,
1981; Iida *et al.*, 1983).

(3) IS elements are active agents of DNA reorganization and dramatically
alter the genealogical possibilities of adjacent sequences. That is, the

Table 13.2. *Spontaneous mutants of* E. coli *lacking galactokinase activity*

17/77 pleiotropic: 11 kinase⁻ transferase⁻, 6 kinase⁻ transferase⁻epimerase⁻
2 deletions
2 *galT*::IS1
9 *galT*::IS4
1 *galE*::IS (unclassified insertion)
3 not fully characterized

insertion of an IS element can make a locus more likely to alter in predictable ways. Taking the *gal* operon as an example, I will cite three illustrations of this point. First, Reif & Saedler (1975) showed that the presence of an IS1 element in the promoter region increased the probability of deletion per cell division of adjacent coding sequences by over three orders of magnitude at 30 °C. Second, the presence of an IS2 element in the promoter region permits the appearance of strains expressing adjacent coding sequences constitutively as a result of rearrangements in the IS2 sequences (Ghosal, Gross & Saedler, 1979; Besemer, Görtz & Charlier, 1980; Ahmed, Bidwell & Musso, 1980). Third, the presence of IS1 elements on either side of the *galE* cistron allows it to transpose into plasmids and be transferred extrachromosomally to other cells (unpublished observations). This last result was predictable from the observation that many different segments of DNA flanked by IS1 repeats are transposable elements similar to IS1 itself (MacHattie & Jackowski, 1977; Kleckner, 1981; Iida *et al.*, 1983).

Several questions arise about the role of IS elements as major agents of spontaneous variation in *E. coli*. Is *E. coli* (or, more generally, prokaryotic heredity) exceptional in this regard? The answer is clearly 'No', for studies of spontaneous variants with altered coding sequence regulation in yeast (Roeder & Fink, 1983), morphological, homeotic and lethal mutants of *Drosophila* (Rubin, 1983; W. Bender, personal communication), and corn plants with novel pigmentation patterns (McClintock, 1951, 1965; Peterson, 1960; Neuffer, 1966) all show them to result chiefly from the action of transposable elements.

A second question concerns the biochemistry of IS element activity. Is it just a by-product of other fundamental cellular processes or are there specific mechanisms that catalyze transposition and other changes? Here the answer is more complex but not in agreement with the notion of transposable elements as passive components of the genome. Much of the genetic activity of IS elements is understandable as the consequence of a particular kind of recombination event that involves DNA replication (Shapiro, 1979; Cohen & Shapiro, 1980). This means that transposition and related rearrangements involve many biochemical reactions (for example, the synthesis of hundreds or thousands of phosphodiester bonds) and almost certainly utilize much of the normal cellular replication apparatus to carry them out. However, alteration of the internal structure of IS and other transposable elements renders them unable to carry out recombination events or to do so in an abnormal way. These kinds of experiments show that transposable elements encode transcripts and polypeptides and contain internal signals which are either essential for replicative recombination or regulate its timing (Kleckner, 1981, 1983; Machida *et al.*, 1982; W. Reznikoff, personal communication).

A third question concerns the specificity of transposable element-mediated variation. How predictable (i.e. nonrandom) is it with respect to timing and the nature of the DNA sequences involved in reorganization? Due to the scope of the problem (even in a simple prokaryotic genome), the answer is only partial but positive where appropriate data exist. (Here I am discussing only transposable elements, but there are other mobile elements in bacterial cells which are important agents of variation. These include temperate bacteriophages which display very elaborate regulation of timing of recombinational activity and show well-defined sequence specificities (Campbell, 1983).) Some elements regulate the transcription of functions needed for replicative recombination. Derepression of these functions by mutation, intercellular transfer or environmental manipulations increases the rate of recombination by orders of magnitude for a discrete period until repression is reestablished (see Bukhari *et al.*, 1977, Calos & Miller, 1980, and Kleckner, 1981, for summaries of the relevant data). As far as sequence specificity is concerned, there is a hierarchy of determinants. For all transposable elements, the terminal sequences are extremely specific for replicative recombination; this means that one breakpoint in each rearrangement is fixed and also that the position of the elements (hence of their termini) is an important determinant of what rearrangements can occur. The terminal sequences of transposable elements can show various specificities for the 'target' sequences elsewhere in the genome to which they become attached (in other words, for the second breakpoint in each rearrangement). A detailed review of the relevant experimental observations is not appropriate here, but I will list some of the features which have been shown to influence recombination of specific genomic regions with different transposable elements: richness in AT basepairs, presence of sequences that show homology with element termini, presence of short oligonucleotides that match closely to a preferred symmetrical substrate, and the presence of other transposable elements in the target replicon. This last feature is particularly intriguing because of the way one element can influence a series of events. Transposable elements can have both stimulatory and inhibitory effects on each other, in some cases serving as hot-spots for recombination with another element and in other cases blocking recombination over regions of tens of thousands of basepairs (e.g. Robinson, Bennet & Richmond, 1977; Grinsted *et al.*, 1978).

We are still far from a comprehensive understanding of the elaborate regulatory interactions which govern when and where transposable elements will act to reorganize a prokaryotic genome. But these interactions do exist and must be considered in theories of prokaryotic evolution. There are many observations which show that the rate of particular rearrangements in a bacterial population is a function of genetic constitution, physiological state and genetic history (i.e. when various elements entered the cell in the population). It is, therefore, unrealistic to postulate any inherent background rate of spontaneous variation.

Spontaneous mutation in *Drosophila melanogaster* – hybrid dysgenesis

Much of the importance of *D. melanogaster* as an experimental organism for working out details of transmission genetics lay in the stability of its genome. Chromosome rearrangements and new allelic forms of specific loci reportedly arose rarely in the laboratory without mutagenic treatment, thus permitting the construction of reliable cytogenetic maps. This reliability in the laboratory gave rise to the idea that spontaneous variation was infrequent and sporadic. (We might profitably reflect that the need for genomic stability by most cytogeneticists influenced their observations and reports.) It was therefore a surprise to find that chromosomes from wild *D. melanogaster* populations very often caused a high level of genetic instability when introduced into laboratory cultures. The instabilities included visible and recessive lethal mutations, chromosome rearrangements, and recombination in the male germ line (frequently premeiotic and non-reciprocal). Generally, a majority of the chromosomes from wild flies carried 'mutator' determinants, and genetic instability was often accompanied by sterility of the hybrid female progeny from crosses between wild and laboratory flies. Most of these initially confusing observations are now understood to reflect a set of phenomena called 'hybrid dysgenesis' (Kidwell, Kidwell & Sved, 1977; Engels, 1980; Bregliano *et al.*, 1980; Bregliano & Kidwell, 1983). Because hybrid dysgenesis results from the presence in natural populations of transposable elements that are agents of genomic reorganization, I think it is relevant to summarize the current state of knowledge. The experimental evidence is given in the reviews cited above.

(1) Dysgenic phenomena (spontaneous variation, distortion of chromosome transmission ratios, female sterility) occur when a sperm carrying chromosomes from wild populations fertilizes the egg of a laboratory strain.

The reciprocal cross generally shows little or no dysgenesis. There are, therefore, two components to the creation of hybrid dysgenesis: one contributed by the sperm and one by the egg.

(2) The sperm's contribution to hybrid dysgenesis consists of one or more transposable elements present in the chromosomes of wild populations. There are actually at least two independent sets of such elements: I ('inducer') factors and P ('paternal') factors. There are differences in the behaviour of I and P factors. For example, I factors are active only in the female germ line whereas P factors are active both in the male and female germ line, and the female sterility induced by these factors appears at distinct stages of the reproductive cycle. Nonetheless, the two systems are similar in most salient features and will be considered together. In the germ line of the dysgenic hybrids, I and P factors transpose to new chromosomal locations, thereby 'contaminating' chromosomes from the laboratory strain and sometimes inducing detectable mutations. Among loci determining visible phenotypes on the X chromosome, P factors appear to have a special affinity for the *singed* locus, and P factor activity in nature has been observed in the form of *singed* mutants in wild populations (Golubovsky, 1978). I and P factors also serve as hot-spots for chromosome fragmentation, rearrangements and non-homologous recombination events. Because of their transposability, I and P factors are not inherited from dysgenic hybrids in a Mendelian fashion. They can be mapped, however, because they do not transpose at high rates in the reciprocal non-dysgenic hybrids formed by crossing wild females and laboratory males. P factors have been identified with a 2900 basepair DNA repeat element, and certain *white* locus mutations that arose in dysgenic hybrids contain intact or partial P factors (Rubin, 1983).

(3) The egg's contribution to hybrid dysgenesis is a property known as 'cytotype' because it is a cellular property and not directly encoded by the chromosomes. In the wild population, the eggs have I or P cytotype and do not permit I or P factor activity. Hence dysgenic phenomena are rare in wild populations homogeneous for the presence of these elements. In the laboratory strains, the eggs have R ('reactive') or M ('maternal') cytotype which permits I or P factors to reorganize chromosomes and induce female sterility. Cytotype is inherited in a matrilineal fashion that differs from both chromosomal and cytoplasmic inheritance. The cytotype of the eggs in a female fly will generally resemble that of her mother and grandmother, *independent of the presence or absence of I or P factors in her chromosomes*. In other words, an egg can have R or M cytotype with I or P chromosomes or, conversely, I or P cytotype with R or M chromosomes.

While the egg cytotype eventually comes to correspond to the female's chromosomal constitution, the discordancy between them can be maintained for at least ten generations. Cytotype is also subject to environmental effects; ageing and temperature regimes can reduce how strongly an R or M female's eggs will respond to the introduction of I or P chromosomes.

(4) In addition to cytotype control of I and P factor activity, there are also developmental controls. No somatic variegation or instability is detectable in flies where germ-line changes occur in virtually every cell lineage.

(5) Examination of *D. melanogaster* strains collected from the wild at different periods indicates that I factors entered wild populations in the Americas early in this century and were present in all of them worldwide by the 1970s. P factors appear for the first time about 1950 (again in the Americas) and have almost completed their spread throughout the world (Bregliano & Kidwell, 1983).

There are two important conclusions from the consideration of hybrid dysgenesis which I would like to emphasize. First, variation involving chromosome rearrangements is not an individual process in this system. It requires an interaction that is both cellular (sperm and egg) and populational. Second, the rate of genomic variation can change dramatically within a single generation and (because of the matrilineal inheritance of cytotype) remain elevated for several generations until cytotype changes. At some loci, the mutation rate can increase by more than a thousand-fold as a consequence of hybrid dysgenesis (Simmons *et al.*, 1980). Thus, in a sexually reproducing metropolitan species like *D. melanogaster*, just as in *E. coli*, there can be no fundamental underlying rate of spontaneous variation independent of the history of any particular population.

Variation and genetic engineering

If we consider how present-day genetic engineers set out to manipulate genomes for specific purposes, we notice some thought-provoking parallels with the normal cellular processes of variation as we have seen them to occur in *E. coli* and *D. melanogaster* (which are by no means unique in this respect).

(1) Genetic engineers and cells use enzymatic tools to reorganize DNA molecules. They also use specific substrates for those enzymes – for example, oligonucleotide linkers in the Eppendorf tube and transposable elements in the cell. The characteristics of the enzyme–substrate systems

employed determine to a large degree the nature of the changes that can be made.

(2) Because there are multiple problems to be solved in each strain construction, the genetic engineer has to follow certain pathways. In order to synthesize a mammalian protein in bacteria, for example, it is generally necessary first to clone the coding sequence in an appropriate vector, then subclone it in another 'expression' vector, and finally optimize the coding sequence with respect to transcription and translation signals. It may also be necessary or desirable to fuse the mammalian coding sequence with a bacterial coding sequence in order to stabilize the product or have it secreted into the growth medium. These sequential processes resemble the variational genealogies determined by the positions of specific substrates for DNA reorganization activities. One good analogy is the creation of different transposons by the movement of IS1 near specific determinants and their subsequent insertion into plasmids. Another analogy is the creation of constitutive *gal* operons by IS2 transposition into the promoter region followed by internal reorganization of IS2 sequences.

(3) Genetic engineers use different enzymes at different times according to the specific changes they wish to make. Similarly, living cells control the activity of their systems for genomic reorganization. Indeed, this is essential, for without such regulation they would not have a functioning hereditary apparatus. For instance, cytotype control of I and P factors is necessary to the survival and reproduction of wild *D. melanogaster* populations. Similarly, successful genetic engineers need to stabilize their specially constructed strains to make use of them. Although the preceding discussion has dealt with the germ-line aspects of variation, it is important to remember that there are many well-documented cases of precisely controlled genomic reorganization in somatic development, such as chromosome fragmentation in the formation of the ciliate macronucleus (Nanney, 1980), chromatin diminution in *Ascaris* (Boveri, 1892) and *Cyclops* (Beermann, 1977), recombination of immunoglobulin coding determinants in lymphocyte differentiation (Tonegawa *et al.*, 1980), and activation of pigment loci in growth of corn kernels (McClintock, 1965).

What are the implications of these parallels between genetic engineering and cellular processes of variation for the formulation of theories about evolution? I think the major points are what they teach us about the specificity and timing of changes in hereditary programs, and this is directly relevant to basic questions about the direction and tempo of the evolutionary process. There is genealogical information in genomes. Not all possible hereditary changes are equally probable, and the particular genetic constitution of any cell (its capacity to synthesize reorganization activities, the genomic distribution of substrates for those activities) conditions the nature of variation that may occur to give rise to a daughter cell with new heritable characteristics. Whether or not a particular variation occurs within a given cell generation depends upon regulatory factors that control reorganization activities, often in response to outside stimuli such as radiation, viral infection, cell hybridization or intercellular transfer of genetic elements. Because some of the agents of variation are themselves hereditary determinants encoding regulatory functions, variation is subject to both positive and negative feedback networks. Thus, there can be periods of regular genomic reproduction interspersed with periods of rapid change. Hybrid dysgenesis in *D. melanogaster* illustrates this point and exemplifies how the agents of variation can initiate a process of reproductive isolation.

In many cases, the differences between species appear to have more to do with genome organization (chromosome structure, ploidy, distribution and structure of repetitive elements, sex determination) than with adaptation to different ecological niches (for examples, see White (1978) and Singer (1982)). Recombinant DNA technology and sequencing are deepening our knowledge of patterns in interspecific variability. Now that the methods for directly characterizing genomic differences and similarities have become so powerful, it may prove useful to focus on the role of variation, rather than selection, in the origin of species.*

Further Reading

McClintock, B. (1965). The control of gene action in maize. *Brookhaven Symposia in Biology*, **18**, 162–84.

Hanawalt, P. & Setlow, R. B. (eds) (1975). *Molecular Mechanisms for Repair of DNA*, part A. New York: Plenum Press.

Bukhari, A. I., Shapiro, J. A. & Adhya, S. L. (eds) (1977). *DNA Insertion Elements, Plasmids and Episomes*. Cold Spring Harbor: Cold Spring Harbor Laboratory.

Shapiro, J. A. (ed.) (1983). *Mobile Genetic Elements*. New York: Academic Press.

* Darwin himself foresaw this in one of his insertions into the sixth edition of *The Origin of Species* when he wrote of '...variations which seem to us in our ignorance to arise spontaneously. It appears that I formerly underrated the frequency and value of these latter forms of variation, as leading to permanent modifications of structure independently of natural selection.' (Chapter XV, p. 395).

References

Adhya, S. & Shapiro, J. A. (1969). The galactose operon of *E. coli* K-12. I. Structural and pleiotropic mutations of the operon. *Genetics,* **62,** 231–249.

Ahmed, A., Bidwell, K. & Musso, R. (1980). Internal rearrangements of IS2 in *Escherichia coli. Cold Spring Harbor Symposium on Quantitative Biology,* **45,** 141–51.

Bachmann, B. J. & Low, K. B. (1980). Linkage map of *Escherichia coli* K-12, edition 6. *Microbiological Reviews,* **44,** 1–56.

Beermann, S. (1977). The diminution of heterochromatic chromosomal segments in *Cyclops* (Crustacea, Copepoda). *Chromosoma* (Berl.), **60,** 297–344.

Besemer, J., Görtz, G. & Charlier, D. (1980). Deletions and DNA rearrangements within the transposable DNA element IS2. A model for the creation of palindromic DNA by DNA repair synthesis. *Nucleic Acids Research,* **8,** 5825–33.

Boveri, Th. (1982). Über die Entstehung des Gegensatzes zwischen den Geschlechtszellen und den somatischen Zellen bei *ascaris. S. B. Ges. Morph. Physiol. Münch.,* **8,** 114–25.

Bregliano, J. C. & Kidwell, M. (1983). Hybrid dysgenesis determinants. In *Mobile Genetic Elements,* ed. J. A. Shapiro, pp. 363–410. New York: Academic Press.

Bregliano, J. C., Picard, G., Bucheton, A., Pelisson, A., Lavige, J. M. & L'Heritier, P. (1980). Hybrid dysgenesis in *Drosophila melanogaster. Science,* **207,** 606–11.

Bukhari, A. I., Shapiro, J. A. & Adhya, S. L. (eds) (1977). *DNA Insertion Elements, Plasmids and Episomes.* Cold Spring Harbor: Cold Spring Harbor Laboratory.

Calos, M. P. & Miller, J. H. (1980). Transposable elements. *Cell,* **20,** 579–95.

Campbell, A. M. (1983). Bacteriophage λ. In *Mobile Genetic Elements,* ed. J. A. Shapiro, pp. 65–104. New York: Academic Press.

Cohen, S. N. & Shapiro, J. A. (1980). Transposable genetic elements. *Scientific American,* **242,** 36–45.

Cox, E. C. (1976). Bacterial mutator genes and the control of spontaneous mutation. *Annual Review of Genetics,* **10,** 135–56.

Cox, E. C. & Horner, D. L. (1982). Dominant mutators in *Escherichia coli. Genetics,* **100,** 7–18.

Devoret, R. & George, J. (1967). Induction indirecte du prophage λ par le rayonnement ultraviolet. *Mutation Research,* **4,** 713–34.

Duncan, B. K., Rockstroh, P. A. & Warner, H. R. (1978). *Escherichia coli* K-12 mutants deficient in uracil-DNA glycosylase. *Journal of Bacteriology,* **134,** 1039–45.

Engels, W. R. (1980). Hybrid dysgenesis in *Drosophila* and the stochastic loss hypothesis. *Cold Spring Harbor Symposium on Quantitative Biology,* **45,** 561–5.

Fabre, F. & Roman, H. (1977). Genetic evidence for inducibility of recombination competence in yeast. *Proceedings of the National Academy of Sciences, USA,* **74,** 1667–71.

Fiandt, M., Szybalski, W. & Malamy, M. H. (1972). Polar mutations in *lac, gal* and phage λ consist of a few IS-DNA sequences inserted with either orientation. *Molecular and General Genetics,* **119,** 223–31.

Ghosal, D., Gross, J. & Saedler, H. (1979). DNA sequence of IS2-7 and generation of mini-insertions by replication of IS2 sequences. *Cold Spring Harbor Symposium on Quantitative Biology,* **43,** 1193–6.

Glickman, B. & Radman, M. (1980). *Escherichia coli* mutator mutants deficient in methylation-instructed mismatch correction. *Proceedings of the National Academy of Sciences, USA*, **77**, 1063–7.

Golubovsky, M. D. (1978). Two types of instability of singed alleles isolated from populations of *D. melanogaster* during mutation outburst in 1973. *Drosophila Information Services*, **53**, 171.

Grinsted, J., Bennett, P. M., Higginson, S. & Richmond, M. H. (1978). Regional preference of insertion of Tn501 and Tn801 into RP1 and its derivatives. *Molecular and General Genetics*, **166**, 313–20.

Hall, R. M. & Brammar, W. J. (1973). Increased spontaneous mutation rates in mutants of *Escherichia coli* with altered DNA polymerase III. *Molecular and General Genetics*, **121**, 271–6.

Hayes, W. (1968). *The Genetics of Bacteria and Their Viruses*. 2nd edn. Oxford: Blackwell.

Iida, S., Meyer, J. & Arber, W. (1983). Prokaryotic IS elements. In *Mobile Genetic Elements*, ed. J. A. Shapiro, pp. 159–221. New York: Academic Press.

Jacob, F. & Wollman, E. (1961). *Sexuality and the Genetics of Bacteria*. London: Academic Press.

Kato, T. & Shinoura, Y. (1977). Isolation and characterization of mutants of *Escherichia coli* deficient in induction of mutations by ultraviolet light. *Molecular and General Genetics*, **156**, 121–31.

Kidwell, M. G., Kidwell, J. F. & Sved, J. A. (1977). Hybrid dysgenesis in *Drosophila melanogaster*: a syndrome of aberrant traits including mutation, sterility and male recombination. *Genetics*, **86**, 813–33.

Kleckner, N. (1981). Transposable elements in prokaryotes. *Annual Review of Genetics*, **15**, 341–404.

Kleckner, N. (1983). Transposon Tn10. In *Mobile Genetic Elements*, ed. J. A. Shapiro, pp. 261–98. New York: Academic Press.

Konrad, E. B. (1978). Isolation of an *Escherichia coli* K-12 *dnaE* mutation as a mutator. *Journal of Bacteriology*, **133**, 1197–202.

Little, J. W., Edmiston, W. H., Pacelli, L. Z. & Mount, D. (1980). Cleavage of the *Escherichia coli* lexA protein by the recA protease. *Proceedings of the National Academy of Sciences, USA*, **77**, 3225–9.

MacHattie, L. A. & Jackowski, J. B. (1977). Physical structure and deletion effects of the chloramphenicol resistance element Tn9 in phage lambda. In *DNA Insertion Elements, Plasmids and Episomes*, ed. A. I. Bukhari, J. A. Shapiro & S. L. Adhya, pp. 219–28. Cold Spring Harbor: Cold Spring Harbor Laboratory.

Machida, Y., Machida, C., Ohtsubo, H. & Ohtsubo, E. (1982). Factors determining frequency of plasmid cointegration mediated by insertion sequence IS1. *Proceedings of the National Academy of Sciences, USA*, **79**, 277–81.

McCann, J., Spingarn, N., Kobori, J. & Ames, B. (1975). Detection of carcinogens as mutagens: Bacterial tester strains with R factor plasmids. *Proceedings of the National Academy of Sciences, USA*, **72**, 979–83.

McClintock, B. (1951). Chromosome organization and genic expression. *Cold Spring Harbor Symposium on Quantitative Biology*, **16**, 13–47.

McClintock, B. (1965). The control of gene action in maize. *Brookhaven Symposia in Biology*, **18**, 162–84.

Miller, H. I., Kirk, M. & Echols, H. (1981). SOS induction and autoregulation of the *himA* gene for site-specific recombination in *Escherichia coli*. *Proceedings of the National Academy of Sciences, USA*, **78**, 6754–8.

Monk, M. (1967). Observations on the mechanism of indirect induction by mating with ultraviolet *coll* donors. *Molecular and General Genetics*, **100**, 264–74.

Mortelmans, K. E. & Stocker, B.A.D. (1976). Ultraviolet light protection, enhancement of ultraviolet light mutagenesis, and mutator effect of plasmid R46 in *Salmonella typhimurium*. *Journal of Bacteriology*, **128**, 271–82.

Nanney, D. L. (1980). *Experimental Ciliatology*. New York: John Wiley and Sons.

Neuffer, M. G. (1966). Stability of the suppressor element in two mutator systems at the *A1* locus in maize. *Genetics*, **53**, 541–9.

Peterson, P. A. (1960). The pale green mutable system in maize. *Genetics*, **45**, 115–33.

Radman, M. (1974). Phenomenology of an inducible mutagenic DNA repair pathway in *Escherichia coli*: SOS repair hypothesis. In *Molecular and Environmental Aspects of Mutagenesis*, ed. L. Prakash, F. Sherman, M. Miller, C. Lawrence & H. W. Tobin, pp. 128–42. Springfield, Ill.: Charles C. Thomas.

Reif, H.-J. & Saedler, H. (1975). IS1 is involved in deletion formation in the gal region of *E. coli* K12. *Molecular and General Genetics*, **137**, 17–28.

Roberts, J., Roberts, C. & Craig, N. (1978). *Escherichia coli* recA gene product inactivates phage λ repressor. *Proceedings of the National Academy of Sciences, USA*, **75**, 4714–18.

Robinson, M. K., Bennet, P. M. & Richmond, M. H. (1977). Inhibition of TnA translocation by TnA. *Journal of Bacteriology*, **129**, 407–17.

Roeder, G. S. & Fink, G. (1983). Transposable elements in yeast. In *Mobile Genetic Elements*, ed. J. A. Shapiro, pp. 299–327. New York: Academic Press.

Rubin, G. M. (1983). Dispersed repetitive DNAs in *Drosophila*. In *Mobile Genetic Elements*, ed. J. A. Shapiro, pp. 329–62. New York: Academic Press.

Shapiro, J. A. (1967). The structure of the galactose operon in *Escherichia coli* K-12. Ph.D. Thesis, University of Cambridge.

Shapiro, J. A. (1969). Mutations caused by the insertion of genetic material into the galactose operon of *Escherichia coli*. *Journal of Molecular Biology*, **40**, 93–105.

Shapiro, J. A. (1979). Molecular model for the transposition and replication of bacteriophage Mu and other transposable elements. *Proceedings of the National Academy of Sciences, USA*, **76**, 1933–7.

Simmons, M. J., Johnson, N. A., Fahey, T. M., Nellett, S. M. & Raymond, J. D. (1980). High mutability in male hybrids of *Drosophila melanogaster*. *Genetics*, **96**, 479–90.

Singer, M. (1982). Highly repeated sequences in mammalian genomes. *International Review of Cytology*, **76**, 67–112.

Starlinger, P. & Saedler, H. (1976). IS-elements in microorganisms. *Current Topics in Microbiology and Immunology*, **75**, 111–52.

Szybalski, W. (1977). IS elements in *Escherichia coli*, plasmids and bacteriophages. In *DNA Insertion Elements, Plasmids and Episomes*, ed. A. I. Bukhari, J. A. Shapiro & S. L. Adhya, pp. 583–90. Cold Spring Harbor: Cold Spring Harbor Laboratory.

Tonegawa, S., Sakano, H., Maki, R., Traunecker, A., Heinrich, G., Roeder, W. & Kurosawa, Y. (1980). Somatic reorganization of immunoglobulin genes during lymphocyte differentiation. *Cold Spring Harbor Symposium on Quantitative Biology*, **45**, 839–58.

Walker, G. C. & Dobson, P. P. (1979). Mutagenesis and repair deficiencies of *Escherichia coli* umuC mutants are suppressed by the plasmid pKM101. *Molecular and General Genetics*, **172**, 17–24.

Watson, J. D. & Crick, F. H. C. (1953). Genetical implications of the structure of desoxyribose nucleic acid. *Nature*, **171**, 964–7.

Witkin, E. (1976). Ultraviolet mutagenesis and inducible DNA repair in *Escherichia coli*. *Bacteriological Reviews*, **40**, 869–907.

Weinstock, G., McEntee, K. & Lehman, I. R. (1979). ATP-dependent renaturation of DNA catalyzed by the *recA* protein of *Escherichia coli*. *Proceedings of the National Academy of Sciences, USA*, **76**, 126–30.

White, M. J. D. (1978). *Modes of Speciation*. San Francisco: Freeman.

EVOLUTION OF WHOLE ORGANISMS

14

Gene, organism and environment

R. C. LEWONTIN

The modern theory of evolution is, as is so often said, a fusion of the two great insights of nineteenth century biology: Darwin's realization that the variation among species arises from the conversion of variation between individuals within species, and Mendel's discovery of the segregation of discrete factors as the basis for the inheritance of differences between individuals. We are constantly reminding ourselves and others that the immense progress made in biology in the present century rests firmly on these two major discoveries of a previous time. What is not always appreciated, however, is that the legacies of Darwin and Mendel are also responsible for certain difficulties in biology, difficulties that prevent us from some kinds of further progress and which keep us locked into a rigid framework of thought about the development and evolution of organisms. These difficulties arise, ironically, from the very source of Mendel and Darwin's success as biologists, their separation of internal from external forces acting on organisms. For Mendel the internal 'factors' were the causes of the form of the organism and were, in the end, the proper objects of study. The 'factors', what we now call 'genes', were the subjects and the organisms the objects of developmental forces. From this view of gene as the cause of organism has flowed the entire corpus of modern mechanical and molecular genetics. For Darwin, the external world, the environment, acting on the organism was the cause of the form of organisms. The environment, the external world with its autonomous properties, was the subject and the organism was, again, the object acted upon. In Darwinism, the organism is the interaction of two causal sequences, autonomous in their dynamics. Internal forces produced the variation among organisms, and autonomous external forces moulded the species on the basis of these autonomous internally caused variations. The essence of Darwin's account of evolution was the separation of causes of *ontogenetic* variation, as coming from internal factors, and causes of *phylogenetic* variation, as being imposed from the external environment by way of internal selection. (I gloss over Darwin's later flirtation with Lamarckism, since the proposition that the environment specifically engenders heritable adaptive variation is at total variance with the Darwinian mechanism of evolution.) It is from this view of environment as the cause of organisms that the entire corpus of modern evolutionary biology arises.

We cannot appreciate fully the nature of the change in biology wrought by Mendel and Darwin unless we understand the historical importance of the objectification of the organism. Descartes' metaphor of the organism as machine had virtually no impact on biology for 200 years. So, for example, even Harvey's mechanical description of the circulation of blood was not really accepted until the beginning of the nineteenth century, and Caspar Friedrich Wolff's epigenetic theory and his remarkably modern distinction between genotype and phenotype had no effect in embryology until the *Entwicklungsmechanik* of the latter part of the last century. Biology lacked clear notions of separable causes and effects and, more important, a systematic commitment to the analysis of biological systems along mechanistic lines. Lamarck's view that information from the external world could become permanently incorporated in organisms and their progeny through the mediation of a living being's needs and will was quintessentially representative of pre-Darwinian biology. As Charles Gillespie (1959) so cogently puts it: 'Larmarck's theory of evolution belongs to the contracting and self-defeating history of subjective science, and Darwin's to the expanding and conquering history of objective science'.

By making organisms the *objects* of forces whose *subjects* were the internal heritable factors and the external environment, by seeing organisms as the *effects* whose *causes* were internal and external autonomous agents, Mendel and Darwin brought biology at last into conformity with the epistemological meta-structure that already character-ized physics since Newton and chemistry since Lavoisier. This change in world view was absolutely essential if biology was to progress by making contact with physical science and by becoming quantitative and predictive. The mechanistic reductionism and the clear separation of internal and external were as necessary in the nineteenth century for the creation of a scientific biology as Newton's ideal bodies and perfect determinism were for the physics of the seventeenth. But we must not confuse the historically determined necessity of a particular epistemological stance at one stage in the development of a science with a perfect model that will guarantee all future progress. On the contrary, the very progress made possible by certain revolutionary formulations may lead eventually to results that are in contradiction with those earlier formulations and which can be resolved only by their reexamination. Yet those reexminations are themselves rooted in the past formulations. Newtonian mechanics and classical optics were in serious contradiction with the newer observations of physics at the end of the nineteenth century, but anyone familiar with the development of the Special Theory of Relativity immediately sees how that theory, in one aspect a negation of Newtonian principles, is built entirely on a Newtonian framework and could never have been developed in the absence of classical physics.

As in physics, so too in biology. As time has passed, Mendel's view of organisms as the manifestation of autonomous internal 'factors' with their own laws, and Darwin's view of organisms as passive objects moulded by the external force of natural selection, have become increasingly in contradiction with the known facts of developmental and population biology. But the situation in biology has developed rather differently than in physics. All of physical science seemed blocked by the repeated conflict between classical formulations and the new observations of radioactivity, of light, and of astronomy. Without relativity and quantum theory, physics and theoretical chemistry would have ground to a halt. On the other hand, biology, far more diverse in its subject matter, far more loosely tied together into a coherent science, has undergone a radically uneven development. Some branches such as molecular biology have made extraordinary progress by concentrating on just those questions for which the simple mechanical reductionism of the nineteenth century is the perfect epistemology. Developmental biology, the study of cognition and memory, and evolutionary biology, on the other hand, have profited only marginally from these rapid advances. Rather, they are stalled by their attempt to use outdated concepts to confront a rich phenomenology to which these concepts clearly do not apply. Evolutionary biology suffers particularly because it is the nexus of all other biological sciences, so that a lack of progress in developmental biology, in ecology, in behavioural science, all are fatal to a proper understanding of evolution.

Specifically, evolutionary biology must confront two issues about the forms of organisms. One is the ontogenetic process by which the sequence of forms that comprise an individual's life history come into being. The second is the phylogenetic process by which species as collective entitites form and change based on the variations among the individuals that make them up. Classical, post-Darwinian, post-Mendelian biology has settled on two metaphors through which the processes are seen. The first, ontogenetic, process is seen as an *unfolding* of a form, already latent in the genes, requiring only an original triggering at fertilization and an environment adequate to allow 'normal' development to continue. The second, phylogenetic, process is seen as *problem and solution*. The environment 'poses the problem'; the organisms posit 'solutions', of which the best is finally 'chosen'. The organism proposes; the environment disposes. These two metaphors are simply the forms of the original Mendelian view that internal factors make the individual organism and the Darwinian view that external forces determine the collectivity. In the balance of this chapter, I want to make clear why these two metaphors are wrong. Individual development is not an unfolding, and evolution is not a series of solutions to present problems. Rather, genes, organisms, and environments are in reciprocal interaction with each other in such a way that each is both cause and effect in a quite complex, although perfectly analysable, way. The known facts of development and of natural history make it patently clear that genes do not determine individuals nor do environments determine species.

Gene, environment and organism in development

I will begin with the obvious. It is well known that Mendel solved the basic problem of the laws of inheritance by investigating the heredity of very special sorts of differences, those in which there was a determinate correspondence between genotype and phenotype. Given the genotype, the phenotype corresponding to it was unambiguously defined, at least under the condition of Mendel's experimental garden. Indeed, it is the essence of the Mendelian methodology that one can read off the genotype given either the individual's phenotype or, in the case of complete dominance, the phenotypes of the progeny of a single test cross. It is not always explicitly pointed out that all of modern biochemical, molecular, and developmental genetics also has depended for progress on finding genetic differences that make clear-cut, non-overlapping phenotypic classes under easily controlled conditions. It is amusing to contemplate where bacteriophage genetics (and, *a fortiori*, all of molecular genetics) would be now if Benzer's bacteriophage mutants had only differed from each other by 1% in the number of progeny phage they produced on restrictive hosts. One consequence of this methodological history is that textbooks of genetics (and of evolution) consistently describe organisms as 'determined' by their genes. It is a short step to seeing organisms as 'lumbering robots' controlled by their genes 'body and mind'. It is not hard to see why such unbiological rubbish is taken seriously. Yet, the vast majority of morphological, behavioural, and physiological differences among individuals do not 'Mendelize'. It is simply not possible to read off the genotypic differences between tall and short individuals of *Rumex acetosella* from their individual phenotypes or from any number of controlled test crosses involving them or their relatives, not to speak of the genotypic difference between faithful spouses and philanderers. The reaction of many evolutionists (encouraged, again, by textbooks) to this obvious fact is to consign such characters to 'polygenic control', invoking the multiple factor hypothesis, and thus saving the basic model that genes determine organisms. But quantitative variation of a character is not *prima facie* evidence that it is influenced by many genes. Single gene mutations affecting eye shape, wing variation and bristle number in *Drosophila*, or enzyme structure in humans, all show quantitative variation in phenotype and, unless care is taken to control the condition of development, considerable overlap between genotypes in their phenotypic distributions. The fundamental general fact of phenogenetics is that the phenotype of organisms is a consequence of non-trivial interaction between genotype and environment during development. All that genes ever do is to specify a norm of reaction over environments. Moreover, fitness too is a phenotype and varies from environment to environment, both because other aspects of the phenotype develop differently in different environments, and because a given shape or behaviour or physiology will confer different fitnesses in different environments. During the 15 years between 1950 and 1965, population geneticists, largely under the influence of plant and animal breeders and of Schmalhausen's seminal work, *Factors of Evolution* (1949), devoted considerable attention to the environmental contingency of phenotype and to the effects of selection in different environmental regions (see, for example, Lerner (1954), Dobzhansky & Spassky (1944), Falconer (1960), and Robertson (1960)), but this work passed out of fashion with the advent of molecular population genetics. Indeed, except for the famous work of Clausen, Keck & Heisey (1958), no study of the norms of reaction of naturally occurring heterozygous genotypes has appeared until the experiments of Gupta on *Drosophila* (Gupta & Lewontin (1982)). Thus, the basic data for judging the effects of selection in particular genotypes are simply lacking. The little that is known shows clearly that the developmental responses of different genotypes to varying environments are non-linear and do not allow the simple ordering of genotypes along a one-dimensional scale of phenotype. Norms of reaction cross each other so that no genotype gives a phenotype unconditionally larger, smaller, faster, slower, more or less different than another. These well-known facts seem, however, to have made no impact on evolutionary theorists who continue to speak about selection for a character and about genes that are selected because they produce that character.

A second, less well-known, feature of development is that phenotype is not given even when the genotype and the environment are completely specified. There is a significant effect of 'developmental noise' (Waddington (1957)) in producing phenotype. The two sides of a *Drosophila* have the same genotype, and no reasonable definition of environment will allow that the left and right sides of a pupa developing halfway up the side of a glass milk bottle in the laboratory are in different environments. Yet, the number of sterno-pleural bristles and the number of eye facets differ between the two sides of an individual fly. Small events at the level of thermal noise acting during cell division and differentiation have large effects on the final developmental outcome.

It is only the beginning of understanding to agree that internal and external factors contribute to phenotype. The reaction to that knowledge is usually to attempt to assign the relative weights to genotype, environment and developmental noise, as for example in a heritability study or some other form of the analysis of variance. The second fact that evolutionary biologists must cope with is that the influence of each factor depends upon the influence of the other factors so that no assignment of fixed weights to genetic, environmental and noise components of variation is possible (Lewontin, 1974). There are no more important experimental results for evolutionary biology than those of Rendel (1967) on canalization, yet the early impact of these findings on evolutionary theory seems now to be forgotten. What Rendel and others have shown is the *reciprocity* of effects of genetic state on environmental sensitivity and of environmental state on genetic sensitivity of the developing organism. Moreover, the genotype is itself hierarchically organized so that the environmental sensitivity of a given gene substitution will depend upon genes at other loci. The consequence is that selection can change the average expression of a trait, independently from the variation of that expression in response to variable environment and to developmental noise. Thus, by selection, an organism can be made developmentally insensitive or highly sensitive to perturbations of its genotype, its environment, developmental accidents, or any combination of these. Internal and external facts not only play a role in development, but each determines the role played by the other.

Thus far, my description of development is still in terms of the organism as the object, the effect of causes both internal and external to it, although the causes themselves interact with each other. The final step in the integration of developmental biology into evolution is to incorporate the organism as itself a *cause* of its own development, as a mediating mechanism by which external and internal factors influence its future. To describe phenotype as the consequence of gene, environment, and accident leaves out of account entirely the element of *temporal order* which is of the essence in a developmental process. The organism's phenotype is in a continual state of change from fertilization to death. The phenotype at any instant is not simply the consequence of its genotype and current environment, but also of its phenotype at the previous instant. That is, development is a first order Markov process in which the next step depends upon the present state. The temporal order of environments is thus a critical, but not a sufficient, prediction of future development. If a *Drosophila* adult hatches from the pupal case as an unusually small fly, it does not matter what the contributing causes were to making it small, it now has an unusually high surface to volume ratio, which will play an important role over the rest of its life history. Small changes in ambient temperature or in its own activity will be felt by it in a way different from its larger sibs and will have different effects on its reproductive rate. The way in which its genotype and its future environmental sequence influence the fly are themselves *effects* of the organism as *cause*. The organism is not simply the object of developmental forces, but is the subject of these forces as well. Organisms as entities are one of the causes of their own development.

Organism and environment in evolution

Our usual description of evolution by natural selection is framed in terms of the process of *adaptation*. A species' environment exists and changes as a consequence of some autonomous forces outside the species itself. This outside world poses problems for the species, problems of acquiring space, consumables, light and individuals of the opposite sex. Those most successful in solving the problems, because, by chance, their morphologies, physiologies, and behaviours make them mechanically the best fit to do so, leave the most offspring and thus the species adapts. This view of evolution, however, has certain paradoxical features (Lewontin, 1978). One is that all extant species are said to be already adapted to their environments. A good deal of evolutionary biology is taken up with demonstrating that their features represent optimal solutions to environmental problems. What then is the motive power of further evolution? The solution proposed by Van Valen (1973) is that the environment is constantly moving and that species are simply running to keep up. In that case, it is the autonomous forces of environmental change that govern the rate of evolution, and we would be well advised to study the laws of environmental rather than organismic change if we want to understand what has been happening. More paradoxical is the necessity of defining environments without organisms. To make the metaphor of adaptation work, environments or ecological niches must exist before the organisms that fill them. There must be a preferred, denumerable set of combinations of factors that make 'environments' and a non-denumerable infinity of combinations of factors that are not. The history of life is then the history of the coming into being of new forms that fit more and more closely into these preexistent niches. But what laws of the physical universe can be used to pick out the possible environments waiting to be filled? In fact, we only recognize an 'environment' when we see the organism whose environment it is. Yet so long as we persist in thinking of evolution as adaptation, we are trapped into an insistence on the autonomous existence of environments independent of living creatures. In fact, we should be able to list the environments on Mars, being careful all the while not to be influenced by our knowledge of earthly life! If, on the other hand, we abandon the metaphor of adaptation, how can we explain what seems the patent 'fit' of organisms and their external worlds? Fish have fins, and not only fish but whales, seals, penguins and even sea snakes have some sort of flattened body part that the animal uses for swimming. Moles have long claws as do anteaters; birds and bats have wings, and so on. The marvellous fit of organisms to their environments seems obvious. What is left, then, but some concept of a progressive fitting of organisms to predetermined adaptive peaks?

What is left out of this adaptive description of organism and environment is the fact, clear to all natural historians, that the environments of organisms are made by the organisms themselves as a consequence of their own life activities. How do I know that stones are part of the environment of thrushes? Because thrushes break snails on them. Those same stones are not part of the environment of juncos who will pass by them in their search for dry grass with which to make their nests. Organisms do not adapt to their environments; they construct them out of the bits and pieces of the external world. This construction process has a number of features:

(1) Organisms determine what is relevant. While stones are part of a thrush's environment, tree bark is part of a woodpecker's, and the

undersides of leaves part of a warbler's. It is the life activities of these birds that determine which parts of the world, physically accessible to all of them, are actually parts of their environments. Moreover, as organisms evolve, their environments, perforce, change. All animals are covered with a thin boundary layer of warm moist air as a consequence of their metabolism. Small ectoparasites may be completely immersed in that boundary layer which thus determines their environmental temperature and humidity. But if natural selection should increase the body size of these parasites, they may emerge through the layer into the stratosphere of a colder drier world. It is the genes of sea-lions that make the sea part of their environment and the genes of lions that make the savannah part of theirs, yet those genes are descended from a common carnivore ancestor.

(2) Organisms alter the external world as it becomes part of their environments. All organisms consume resources by taking up minerals, by eating. But they may also create the resources for their own consumption, as when ants make fungus farms, or trees spread out leaves to catch sunlight. They may create an environment more hospitable to their own species, as for example when beavers raise the water level of a pond, but, in contrast, white pine in New England creates a dense shade that prevents its own reseeding. Ecological succession is precisely the history of self-destruction of species by alterations of their own environment.

(3) Organisms transduce the physical signals of the external world. Changes in external temperature are not perceived by my liver as thermal changes but as alterations in the concentrations of certain hormones and ions. The photon energy impinging on my retina and the vibrational energy at my ear drums when I see and hear a rattlesnake are immediately changed through the mediation of my central nervous system into changes in adrenalin concentration. But this change is in part a consequence of my biology, since another rattlesnake would presumably react rather differently.

(4) Organisms create a statistical pattern of environment different from the pattern in the external world. Organisms, by their life activities, can damp oscillations, for example in food supply by storage, or in temperature by changing their orientation or moving. They can, on the contrary, magnify differences by using small changes in abundance of food types as a cue for switching search images. They can also integrate and differentiate. Plants may flower only when a sufficient number of days above a certain temperature have been accumulated. *Cladocera* can change from asexual to sexual reproduction in response to a large alteration in oxygen supply, temperature or food supply, irrespective of the absolute

level. Even the period of external oscillation can be modulated, as when cicades count a prime number of seasonal fluctuations.

It might be objected that the notion of organisms constructing their environments leads to absurd results. After all, hares do not sit around constructing lynxes! But in the most important sense they do. First, the biological properties of lynxes are presumably in part a consequence of selection for catching prey of a certain size and speed, i.e. hares. Second, lynxes are not part of the environment of moose while they are of hares, because of biological differences between moose and hares. Then what about laws of physical nature? Organisms did not pass the law of gravity. Yet, whether gravitation is an aspect of the environment of an organism depends upon biological properties of the organism, for example size. Bacteria are 'outside' gravity, but their very small size makes them subject to a very different physical phenomenon, Brownian motion of molecules, which larger organisms do not notice.

The metaphor of construction rather than adaptation leads to a different formulation of natural selection and evolution. On the adaptive view evolution can be represented formally as a pair of differential equations. The first, describing the change in organisms, O, as a function of organism and environment, E, $dO/dt = f(O, E)$, and a second law of the autonomous change of environment, $dE/dt = g(E)$. A constructionist view makes this into a pair of *coupled* differential equations in which organism and environment coevolve, each as a function of the other, $dO/dt = f(O, E)$, and $dE/dt = g(O, E)$.

The parallel in population genetics is between frequency-independent and frequency-dependent selection. In the first case, one postulates a set of adaptive peaks which may or may not change in time as a consequence of external forces, and a set of gene frequencies that change in response to the potential field represented by the adaptive peaks. In the second case, the location and existence of the peaks are themselves functions of the genetic composition of the evolving population. The model of the first process is climbing a mountain peak, of the second walking on a trampoline. While population geneticists usually model selection as a frequency-independent process, adding frequency dependence as an added complication of marginal interest, the actual situation is the reverse. Most selective processes are frequency-dependent, as for example are any processes in which fitness depends on relative position in an ordered series, or in which there is competition for resources in short supply with different relative success of different types.

A problem seems to be posed for the constructionist view by the phenomenon of convergence. If environments do not preexist, how are we to explain the remarkable similarities that evolved independently in different groups? Not only do animals as unrelated as fish, mammals, and birds all have fin-like appendages when they are aquatic, but a whole marsupial fauna developed with 'wolves', 'moles', 'mice' and 'cats' (although no ungulates, marine forms, elephants, bats, or horses). The error is to suppose that because organisms construct their environments they can construct them arbitrarily in the manner of a science fiction writer constructing an imaginary world. The coupled equations of coevolution of organism and environment are not unconstrained. Some pathways through the organism–environment space are more probable than others, precisely because there are real physical relations in the external world that constrain change. The construction of an environment that includes living in a fluid medium of a certain density and viscosity places certain quite loose constraints on morphology. Either the organism will be sessile, as many are, and wait for food, or it will be vagile and chase it. If it is vagile it may propel itself by jet action like the squid or by umbrella sculling like the jelly-fish or by flattened appendages. But even in the last case, there are those who undulate up and down (whales), side to side (sharks), fly in the water (rays), beat tiny wings rapidly (sea horses), and a variety of other fin movements. Where there is strong convergence is in certain marsupial–placental pairs, and this should be taken as evidence about the nature of constraints on development and physical relations, rather than as evidence for pre-existing niches.

The nagging problem of adaptation has always been what one imagines the progression of events to be during the process itself, since we always observe only the finished product. If a seal's flippers are an adaptation to water, at what stage in the evolutionary history of seals did swimming in water become the 'problem' which seals 'solved' by losing their legs? No one imagines that a whole group of terrestrial carnivores simply plunged into the water one day, experiencing a new major adaptive problem, and then proceeded to adapt to it by the usual route of natural selection for small increases in flipper-like morphology. Nor, alternatively, can we say that swimming has always been a major problem for carnivores. There is, in fact, no reason why we should not, *a priori*, reverse the entire scenario. Perhaps an early pinniped ancestor acquired slightly flipper-like appendages for an entirely different reason – genetic drift or some pre-adaptation. Partly aquatic life then became an *opportunity* rather than a problem. This is not a very satisfactory theory, at least as a typical one for evolution, but for precisely the same reason that the standard adaptive story is unacceptable. Both uncouple organism and environment in such a way that we must regard one as undergoing significant autonomous change and the other as responding.

Concentrating as we always have on the problem of how the organisms change under natural selection, we have neglected to ask seriously how the environment to which the organism is supposedly responding has come to be a problem in the first place. Presumably the non-aquatic carnivore ancestors of the Pinnipedia slowly incorporated the water as a more and more energetically significant aspect of their environment while their morphologies and physiologies changed to make that appropriation more energetically rewarding. Were a complete reconstruction possible, we would be unable to find the moment at which swimming was for the first time a 'problem' to be 'solved' by the animal.

Organisms, then, both make and are made by their environment in the course of phylogenetic change, just as organisms are both the causes and consequences of their own ontogenetic development. The alienation of internal and external causes from each other and of both from the organism, seen simply as passive result, does not stand up under even the most casual survey of our knowledge of development and natural history. It is a tribute to the power of long-held ideology that the study of evolution continues to lean so heavily on an impoverished view of the relation between gene, environment, and organism. If the hundredth anniversary of the death of Darwin is not to mark the death of Darwinism, we need to struggle for its transfiguration.

Further reading

Lewontin, R. C. (1974). The analysis of variance and the analysis of causes. *American Journal of Human Genetics*, **26**, 400–11.
Lewontin, R. C. (1978). Adaptation. *Scientific American*, **239**, (9), 156–69.

References

Clausen, J., Keck, D. D. & Heisey, W. W. (1958). Experimental studies on the nature of species, Vol. 3: Environment responses of climatic races of *Achillea*. *Carnegie Institution of Washington Publication 581*, 1–129.
Dobzhansky, Th. & Spassky, B. (1944). Manifestation of genetic variants in *Drosophila pseudoobscura* in different environments. *Genetics*, **29**, 270–90.
Falconer, D. S. (1960). Selection of mice for growth on high and low planes of nutrition. *Genetical Research, Cambridge*, **1**, 91–113.

Gillespie, C. C. (1959). Lamarck and Darwin in the history of science. In *Forerunners of Darwin*, ed. B. Glass, O. Temkin & W. L. Straus, p. 265–91. Baltimore: The Johns Hopkins Press.

Gupta, A. P. & Lewontin, R. C. (1982). A study of reaction norms in natural populations of *D. pseudoobscura*. *Evolution*, **36**, 934–48.

Lerner, I. M. (1954). *Genetic Homeostasis*. Edinburgh: Oliver and Boyd.

Lewontin, R. C. (1974). The analysis of variance and the analysis of causes. *American Journal of Human Genetics*, **26**, 400–11.

Lewontin, R. C. (1978). Adaptation. *Scientific American*, **239**, (9), 156–69.

Rendel, J. M. (1967). *Canalization and Gene Control*. London: Academic Press.

Robertson, F. W. (1960). The ecological genetics of growth in *Drosophila*. II. Selection for large body size on different diets. *Genetical Research, Cambridge*, **1**, 305–18.

Schmalhausen, I. I. (1949). *Factors of Evolution: The Theory of Stabilizing Selection*. Philadelphia: Blakeston.

Van Valen, L. (1973). A new evolutionary law. *Evolutionary Theory*, **1**, 1–30.

Waddington, C. H. (1957). *The Strategy of the Genes*. London: Allen and Unwin.

15

Population genetics and ecology: the interface

E. NEVO

The problem: the nature and evolution of allozyme polymorphism

Ecological genetics: the dialogue between environment and organism

Evolutionism, the world view that considers constant change a central process of nature, always regarded organic evolution as a coevolving dialogue between organism and environment. Lamarck, the forerunner of Darwin, emphasized adaptation as a central concept of biological evolution despite his wrong explanatory model of its origin (Mayr, 1972). Darwinian evolutionism introduced the idea of individual variation as central to the process of environmental selection, the natural nonrandom architect of adaptational design. Darwinism holds that evolutionary change results from the interplay between variation and selection through differential survival and reproduction of individual variants.

Modern evolutionary theory is based on the interdisciplinary structure between population genetics and ecology, empirically hybridized in the science of ecological genetics (Ford, 1971). The main concern of ecological genetics is the origin and evolution of adaptations, attempting to evaluate the ecological forces that mould the genetic structure and evolution of populations and species. Nevertheless, a true theoretical integration of genetics and ecology is still a future challenge (Sammeta & Levins, 1970; Levin, 1978) despite some insights (i.e. Brussard, 1978; Roughgarden, 1979). A major problem is that the conceptual frameworks of genetics and ecology are not easily matched. For example, what is the ecological significance of genetic population pattern? Or, how can the high order complexities of ecology and genetics, environmental heterogeneity and multilocus structure, respectively, be meaningfully associated? Likewise, to what extent are allele frequency configurations determined by deterministic, ecological and population parameters as against random factors such as founder effects, initial conditions, bottle-necking, sampling and random drift fluctuations?

The spectacular developments of molecular biology allowed the analysis of individual genetic variation, and hence the study of genetic structure of populations and species (reviewed in Lewontin, 1974; Nei, 1975; Ayala, 1976; Wright, 1969, 1978; Ewens, 1979; Brown, 1979; Hartl, 1980; Wills, 1981). Thus ecological genetics shifted largely from visual to biochemical polymorphisms and concentrated on analysing the ecological variables which maintain and differentiate enzyme polymorphism and the relationship of the latter with phenotypic diversity. Yet the relative importance of the forces interacting in genetic population differentiation, i.e., population size and structure, mating pattern, mutation, recombination, migration, selection and genetic drift, remain now as enigmatic and controversial as ever, at the molecular level. Ironically, the great genetic variation found in nature (reviewed in Nevo, 1978) caused more confusion than elucidation of the relative roles of these forces. The problem is central to neo-Darwinism: what proportion of genetic diversity in nature is the basis of adaptive evolutionary change (Lewontin, 1974)?

Evolution of genetic polymorphism: the selectionist and neutrality schools
Genetic variation in nature can be increased, decreased, or stabilized by a combination of random and nonrandom factors. Mutation and migration introduce variation into populations, while sexual recombination and hybridization increase the genotypic array, and random drift and purifying selection decrease variation. Currently there are two major rival theories for explaining genetic diversity in nature (Lewontin, 1974). The selectionist theory proposes a variety of selective mechanisms that can actively maintain genetic polymorphism (see detailed list and references in Hartl, 1980, p. 225). Most of these are varieties of balancing selection mechanisms in which one homozygote is favoured under some conditions (niche, habitat, season, density, frequency, life cycle, fitness components, gametes, sexes, groups, etc.), but disfavoured under others. The most likely general selective mechanisms that can maintain genetic polymorphism are spatially and/or temporally heterogeneous environments and epistasis (Hartl, 1980). Selectionists believe that a large part of protein polymorphism is maintained by some form of, or a combination of, balancing selection mechanisms (e.g., Milkman, 1976; Ayala, 1977; Clarke, 1979; Wills, 1981). Finally, the presumed constancy of molecular evolutionary rates which seemingly support neutrality (Kimura, 1979b) is also explainable on ecological grounds (Van Valen, 1974).

The neutral theory of molecular evolution (Kimura, 1968; Kimura & Ohta, 1971 and modifications in Kimura, 1979a) holds that the laws of phenotypic and molecular evolution are different. While Darwinian positive environmental selection acts mainly on phenotypes through polygenes, it 'cares little how those phenotypes are determined by genotypes' (Kimura, 1979b). At the genotypic level most intraspecific variability and evolutionary change are caused not by selection but by random drift of mutant genes that are selectively equivalent in functional efficiency. The neutral theory allows for negative or purifying selection of deleterious mutations and also for selection of a small proportion of advantageous mutations which cause slow adaptive and progressive phenomena (Nei, 1975), hence are primarily noticeable over geological time. However, most molecular diversity in nature is considered nonselective and is maintained in populations through mutational input and random extinction. Polymorphism is simply an incidental and largely unimportant phase of molecular evolution. Neutralists believe that the level of observed gene diversity in nature is in rough agreement with the expected value for neutral mutations, when mutation rate is inferred from the rate of amino-acid substitutions.

Obviously, the neutral theory, original or modified, continues the heritage of the classical position for which variation is constantly being removed by purifying selection and genetic drift, rather than being preserved in populations as asserted by the balance school (Lewontin, 1974). It remains controversial what proportion of genetic variation is attributable to selection and what proportion to random factors.

The theoretical testing of neutral models

A critical analysis of the testing of neutral models is given by Ewens (1979, and references therein). The implicit assumption of all tests involves panmixia and stationarity, without consideration of geographical subdivision. The analysis is based on three different mathematical models: the *charge state*, the *infinite allele* and the *infinite site* models. Caution should be exercized in matching the mathematical model and data set. Ewens (1979) concludes that a variety of practical circumstances is testable under the more complete testing theory existing for the infinite allele model, and that so far in all tests neutrality is still rejected.

Nei and his coworkers (Nei, 1975; Nei, Fuerst & Chakraborty, 1976) have used the relationship between the mean (*He*) and variance v(*He*) of gene diversity among loci for testing the null hypothesis of neutral mutations. Nei and coworkers were not able to reject the neutral hypothesis in the above tests but admitted that these tests were not always powerful.

M. Nei and D. Graur (personal communication) concluded 'that available data on protein polymorphism are most easily explained by a modified form of neutral mutations hypothesis in which the effects of bottle-necks and fluctuating selection are taken into account'. However, their analysis completely ignores the interactions between genotype, environment and population numbers (Ayala, 1968).

Finally, various less formal models were suggested to test the neutral theory. These include among others, tests correlating various genetic observations with environmental phenomena (see Ewens, 1979, p. 281 and his references and Nevo, Zohary, Brown & Haber, 1979, etc.). As indicated by Ewens (1979), such informal tests are often better than the formal procedures, especially in ecological circumstances where a useful mathematical theory is unlikely.

If current tests appear to have insufficient power to resolve the neutrality question there are other promising alternatives. A possible approach is the development of a theory of genotype–phenotype interaction with selection on the phenotype (Lewontin, 1974) which may then allow direct correlation with the environment. This, as pointed out by Ewens & Feldman (1976) 'clearly involves more ecology than population geneticists have been willing to use'.

Genetic–environmental correlations across species: objective and rationale

The present study is an attempt to test the idea that ecological forces and population structure are substantial in the genetic differentiation of populations and species. The methodology used in this study is to explore direct correlation between genetic, ecological and biotic profiles involving many shared loci in many populations and unrelated species, varying in biological parameters and historical record but distributed *over a shared, diverse and stressful ecological background*. The major question is, what proportion of allozyme variation in natural populations may be adaptive? If parallel, generalized and predictive genetic patterns, emerge across distinctly unrelated species, then selection must be a major differentiating force maintaining allozyme polymorphisms in natural populations. Israel is suitable for this analysis owing to its extremely variable ecological background over very short distances.

The evolutionary play: the ecological theatre and genetic participants

The ecological background structure of Israel

Physical factors

The physical geography of Israel is distinguished by substantial geomorphologic variety and climatic diversity within a small area. Geomorphologically, Israel is characterized by mountains and basins, plains and highlands. Geologically, it consists mostly of sedimentary rocks, primarily Mesozoic and Cenozoic limestones, dolomites, chalks, marls and sandstones, while Precambrian magmatic rocks characterize the southern Elat region and the flanking mountains of Sinai. This background weathers into hundreds of soil varieties grouped into 24 soil types. Climatically, the country is divided into two distinct intergrading parts, the northern mesic Mediterranean and southern xeric regions, by one of the world's major climatic borders – the border of aridity. The combined diverse geological and climatic background causes significant ecological variation over remarkably short distances.

Biotic factors

The Near East in general and Israel in particular are extremely diverse biogeographically, including Mediterranean, European, Asiatic and African plant and animal communities. Relatively few species occur across the country in both mesic and xeric environments. Most species belong to the Mediterranean, steppic or desert territories. The Mediterranean territory has a mean yearly minimum rainfall of about 350 mm and the dominant soil types are terra rossa and rendzina. Maquis and forests of the Quarcetea class form the climax vegetation. Deforestation resulted in dwarf shrub formations called garrigue and batha. The steppic or Irano–Turanian territory comprises a narrow strip between the Mediterranean and Saharo–Arabian desert territory, characterized by 150–350 mm rainfall and grey steppe and loess-like soil. No arboreal climax exists and dwarf shrub formations predominate. The Saharo–Arabian desert territory is the largest in extent, occupying about half the area of Israel including the Judean and Negev deserts, as well as the Sinai peninsula. Mean annual rainfall ranges between 25 and 150 mm. Vegetation is rare and patchy, largely limited to seasonal stream beds and comprising about 80% annuals and a few scattered shrub formations. Detailed discussions and illustrations of the climatic and biotic factors appear in the *Atlas of Israel* (1970), and in Zohary (1973). This varied physical and biotic background combined with historical factors result in rich biological diversity suitable for analysing the genetic differentiation of populations and species.

Table 15.1. *Ecogeographical, life history, and genetic variables for 38 species*

Species	Number of individuals	popula-tions	loci	A	P–1%	P–5%	H	He
PLANTS								
1 *Hordeum spontaneum* (1)	1203	28	12	1.512	0.289	0.238	0.004	0.096
2 *Triticum dicoccoides* (2)	457	12	13	1.404	0.340	0.275	0.004	0.099
MOLLUSCA								
3 *Buliminus labrosus* (3)	140	2	15	1.414	0.310	0.274	0.070	0.075
4 *B. diminutus* (4)	20	1	15	1.067	0.067	0.067	0.017	0.019
5 *Sphincterochila cariosa* (5)	64	2	14	1.357	0.286	0.214	0.058	0.061
6 *S. fimbriata* (5)	97	3	14	1.857	0.548	0.452	0.162	0.186
7 *S. prophetarum* (5, 6)	178	4	14	2.107	0.518	0.464	0.088	0.191
8 *S. zonata* (5, 6)	110	2	14	1.714	0.429	0.357	0.145	0.124
9 *S. aharonii* (5)	60	2	14	1.250	0.250	0.107	0.044	0.061
10 *Theba pisana* (7)	262	8	13	1.543	0.366	0.336	0.088	0.136
11 *Trochoidea setzenii* (8)	97	3	14	1.784	0.416	0.293	0.106	0.136
12 *T. erkelii* (8)	29	1	14	1.846	0.538	0.538	0.085	0.148
INSECTA								
13 *Gryllotalpa gryllotalpa* (2n = 19) (4)	414	11	14	1.592	0.470	0.234	0.025	0.069
14 *G. gryllotalpa* (2n = 23) (4)	22	1	14	1.214	0.214	0.214	0.011	0.053
15 *G. africana* (2n = 23) (4)	34	1	8	1.125	0.125	0.000	0.004	0.004
16 *Grylloides hebraeus* (4)	34	1	13	1.462	0.308	0.077	0.022	0.046
17 *Gryllus bimaculatus* (4)	34	1	13	2.000	0.615	0.462	0.095	0.137
18 *Periplaneta americana* (4)	60	1	10	1.200	0.200	0.100	0.000	0.019
19 *Blatella germanica* (4)	30	1	12	1.250	0.250	0.083	0.014	0.027
20 *Dociostaurus curvicercus* (9)	216	7	9	1.778	0.524	0.349	0.057	0.141
21 *D. genei* (9)	91	3	8	2.083	0.750	0.625	0.095	0.185
22 *D. sp.* ('sands') (9)	30	1	8	1.500	0.375	0.125	0.012	0.070
AMPHIBIA								
23 *Hyla arborea* (10)	218	8	11	2.136	0.602	0.455	0.139	0.157
24 *Bufo viridis* (11)	507	11	11	1.736	0.504	0.372	0.131	0.163
25 *Rana ridibunda* (12)	340	11	11	1.413	0.306	0.273	0.070	0.097
26 *Pelobates syriacus* (13)	61	3	13	1.154	0.154	0.103	0.060	0.046
REPTILIA								
27 *Agama stellio* (14)	242	9	12	1.509	0.417	0.287	0.084	0.106
MAMMALIA								
28 *Acomys cahirinus* (2n = 38) (15)	133	5	12	1.219	0.188	0.188	0.041	0.069
29 *A. cahirinus* (2n = 36) (15)	39	2	13	1.125	0.125	0.125	0.035	0.050
30 *A. russatus* (15)	51	3	12	1.228	0.194	0.194	0.009	0.068
31 *Gerbillus allenbyi* (16)	103	3	13	1.308	0.179	0.154	0.038	0.064
32 *G. pyramidum* Negev (16)	5	1	12	1.000	0.000	0.000	0.000	0.000
33 *G. pyramidum* Coastal Plain (16)	27	1	12	1.000	0.000	0.000	0.000	0.000
34 *Mus musculus* (17)	60	5	11	1.345	0.327	0.291	0.053	0.096
35 *Spalax ehrenbergi* (2n = 52) (18, 19)	70	(b)	10	1.200	0.100	0.100	0.016	0.056

No. of loci	A	P–1%	P–5%	H	He	Gr	Hr	Pt	Ar	Pz	Bs	Mo	Gf	Lo	Gl	Fe	Ht	Do	Or	Tr
			Genetic indices based on all loci							Life history and ecological characteristics (a)										
28	1.483	0.298	0.238	0.003	0.099	3	2	2	4	3	3	1	1	1	1	2	2	2	1	3
50	1.327	0.250	0.203	0.002	0.071	3	1	1	3	3	3	1	1	1	1	2	2	2	1	3
31	1.364	0.273	0.218	0.040	0.048	3	1	3	3	2	1	2	1	3	2	2	2	2	1	2
30	1.067	0.067	0.067	0.010	0.017	2	1	1	1	1	1	2	1	3	2	2	2	2	1	2
28	1.286	0.232	0.161	0.043	0.053	3	1	3	3	3	1	2	1	3	2	2	2	2	1	2
27	1.654	0.423	0.359	0.104	0.139	3	1	2	2	3	1	2	1	3	2	2	2	2	1	2
29	1.889	0.523	0.410	0.074	0.157	2	1	2	1	3	1	2	1	3	2	2	2	2	1	2
29	1.724	0.440	0.300	0.079	0.100	2	1	2	1	3	1	2	1	3	2	2	2	2	1	2
29	1.228	0.228	0.123	0.043	0.055	1	1	2	3	2	1	2	1	3	2	2	2	2	2	2
18	1.650	0.460	0.424	0.105	0.166	3	2	3	3	3	1	2	1	1	1	2	2	2	2	2
27	1.564	0.314	0.235	0.064	0.097	3	2	2	2	3	1	2	1	3	2	2	2	2	1	2
22	1.773	0.500	0.409	0.056	0.131	2	1	1	1	2	1	2	1	3	2	2	2	2	1	2
21	1.756	0.568	0.376	0.030	0.116	3	1	2	3	3	1	3	1	2	1	2	1	3	1	1
20	1.200	0.200	0.150	0.010	0.039	1	1	1	1	1	1	3	1	2	1	2	1	3	2	1
15	1.067	0.067	0.000	0.002	0.002	3	1	1	1	2	1	3	1	2	1	2	1	3	1	1
26	1.385	0.308	0.077	0.021	0.041	2	1	2	3	2	1	3	1	2	1	2	2	3	1	1
25	1.840	0.560	0.360	0.055	0.110	3	2	2	3	2	1	3	1	2	1	2	2	3	1	1
20	1.150	0.150	0.100	0.050	0.025	4	1	2	3	2	1	3	1	2	1	2	2	3	1	2
19	1.211	0.211	0.053	0.012	0.021	4	1	2	3	2	1	3	1	2	1	2	2	2	1	2
19	1.803	0.553	0.416	0.074	0.160	2	2	2	3	3	1	3	1	1	1	2	2	3	1	2
18	1.883	0.641	0.462	0.057	0.135	3	2	2	2	3	1	3	1	1	1	2	2	3	1	2
16	1.750	0.500	0.313	0.055	0.138	1	1	2	3	3	1	3	1	1	1	2	2	3	2	2
27	1.620	0.394	0.208	0.072	0.080	3	2	3	3	2	2	3	2	3	3	2	2	3	1	1
26	1.664	0.435	0.320	0.130	0.148	4	2	3	4	2	2	3	2	3	3	3	2	3	1	1
28	1.409	0.328	0.256	0.069	0.092	4	2	3	3	2	2	3	2	3	3	2	2	3	1	1
32	1.076	0.076	0.043	0.028	0.021	2	1	1	3	2	2	2	1	3	3	3	1	2	1	1
25	1.409	0.333	0.236	0.066	0.082	3	2	3	4	2	3	4	2	3	3	2	2	1	1	1
35	1.184	0.172	0.143	0.028	0.050	3	1	3	4	2	2	4	3	2	1	1	2	3	2	2
35	1.164	0.164	0.119	0.026	0.050	1	1	3	1	2	2	4	3	2	1	1	2	3	1	2
33	1.126	0.115	0.092	0.005	0.026	2	1	2	1	2	2	4	3	2	1	1	2	3	1	2
29	1.207	0.149	0.080	0.019	0.033	1	1	2	3	2	2	4	3	2	1	1	2	3	1	2
28	1.071	0.071	0.071	0.000	0.025	3	1	2	2	2	2	4	3	2	1	1	2	3	1	2
28	1.071	0.071	0.071	0.004	0.021	1	1	2	3	2	2	4	3	2	1	1	2	3	2	2
37	1.265	0.243	0.222	0.058	0.075	4	2	3	3	2	2	4	3	2	1	2	2	3	1	2
25	1.160	0.120	0.120	0.035	0.046	1	1	3	3	2	3	3	2	2	1	1	1	3	2	1

Table 15.1 (*cont.*)

Species	Number of			Genetic indices based on shared loci				
	indi-viduals	popula-tions	loci	A	P–1 %	P–5%	H	He
36 S. ehrenbergi (2n = 54)								
(18, 19)	119	(b)	10	1.400	0.300	0.100	0.011	0.028
37 S. ehrenbergi (2n = 58)								
(18, 19)	97	(b)	10	1.300	0.200	0.200	0.034	0.091
38 S. ehrenbergi (2n = 60)								
(18, 19)	112	(b)	10	1.400	0.200	0.200	0.053	0.102
Sum	5866	162						
Means	154		12.1	1.461	0.315	0.235	0.052	0.086
Standard errors	35			0.051	0.029	0.025	0.007	0.009

(*a*) Abbreviations and codes: A, mean number of alleles per locus; *P–1*%, proportion of polymorphic loci, criterion 1% for polymorphism; *P–5*%, proportion of polymorphic loci, criterion 5% for polymorphism; *H*. mean heterozygosity per individual per locus; *He*, expected heterozygosity under panmixia (Nei, 1975); Gr, geographical range: 1 = endemic; 2 = narrow; 3 = regional; 4 = widespread; Hr, habitat range: 1 = habitat specialist; 2 = habitat generalist; Pst (Pt), population structure: 1 = isolate; 2 = patchy; 3 = continuous; Ar, aridity: 1 = arid; 2 = subarid; 3 = mesic; 4 = mesic and arid; Psz (Pz), population size: 1 = small; 2 = medium; 3 = large; Bs, body size: 1 = small; 2 = medium; 3 = large; Mo, adult mobility: 1 = sedentary; 2 = very low; 3 = low; 4 = high; Gf, dispersal of youngsters: 1 = low; 2 = medium; 3 = high; Lo, longevity: 1 = 1 year; 2 = up to 3 years; 3 = > 3 years; Gl, generation length: 1 = 1 year; 2 = 1.5 to 3 years; 3 = > 3 years; Fe, number of descendants: 1 = < 10; 2 = < 100; 3 = > 100; Ht, habitat type: 1 = subterranean habitat; 2 = others; Do, dormancy: 1 = in winter; 2 = in summer; 3 = no dormancy; Or, origin of species or population: 1 = old; 2 = recent; Tr, territoriality: 1 = territorial; 2 = nonterritorial; 3 = not applicable.
(*b*) *Spalax* data were collected from many localities, but only averages have been used.

Genetic variables

The review summarizes a re-analysis (Nevo, in preparation) of previously published and some unpublished studies of 38 plant and animal species in Israel, involving 162 populations and 5866 individuals each tested on the average at 26 loci (range 15–50) (Table 15.1). However, while some results involve also the nonshared loci among species, the major comparative analysis involves metabolically homologous (i.e., catalytically similar) enzymes encoded by 15 gene loci shared by most species. These include: α-glycerophosphate dehydrogenase (α*Gpd*); hexokinase (*Hk*); malate dehydrogenases, 2 loci (*Mdh-1,2*); phosphoglucose isomerase (*Pgi*); phosphoglucomutase (*Pgm*); 6-phosphogluconate dehydrogenase (*6Pgd*); aspartate aminotransferase (*Aat*); esterases, 2 loci (*Est-X,Y*); leucine amino peptidase (*Lap*); peptidases, 2 loci (*Pept-X,Y*); superoxide dismutase (*Sod*); and general protein (*Gp*). Since it is impossible to match homologous peptidases and esterases across different genera and higher taxa, we have consistently

Genetic indices based on all loci						Life history and ecological characteristics (a)														
No. of loci	A	P–1%	P–5%	H	He	Gr	Hr	Pt	Ar	Pz	Bs	Mo	Gf	Lo	Gl	Fe	Ht	Do	Or	Tr
25	1.280	0.240	0.200	0.016	0.024	1	1	3	3	2	3	3	2	2	1	1	1	3	2	1
25	1.200	0.160	0.120	0.037	0.057	1	1	3	3	2	3	3	2	2	1	1	1	3	2	1
25	1.360	0.280	0.160	0.069	0.065	1	1	3	3	2	3	3	2	2	1	1	1	3	2	1
26.6	1.403	0.293	0.208	0.043	0.074															
	0.044	0.027	0.021	0.005	0.008															

References
(1) Nevo, Zohary, Brown & Haber (1979)
(2) Nevo, Golenberg, Beiles, Brown & Zohary (1982)
(3) Nevo, Bar-El, Beiles & Yom-Tov (1982)
(4) Nevo (1978)
(5) Nevo, Bar-El & Bar (1983)
(6) Nevo, Bar-El, Beiles & Yom-Tov (1982)
(7) Nevo & Bar (1976)
(8) Nevo, Bar-El, Bar & Beiles (1981)
(9) Nevo, Alkalay & Blondheim (in preparation)
(10) Nevo & Yang (1979)

(11) Dessauer, Nevo & Chuang (1975)
(12) Nevo & Yang (1982)
(13) Nevo (1976)
(14) Nevo (1981)
(15) Nevo (1982a)
(16) Nevo (1982b)
(17) Bonhomme et al. (1982)
(18) Nevo & Shaw (1972)
(19) Nevo & Cleve (1978)

chosen for each of the 38 species analysed the two nonhomologous, most polymorphic, peptidases and esterases. The same was done in all cases where intergeneric homology was unknown.

The genetic data were all derived by routine standardized horizontal starch gel electrophoresis (Selander *et al.*, 1971) yielding conservative but evolutionarily meaningful comparative estimates of genic variation across all species. The original gene frequencies used in this review appear in the individual studies cited in the table. Five measures of intrapopulation genetic indices (GIs) were recalculated for each population and species based on 15 *shared* loci and are defined, symbolized and given along with GIs of all loci at the bottom of the table. They include indices of polymorphisms (criterion 1% and 5% – P–1%, P–5%), heterozygosity (H), gene diversity or expected heterozygosity under panmixia calculated from allelic frequencies, (He), and allele diversity (A) (see discussion of the properties of these measures in Nei, 1975; and in Karlin & Feldman, 1981).

Fig. 15.1. The geographic distribution of 162 tested populations at 93
localities in Israel.

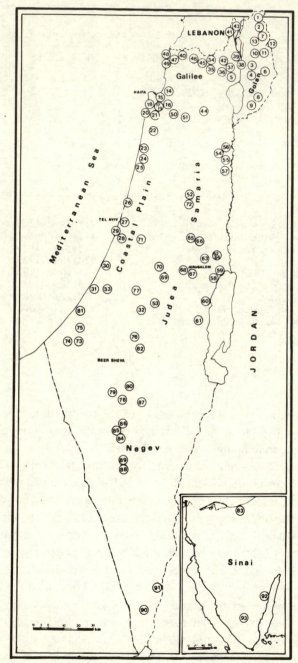

The sensitivity of electrophoresis as a detector of genetic variation was discussed by Ramshaw, Coyne & Lewontin (1979). Sequential electrophoresis, involving variable pH and gel concentration, seems to detect most amino-acid substitutions. However, the studies reported here were standardized and routine hence the resolution is probably about one third of the genetic variation at these loci within populations, and each mobility class, or electromorph (King & Ohta, 1975) may contain more than one allele. Thus the levels of genetic variation reported here are conservative but comparable.

Ecogeographical variables of localities
This review includes 162 populations sampled in 93 localities throughout Israel and some parts of Sinai (Fig. 15.1). The analysis involves 17 ecogeographical variables comprising geographical and climatic parameters used in the analysis. All variables are defined and symbolized in Fig. 15.4; their values are from maps in the *Atlas of Israel* (1970), and from records of the Israel Meteorological Service. The climatological records are multiple year averages derived from the closest meteorological station to the collecting sites.

Biotic variables of species
The 15 biotic, ecological and life history or populational variables of each of 38 species analysed are defined, symbolized, categorized and appear as 15-variable profiles in Table 15.1. Categorization was based on the individual papers in the table and their cited literature, as well as on field experience.

The evidence: ecological and biotic patterns and correlates of genetic diversity

Univariate analysis of genetic variation

Considerable variation in levels of gene diversity existed within and between species and among the loci tested. The means of overall gene diversity (*He* and *P–1%*) and *He* of 2 loci (*Pgm* and *Lap*) for each category of eight biotic factors appear in Fig. 15.2. Significant differences among higher taxa appear primarily in individual loci (*Pgm* and *Lap*) rather than in the GIs, except in mammals which displayed a significant reduction in *P–1%*.

Our results indicate that generalist and overground species and those with large population size, high fecundities, and low gene flow harbour

Fig. 15.2. Levels of mean genic diversity (*a*), polymorphism (*b*), genic diversity of PGM (*c*) and genic diversity of LAP (*d*). N, no. of species represented. Levels of significance are given. Note that LAP was monomorphic in several categories, and in these cases the sample size (N) is given unenclosed.

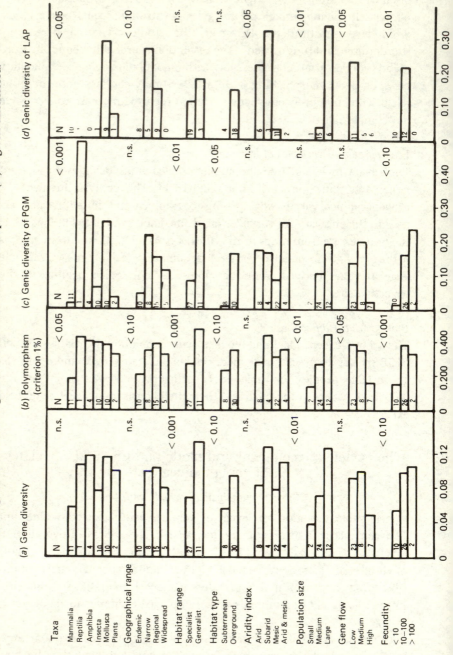

more genetic variation than do species with contrasting characteristics. Reduction in genetic diversity was evident in endemic species, as in plants (Hamrick, Linhart & Mitton, 1979) and in mammals (Nevo, 1982*d*). The genetic pattern of generalist–specialist species is in accord with that reported for 242 species (Nevo, 1978) and 140 mammalian species (Nevo, 1982*d*). Likewise, the pattern of overground versus underground species is in line with that reported for small mammals (Nevo, 1979) and for the 140 mammalian species.

The pattern described above highlights the importance of ecological factors in genetic differentiation. Notably, species with large populations were often associated with widespread distribution, high fecundity, habitat generalism, overground life, and patchy or continuous population distribution, all factors associated with high levels of genetic variation. Theoretically, habitat dimensionality is an important determinant of effective population size (Kimura & Ohta, 1971). Since animal numbers are regulated by the genetic composition of the population and by environmental factors (Ayala, 1968), the level of genetic diversity is related to environmental heterogeneity.

Our species sample was insufficient to resolve whether overall gene diversity (*He*) varies significantly among the categories of the remaining biotic factors which appear in the table, except in longevity.

Multivariate analysis of genetic variation

Discriminant analysis

In order to characterize, classify, and distinguish statistically between various groupings, data were analysed by stepwise discriminant analysis (Nie *et al.*, 1975). The results for three ecological and three populational factors of the 38 species appear in Fig. 15.3. In general, the analysis discriminated significantly ($P < 0.01$) between the ecological categories of geographical range, habitat range and aridity. Correctly classified groupings ranged between 81% and 90%. Likewise, the populational categories of population structure, population size and fecundity were discriminated significantly and correctly classified between 76% and 90%.

Smallest space analysis

Genetic diversity of species: biotic patterns. The multivariate pattern of the data was analysed by means of Smallest Space Analysis (SSA) which is a nonmetric technique designed to plot the intercorrelations of multivariate data in a space of minimal dimensionality (Guttman, 1968; Levy, 1981). The SSA treats each variable as a point in a Euclidean space in such a way

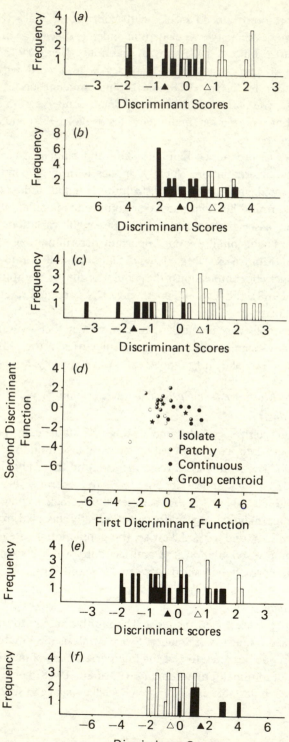

that, the higher the correlation between two variables, the closer they are in the space. The space of smallest dimensionality represents an inverse relationship between the observed correlations and the geometrical distances.

The multivariate pattern of the 38 species was represented graphically by three SSA-I diagrams (Fig. 15.4*a–c*) based on the shared and nonshared genetic indices (GIs) and gene diversity of the 15 polymorphic loci. SSA-Ia involved all 15 biotic factors (ecological, populational, life-history) while SSA-Ib involved only ecological, and SSA-Ic only populational parameters. Proximity of points in the diagrams reflected higher correlations, as is seen by the contiguity of all shared and nonshared genetic indices in all three diagrams. Notably, the genetic variables were related to the biotic structures. High genic diversity (*He*) in five loci (*Mdh-2, 6Pgd, Est-X,Y Gp*) was associated with the highest categories of aridity index, i.e., mesic environments, while another ten loci and all GIs were more genetically diverse in xeric regions. Likewise, high *He* of six loci (*Pgi, Pept-X,Y, Aat, Pgm, Gp*) and high variation in GIs was associated with widespread geographic range, habitat generalism, and overground habitats, while the opposite was largely true for eight other loci (Fig. 15.4*a, b*). Regional differentiation in the SSA diagram was also associated with population factors. Thus, high *He* in seven loci and GIs were related to large population size, patchy and isolated population structure, low gene flow, high fecundity and small body size, while the opposite was true for the other eight loci (Fig. 15.4*c*). Notably, the 14 enzymatic loci appear in two circles following the Gillespie–Kojima (1968) classification: an outer circle involving seven specific (glucose metabolizing) enzymes, and an inner circle involving seven nonspecific enzymes, indicating higher intercorrelations in the second group.

Fig. 15.3. Histograms of discriminant scores of the 38 species represented by ecological and populational parameters. Filled and open triangles, group centroids. (*a*) Geographical range. Discriminant variables: *P1%, A, 6 Pgd, Est-X.* 81% correctly classified, $P = 0.0095$. Filled columns, endemic or narrow; open columns, regional or widespread. (*b*) Habitat range. Discriminant variables: *Est-X, Pgm.* 85% correctly classified, $P = 0.0002$. Filled columns, specialist; open columns, generalist. (*c*) Aridity. Discriminant variables: *Est-X, He, P1%, Est-Y.* 90% correctly classified, $P = 0.0003$. Filled columns, arid; open columns, mesic. (*d*) Population structure. Discriminant variables: *Est-X, Pgi, Pept-X, 6-Pgd.* 77% correctly classified, $P = 0.0045$. (*e*) Population size. Discriminant variables: *He.* 76% correctly classified. $P = 0.0145$. Filled columns, small and medium; open columns, large. (*f*) Fecundity. Discriminant variables: *P1%, Est-X, 6-Pgd, A, P5%,* 90% correctly classified, $P = 0.002$. Filled columns < 10 descendants; open columns, > 10 descendants.

Fig. 15.4. Smallest Space Analyses on the 38 species (SSA–I a–c) and on 162 populations (SSA–I d–f). Genetic indices, GIs (1–5 based on 15 shared loci; 6–10 on all loci); He of 15 individual loci (11–25); ecological and life history variables (solid black circles); region of all genetic indices, GIs (dashed lines); outer circle connecting enzymes utilizing internal metabolites (11–16; dotted and dashed lines); inner circle connecting enzymes utilizing external metabolites. (17–24; dotted lines). Categorization of biotic factors appears in Table 15.1. All climatic factors and values are means from the *Atlas of Israel* (1970) and multiple year records of the Meteorological Service of Israel. Climatic and geographical variables

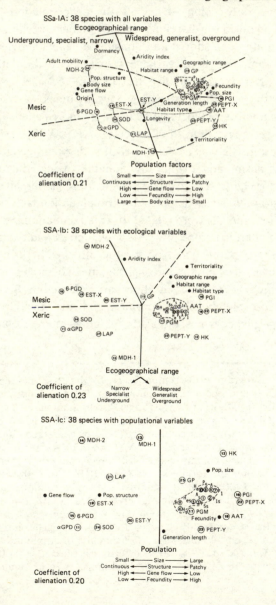

with their abbreviations: Longitude (Ln); Latitude (Lt); Altitude (Al); Mean annual temperature (Tm); Mean August temperature (Ta); Mean January temperature (Tj); Mean seasonal temperature difference (August minus January, Td); Mean daily difference of temperature (Tdd); Mean no. of tropical days (Trd); Mean annual rainfall (Rn); Mean annual no. of rainy days (Rd); Mean annual humidity (Huan); Mean humidity at 14 hr (Hu$_{14}$); Humidity index after Thornthwaite (Th); Mean no. of Sharav days (Sh); Mean no. of dew nights in summer (Dw); Mean annual evaporation (Ev).

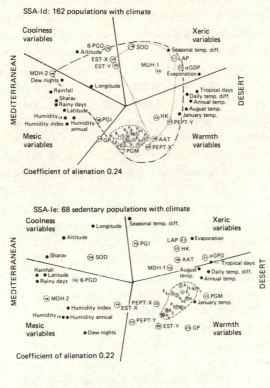

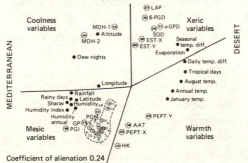

Genetic diversity of populations: climatic patterns. The multivariate pattern of the 162 populations was represented graphically by three SSA-I diagrams (Fig. 15.4*d*–*f*). Two regions emerged reflecting Israel's sharp climatic division. These were Desert and Mediterranean each subdivided by humidity variables (Hu14, Huan, Rd, Rn, Th), cooling variables (indicated by altitude, A1), xeric variables (Ev, Td), and warmth variables (Tj, Tm, Tq), all symbols defined in Fig. 15.4. In the diagram of all 162 populations (Fig. 15.4*d*) the shared and nonshared GIs appeared together in the desert warm subregion, indicating higher genetic variation in most populations towards the peripheral deserts, a trend climaxed in sedentary organisms (Fig. 15.4*e*). Genic diversity estimates for each locus (*He*) are located regionally in accord with climate. Thus, *Pgi* appeared in the Mediterranean mesic subregion: *6Pgd* and *Est-X,Y* in the Mediterranean cool subregion, while *Mdh-2* characterizes the entire Mediterranean region; *Lap*, α*Gpd*, *Mdh-1* and *Sod* in the desert xeric subregion, and *Hk*, *Pept-X,Y*, *Aat*, *Pgm* in the desert warm subregion. Enzymes were largely organized in two circles as previously described. SSA-I was computed in two, three and four dimensions. The goodness of fit of the diagram to the data set, represented by the coefficient of alienation, was 0.24, 0.15, 0.11, for two, three and four dimensions. Thus, even in two dimensions the fit of the data in the diagram was good. Adding dimensions does not add to the fit of the regionality to the content.

The SSA analysis was also conducted by subdividing the 162 populations into 68 sedentary organisms (plants and landsnails; Fig. 15.4*e*) and 94 nonsedentary organisms (vertebrates and insects; Fig. 15.4*f*). In sedentary organisms the GIs and two groups of loci (*Pgi*, *Lap*, *Hk*, *Aat*, *Mdh-1*, α*Gpd* and *Pgm*, *Gp*, *Pept-X,Y*, *Est-Y*) appeared in the desert, whereas another group (*Mdh-2*, *6Pgd*, *Sod*) appeared in the Mediterranean region. In nonsedentary organisms (Fig. 15.4*f*) the GIs and three loci (*Pgi*, *Pgm*, *Gp*) appeared in the Mediterranean mesic subregion whereas in sedentary organisms they appeared in the desert. Likewise, *6Pgd* and *Sod* were here located in the desert, while *Mdh-1,2* were associated with the Mediterranean cool subregion. Finally, six loci (*Lap*, α*Gpd*, *Pept-X,Y*, *Hk*, *Aat*) appeared consistently in all three SSA diagrams in the desert region.

Multiple regression analysis
Genetic diversity of species: biotic correlates. A first test for the best predictors of genic diversity (*He*) and the 15 shared loci of the 38 species (Nevo, in preparation) was conducted by stepwise multiple regression analysis, (Nie *et al.*, 1975) employing all 20 biotic and taxonomic variables or their ecological, populational and taxonomic subdivisions as independent

variables. In general, biotic parameters explained significantly a substantial amount of the genetic variance in *He* (Fig. 15.5) and in *He* of each of the 15 loci except *6-Pgd*. Thus for *He*, 0.58 of the variance was explained significantly $(P < 0.001)$ by four-variable combinations involving all variables. Similarly, the ecological, populational and taxonomic variables separately explained significantly 0.36, 0.43 and 0.21 of the variance in *He*, respectively. The same pattern was obtained for 14 out of 15 polymorphic loci analysed. Explanations averaged 0.50 and ranged from 0.28 in *Aat* to 0.75 in *Lap*. In sum, biotic factors explained a substantial amount of genetic variability. Notably, ecological, population, and taxo-nomic variables each explained alone more than a third, and together, more than half of the genetic variance.

Since problems arise in selection of the independent variables, and in our data set which may deviate from normality, both the level and extent of the significance found may be questioned and spuriousness invoked (Levins & Lewontin, 1980). To meet this criticism we randomized our data set and ran 10 different random sets by the same multiple regression program. The results suggested that a large amount of the significance reported here is

Fig. 15.5. Histograms displaying the explained levels of variance in gene diversity (He) represented by coefficients of multiple regression (R^2) of 38 species in Israel. The independent variables include all 20 biotic factors, 5 ecological, 5 populational and 5 taxonomic (Mammalia, Amphibia, Mollusca, Plants, Insecta) factors. Levels of significance: overall, ecological and populational factors, $P < 0.001$; taxonomic factors, $P < 0.05$.

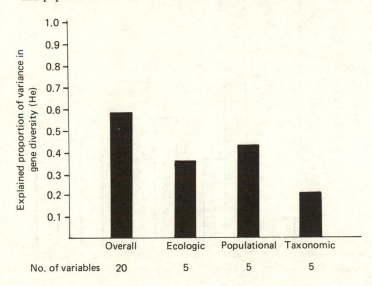

not spurious. For example, in the ecological subdivision the random simulations were less frequently significant at a lower level, and differed in combination of the independent variables. We concluded that although the real significance may be lower than that reported, and some spuriousness may be included, it appears that most of the correlations found are real and biologically meaningful.

Genetic diversity of populations: climatic correlates. A second MR study was conducted on the 162 populations employing genetic variables as dependent, and climatic and geographic as independent variables. In general, 80% of the GIs and 60% of the loci tested were significantly explained, though at a low proportion ($R^2 = 0.03-0.23$), by one to three variable combinations of water availability (*He*, *6-Pgd*, *Aat*), temperature (*Lap*), a combination of both (*Mdh-2*, *Est-X*, *Pgm*, *αGpd*), or a combination of climatic and geographical variables (*Pept-Y*, *Pgi*, *Gp*). The relatively low, yet significant,

Fig. 15.6. Histograms displaying the explained levels of variance in gene diversity (He) represented by coefficients of multiple regression (R^2) of 162 populations of Israel. The independent variables include 3 geographic and 10 out of the 14 climatic factors that appear in Fig. 15.4 *d–f.* The histograms represent overall, climatic, geographic and taxonomic subdivisions. Levels of significance: $P \leqslant 0.05$: overall, steppic, Amphibia, Mollusca; $P < 0.01$: mountains, plants; $P < 0.01$: Mediterranean, Coastal Plain.

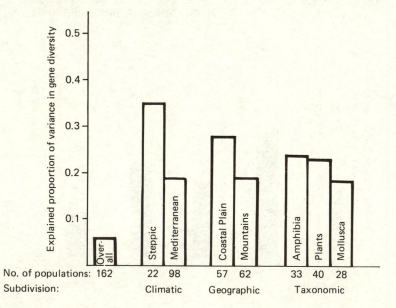

correlations of genetic variables with climate reflected the mixed and sometimes opposed species patterns. To overcome this problem, we ran a series of MR studies on subsets of our data subdivided into taxonomic, ecological, and population parameters, thereby increasing distinctly the level of explanation (Fig. 15.6).

The same procedure raised substantially the explanation of the variance of *He* of 14 loci by climate (*Pept-X,Y, Mdh-1,2, 6-Pgd. Est-X,Y, Aat, Pgm, Pgi, αGpd, Lap, Sod, Gp*). Only *Hk* remained unexplained even after subdivision, possibly owing to its fewer entries. The subdivided entries explained generally from about a fifth to a third of the genetic variance in *He*, only partly overlapping with the range of loci explained in the previous analysis of 38 species. Therefore, in sum, the combined MRs of biotic and climatic factors explained a substantial amount of the genetic variance in the 38 species and 162 populations.

Genetic differentiation in sedentary versus vagile species

The degree of genetic differentiation within and between populations of vagile species (insects and vertebrates) as compared with sedentary species (plants and landsnails) was analysed by *Dst*, or the average gene diversity between populations (Nei, 1973). This technique extends the Wright

Fig. 15.7. Genetic differentiation of populations in sedentary versus vagile species, following Nei (1973) Dst analysis. Ht = total gene diversity; Hs = mean within population diversity; Dst = Ht–Hs. Filled columns, 10 sedentary species; open columns, 13 vagile species.

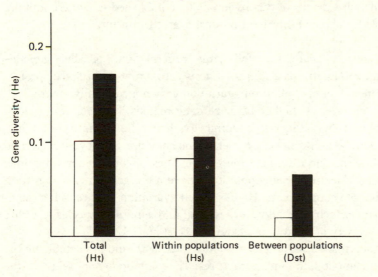

fixation index analysis *Fst* (Wright, 1965) to multiallelic structures. The results presented in Fig. 15.7 indicated distinct differential patterns between the two groups. Of the total genetic variation found in vagile organisms 19% (0.020/0.103) existed between populations as against 39% (0.066/0.170) in sedentary ones, i.e., genic divergence was about twicefold in the latter (Wilcoxon test, $P < 0.01$). This pattern was in line with that found for inbreeding plants (Brown, 1979), sedentary animals and snails (Selander & Kaufman, 1975), and vagile organisms, including man (Nei, 1975). This differential pattern reinforces the populational and ecological implications in genetic differentiation.

Testing neutrality by distribution of single loci heterozygosity

We have applied the neutrality test (Nei *et al.*, 1976) to the data discussed here and present the results in Fig. 15.8. Obviously, most points appeared below the lowest theoretical expected curve which was based on the stepwise mutation model. Out of 38 species, 27 showed lower variance in gene diversity (*He*), or higher *He* values, than the theoretical expectation. The probability of obtaining 27 or more points below the curve was $P = 0.007$. A sign test, assuming $P = 0.50$ below the curve, indicated a significant deviation from the expected ($P < 0.02$). Thus the null hypothesis of neutral mutations in our case was rejected. This conclusion would have been reinforced had our calculations been conducted on a steeper theoretical curve based on the more realistic stepwise mutation model assuming varying mutation rate among loci.

The detailed results of this review will be published elsewhere and are available in tabular form upon request from the author.

Our results indicate the following general and parallel genetic–environmental patterns and associations across many unrelated species.

(1) The amounts of allozymic variation vary nonrandomly among loci, populations, species, higher taxa, climatic regions, habitats, and species varying in geographic range, longevity, generation length, fecundity, population size and structure, dispersal and body size. Species characterized by generalism, large ranges, high fecundities, a long generation time, overground life and large populations, have more genetic variation than their opposite counterparts. Higher genetic variation implies richer genotypic array which may allow widespread and generalist species to utilize a wider variety of microhabitats (Hamrick *et al.*, 1979).

(2) Allozymic variation is partly correlated with and predictable by the physical climatic and biotic structures. The climatic factors relate to the

availability of water and to the temperature regime in accord with the climate of Israel which is divided distinctly into mesic and xeric regimes. Ecological and population factors each explain about a third and together more than half of the genetic variance. However, significant inter-correlations occur both within and between the ecologic, population and taxonomic categories, preventing the separate assessment of each category and its contribution to genetic diversity. While each category appears substantial, their intercorrelations seem to represent a confounded structure of nature in which the various forces interact rather than operate separately. This is of course also true for historical factors which affect population parameters. The disentanglement of some of these inter-correlations and the assessment of the separate contribution of the

Fig. 15.8. Relationship between the mean and variance among loci of gene diversity (He) of the 38 species, as a test of the neutral model (Nei *et al.*, 1976.)

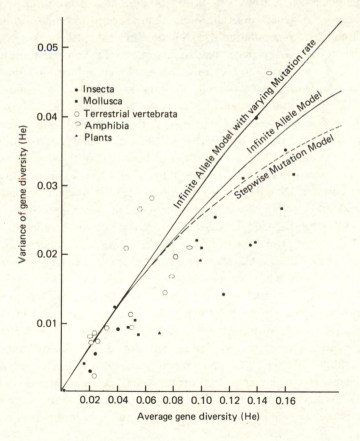

ecological factors necessitate controlled laboratory experimentation (Nevo, Lavie & Ben-Shlomo, 1982*a*).

(3) In most species that penetrate the xeric regions, either plants (*Hordeum spontaneum*), invertebrates (*Sphincterochila* landsnails), or vertebrates (*Agama, Acomys, Spalax*), the levels of polymorphism and heterozygosity increase towards the desert. Exceptions to this pattern are frogs, toads and mole crickets which are associated with water at least in part of their life cycle.

The theory: evolutionary significance of allozyme polymorphisms

Natural selection, genetic drift, and allozyme polymorphism

The parallel genetic patterns across many different species sharing similar ecologies suggest that positive Darwinian selection is massively involved in genetic differentiation of populations. Random factors are unlikely to be prime movers of genetic differentiation in most species and genes studied here. Neither current population size (N) nor historical bottle-necking can alone explain their genetic structure, in contrast to the conclusion drawn for 72 species analysed by Nei & Graur (personal communication). While N does contribute to the level of He in our data (Fig. 15.2) its contribution is at best similar to that of other intercorrelated population and ecological factors. In most species analysed here, even in the relatively low frequency populations of *Spalax* (Nevo, Heth & Beiles, 1982) the actual N in the entire species runs in each at least in the order of hundreds of thousands, and at most in millions of individuals. Assuming as did Nei & Graur a high correlation between N and effective population size (Ne), or even assuming a small ratio of 0.1 or less for Ne/N as calculated for some insects (Nei & Tajima, 1981), most species discussed here are less likely to be affected by genetic drift, either at present or in the past. Massive bottle-necking due to Pleistocene glaciations (Martin & Wright, 1976; Nei & Graur, personal communication), cannot explain their low but varied He values.

The analysis of the fossil record in Israel and the Levant (Tchernov, 1968, 1982), indicates that despite a general desiccation trend in the Levant during the Upper-Pleistocene and Holocene periods, the current Mediterranean region remained mesic without drastic perturbations towards aridity. This implies that past severe and extensive bottle-necking can not be invoked to explain major current genetic patterns. The climatic change in the Levant, from the colder and wetter Würm to the warmer and drier post-periglacial conditions, did not cause an overall faunal and floral turnover but resulted in morphogenetic changes due to increasing aridity and heat such as decreasing size in mammals following Bergmann's Rule *in time* (Tchernov, 1982). These are precisely the climatic factors explaining the genic differentiation in space discussed here. Thus the interaction of ecologic, population and historic factors best explains the genetic structure of our species.

The neutralist–selectionist controversy is yet unresolved. It will certainly not be settled by any single study, no matter how exhaustive, whether correlative or biochemical and physiological. In fact, it may never be resolved completely since the sharp dichotomy of either school may be unrealistic. The real problem is how much adaptiveness is implicated at the molecular level. Even this target may be unrealistic since as Wills (1981) noted 'it seems unlikely that we will ever be able to determine what proportion of all alleles in a population is neutral and what proportion is subject to selection'. Moreover, the issue was oversimplified and the sharp dichotomy between neutral and selected alleles is artificial, primarily since alleles may change direction of selection not only with the environment but also with the genetic background (Hartl & Dykhuizen, 1981) and hierarchy of selective units (Bargiello & Grossfield, 1979), analogous to Wright's shifting balance theory of evolution (Wright, 1970). In any event, our understanding of the relative roles of stochastic and deterministic factors at the molecular level depends on extensive and intensive field and critical laboratory studies of single and multi-locus systems in many species with full consideration of the ecological, biochemical, physiological and historical factors.

Critical analysis of the methodology of genetic-environment correlations

The problem of discriminating between selective and neutral models of electrophoretic profiles is often ambiguous and difficult (Ewens, 1979; Hartl, 1980). Notably, neutrality tests involve the largely unknown estimates of mutation rates, selection coefficients, population sizes and migration rates. While these estimates are hardly known or attainable in most natural populations it appears plausible to explore quantifiable genetic–environment associations (i.e., Sammeta & Levins, 1970; Bradshaw, 1971; Ford, 1971; Bryant, 1974; Nevo, 1978; Nevo *et al.*, 1979; Hamrick *et al.*, 1979), though they need additional evidence to support a causal mechanism and evaluate the proportion of selective and nonselective causes through fitness tests (see Hedrick, Ginevan & Ewing, 1976; Felsenstein, 1976 for critical reviews of visual and allozyme polymorphisms and their environmental correlates). In particular, temperature and water factors essential for life processes are correlated directly and indirectly with many important variables affecting life, as well as with niche and environ structures (Patten, 1981) and are easily quantifiable.

Genetic–environmental correlations seem therefore a powerful methodology if conducted and evaluated properly. This is particularly true if correlations are sought between allele frequencies at homologous loci in closely related sympatric species (i.e., Clarke, 1975; Harrison, 1977) and between allele frequencies and metabolically homologous shared loci in many unrelated species, as in this study.

The unequivocal demonstration of the adaptive significance of an enzyme polymorphism depends on interrelated multidisciplinary evidence. The latter must show that the enzyme phenotypic diversity varies with the environment and involves *in vivo* biochemical and physiological differences among allozyme variants that contribute to fitness. This has been demonstrated for leucine amino peptidase (*Lap*) polymorphism in the marine bivalve *Mytilus edulis* by Koehn and colleagues (Koehn, 1978; Koehn & Immerman, 1981). Yet such a comprehensive approach depends on extensive efforts in several experimental fields, finally demonstrating a highly probable adaptiveness in a specific case rather than substantiating a general pattern. Furthermore, it will usually prove very difficult to obtain all the experimental information an ideal analysis might require (Johnson, 1979). Recall that the neutral theory never questioned specific examples of adaptive evolution at the molecular level. Neutralists believe that adaptive evolution occurs by positive Darwinian selection only at a small proportion of genes, while most molecular evolutionary change is propelled by random drift (Nei, 1975; Kimura, 1979*b*).

Genetic–environmental correlations demonstrate inferentially the adaptive significance of enzyme polymorphisms. Yet, if conducted on many loci and over many populations and species sharing an ecological background, as in this study, they may roughly quantify the proportion of genic diversity and/or gene substitutions involved in adaptive evolution, at least of structural genes coding for soluble enzymes. Remarkably, most loci analysed here are similarly correlated with the environment in many unrelated species. Thus the assumption of the neutral theory that only a small proportion of genes are involved in adaptive molecular evolution is inconsistent with our data. Furthermore, since these correlations may be viewed as independent evolutionary trials due to the absolute reproductive isolation among most taxa tested, the underlying shared ecology appears to suggest the causative mechanism of genetic differentiation. Clearly, the correlations found are descriptive rather than deterministic. They do not establish directly causal relations between climatic and genetic variation. Yet their consistency, over many species, strongly suggests that they score indirectly useful, presumably genuine, associations between climate and allozymes. These are presumably mediated at least partly by climatic selection, apparently directed at allele frequencies of the specific loci studied, rather than to linked loci. Other correlates of genic diversity (i.e., quaternary structure, subunit molecular weight) do not exclude the environmental molecular correlations analysed here.

The methodology of genetic–environmental correlations was extensively criticized due to its supposed nonexistence, nonconvincingness, spuriousness, noncausality, nongenetic mechanisms, inferentiality, etc. (Schnell & Selander, 1981; Nei & Graur, and others). While caution must be exercised (Levins & Lewontin, 1980), overdue pessimism is unwarranted. Taken individually, any of the cases analysed here could be dismissed as a mere and partly spurious genetic–environmental correlation. However, these observations collectively constitute coherent corroborating evidence of parallel *genetic–environmental* associations in many unrelated species differing in their biological and historical records but sharing a similar varying and stressful ecological background. The demonstration of such genetic parallelism in mostly long-closed genetic systems which do not exchange genes, strongly suggests that a substantial amount of seemingly adaptive differentiation of enzyme polymorphisms in nature is mediated by natural selection as an organismal response to climatic challenges. While stochasticity, founder effects, genetic drift, sampling, neutrality, initial conditions and constraints may be secondarily involved they are unlikely to primarily explain, singly or in combination, parallel genetic patterns on such an extensive scale. The long but promising road is now open for critically testing the allozymes analysed here, to evaluate *directly* their hypothesized causal environmental relations through biochemical kinetics and physiological function, and assess their presumed differential contribution to fitness. The prospective results are expected to unveil the underlying genetic mechanism(s) and the total interaction of forces causing differentiation and maintenance of allozyme polymorphisms.

Biochemical and physiological studies of allozyme polymorphism

Numerous studies demonstrated *in vitro* kinetic differences between isozymes (many articles in Markert, 1975) and allozyme variants (Koehn, 1978; Place & Powers, 1979; for detailed references see Nevo, 1982c). Although *in vitro* kinetic differences between allozymes could indicate that selection is operating, only the demonstration of *in vivo* physiological differences that ultimately determine fitness may substantiate the selection hypothesis (Lewontin, 1974). Several studies have indeed described such *in vivo* differences affecting fitness (Koehn, 1978; Cavener & Clegg, 1981; DiMichele & Powers, 1982; Burton & Feldman, 1982). Finally, environmental and physiological stresses may unveil fitness differences among allozyme variants of several enzymes such as amylase, alcohol dehydrogenase, glucose-6-phosphate dehydrogenase and 6-phosphogluconate dehydrogenase, lactate dehydrogenase, aminopeptidase-I, phosphoglucomutase and phosphoglucose isomerase (see references in Nevo, 1982*c*).

In sum, this study corroborates previous ones (Nevo, 1978; Hamrick *et al.*, 1979) in suggesting a general pattern in nature. Genic diversity of allozymes varies nonrandomly among loci, populations, species and habitats and is at least partly correlated with and propelled by environmental heterogeneity in space and time as predicted by Darwinian evolutionary theory. Natural selection, primarily through spatiotemporally varying environments and epistasis, appears to be an important determinant of genetic differentiation thereby maintaining genetic polymorphisms in nature which are largely adaptive and provide the basis of adaptive evolutionary change. The correlative approach, if conducted properly, may provide clues concerning the relative roles and interaction of selective and nonselective forces acting at the ecological–genetic interface. The environmental associations can then suggest promising and critical biochemical, physiological and biological experiments to unveil potential differential fitness components of allozymes and establish unequivocally their adaptive nature. These combined field and laboratory observations and experiments may help to distinguish between alternative hypotheses about the origin and evolution of allozyme and DNA polymorphisms in nature and their roles in adaptation and speciation.

Theoretical overview and prospects

The idea of positive correlation between genetic and environmental variation, the niche width variation hypothesis, is widespread in evolutionary biology (Levene, 1953; Levins, 1968). It was tested critically and confirmed for karyotypic, electrophoretic (Powell, 1971; McDonald & Ayala, 1974; Powell & Wistrand, 1978) and quantitative traits (Mackay, 1981 and references therein). It was also supported by allozyme studies comparing habitat specialist and generalist species (Nevo, 1978, 1982*d* and this study) and by life history analysis (Hamrick *et al.*, 1979). Theoretically, the existence of a protected polymorphism is more likely in more heterogeneous environments (Karlin, 1982*a*). Likewise, models with spatial heterogeneity and limited migration may be more effective in maintaining genetic polymorphism than alternative models with less environmental heterogeneity and higher migration (Karlin & McGregor, 1972). Furthermore, theory predicts that a mixture of underdominance, directional, and overdominant spatially varying selection can produce a wide variety of stable polymorphic and/or fixation states, and spatial is more effective than temporal variation in protecting polymorphisms (Karlin, 1982*a*).

The two-niche Levene model (1953) has been extended recently to multilocus and multiniche structures (reviews in Hedrick *et al.*, 1976; Felsenstein, 1976; Hedrick, Jain & Holden, 1978; Karlin, 1979, 1982*a*, *b*). Gillespie and colleagues (Gillespie, 1974*a*, *b*, 1977; Gillespie & Langley, 1974) concluded on theoretical grounds that allozymic polymorphism is primarily due to selection acting on environmental variation in gene function. Genetic variation will be more likely in spatio-temporally more variable environments than in constant ones. Furthermore, heterozygote intermediacy plus random environmental fluctuations are sufficient elements to explain genetic variability, and the conditions for polymorphism in heterogeneous environments are less stringent than those of overdominant selection in multiple allelic systems, as also concluded by Lewontin, Ginzburg & Tuljapurkar (1978). Theoretically, non-overdominant stable equilibrium may be generated even in a single niche if selection coefficients vary from generation to generation in a specific manner in infinitely large populations (Dempster, 1955; Haldane & Jayakar, 1963; Karlin & Levikson, 1974).

The maintenance of genetic polymorphism under various natural structured viability regimes versus random fitness assignments was theoretically compared by Karlin (1981). An increased likelihood for a globally stable equilibrium is predicted for the more structured viability models. Accordingly, 'if observed allele frequency data exhibit a reasonably consistent common set over different populations and epochs, the contingencies for a structured selection mechanism may be relevant. On the other hand, where the allele frequency observations vary significantly in space or time with few segregating alleles in any particular sample, an explanation of the observed variability based on fitness interactions is unlikely. Other forces, such as migration, population structure, mating pattern, genetic frequency and/or ecological density factors and strong randomizing recombination interactions, may be important'. We conclude that the data analysed in this study exhibit a reasonably consistent common set over different populations, species and areas, and hence are primarily explained by a structured selection mechanism as predicted by theory.

My deep gratitude is extended to all my colleagues, too numerous to enlist here, who collaborated in the continuous efforts to understand genetic differentiation in nature. Thanks are due to L. Guttman, who greatly assisted in the multivariate Smallest Space Analysis and its application to our data, and to S. Heiberman, who generously advised in the Multiple Regression Analysis. Special thanks are due to A. Beiles for many stimulating conversations, arguments, discussions and the indispensable statistical analyses conducted throughout this study. Thanks for constructive comments on the manuscript are due to D. Adler, R. Ben-Shlomo, S. Karlin and M. W. Feldman. This study was supported by grants from the United States–Israel Science Foundation (BSF), Jerusalem, Israel.

Further reading

Lewontin, R. C. (1974). *The Genetic Basis of Evolutionary Change.* New York: Columbia University Press.

Brussard, P. F. (1978). *Ecological Genetics: The Interface.* New York: Springer Verlag.

Hartl, D. L. (1980). *Principles of Population Genetics.* Sunderland, Mass.: Sinauer Assoc. Inc.

References

Atlas of Israel (1970). Surveys of Israel. Jerusalem: Ministry of Labour; Amsterdam: Elsevier.

Ayala, F. J. (1968). Genotype, environment and population numbers. *Science,* 16, 1453–9.

Ayala, F. J. (ed.) (1976). *Molecular Evolution.* Sunderland, Mass.: Sinauer Assoc. Inc.

Ayala, F. J. (1977). Protein evolution in different species: is it a random process? In *Molecular evolution and polymorphism,* ed. M. Kimura, pp. 73–102, Mishima, Japan: National Institute of Genetics.

Bargiello, T. & Grossfield, J. (1979). Biochemical polymorphisms: the unit of selection and the hypothesis of conditional neutrality. *Biosystems,* 11, 183–92.

Bonhomme, F., Catalan, J., Britton-Davidian, J., Chapman, V. R., Moriwaki, K., Nevo, E. & Thaler, L. (1982). Biochemical diversity and evolution in the genus *Mus. Biochemical Genetics* (in press).

Bradshaw, A. D. (1971). Plant evolution in extreme environments. In *Ecological Genetics, and Evolution.* New York: Appleton-Century-Crofts.

Brown, A. H. D. (1979). Enzyme polymorphism in plant populations. *Theoretical Population Biology,* 15, 1–42.

Brussard, P. F. (ed.) (1978). *Ecological Genetics: The Interface.* New York: Springer Verlag.

Bryant, E. H. (1974). On the adaptive significance of enzyme polymorphisms in relation to environmental variability. *American Naturalist,* 108, 1–19.

Burton, R. S. & Feldman, M. W. (1982). Physiological and fitness effects of an allozyme polymorphism: Glutamate-Pyruvate transaminase and response to hyperosmotic stress in the copepod *Tigripus californicus. Biochemical Genetics* (in press).

Cavener, D. R. & Clegg, M. T. (1981). Evidence for biochemical and physiological differences between enzyme genotypes in *Drosophila melanogaster*. *Proceedings of the National Academy of Science, USA*, **78**, 4444–7.

Clarke, B. (1975). The contribution of ecological genetics to evolutionary theory: detecting the direct effects of natural selection on particular polymorphic loci. *Genetics Supplement*, **79**, 101–13.

Clarke, B. (1979). The evolution of genetic diversity. *Proceedings of the Royal Society of London, B*, **205**, 453–74.

Dempster, E. (1955). Maintenance of genetic heterogeneity. *Cold Spring Harbor Symposium of Quantitative Biology*, **20**, 25–32.

Dessauer, H. C., Nevo, E. & Chuang, K. C. (1975). High genetic variability in an ecologically variable vertebrate, *Bufo viridis*. *Biochemical Genetics*, **13**, 651–61.

DiMichele, L. & Powers, D. A. (1982). LDH-B genotype-specific hatching times of *Fundulus heteroclitus*. *Nature*, **296**, 563–4.

Ewens, W. J. (1979). *Mathematical Population Genetics*. Berlin: Springer Verlag.

Ewens, W. J. & Feldman, M. W. (1976). The theoretical assessment of selective neutrality. In *Population Genetics and Ecology*, ed. S. Karlin & E. Nevo, pp. 303–37. New York: Academic Press.

Felsenstein, J. (1976). The theoretical population genetics of variable selection and migration. *Annual Review of Genetics*, **10**, 253–81.

Ford, E. B. (1971). *Ecological Genetics*, 3rd edn. London: Chapman & Hall.

Gillespie, J. H. (1974*a*). Polymorphism in patchy environments. *The American Naturalist*, **108**, 145–51.

Gillespie, J. H. (1974*b*). The role of environmental grain in the maintenance of genetic variation. *American Naturalist*, **108**, 831–6.

Gillespie, J. H. (1977). A general model to account for enzyme variation in natural populations. III. Multiple Alleles. *Evolution*, **31**, 85–90.

Gillespie, J. H. & Kojima, K. (1968). The degree of polymorphism in enzymes involved in energy production compared to that in nonspecific enzymes in two *D. ananassae* populations. *Proceedings of the National Academy of Sciences, USA*, **61**, 582–5.

Gillespie, J. H. & Langley, C. H. (1974). A general model to account for enzyme variation in natural populations. *Genetics*, **76**, 837–48.

Guttman, L. (1968). A general nonmetric technique for finding the smallest coordinate space for a configuration of points. *Psychometrika*, **33**, 469–506.

Haldane, J. B. S. & Jayakar, S. D. (1963). Polymorphism due to selection of varying direction. *Journal of Genetics*, **58**, 237–42.

Hamrick, J. L., Linhart, Y. B. & Mitton, J. B. (1979). Relationships between life history characteristics and electrophoretically detectable genetic variation in plants. *Annual Review Ecology & Systematics*, **10**, 173–200.

Harrison, R. G. (1977). Parallel variation at an enzyme locus in sibling species of field crickets. *Nature*, **266**, 168–70.

Hartl, D. L. (1980). *Principles of Population Genetics*, Sunderland, Mass.: Sinauer Assoc. Inc.

Hartl, D. L. & Dykhuizen, D. E. (1981). Potential for selection among nearly neutral allozymes of 6-phosphogluconate dehydrogenase in *Escherichia coli*. *Proceedings of the National Academy of Sciences, USA*, **78**, 6344–8.

Hedrick, P. W., Ginevan, M. E. & Ewing, E. P. (1976). Genetic polymorphism in heterogeneous environments. *Annual Review Ecology and Systematics*, **7**, 1–32.

Hedrick, P., Jain, S. & Holden, L. (1978). Multilocus systems in evolution. *Evolutionary Biology*, **11**, 101–85.

Johnson, G. (1979). Genetic polymorphism among enzyme loci. In *Physiological Genetics*, ed. J. G. Scandalios. New York: Academic Press.

Karlin, S. (1979). Principles of polymorphism and epistasis for multilocus systems. *Proceedings of the National Academy of Sciences, USA*, **76**, 541–5.

Karlin, S. (1981). Some natural viability systems for a multiallelic locus: a theoretical study. *Genetics*, **97**, 457–73.

Karlin, S. (1982a). Classifications of selection-migration structures and conditions for a protected polymorphism. *Evolutionary Biology*, **14**, 61–204.

Karlin, S. (1982b). *Theoretical Population Genetics*. New York: Academic Press.

Karlin, S. & Feldman, M. W. (1981). A theoretical and numerical assessment of genetic variability. *Genetics*, **97**, 475–93.

Karlin, S. & Levikson, B. (1974). Temporal fluctuations in selection intensities: Case of small population size. *Theoretical Population Biology*, **6**, 383–412.

Karlin, S. & McGregor, J. L. (1972). Application of method of small parameters to multiniche population genetic models. *Theoretical Population Biology*, **3**, 186–209.

Kimura, M. (1968). Evolutionary rate at the molecular level. *Nature*, **217**, 624–6.

Kimura, M. (1979a). Model of effectively neutral mutations in which selective constraint is incorporated. *Proceedings of the National Academy of Science, USA*, **76**, 3440–4.

Kimura, M. (1979b). The neutral theory of molecular evolution. *Scientific American*, **241**, 98–126.

Kimura, M. & Ohta, T. (1971). *Theoretical Aspects of Population Genetics*. Princeton: Princeton University Press.

King, J. L. & Ohta, T. H. (1975). Polyallelic mutational equilibria. *Genetics*, **79**, 681–91.

Koehn, R. K. (1978). Physiology and biochemistry of enzyme variation: The interface of ecology and population genetics. In *Ecological Genetics: The Interface*, ed. P. Brussard, pp. 51–72. New York: Springer Verlag.

Koehn, R. K. & Immerman, F. W. (1981). Biochemical studies of aminopeptidase polymorphism in *Mytilus edulis*. I. Dependence of enzyme activity on season, tissue and genotype. *Biochemical Genetics*, **19**, 1115–42.

Levene, H. (1953). Genetic equilibrium when more than one ecological niche is available. *American Naturalist*, **87**, 331–3.

Levin, S. A. (1978). On the evolution of ecological parameters. In *Ecological Genetics: The Interface*, ed. P. Brussard, pp. 3–26. New York: Springer Verlag.

Levins, R. (1968). *Evolution in Changing Environments*. New Jersey: Princeton University Press.

Levins, R. & Lewontin, R. (1980). Dialectics and reductionism in ecology. *Synthese*, **43**, 47–78.

Levy, S. (1981). Lawful roles of facets in social theories. In *Multidimensional Data Representation: When and Why*, ed. I. Borg. Ann Arbor: Mathesis Press.

Lewontin, R. C. (1974). *Genetic Basis of Evolutionary Change*. New York: Columbia University Press.

Lewontin, R. C., Ginzburg, L. R. & Tuljapurkar, S. D. (1978). Heterosis as an explanation for large amounts of genic polymorphism. *Genetics*, **88**, 149–69.

Mackay, T. F. C. (1981). Genetic variation in varying environments. *Genetical Research, Cambridge*, **37**, 79–93.

Markert, C. L. (ed.) (1975). *Isozymes*. Volumes I–IV. New York: Academic Press.

Martin, P. S. & Wright, H. E. (1976). *Pleistocene extinctions*. New Haven: Yale University Press.

Mayr, E. (1972). Lamarck revisited. *Journal of the History of Biology*, **5**, 55–94.

McDonald, J. F. & Ayala, F. J. (1974). Genetic response to environmental heterogeneity. *Nature*, **250**, 572–4.

Milkman, R. D. (1976). Selection is the major determinant. *Trends in Biochemical Sciences*, **1**, N152–4.

Nei, M. (1973). Analysis of gene diversity in subdivided populations. *Proceedings of the National Academy of Sciences, USA*, **70**, 3321–3.

Nei, M. (1975). *Molecular Population Genetics and Evolution*. Amsterdam: North Holland Publ. Co.

Nei, M., Fuerst, P. A. & Chakraborty, R. (1976). Testing the neutral mutation hypothesis by distribution of single locus heterozygosity. *Nature*, **262**, 491–3.

Nei, M. & Tajima, F. (1981). Genetic drift and estimation of effective population size. *Genetics*, **98**, 625–40.

Nevo, E. (1976). Genetic variation in constant environments. *Experientia*, **32**, 858–9.

Nevo, E. (1978). Genetic variation in natural populations: patterns and theory. *Theoretical Population Biology*, **13**, 121–77.

Nevo, E. (1979). Adaptive convergence and divergence of subterranean mammals. *Annual Review Ecology and Systematics*, **10**, 269–308.

Nevo, E. (1981). Genetic diversity and climatic selection of the lizard *Agama stellio* in Israel and Sinai. *Theoretical and Applied Genetics*, **60**, 369–80.

Nevo, E. (1982*a*). Genetic differentiation and speciation in spiny mice, *Acomys*. *Proceedings of the 3rd International Theriological Congress, Helsinki*: August, 1982.

Nevo, E. (1982*b*). Genetic structure and differentiation during speciation in fossorial gerbil rodents. *Mammalia*, (in press).

Nevo, E. (1982*c*). Adaptive significance of protein variation. *Proceedings of the British Systematics Association Symposium*. Adaptive and Taxonomic Significance of Protein Variation, York: 13–15 July, 1982. Academic Press.

Nevo, E. (1982*d*). Ecological and populational correlates of allozyme polymorphisms in mammals. *Proceedings of the 3rd International Theriological Congress, Helsinki*: August, 1982.

Nevo, E. & Bar, Z. (1976). Natural selection of genetic polymorphisms along climatic gradients. In: *Population Genetics and Ecology*, ed. S. Karlin & E. Nevo, pp. 159–84. New York: Academic Press.

Nevo, E., Bar-El, Ch. & Bar, Z. (1983). Genetic diversity, climatic selection and speciation of *Sphincterochila* snails in Israel. *Biological Journal Linnean Society*.

Nevo, E., Bar-El, Ch., Bar, Z. & Beiles, A. (1981). Genetic structure and Climatic correlates of desert landsnails. *Oecologia*, **48**, 199–208.

Nevo, E., Bar-El, Ch., Beiles, A. & Yom-Tov, Y. (1982). Adaptive microgeographic differentiation of allozyme polymorphisms in landsnails. *Genetica*, **59**, 61–7.

Nevo, E. & Cleve, H. (1978). Genetic differentiation during speciation. *Nature*, **275**, 125–6.

Nevo, E., Golenberg, E., Beiles, A., Brown, A. H. D. & Zohary, D. (1982). Genetic diversity and environmental associations of wild wheat. *Triticum dioccoides*, in Israel. *Theoretical and Applied Genetics*, **62**, 241–54.

Nevo, E., Heth, G. & Beiles, A. (1982). Population structure and evolution in subterranean mole rats. *Evolution*.

Nevo, E., Lavie, B. & Ben-Shlomo, R. (1982). Selection of allozyme polymorphisms in marine organisms: pattern theory and application. *Proceedings, 4th International Congress on Isozymes*. In *Isozymes: Current Topics in Biological and Medical Research*, ed. M. Siciliano; New York: Liss Publ.

Nevo, E. & Shaw, C. (1972). Genetic variation in a subterranean mammal, *Spalax ehrenbergi. Biochemical Genetics*, **7**, 235–41.

Nevo, E., Zohary, D., Brown, A. H. D. & Haber, M. (1979). Genetic and environmental associations of wild barley, *Hordeum spontaneum*, in Israel. *Evolution*, **33**, 815–33.

Nevo, E. & Yang, S. Y. (1979). Genetic diversity and climatic determinants of tree frogs in Israel. *Oecologia*, **41**, 47–63.

Nevo, E. & Yang, S. Y. (1982). Genetic diversity and ecological relationships of marsh frog populations in Israel. *Theoretical & Applied Genetics*, **63**, 317–30.

Nie, N. H., Hull, C. H., Jenkins, J. G., Steinbrenner, K. & Bent, D. H. (1975). *SPSS: Statistical Package for the Social Sciences*. 2nd. edn. New York: McGraw Hill Book Co.

Patten, B. C. (1981). Environs: The superniches of ecosystems. *American Zoologist*, **21**, 845–52.

Place, A. R. & Powers, D. (1979). Genetic variation and relative catalytic efficiencies: Lactate dehydrogenase B allozymes of *Fundulus heteroclitus. Proceedings of the National Academy of Sciences, USA*, **76**, 2534–8.

Powell, J. R. (1971). Genetic polymorphism in varied environments. *Science*, **174**, 1035–6.

Powell, J. R. & Wistrand, H. (1978). The effect of heterogeneous environments and a competitor on genetic variation in *Drosophila. American Naturalist*, **112**, 935–47.

Ramshaw, J. A. M., Coyne, L. A. & Lewontin, R. C. (1979). The sensitivity of gel electrophoresis as a detector of genetic variation. *Genetics*, **93**, 1019–37.

Roughgarden, J. (1979). *Theory of Population Genetics and Evolutionary Ecology: An Introduction*. New York: McMillan.

Sammeta, K. P. V. & Levins, R. (1970). Genetics and ecology. *Annual Review of Genetics*, **4**, 469–88.

Schnell, G. D. & Selander, R. K. (1981). Environmental and morphological correlates of genetic variation in mammals. In *Mammalian Population Genetics*, ed. M. H. Smith & J. Joule, pp. 60–99. Athens: The University of Georgia Press.

Selander, R. K. & Kaufman, D. W. (1975). Genetic structure of populations of the brown snail (*Helix aspersa*). I. Microgeographic variation. *Evolution*, **29**, 385–401.

Selander, R. K., Smith, M. H., Yang, S. Y., Johnson, W. E. & Gentry, G. B. (1971). Biochemical polymorphism and systematics in the genus *Peromyscus*. I. Variation in the old field-mouse (*Peromyscus polionotus*). *University of Texas Publ. 7103*, 49–90.

Tchernov, E. (1968). Succession of rodent faunas during the Upper Pleistocene of Israel. Hamburg and Berlin: Mammalia Depicta. Paul Parey.

Tchernov, E. (1982). Faunal turnover and extinction rate in the Levant. In *Pleistocene Extinction*, ed. P. S. Martin.

Van Valen, L. (1974). Molecular evolution as predicted by natural selection. *Journal of Molecular Evolution*, **3**, 89–101.

Wills, C. (1981). *Genetic Variability*. Oxford: Clarendon Press.

Wright, S. (1965). The interpretation of population structure by *F*-statistics with special regard to systems of matings. *Evolution*, **19**, 355–420.

Wright, S. (1970). Random drift and the shifting balance theory of evolution. In *Mathematical Topics in Population Genetics*, ed. Kojima, pp. 1–31, New York: Springer Verlag.

Wright, S. (1969). Evolution and the Genetics of Populations. Vol. 2. *The Theory of Gene Frequencies*. Chicago: The University of Chicago Press.

Wright, S. (1978). Evolution and the Genetics of Populations. Voi. 4. *Variability Within and Among Natural Populations*. Chicago: The University of Chicago Press.

Zohary, M. (1973). *Geobotanical Foundations of the Middle East*. Stuttgart: Fischer Verlag, pp. 738.

16

A Darwinian plant ecology

JOHN L. HARPER

'...on a piece of ground 3 feet long and 2 feet wide, dug and cleared, and where there could be no choking from other plants, I marked all the seedlings of our native weeds as they came up, and out of 357 no less than 295 were destroyed, chiefly by slugs and insects. If turf which has long been mown, and the case would be the same with turf closely browsed by quadrupeds, be let to grow, the more vigorous plants gradually kill the less vigorous, though fully grown plants; thus out of 20 species growing on a little plot of mown turf (3 feet by 4 feet) 9 species perished, from the other species being allowed to grow up freely' (*The Origin of Species*).

It is difficult to detect any direct influence of Darwin's writings on the development of the main stream of plant ecology. The extreme reductionist approach that is represented in the above quotation, and is apparent again and again in his writings, not only in *The Origin of Species* but in many of his later books, is conspicuous by its absence from early plant ecological texts and is barely represented in the ecological literature until towards the middle of the twentieth century. The approach that involved marking individual plants or seedlings in the field, tracing the fate of individual leaves as they are pulled down earthworm burrows, the behaviour of tendrils as they touch a support, the fate of insects as they land on a *Drosera* leaf, or recording the number of seeds at the bottom of an earthworm burrow, represented a reductionist level of concentrated observation that contrasted with the geographical view of vegetation with which Warming and others set the early direction of plant ecology. There is no way in which Darwin can be regarded as a parent of the science of plant ecology. Nothing illustrates this point so clearly as the fate of a paper published in 1874 by C. Nägeli entitled *Verdrängung der Pflanzenformen durch ihre Mitbewerber*. This paper was directly stimulated by *The Origin of Species* and in it Nägeli developed mathematical models that describe the interaction between populations of two species of plants. His models attempt to describe in formal mathematical terms the replacement of a population of one species by another and also situations in which pairs of species persist together as stable, mixed populations. Nägeli's models included density-dependent and frequency-dependent situations and came very close to providing a formal description of the niche. It might have been expected that such a paper from Nägeli, who was one of the most distinguished botanists in Europe at that time, would have had immense impact on the early development of plant ecology. Instead, it appears to have been wholly ignored for 60 years until it was mentioned briefly by Gauze (1934) in his book *The Struggle for Existence*.

Plant ecology developed not as a study of the factors affecting the lives and deaths of individual plants and their parts but as a study of the distribution of vegetation types and of particular species. It included also the description of those specialized features of morphology and physiology that distinguish species and might (often by more or less inspired guesswork) be said to account for the differences in their distribution. That these features were often called 'adaptations' did nothing to explain them. Much of the early science of plant ecology sought for correlations between vegetation and physical, not biotic, factors in the environment, most particularly temperature, water supply and soil types. These forces have been described as 'Wallacian' (Harper, 1977), because they represent those agents of natural selection that were of more concern to Wallace than to Darwin in accounting for how organisms are as they are and behave as they do. The role of biotic (Darwinian) forces in determining the distribution and abundance of species was largely neglected except by token reference to grazing animals. The essentially Darwinian forces of struggle for existence, involving competitive interactions between members of the same species and between different species, played a negligible part in the interpretation of natural vegetation.

The Darwinian approach, involving reductionist concentration on individual plants and the hazards that they experience, particularly the interference from their neighbours, entered the science of plant ecology in the late nineteen twenties in simple experiments involving mixed populations of two or more species and I have described elsewhere (Harper, 1967) the curious piece of history in which three leading ecologists, Sukatschev in Russia, Clements in the United States and Tansley in Britain, all made simple competition experiments involving deliberately-sown plant populations. None of these authors continued with this type of study and all returned to essentially descriptive studies of vegetation. It was as if they had found it too difficult to bridge the gap between simple experimental systems and the complexity of nature – as if the reductionism of the experimental method lost the holist qualities of the integrated complex whole that these distinguished ecologists saw in natural vegetation. Both Clements' *Community as an Organism* and Tansley's *Ecosystem* can be represented as a retreat to community holism after a brief flirtation with an organismal, Darwinian ecology.

A piece of ecological history that remains to be fully researched was the decision by a number of individuals, many apparently working in isolation from each other, to establish, like Darwin, permanent plots within which the fate of individual plants could be recorded over time. The great classic amongst such observations is the work of Carl Olaf Tamm (1948, 1972) who marked out permanent quadrats in woodland and grassland near his country home and as a hobby mapped, remapped and continues to map the plants within them. Tamm attributes his decision to embark on this long-term programme to stimulus from Romell, a distinguished Swedish ecologist and soil scientist. Other permanent quadrats were set up by Forrest Shreve (1915) at the Desert Laboratory of the Carnegie Institute of Washington at Tucson, Arizona, and it appears to be through a colleague of Shreve, W. A. Cannon, that T. G. B. Osborn was stimulated in 1926 to set up permanent quadrats in heavily used shrub land and a reserve released from grazing at Koonamore in S. Australia (Osborn, Wood & Partridge, 1935; O. B. Williams & Mott, 1981). There is also a European tradition that precedes the work of Tamm, associated with the name of Bogdanovskaya–Gienef (1926) who seems to have been a pupil of Sukatschev and to have influenced the reductionist approach to ecology later developed by Rabotnov, Uranov and their pupils in Russia. She reported her two-year study of four 20×20 cm quadrats. Her publications list appears to consist of only two papers, but she may have had a greater influence than this suggests (J. White, personal communication). The peculiar quality of Tamm's work was that he recorded individual plants, shoots or rosettes, within his populations and this enabled him to follow the fates of individual plant units (often tillers or ramets) rather than to study the grosser vegetational change that was the aim of many others who set up permanent quadrats. The data obtained from Tamm's studies revealed, for the first time, the magnitude of the flux that underlay the apparent stability of many plant communities and (after Darwin) gave the first real insight into plant community dynamics dominated by establishment and deaths. Vegetation is composed of the few plants that survive and grow: to explain that vegetation it may be more important to study the many that die.

In many of Tamm's populations it can be shown (Harper, 1967) that the plants originally present were progressively lost and that the rate of loss was remarkably constant. Half-lives could be calculated for the populations – differing from species to species but constant over the years. Such data made it clear that (i) it was realistic and profitable to study individual species, even within a complex community and (ii) that the dynamics of the populations appeared to be largely independent of year-to-year fluctuations in climate. Later studies of permanent quadrats in grassland by Sagar, Sarukhán, Hawthorn and others (see Harper, 1977) extended over fewer years but involved repeated observations within each year. They showed that the death risks to plants in populations within years were commonly greatest when the survivors were growing fastest – not during the periods when the anthropomorphically-minded botanist would regard the physical environment as 'harsh'. The likely explanation is that it is biotic pressure from competing neighbours rather than 'harshness' of the physical environment that is the prime cause of death of most plants and perhaps of natural selection. Most of these detailed demographic studies have been made with pasture or woodland systems in northern temperate regions and it could well be that in arid zones, and some other extreme environments, biotic pressures are less dominant and then climatic factors may play the major role in killing plants and in natural selection.

At a Symposium organized to pay tribute to Darwin a hundred years after his death, it is perhaps permitted to expand on one tiny fragment of ecological research that may truly be said to have been directly stimulated by Darwin's own writings, in particular by the paragraph quoted as heading to this paper.

Early in the 1960s the decision was made to concentrate a number of intensive ecological studies based at Bangor, North Wales, on a single, small (1 ha) field of permanent grassland, part of the College Farm at Henfaes, Aber, near Bangor. The field was chosen because it was superficially very dull – lacking any obvious heterogeneity of contour and relatively homogeneous in soil properties. The field had not been ploughed for at least 80 years (probably not for more than 150 years) and there is no record of fertilizer or herbicide ever having been applied. A general description of the field is given in Turkington & Harper (1979a) and more detailed analysis of the soil and flora in Turkington (1975).

Many graduate students and overseas visitors have worked on aspects of the ecology of this field. In many cases the studies have concentrated on populations of a single species within 1 m² quadrats, though sometimes the reductionist level of study has been yet smaller – a scale of 1 cm² (Thorhallsdottir, 1983). The concentration of effort within one small field, and on small quadrats within it, gives the studies high precision and high relevance but with an absolute sacrifice of generality. We do not know whether most of what we have observed in this field can be generalized to other fields or, indeed, to less intensively studied parts of the same field.

The vegetation of the field was analysed by ordination and correlation techniques which showed that only a minor part of the variation in species distribution could be accounted for by underlying edaphic factors, though in the peripheral areas of the pasture the presence of hedges and trees accounted for significant changes in the vegetation – e.g. *Dactylis glomerata* occurred mainly in or close to the shade of the trees. Two common species, *Lolium perenne* and *Trifolium repens* were usually positively associated in their microdistribution and negatively associated with other species. The structure of the vegetation was interpreted as determined by regeneration cycles directed by *T. repens* and *L. perenne* – the species themselves appeared to be the prime determinants of each others distribution (Turkington and Harper, 1979a, b).

The most intensive studies on the field have been made on populations of three species of *Ranunculus* and on *Trifolium repens*. Some of the work is still in midstream and consequently some material referred to in this paper represents an interim report.

Two of the species present in the field are of particular interest, both to population biologists interested in the manner in which the numbers of plants are regulated and to evolutionists concerned with the extent and significance of natural variation. The two species are *Ranunculus repens* and *Trifolium repens*. Both possess the property of clonal growth by which the product of a single zygote forms a spreading clone of rooted nodes capable of vegetative extension through the sward. Such plants have the potential for a single clone or genet (product of a single zygote) to dominate large areas of the vegetation. Indeed, it would be theoretically possible for one genet of *Trifolium repens* or *Ranunculus repens* to have occupied the whole of the Henfaes field. In other plants with such clonal growth it is known that the product of a single zygote may indeed occupy considerable areas of land. An extreme example is bracken, *Pteridium aquilinum*; Oinonen (1967) showed that individual clones in Finland were up to 1440 years old and one old clone extended over an area of 474×292 m. There are other cases in which clone-forming species are known to produce large areas of genetic monotony, e.g. *Festuca rubra* (Harberd, 1961) and *Holcus mollis* (Harberd, 1967). It might be expected that where such clonal growth is possible, the struggle for existence over long periods of stable management would lead to the local dominance of single clones – those that had succeeded in a struggle for existence with others. The population dynamics and genetics of these two species in the permanent pasture at Henfaes seemed to offer the opportunity to study natural selection in action – Hutchinson's 'ecological theater and evolutionary play'.

Population dynamics of *Ranunculus repens*

The detailed demographic studies of *Ranunculus repens* made by Sarukhán in the field at Henfaes and followed by unpublished studies by Soane made it possible to give quantitative measures to the population dynamics of this species. These are summarized in the life-cycle diagram of Fig. 16.1. Populations of this species showed vigorous vegetative growth, and most new rosettes recruited to the population appeared as ramets (clonal replicates) from existing rosettes. The input of new seedlings was small and the number eventually contributing to mature rosettes in the pasture seemed insignificant in comparison with the contribution from clonal growth. In comparison with the other two species of *Ranunculus*, *R. acris* and *R. bulbosus* present in the pasture, seed production and the numbers of seedlings observed was very small. Plants of *Ranunculus repens* produced

Fig. 16.1. The population dynamics of *Ranunculus repens* in a permanent grassland field in North Wales. The diagram represents average population fluxes within 1 m² quadrats. (From Sarukhán, 1971.)

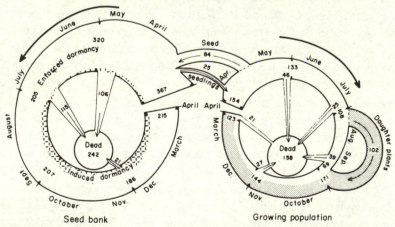

Seed bank   Growing population

R repens

on average less than one seed per rosette (cf. *ca.* 10 for *R. acris, ca.* 15 for *R. bulbosus*). In the studied quadrats, 25 seedlings of *Ranunculus repens* emerged per metre square from the long-lived bank of seeds in the soil, in contrast with 176 of *Ranunculus acris* and 95 of *Ranunculus bulbosus*.

The dynamics of *Ranunculus repens* populations was followed over four years during which new seedlings that died and those that survived to form rosettes were recorded. The development of clones from the rosettes present at the beginning of the study was also recorded in detail so that the clonal parent was known for almost every rosette present in the quadrats at the end of the study. These data allowed Soane & Watkinson (1979) to build a computer simulation model to examine the relationship between the flux of ramets, the recruitment of seedlings and the diversity of genets within

Fig. 16.2. Family survivorship curves (filled symbols, observed; open symbols, simulated) for the founder members of three populations of *Ranunculus repens* with 95% confidence limits that the observed result in the field is consistent with the sample mean and standard deviation obtained from the relevant set of replicated simulations.

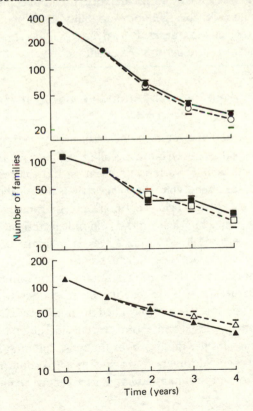

the populations. The computer models simulated the actual flux of ramets in each of eight studied populations and followed the fates of ramets and families of ramets *assuming no selection* between families. The real and simulated changes in the populations are shown in Fig. 16.2. Agreement is extremely close and provides little evidence for selection between families or against new seedling recruits. In the simulation model, in the absence of selection, the number of original families present in the population declined at an approximately exponential rate. With the passage of time the contribution that these families made to the total genetic diversity of the population became subordinate to the seedling recruits: although the number of seedlings appears to be very small, their contribution to the total number of genetic individuals in the population is clearly significant in determining the number of clones or genets that are present. In the absence of selection, the observed small numbers of seedling recruits would apparently be sufficient to maintain potentially high genet diversity within such a vigorously clonal plant population. Such genetic diversity was indeed present, because there was visible genetic polymorphism within the populations and it was shown (M. J. Lawrence, personal communication) that there was considerable genetic variability both in quantitative characters and in polymorphism at two enzyme loci in populations immediately adjacent to the permanent quadrats.

Population dynamics and genetic variation in *Trifolium repens*

Analysis of the population dynamics and associated genetics of *Trifolium repens* in the same pasture is even more revealing. *Trifolium repens* is unusual in that much of the genetic variation present within the populations is easily recognised in phenotypic differences that are visible or easily determined on plants in the field. Leaf mark polymorphism is one such property. A variety of white leaf marks is found in natural populations and these are represented by multiple alleles at a single locus. In mid-summer, when the marks are most fully expressed, the genotypes of most plants can be identified in the field and an estimate of the number of clones present within an area can be made: it will, of course, be a minimal estimate. Cahn & Harper (1976a) determined the number of clones present within 10×10 cm quadrats and found, to considerable surprise, that between 3 and 4 clones per quadrat was the most common situation in the field. This level of genetic diversity at a fine scale was confirmed in other permanent grasslands in Britain including Port Meadow, Oxford, which is at least 896 years old! Clearly, single clones did not dominate patches even at this very fine scale.

More recently, Trathan has identified genetic individuals with more precision using isoenzyme analysis. He finds 48–50 distinct genotypes present per metre square. The various clones weave amongst and intermingle with each other and amongst grasses and associated herbs. The degree of intermingling may itself reflect the growth form of stoloniferous species. Both *Ranunculus repens* and *Trifolium repens* have 'guerilla' growth forms in contrast to the predominantly 'phalanx' forms of the associated grass species. By 'guerilla' growth form is implied one that is continually wandering amidst associated vegetation, creeping into new and escaping from old patches in the community. In contrast, 'phalanx' growth forms develop a structure of tightly packed shoots (most of the pasture grasses and some pasture dicots such as *Bellis perennis*) (Lovett Doust, 1981). Not only has *Trifolium repens* a guerilla growth form but its guerilla character is exaggerated when it is growing with grasses; its branching then tends to be reduced and growth is concentrated in linear extension. Instead of a genet locally consolidating its occupancy of a site, individual stolons wander as linear extensions of the genet into surrounding vegetation. This growth form itself maximises the chance that genets will intermingle: it maximizes the role of interspecific contacts in the life of the genet.

Darwin commented on the growth of such plants and the ways in which they penetrate amongst other vegetation. He describes his own (very Darwinian!) experiment in which he allowed the stolons of *Saxifraga sarmentosa* (a classic 'guerilla' growth form) to encounter an artificial vegetation that he had constructed: 'Many long pins were next driven rather close together into the sand, so as to form a crowd in front of... two thin lateral branches; but these easily wound their way through the crowd. A thick stolon was much delayed in its passage; at one place it was forced to turn at right angles to its former course; at another place it could not pass through the pins, and the hinder part became bowed; it then curved upwards and passed through an opening between the upper part of some pins which happen to diverge; it then descended and finally emerged through the crowd' (Darwin, 1880).

Another approach to the study of variation within populations of white clover in the Henfaes field was made by Burdon (1980). He sampled 50 white clover clones from a grid covering the whole field. He multiplied the clones in the glasshouse and screened them for a variety of characters (Fig. 16.3). These included a number with simple Mendelian inheritance and a number of characters of agronomic importance with polygenically controlled expression. He was able to use these characters to produce identity diagrams that distinguished and 'finger-printed' each clone. The 50 clones differed on average from one another in 3.3 vegetative characters. If floral characters were included in the comparison the average difference between clones was 5.4 characters. (One pair of clones differed in 13 statistically significant and apparently independent respects!) Many of the characters considered had been shown by other workers to be of selective importance in white clover or another species of *Trifolium* (e.g. Cahn & Harper (1976b) had presented evidence suggesting that sheep selected between leaf marks; Dirzo & Harper (1982a) and others have shown that slugs select between cyanogenic and acyanogenic forms; Black (1960) had shown the selective value of long petioles).

The polymorphisms in the Henfaes population were not exhausted in Burdon's study. It has now been shown that variation in relative growth rates can be added to the list (Burdon & Harper, 1980). Natural populations of white clover are polymorphic for incompatibility alleles and for the ability to form nitrogen fixing symbioses with strains of *Rhizobium* but these have not been looked at at Henfaes. Berrington (personal communication) has recently shown that clones within the field differ in their ability to form endotrophic mycorrhizal associations. The polymorphism that is found within white clover populations in the field represents therefore a subtle (or? random) variety of unique associations of apparently selectively important properties. The imagination of the most extreme selectionist is stretched to breaking point by such a situation.

Fig. 16.3. (*a*) Individuality diagram for fifty clones of *Trifolium repens* sampled from the field of permanent grassland described in text. (*b*) The characters used in the individuality diagram. (From Burdon, 1980.)

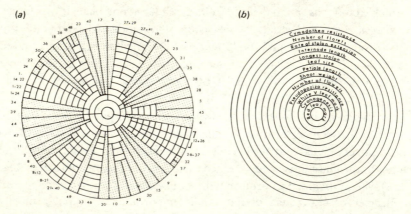

Evidence for local specialization within white clover

Plants that are capable of clonal growth offer peculiar opportunities for testing the extent to which particular genets are locally specialized. A plant may be sampled from the field, multiplied clonally and the clonal products (genetic identities) can then be reinserted into the field, both in the places from which the clones originally came and into other places. Plants of the same genotype can then be tested in different environments. Turkington made such an experiment with white clover in the field at Henfaes (Turkington & Harper, 1979*b*). Clones of white clover were sampled from within patches in the field dominated by each of four common grasses, *Lolium perenne*, *Holcus lanatus*, *Agrostis tenuis* and *Cynosurus cristatus*. The clones were multiplied in the glasshouse and then transplanted back into patches of the field dominated by the four grass species. The performance of the transplants was measured by vegetative growth expressed as dry weight at harvest after twelve months. The results are shown in Fig. 16.4. Over the whole experiment clones of clover that were returned to their original grass associate made more growth than those introduced to alien sites (significant at $P < 0.001$). The clover clones had also been introduced into sites from which the existing vegetation had been denuded by treatment

Fig. 16.4. The dry weight of plants of *Trifolium repens* from a permanent grassland sward, sampled from patches dominated by four different perennial grasses and grown in all combinations of mixture with the four grass species. Clover 'types': At, *Agrostis tenuis*; Cc, *Cynosurus cristatus*; Hl, *Holcus lanatus*; Lp, *Lolium perenne*. (From Turkington & Harper, 1979*b*.)

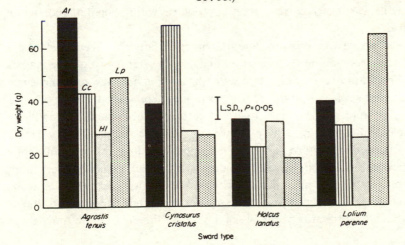

with the paraquat herbicide. Some of the 'principal diagonal effect' remained, though the difference in yield between clones returned to their site and those returned to alien sites was now significant only at $P < 0.05$. Turkington made a further experiment in which he introduced the four clone types of white clover into pure swards of the four grass species that had been sown on soil sampled from the experimental field. This part of the experiment was designed to remove possible place-to-place variations in soil conditions in the studied field from the comparison. The yield of clones grown with the grass species from which they had originally been sampled again exceeded that made when they were grown in an 'alien' sward, $P < 0.00001$. It is difficult to interpret the results of this experiment as representing anything but evidence of precise specialization of clover clones in their ability to grow in association with particular grass neighbours. It suggests that, within the pasture, strains of white clover have been selected by competitive interaction with associated grasses and that different species of grass contribute to the diversifying or disruptive selection operating upon the population of white clover. Some, at least, of the variation within the white clover populations thus appears to be directly interpretable in terms of attributes contributing to present fitness. It is a very Darwinian interpretation to suggest that the grass neighbours may be primary forces selecting and diversifying the clover populations. Hill (1976) grew a *single* clone of white clover with a variety of clones of *Lolium perenne*. Quite distinct phenotypic modifications were elicited from the clone by the different ryegrass strains. If different strains of *Lolium perenne* produce different phenotypes from the same clone of white clover, it is difficult to escape the conclusion that different species of grass are even more likely to exert different selective pressures within populations of white clover.

It is not easy to measure and describe just how the different clones of white clover differ in their reaction to different neighbouring grass species or forms. Survivorship and dry matter production are very gross measures of a plant's reaction to different types of neighbour. We have, at present, no real indication of the manner in which ecological compatibility (ecological combining ability) between particular strains of white clover and particular pasture grasses is accomplished. It may represent subtle differences in growth cycle or growth form or more complex interactions involving the soil microflora, perhaps the mycorrhizae. There may be subtleties of interaction below the soil surface of which we know little or nothing. The hazards in the life of a plant in the field are not only those of competition from neighbours, though it may be these that are the most relentless.

Hazards in the life of a plant in the field

Some of the hazards to the life of a plant in the field can be measured by studying the fate of individual leaves or flowers or seeds. An attempt to catalogue and quantify these hazards in the field at Henfaes was made by Peters (1980) who included white clover amongst the species that he studied. He marked young leaves as they began to expand and then followed their fate by repeated observation. From this he could obtain survivorship curves for cohorts of leaves born in the same time period and record some of the causes of death or damage within the populations (Fig. 16.5). Some leaf predators leave tell-tale records of their activity. In particular, grazing molluscs leave characteristic erosions from a leaf edge; birds, particularly the wood-pigeon, feed on clover and often leave characteristic beak-marks; weevils remove circles of tissue, often leaving

Fig. 16.5. The proportions of the leaf population of *Trifolium repens* suffering predation by slugs (triangles), weevils (circles), and sheep (squares) in different seasons on a field of permanent grassland described in text. (Modified from Peters, 1980.)

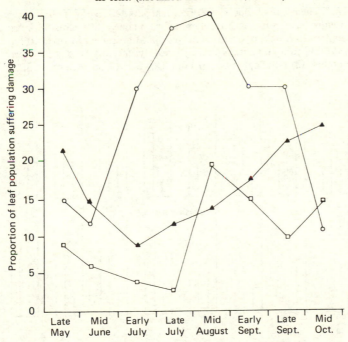

the upper epidermis intact; sheep (causing damage probably indistinguishable from that caused by rabbits) remove whole leaves, leaving torn petioles
or leave their bite marks on the leaflets that remain. Tracking the fate of
individual leaves immediately reveals a number of other hazards. Leaves
may be submerged under a dropping of dung, soused in a downpour of
urine, trodden on, pulled into the ground by earthworms (further shades
of Darwin) or damaged by frost. The frequency of these various events in,
or ending, the life of a leaf are shown in Table 16.1. Grazing animals are,
however, the major hazards for a leaf though all leaves in a pasture are
not equally at risk. Weevils bite holes in the leaves of white clover and of
Ranunculus species but the relative severity of attacks on *Trifolium* and
Ranunculus change through the season, *Trifolium* being more attacked in
early, and *Ranunculus* in late, summer. Fig. 16.6 shows the relative
proportion of the leaf population that suffered damage from leaf grazers
in populations of *T. repens* and *R. repens*. Surprisingly, the leaves of white
clover suffered proportionately much less from grazing by sheep than did
those of *R. repens*, but leaves of *T. repens* suffered much more from both
slugs and weevils (except towards the end of the growing season).

Grazing by molluscs figured so strongly among the hazards to a clover

Fig. 16.6. The *relative* proportions of the leaf population of *Trifolium repens*
and *Ranunculus repens* suffering damage from sheep (Sh), slugs (Sl), and
weevils (W) in a field of permanent grassland described in text. (*a*) late
May, (*b*) mid-June, (*c*) early July, (*d*) late July, (*e*) mid-August, (*f*) early
September, (*g*) late September, (*h*) mid-October. (From Peters, 1980.)

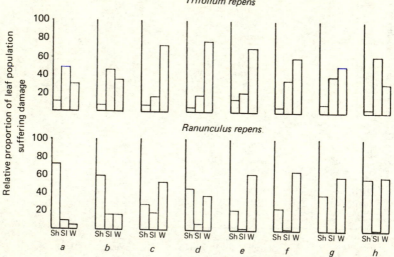

Table 16.1. *Hazards in the life of leaves in a permanent grassland site (excluding those damaged by grazing)*

Values shown are the number of leaves affected out of 1000 sampled; values in parentheses represent leaves that died as a result of the hazard. Leaves receiving urine all died if there was drought or bright sunshine – at other times it was not possible to determine which leaves had or had not received urine. Similarly leaves that had been trampled could be recorded only if some damage was done. Earthworms pulled leaf tips into their burrows but the leaves were scarcely ever killed. (Data recalculated from Peters, 1980.)

	L. perenne	T. repens	A. millefolium	R. acris	R. repens	R. bulbosus
Trampling	9.7 (9.2)	39.0 (26)	33.9 (27.7)	16.6 (13)	20.2 (14.0)	24.9 (11.7)
Dung	8.1 (8.1)	2.8 (0)	11.3 (8.5)	6.7 (0)	0.8 (0)	0.5 (0)
Urine	3.8 (3.8)	3.7 (3.7)	43.8 (43.8)	10.0 (10.0)	6.4 (6.4)	0 (0)
Earthworms	2.3 (0.1)	27.0 (0)	41.0 (0)	112.0 (0)	150.0 (0)	118 (0)
Pathogens	9.4 (0.0)	2.8 (0)	0 (0)	0 (0)	0 (0)	0 (0)
Frost	0.0 (0.0)	9.3 (4.6)	0 (0)	28 (23)	31.6 (16.0)	30.3 (21.5)

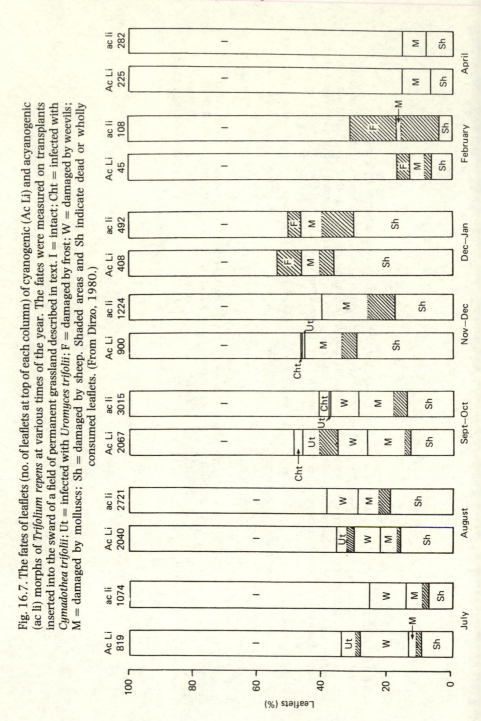

Fig. 16.7. The fates of leaflets (no. of leaflets at top of each column) of cyanogenic (Ac Li) and acyanogenic (ac li) morphs of *Trifolium repens* at various times of the year. The fates were measured on transplants inserted into the sward of a field of permanent grassland described in text. I = intact; Cht = infected with *Cymadothea trifolii*; Ut = infected with *Uromyces trifolii*; F = damaged by frost; W = damaged by weevils; M = damaged by molluscs; Sh = damaged by sheep. Shaded areas and Sh indicate dead or wholly consumed leaflets. (From Dirzo, 1980.)

leaf that it seems reasonable to expect that, in those years when slugs or snails were abundant, they may act as important selective forces within clover populations. White clover is polymorphic for the presence or absence of cyanogenic glucosinolates and is also polymorphic for the beta glucosidases that release HCN from glucosinolates when a leaf is damaged. There is considerable variation in the extent to which cyanogenic properties are expressed and there is some seasonal and perhaps other causes of variation in expression. Nevertheless, it is possible to classify plants in the field by taking leaf samples and performing appropriate tests (see Dirzo & Harper, 1982*a*) and to categorize the plants into four groups Ac Li, Ac li, ac Li, ac li. After a period of exposure to grazing one can then test for the

Fig. 16.8. Contour map made by a trend-surface analysis of the density of active molluscs at Henfaes Field, College Farm, Aber. The values shown by the contour lines are the mean number of molluscs. The mnemonics for mollusc density are: VL, very low; L, low; H, high; VH and VH +, very high. Sampling sites are shown by ●. The central zone of very high slug density and the two isolated zones of high density were mapped directly. They represent patches of *Urtica dioica* which serve as a refuge for slugs. The symbols below each site indicate the locations of the four clover morphs (see Table 16.1): + +, cyanogenic Ac Li; + − glucosidic Ac li; − +, enzymatic ac Li; − −, double recessive ac li. Samples which gave ambiguous results are shown by (A) and are not included in the contingency analysis (see Table 16.1). (From Dirzo & Harper, 1982*a*.)

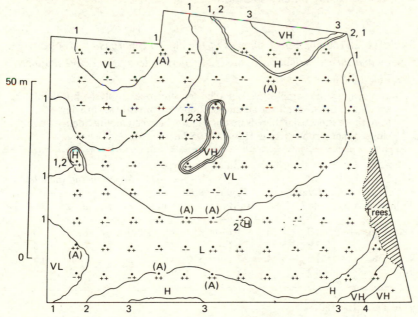

frequency of the various types of damage to clover leaves with the different genotypes and afterwards reconfirm the cyanogenic or acyanogenic status of the plant (Fig. 16.7). There have been many studies of selection by molluscs on cyanogenic and acyanogenic morphs of white clover under laboratory or other controlled conditions. It seemed that it might be possible to relate the variation in the polymorphism to slug density in the Henfaes field. The density of slugs was determined at sites arranged on a grid across the field at Henfaes. Dirzo (1982*b*) used trend-surface analysis to draw contours of mollusc density (see Fig. 16.8) and added some further information from visual inspection of the field. Most notably, he added one big and two small islands (shown in Fig. 16.8 as having high and very high slug densities) which were patches of nettle, *Urtica dioica*. Slugs tend to concentrate in such refuges, presumably because they give protection from desiccation. He then determined the cyanogenesis category of the clover plant nearest to each grid intersection. Table 16.2 shows the distribution of the different glucosinolate morphs of *T. repens* between areas with different densities of active molluscs. There is a clear excess over expectation of cyanogenic forms in the areas of high and very high mollusc density and of acyanogenic forms in the areas of very low mollusc density. It appears that some at least of the variation in the cyanogenesis polymorphism over the field can be explained as local micro-evolution in response to locally patchy selection.

Table 16.2. *The number of samples of the different morphs of* Trifolium repens *and their distribution in areas of different density of active molluscs* (*see Fig. 16.8*)

The cyanogenic samples were pooled for statistical analysis, but details are shown in the last three columns. Values in parentheses are deviations from those expected if there is no correlation between mollusc density and clover type. (From Dirzo & Harper, 1982*a*.)

Mollusc density	Cyanogenic *Ac Li*	Plant 'phenotype' Acyanogenic (all)	Total	Acyanogenic (details) *Ac Li*	*ac Li*	*ac li*
Very high + high	10 (+4.2)	1 (−4.2)	11	0	1	0
Low	29 (+5.3)	16 (−5.3)	45	6	7	3
Very low	10 (−9.5)	27 (+9.5)	37	5	7	15
Total	49	44	93			

$\chi^2 = 18.7; P < 0.001.$

Two strands of evidence, that from the reciprocal transplanting of clover into the neighbourhood of different grasses and that from the study of the distribution of cyanogenic properties and the distribution of slugs, allow an easy interpretation in terms of immediate and *present* selective forces. How far this interpretation can be extended to the whole gamut of characters for which Burdon had shown the populations to be polymorphic must be very doubtful. Certainly almost all of the characters that Burdon lists have been shown to be of selective importance either in clover or in some related species, though sometimes under very specialized circumstances. The scientific method may sometimes mislead. We commonly test for the selective value of a particular feature by holding background variation of both genotype and environment at a minimum. We thereby maximise our chance of demonstrating what we are looking for. The real measure should be whether selection is significant against normal levels of background variation. This is why it was important to test the effects of cyanogenesis and reaction to neighbours *in the field*.

It is difficult to believe that any of the characters examined by Burdon could be selectively neutral. However, the contribution of each property to fitness must vary dramatically from year to year as well as from place to place within the field. Most winters at Henfaes (only 400 m from the sea) are mild and frost is rare. Occasionally, as in the winter of 1982, there is severe frost. Populations of molluscs fluctuate wildly from year to year in North Wales. In some years spring growth of the sward is vigorous and exceeds the capacity of sheep and the other grazing animals to keep it fully grazed. In other years, as in the spring of 1982, a protracted spring drought slowed the growth of the sward and it became tightly grazed. A severe drought is not a common feature on the field but when it occurs it may be in any month from April to September. During the period of our observations we have detected three significant leaf pathogens on white clover in the field, *Uromyces trifolii*, *Cymadothea trifolii* and *Pseudopeziza trifolii*. It seems unlikely that these three diseases attack with equal intensity in all seasons and in all years. In a field that is patchy in space and time, be it ever so small, we may expect that the populations of a species such as white clover will, at any time, reflect selective forces from its past. The genotypic composition of the population may in some cases dimly reflect forces that operated twenty or thirty years ago. Other selective forces may have operated quite recently and left a strong memory or image in the structure of the population's genetics. If this is the case, we would expect to find only a few of the many polymorphisms readily interpretable as responsive to present proximal selective forces. Much of the polymorphism could be transient and, without an even more detailed history of the field, uninterpretable. It is doubtful whether such an explanation of naturally occurring polymorphism could be tested without long-term, detailed recording, not only of the variety of genetic changes occurring within clover populations, but at the same time of a detailed recording of the known hazards in the life of the clover plant over the seasons and the years.

Conclusions

The studies that I have described, concentrated in the field at Henfaes, are now being extended by deliberate experimentation within the field. We are destroying the site as a long-term study on a supposedly stable system by introducing a variety of perturbations such as transplant experiments, the creation of islands for invasion and further perturbations are planned. The study has involved a curious concentration of effort in one very specialized environment. The type of observations that have been made have been quintessentially Darwinian. Another great naturalist, Thoreau, has focussed attention at the same scale: 'Nature will bear the closest inspection. She invites us to lay our eye level with her smallest leaf, and take an insect-view of its plain'. If we are to see evolutionary processes in action in plant communities and the proximal events determining their character we must focus our attention away from an anthropomorphic scale of acres or square metres and onto a scale appropriate to the organisms with which we are concerned. The appropriate scale is determined by the organism and not by us. It will be different for different species. We ask for a plant's eye view of life and death in a sward and hope ultimately to be able to collect these reductionist observations into statements about the population, the species or even possibly the community. I doubt if it is possible to hold the view of Margalef (1968) that 'Relevant evidence does not consist of a massive accumulation of trivia' and reconcile it with his 'Ecology...is the study of systems at a level at which individuals or whole organisms may be considered elements of interaction...'. It was, indeed, from the massive accumulation of trivia and tiny details, that Darwin assembled the evidence for The Origin of Species.

In a volume commemorating Darwin's death, I have tried to show how his way of looking at the behaviour of individual plants in nature can be extended. A hundred years after his death his approach seems more relevant to botanical studies than it has ever been. This part of his intellectual legacy has not yet been fully invested. A part of the legacy, however, ceases to bear interest. He was writing in the Origin for readers most of whom were steeped in Victorian optimism, religion and the romantic movement. It was necessary in 1859 to write about the process of evolution as if it produced the best of all possible worlds, a substitute for the finger of the Almighty at work. If the process of evolution had not been presented in this way (though with careful caveats) it is very questionable whether it could have been accepted so rapidly by Victorian society. It was then appropriate to show how '...from so simple a beginning endless forms most beautiful and most wonderful have been, and are being evolved'. But 'beauty' and 'wonder' are in the eye of the beholder and that eye has itself evolved. The teleology of evolution as a goal-seeking activity persists in indefensible form a hundred years later in the writing of biologists. This particular heritage may be a millstone around the neck of scientific natural history. Most particularly, it harms biology as a means of teaching science to the young (Harper, 1982). Looking back at the variety of studies conducted on plants in the little field at Henfaes, I feel little temptation to explain the behaviour of organisms within it as perfectly fitted for adaptive optima in an ideally evolved ecosystem. Rather, I see the population of plants trapped in narrowly constrained evolved ruts, bearing the battered imprint of recent and not-so-recent selective and other forces. 'I returned, and saw under the sun, that the race is not to the swift, nor the battle to the strong, neither yet bread to the wise, nor yet riches to men of understanding, nor yet favour to men of skill; but time and chance happeneth to them all' (*Ecclesiastes*).

Further reading

Harper, J. L. (1967). A Darwinian approach to plant ecology. Presidential Address, British Ecological Society. *Journal of Ecology*, **55**, 247–70.
Harper, J. L. (1977). *Population Biology of Plants*. New York: Academic Press.
Harper, J. L. (1981). The concept of population in modular organisms. In *Theoretical Ecology. Principles and Application*, ed. R. M. May. 2nd edn. Oxford: Blackwell Scientific Publications.

References

Black, J. N. (1960). The significance of petiole length, leaf area and light interception in competition between strains of subterranean clover (*Trifolium subterraneum* L.) grown in swards. *Australian Journal of Agricultural Research*, **11**, 277–91.
Bogdanovskaya-Gienef, J. D. (1926). On seed regeneration in meadow communities. *Zapiski Leningradskogo sel'sko Khozyaistuen nogo instituta*, Leningrad, **3**, 216–53.

Burdon, J. J. (1980). Intraspecific diversity in a natural population of *Trifolium repens*. *Journal of Ecology*, **68**, 717–35.

Burdon, J. J. & Harper, J. L. (1980). Relative growth rates of individual members of a plant population. *Journal of Ecology*, **68**, 953–7.

Cahn, M. A. & Harper, J. L. (1976*a*). The biology of the leaf mark polymorphism in *Trifolium repens* L. I. Distribution of phenotypes at a local scale. *Heredity*, **37**, 309–25.

Cahn, M. A. & Harper, J. L. (1976*b*). The biology of the leaf mark polymorphism in *Trifolium repens* L. II. Evidence for the selection of leaf marks by rumen fistulated sheep. *Heredity*, **37**, 327–33.

Darwin, C. R. (1880). *The power of movement in plants*. London: John Murray.

Dirzo, R. (1980). Studies on plant-animal interactions: terrestrial molluscs and their food plants. Ph.D. thesis, University of Wales.

Dirzo, R. & Harper, J. L. (1982*a*). Experimental studies on slug-plant interactions. III. Differences in the acceptability of individual plants of *Trifolium repens* to slugs and snails. *Journal of Ecology*, **70**, 101–17.

Dirzo, R. & Harper, J. L. (1982*b*). Experimental studies on slug-plant interactions. IV. The performance of cyanogenic and acyanogenic morphs of *Trifolium repens* in the field. *Journal of Ecology*, **70**, 119–38.

Gauze, G. F. (1934). *The struggle for existence*. Baltimore: Waverly Press.

Harberd, D. J. (1961). Observations on population structure and longevity of *Festuca rubra* L. *New Phytologist*, **60**, 184–206.

Harberd, D. J. (1967). Observations on natural clones of *Trifolium repens* L. *New Phytologist*, **62**, 198–204.

Harper, J. L. (1967). A Darwinian approach to plant ecology. *Journal of Ecology*, **55**, 247–70.

Harper, J. L. (1977). *Population biology of plants*. London: Academic Press.

Harper, J. L. (1982). After description. In *The plant community as a working mechanism*, ed. E. I. Newman. *B.E.S. Special publication Series*, No. 1. pp. 9–25. Oxford: Blackwell Scientific Publications.

Hill, J. (1976). Plasticity of white clover grown in competition with perennial ryegrass. *Annual Report Welsh Plant Breeding Station*, 24–5.

Lovett Doust, L. (1981). Population dynamics and local specialization in a clonal perennial (*Ranunculus repens*). I. The dynamics of ramets in contrasting habitats. *Journal of Ecology*, **69**, 743–55.

Margalef, R. (1968). *Perspectives in Ecological Theory*. Chicago: University of Chicago Press.

Nägeli, C. (1874). Verdrängung der Pflanzenformen durch ihre Mitbewerber. *Sitzungsberichte der Akademie der Wissenschaften München*. **11**, 109–64.

Oinonen, E. (1967). The correlation between the size of Finnish bracken (*Pteridium aquilinum* (L.)), Kuhn clones and certain periods of site history. *Acta foresta fennica*, **83**, 1–51.

Osborn, T. G. B., Wood, J. G. & Partridge, T. B. (1935). On the climate and vegetation of the Koonamore Vegetation Reserve to 1931. *Proceedings of the Linnean Society of New South Wales*, **60**, 392–427.

Peters, B. (1980). The demography of leaves in a permanent pasture. Ph.D. thesis, University of Wales.

Sarukhán, J. (1971). Studies on plant demography. Ph.D. thesis, University of Wales.

Shreve, F. (1915). The vegetation of a desert mountain range as conditioned by climatic factors. *Carnegie Institute of Washington, Publication 217*.

Soane, I. D. & Watkinson, A. R. (1979). Clonal variation in populations of *Ranunculus repens. New Phytologist*, **82**, 557–73.

Tamm, C. O. (1948). Observations on reproduction and survival of some perennial herbs. *Botaniska Notiser*, **3**, 306–21.

Tamm, C. O. (1972). Survival and flowering of some perennial herbs. II. The behaviour of some orchids on permanent plots. *Oikos*, **23**, 23–28.

Thorhallsdottir, T. E. (1983). Dynamics of a grassland community with specific reference to five grasses and white clover. Ph.D. thesis, University of Wales.

Turkington, R. A. (1975). Relationships between neighbours among species of permanent grassland. Ph.D. thesis, University of Wales.

Turkington, R. A. & Harper, J. L. (1979*a*). The growth, distribution and neighbour relationships of *Trifolium repens* in a permanent pasture. I. Ordination pattern and contact. *Journal of Ecology*, **67**, 201–18.

Turkington, R. A. & Harper, J. L. (1979*b*). The growth, distribution and neighbour relationships of *Trifolium repens* in a permanent pasture. IV. Fine scale biotic differentiations. *Journal of Ecology*, **67**, 245–54.

Williams, O. B. & Mott, J. J. (1981). Population biology: a review of the Australian contribution. Abstract of paper presented at the *International Botanical Congress*, Sydney, Australia, August 1981.

17

Irrelevance, submission, and partnership: the changing role of palaeontology in Darwin's three centennials, and a modest proposal for macroevolution

STEPHEN JAY GOULD

Three Centennials

When Cambridge University Press published the Festschrift of the first Darwinian centennial (Seward, 1909), two members of Darwin's inner circle still lived. J. D. Hooker, at 92, contributed a short letter of welcome to the volume. A. R. Wallace, feisty as ever at 86, did not participate. But, a year later (Wallace, 1910), he took some time off in his campaigns for spiritualism, socialism and God, and against vaccination, to publish his last defence of natural selection.

Wallace remained true to the hyperselectionism that out-Darwined Darwin (Gould, 1980), while investing it with his own brand of religiosity and championing a separate status for the human mind. In striking contrast with his own certainty, he expressed disappointment that confusion and disagreement reigned at his partner's century and *The Origin of Species'* half century:

> During the fifty years that have elapsed since the Darwinian theory was first adequately, though not exhaustively, set forth, it has been subject to more than the usual amount of objection and misapprehension both by ignorant and learned critics, by old-fashioned field naturalists, and by the newer schools of physiological specialists (Wallace, 1910, p. 252).

The 1909 centennial may represent the acme of genuine confusion in theories of evolutionary mechanisms (amidst complete confidence in the fact of evolution itself). Conflict still raged about the central issue of whether or not natural selection could act as a creative force in evolutionary change, with non-Darwinians advocating organic response to environmental challenge (neo-Lamarckism), foreordination (phyletic life cycles or classical orthogenesis), or internal driving forces (vitalism) to supply the directed variants that selection might then favour. Yet the one ghost that all these schools thought they had exorcised – discontinuous or saltatory change with its twin spectres of irrationality and chanciness – had made a dramatic reappearance. The very discovery that all had trusted to resolve confusion in their favour – the elucidation of the mechanism of heredity – had reintroduced saltationism in the gentle clothing of evening primroses.

In this light, the centennial Festschrift is a truly remarkable volume. Though born in confusion, it laboured mightily to cover up internal contradictions and to present a hagiography that could only have embarrassed the gentle man whom it honoured. It did not reach artificial agreement by excluding the true opponents of natural selection; rather, these opponents themselves indulged in flights of rhetoric to assimilate Mr Darwin into their camp. Do not all sides invoke the same God before battle?

Hugo De Vries, for example, portrayed Darwin as the father of macromutationism with an extraordinary bit of special pleading based on an evident misreading of a single sentence in *The Origin of Species*: '...the doctrine of natural selection or the survival of the fittest, which implies that when variations or individual differences of a beneficial nature happen to arise, these will be preserved' (quoted by De Vries, 1909, p. 71). De Vries argued that since ordinary, small-scale, continuous variation always exists, it cannot represent what Darwin describes as differences that 'happen to arise'. These must therefore be the large and fortuitous macromutations that, according to De Vries, make new species in a single gulp. And Darwin was at least a closet De Vriesian saltationist all along! A. C. Seward, editor of the volume, considered this rewriting of history so egregious – considering Darwin's deep-seated gradualism and his so strongly and continually expressed denial of evolutionary significance to large variants, or 'sports' – that he took the unusual liberty of inserting a contradictory footnote right within De Vries' text: 'I think it right to point out that the interpretation of this passage from the Origin by Professor De Vries is not accepted as correct by Mr. Francis Darwin or by myself...' (p. 71).

Despite the façade of unity that obeisance to Darwin's name inspired, the articles in Seward's centennial volume well reflect the disarray that existed within evolutionary theory on Darwin's hundredth birthday. Just read de Vries' account of macromutation, Bateson's attack on adaptation, and Weismann's stout defence of strict neo-Darwinism and the 'omnipotence (*Allmacht*) of selection' for the flavour of controversy. Only William Bateson, who never could mince words, honestly admitted the confusion and deep disagreement in concluding (1909, p. 101): 'No one can survey the work of recent years without perceiving that evolutionary orthodoxy developed too fast, and that a great deal has got to come down'.

Yet in this thick and supposedly comprehensive set of tributes, the custodians of all the direct evidence for life's pageant got short shrift. Palaeontology merited only two insipid articles, one on animals and one on plants, by different gentlemen named Scott. W. B. Scott cited some phylogenies, advocated more work and defended the objectivity of pure observation. He even praised Cuvierian creationism because its intellectual emptiness had at least kept 'the infant science in leading-strings until it was able to walk alone', and thus prevented 'a flood of premature generalizations and speculations' (1909, p. 185). D. H. Scott defended the adaptive significance of basic plant anatomy and reiterated Darwin's argument on the imperfection of the geological record to explain the apparent gaps in fossil series.

I do not think that the extreme downplaying of palaeontology in 1909 simply reflected an editorial idiosyncrasy. Palaeontology had played a vigorous role in evolutionary theory for the first two or three decades after 1859 – both in theory (the neo-Lamarckism of Cope and Hyatt, for example) and in empirics (the establishment of phylogenies, and the discovery of such transitional forms as *Archaeopteryx*). But its fortunes then waned along with the entire school of speculative phylogenetics that Haeckel had spawned, and most palaeontologists had returned to their ancient métier of correlating rocks for geologists. It was a mechanistic age, and the torch had passed to those 'newer schools of physiological specialists' mentioned by Wallace.

In sharp contrast, the major Festschrift of the second centennial (for the *Origin* this time), held 50 years later in 1959, featured plenty of palaeontology and precious little disagreement (Tax, 1960). Again, these shifts reflect neither editorial nor national peculiarities, but well record the major movement in evolutionary theory during the intervening 50 years. For 1959 represented the heyday of the hard version of the Modern Synthesis, a movement that brought such traditional fields as palaeontology and systematics under the explanatory aegis of Darwin's original hypothesis of natural selection, finally united fruitfully with the Mendelian genetics of micromutation.

The synthesis had received an indispensable boost from Simpson's brilliant and necessary argument (1944) that the large-scale phenomena of life's pageant could be rendered consistent with Darwinian principles. (I use 'necessary' in the historical, not the logical, sense. As long as palaeontologists continued to insist that macroevolution obeyed different and probably unknowable laws (see Osborn, 1922, p. 141), no satisfactory, unified evolutionary theory could be developed.) Simpson's argument then triumphed at the famous international meeting of geneticists, palaeontologists and systematists that virtually codified the Modern Synthesis at Princeton University in 1947 (Jepsen, Mayr & Simpson, 1949), and led directly to the foundation of the Society for the Study of Evolution and its journal *Evolution*, still the focal point of our profession. D. M. S. Watson, grand old man of British vertebrate palaeontology, brought up a lonely rearguard of one in arguing that certain pervasive trends in amphibians (decrease of cartilage bone and dorso-ventral compression of the skull) and therapsids (decrease of the quadrate and general reduction in depth of all skull parts near the brain case and below the brain itself) could be interpreted neither as mechanically advantageous in themselves nor as correlates of any adaptive change, and might therefore have 'been induced by an internal mechanism' (Watson, 1949, p. 59). But all other palaeontologists presented their adaptationist scenarios and calculations of evolutionary rates in a strictly Darwinian context.

The Synthesis began in an open and pluralistic spirit during the 1930s. Its aim, so clearly displayed in Dobzhansky's founding document of 1937 (see Gould, 1982c), was to render all evolutionary phenomena consistent with the causal mechanisms of modern genetics, whether Darwinian or not (see also Simpson, 1944, but largely completed before World War II, and Mayr, 1942). Yet, throughout the 1940s the Synthesis 'hardened' in its advocacy of a selectionist basis for nearly all important evolutionary change, and underwent a subtle shift away from a methodological concern with explanation by observed genetic mechanisms to a substantive claim for a strict form of Darwinism based on Weismann's virtual *Allmacht* of selection (best seen in comparisons of first and later editions of founding works: Simpson, 1944, 1953; Dobzhansky, 1937, 1951; Mayr, 1942, 1963, for example; see Gould, 1982e and Provine, 1983). As the Synthesis hardened, it also gained a confidence that sometimes verged (in retrospect) on the uncomfortable. Simpson ended a 1950 essay by proclaiming:

This general theory is now supported by an imposing array of paleontologists, geneticists, and other biological specialists. Differences of opinion on relatively minor points naturally persist and many details remain to be filled in, but the essentials of the explanation of the history of life have probably now been achieved (Simpson, 1965, p. 14).

When the world's leading evolutionists converged on Chicago in 1959 for the second centennial, this confidence had reached its pinnacle. Tax's three volumes are a monument to this genuine (if transitory) agreement. Yet, just as Bateson had blown the whistle on the covered disagreement of 1909, one paper challenged, albeit more gently, the true consensus of 1959. E. C. Olson, a leading American vertebrate palaeontologist, expressed his unhappiness with the increasingly dogmatic tone that some synthesists had adopted of late:

The statement is made, in effect, that those who do *not* agree with the synthetic theory do *not* understand evolution and are incapable of so doing, in most cases because they think typologically...Some avid proponents of the synthetic theory would appear to...eliminate as competent students of evolution, because of their inability to understand *the* theory, those who may disagree (Olson, 1960, pp. 526–527; Olson's italics).

Olson then put his finger on the two greatest methodological problems of the synthesis as applied to palaeontology and macroevolution. First, its disconcerting flexibility and resultant capacity to encompass almost any conceivable observation:

The feeling of a slight sense of frustration in the elasticity involved in developing a universal explanation is hard to avoid...There is little or nothing that cannot be explained under the selection theory, and, at present, this theory appears to be unique in this respect (1960, p. 530).

Secondly, Olson recognized that the palaeontological arguments of the synthesis had a peculiar form – they were neither proofs nor exclusions of alternatives, but merely consistency arguments, thereby adding little independent support to the basic theory itself:

This possible danger is amply revealed in some studies of the last decade which seem more concerned with fitting results into the current theory than with evaluation of results in terms of a broader context...They [palaeontological studies] have not revealed inconsistencies, but there is, of course, some question as to whether this is a likely outcome in view of the methodology (1960, pp. 536–7).

Yet, despite his cogent critique of existing methodology, Olson confessed that he had nothing specific to offer in rebuttal. He doubted the syntheticist confidence in extrapolation from micro to macroevolution (p. 537), but admitted (p. 531):

This situation poses a frustrating dilemma for the sincere student who feels from his observations that there is more to evolution than can be studied, tested, and integrated under the synthetic theory, who is confident that real problems exist but also sees no way of making progress toward an understanding by means of the materials that raise the questions in his mind. Few feel that the genetic-selection theory is invalid, but rather consider that there is much evidence that it is not adequate.

To the last sentence, I can only reply, with the benefit of another quarter century, 'amen'. Still, critiques without alternatives never get much of a hearing in happy times, and Olson's warnings inspired no general reassessment.

Thus palaeontology, rendered irrelevant amidst the confusion of the first centennial, contributed greatly to the agreement of the second. But its role, as Olson saw, had become one of subservience. It had relinquished the idea that a direct study of vast times and changes might contribute new principles to a larger synthesis, and had opted for satisfaction that its phenomena *could* be rendered consistent with a body of theory presented as a package by colleagues in other disciplines. One can only hope, if for palaeontological pride alone, that a middle road exists between dismissal and acceptance for testimonial agreement. I believe that we have found this road in time for the third centennial.

The third centennial finds Darwinian theory in healthy turmoil, sometimes misrepresented as death-throes by an uncomprehending (or irresponsible) press, searching to intensify conflict. Legitimate conflict abounds, but we are not witnessing a revival of the anarchic confusion that characterized evolutionary theory at the 1909 centennial. The principle of natural selection seems sound, and proposals for revised and expanded approaches to evolutionary theory retain a Darwinian core. The healthy turmoil includes proposals of two sorts: (1) a widened role for non-adaptation and for chance as a source of evolutionary change, including claims that non-selected features act as important channels for pathways of change (though selection may propel organisms down the permitted paths), and as facilitators in forming pools of cooptable features for future change; (2) attempts to construct a hierarchical theory based on the interaction of selective (and other) forces at numerous levels (from genes to clades) – rather than almost exclusively upon selection among organisms (Eldredge & Cracraft, 1980; Gould, 1982b; Arnold & Fristrup, 1982). I regard critiques of gradualist thought as an important challenge to a pervasive corollary of the theory of natural selection (though Darwin himself viewed the connection as logically entailed), but not as a confutation of the theory itself (Gould, 1982b).

For palaeontologists, challenges to strict Darwinism centre upon one venerable but continuously fascinating issue: can the phenomena of vast times and large amounts of change (macroevolution) be explained as simple extrapolations of observed evolutionary processes acting within populations (microevolution). In his cautionary essay for the second centennial, Olson focussed upon this 'difficulty in extrapolation' and questioned the 'extension of observations [about microevolution] to levels that are incommensurate' (Olson, 1960, p. 533).

I should point out – because this has been a source of frustrating and purely terminological confusion – that I use the terms micro and macroevolution in the purely descriptive sense to designate the phenomenology, whatever its cause, of evolutionary change within versus among species. Others use 'macroevolution' in a restricted causal sense to designate Goldschmidt's saltational theory for the origin of new taxa, thus giving the word itself an anti-Darwinian commitment. In this context, note that Goldschmidt himself, who first popularized these terms, used them in the descriptive sense that I advocate – and then, of course, went on to defend his particular non-Darwinian theory of macroevolutionary mechanisms. Goldschmidt wrote, at the beginning of his major work (Goldschmidt, 1940, p. 8): Macroevolution 'will be used here for the evolution of good species and all the higher taxonomic categories'.

In the rest of this essay, I shall argue that palaeontology has a new role to play at the third centennial – neither irrelevance, nor subservience, but true partnership. For palaeontology is a principal source of independent macroevolutionary theory that cannot be simply extrapolated from the evolutionary processes operating among organisms within populations. But I speak of partnership and not warfare or conflict for a definite reason. Previous claims for an independent macroevolution have embodied the unfortunate and unnecessary methodological claim that such autonomy requires new causal mechanisms necessarily opposed to well-known microevolutionary processes of mutation and selection (or at least declaring their irrelevance). This assumption of warfare is falsely based on the non-hierarchical 'either-or' thinking that characterizes most of science. But when we recognize that genuine novelty also arises, within hierarchical systems, from the different operation of familiar processes upon the material of various levels, then the basis for an independent but complementary theory of macroevolution may be established.

A false and restrictive dichotomy

Few evolutionists realize how integrally bound to the entire history of our subject is the ancient issue of whether macroevolution can be rendered as extrapolated microevolution, or whether it requires explanatory principles in its own right. One might even say that the history of evolutionary thought, in its broadest aspect, *is* the chronicle of this debate. Pre-Darwinian traditions spoke with virtual unanimity of a separate macroevolution different in principle from processes of change in the small. Darwin's uniqueness, and much of his greatness, lies in his proposal of a unified and workable theory based upon an observable microevolution.

I divide the pre-Darwinian traditions of separation into two major schools of thought.

(1) *Ladders and diversions.* The earliest fully formulated evolutionary system, devised by Lamarck, rendered the bush-like pattern of evolution as an interplay of two opposing forces. Life, generated spontaneously in primal simplicity, advanced upon its single upward course, propelled by 'the force that tends, incessantly, to complicate organization'. But if this macro-evolutionary principle of progress generates novelty and produces life's pageant, what causes iterated diversity? Why cannot all organisms be placed on a single ladder? Why do most species exhibit such detailed fine tuning to local environments (eyeless moles and long-necked giraffes)? Lamarck reasoned that a tangential or diverting force must exist, separate in principle from the causes of progress, but the source of microevolutionary adaptation to local circumstance. Organisms are sensitive to l'*influence des circonstances*; they adapt their bodies and behaviours to these local conditions during their lifetimes and pass these changes to their offspring in the form of altered heredity. Lamarck's mechanism for the propagation of diversions – the inheritance of acquired characters – became the 'Lamarckism' of later generations, thereby assigning to oblivion the ladderlike progress that Lamarck himself viewed as more important.

In his later works at least, Lamarck described the two forces of ladders and diversions as separate in principle and inherently opposed. He wrote in 1809:

The state in which we see all animals is, on the one hand, the result of the increasing complexity of organization [*la composition croissante de l'organisation*], which tends to form a regular gradation, and, on the other hand, that of the influence of a multitude of very different circumstances, which continually tend to destroy regularity in the gradation of the increasing complexity of organization.

Thus, the causes of general direction and local adaptation are not only different, but opposed. The microevolutionary process that we can study directly cannot in principle illuminate the larger-scale process that gives life's history its essential character. We are placed in the conceptual dilemma of declaring the observable trivial, and the essential both rigidly separate and probably unobservable.

The first fully elaborated English evolutionary theory, presented by R. Chambers in the *Vestiges of the Natural History of Creation* (1844), followed the Lamarckian tradition of ladders and diversions, with separate and opposed causes. Macroevolutionary progress is an analogue to ontogeny with a unidirectional, humanward propulsion. But disturbing forces divert the flow, arresting general development at a certain stage of universal ontogeny and iterating the products of this arrest into a diverse array of locally adapted species.

(2) *Laws of form*. This second affirmation of an independent macroevolution opposed to the forces of local adaptation arose from the central issue of relationships between form and function (Russell, 1916, remains the indispensable source for this debate). Does form, as Cuvier generally believed, arise from and reflect functions of the parts involved? Is form, in other words, frozen activity? Or does function, as Goethe and E. Geoffroy de Saint-Hilaire maintained, follow the dictates of pre-existing form? If form reflects function and function records local adaptation, then the *Baupläne* of life are directly connected to small-scale, observable ties of organism and environment. An evolutionary continuum based on extra-polated local adaptation might be constructed.

But if form is primary and acts as a constraint upon action of the moment, then local adaptation is minor adjustment within preset possi-bilities, and we must seek a different causal basis for the origin of basic designs. We must search, in other words, for the laws of form. Since these *Baupläne* are discontinuous in life's morphospace, and since they do not record 'adaptive peaks' of function climbed by continuous change, the macroevolutionary process that generates them is probably saltational and unrelated to the minor adaptive fine-tuning (microevolution) that then occurs within their constraints. Thus, adherents to the 'laws of form' tradition, from Geoffroy, to Bateson (1894) who catalogued discontinuities in morphology and sought their formal relations, to D'Arcy Thompson (1917) who developed a wondrously anachronistic Pythagorean mathe-matics to contrast continual modification within and saltational change between *Baupläne* (see Gould, 1971), have generally advocated a macro-evolution separate from, and unilluminated in principle by, the forces of local adaptation.

The problem with both these traditions is despair and unworkability. We can observe and manipulate small-scale changes and local adaptations; they provide us with something evident to do – the *sine qua non* of science as practised. But if such microevolution teaches us nothing about, and is actually opposed to, the larger-scale processes that intrigue us so mightily, then *que faire?* Macroevolution is either too slow or too rare to observe directly. How can we study it at all? Thus, the common property of both ladders and diversions, and laws of form lies in their general unfruitfulness as research strategies. Little enduring, hard-nosed work arose from them, and evolutionary theory remained at an impasse. Enter Darwin.

What was the chief source of Darwin's greatness? He did not invent evolutionary theory. The philosophically radical implications of natural selection, much as they upset both ancient traditions of Western thought and European social realities, lay at too high a level of abstraction to affect directly the daily activities of working scientists. I believe that Darwin's greatest contribution lay in his construction, for the first time, of a workable evolutionary theory.

By asserting a continuum of process, and a reduction of macro-evolutionary events to microevolutionary forces, Darwin proclaimed that the observable and manipulable events of evolution, from artificial selection as practised by breeders to subspecific differentiation in nature, did not stand in opposition to life's pageant as mere fine tuning within stages attained by different forces, but were the actual stuff of evolution – all of evolution. What *can* be studied is, by extension, the essence of the entire process. Darwin understood his methodological contribution perfectly well. He constructed his entire book (Darwin, 1859) as a grand analogy between the palpable process of artificial selection and an inferable natural selection for macroevolutionary events. Why else would the first chapter of such a revolutionary book be largely consecrated to the otherwise arcane subject (for non-fanciers) of pigeon breeding? Darwin's continuum extends from pigeons to phyla. New species arise from the conversion of small-scale, within-population variation to the differences between populations: higher taxa are simply a consequence of more time granted to the same events. The Linnaean hierarchy is a continuum in process.

Darwin's methodology breathed life into evolutionary theory, and made it a science rather than a field of untestable speculation about cosmic directionality. But, in a strict version that later became canonical (though Darwin himself did not espouse it), the continuationist school ended by denying any legitimate independence to macroevolution as a source of theory. And the false dilemma of two unacceptable choices arose: confusion or restriction. Either assert an independence for macroevolution and deny unity to evolutionary theory, or assert continuity and deny that macroevolution offers insight about evolutionary process.

This tradition has persisted ever since Darwin; the false dichotomy continues. In the most widely discussed defence of macroevolution published after the ascendancy of the Modern Synthesis, Richard Goldschmidt (1940, pp. 205–6) argued for new genetic mechanisms opposed in principle to natural selection acting within populations. 'For a long time I have been convinced that macroevolution must proceed by a different genetic method... a pattern change in the chromosomes, completely independent of gene mutations, nay, even of the concept of the gene'. And so Goldschmidt led himself down a garden path lined (as are the gardens of my native Massachusetts during our plague of gypsy moths) with the caterpillars of *Lymantria*. He began with the reasonable notion of 'hopeful monsters' as products of small genetic changes with large phenotypic effects, moved on to a genetic theory of 'systemic mutation' discombobulating the entire genome, and finally denied the corpuscular gene itself, thereby courting derision and dismissal by his evolutionary colleagues (see Gould, 1982*d*).

Even passing suggestions follow the same tradition. When Olson (1960, pp. 6, 533) felt his discomfort with the extrapolationist vision, he too could only express it with the thought that a different genetics must regulate macroevolution, perhaps, he suggested, some aspect of cytoplasmic inheritance.

I persist in thinking, as I argued above, that there is something desperately restrictive and fallacious about this dichotomy. We can have an independent macroevolution without requiring either a new genetics or an opposition to the well-established principles and mechanisms of microevolution. The concept of hierarchy is the point of exit from this historical morass.

Palaeontological expressions of hierarchy, two challenges, and a modest proposal

The part–whole hierarchy of nature forms a conceptual basis for levels, bound by interaction to others, but sufficiently independent to preclude complete explanation by processes acting at other levels (lower ones in the reductionist tradition). Gene, organism, deme, species, and clade represent one imperfect and incomplete display of such a hierarchy. (Ecosystems, and other units treating the interaction of objects that do not form monophyletic sets are incommensurate and do not belong in such a hierarchy.)

Species are the units of macroevolution. Palaeontological phenomena described by differential rates of origin and persistence of species, rather than by differential rates of change within anagenetic lineages, provide a pool of events for an independent macroevolution. These events embody both a 'weak' (but conceptually important) and a strong argument for independent macroevolution.

The weak argument applies to patterns not directly predictable from microevolutionary events, but not including any irreducible process in their explanation. Stebbins & Ayala (1981) advocated this sense of independence in their defence of neo-Darwinism. Patterns well described by stochastic models provide the best cases that palaeontology has offered during the last decade. Van Valen (1973), for example, argued that extinction patterns, as recorded in the fossil record, have the form of radioactive decay curves, implying that, at this level, species (or higher taxa) may be treated as if they had an equal probability of extinction in each interval of time. Raup & Gould (1974) showed that many morphological patterns of apparent order, high correlations among features evolving independently for example, arise at unexpectedly high frequency in random models. Raup *et al.* (1973) and Gould *et al.* (1977) compared shapes of clades in fossils and stochastic simulations to explore whether random models might produce the apparent order observed in the history of diversity. Sepkoski (1978, 1979) has developed a more complex two- or three-phase logistic model for generating plateaus of diversity closely matching the pattern of fossil invertebrate higher taxa.

The best explanation for these random patterns does not involve any new causal (or rather acausal) principle operating at high levels of the hierarchy. Each individual event – the death of a single species for example – may have a conventional cause fully reducible to the Darwinian level of selection upon organisms. But the pattern in large classes of such events may fit random models because the forces regulating individual events are so varied, so independent, and so numerous that their combined effect over all cases is best described in stochastic terms. Nothing in microevolutionary theory predicts whether or not stochastic models will apply. The question must be explored directly in terms of macroevolutionary units and macroevolutionary scales, even though all results along a deterministic–stochastic spectrum may be reducible to microevolutionary processes.

The strong argument for an independent macroevolution identifies causes that emerge at macroevolutionary levels of the hierarchy itself, and are not reducible in principle to Darwinian selection acting upon organisms. These emergent causes are not opposed to, or inconsistent with, conventional microevolutionary causes. Rather, they interact with causes at other levels to yield evolutionary patterns by interaction between levels

(Gould, 1982b). We advocate partnership through hierarchies with levels in bounded independence, not the old 'either-or' dichotomy that viewed any claim for a macroevolutionary cause as necessarily eliminating or contradicting a microevolutionary process.

True species selection (Gould, 1982a, b) provides our clearest example in this category. If species are discrete units both in space and time, as the theory of punctuated equilibrium asserts, then evolutionary trends must be described in terms of their differential success. This differential success might be reducible to microevolutionary causes, and we should not then speak of 'species selection' even though we must describe the trend in terms of species (see Gould, 1982a, and Stanley, 1979 for many examples, but different terminological commitments). Vrba's 'effect hypothesis' lies in this category and provides a fine example (based on the evolution of African bovids) of why even reducible trends cannot be described in terms of microevolutionary units alone (Vrba, 1980, and personal communication).

Differential success, however, is often not reducible to Darwinian selection upon individuals, but operates through direct selection upon such group-level properties as differential rates of speciation (see Hansen, 1978 and Gilinsky, 1981, for potential examples; the causes of enhanced speciation may be reducible, but usually they are not, since they involve population structure, size, migration dependent upon population structure, etc.). Trends powered by high speciation rates in certain subclades are primary candidates for species selection. The rise to dominance of many morphological features may express an accidental tie with species that branch more often, rather than any Darwinian advantage for the form itself.

These ideas of independence through hierarchy have been challenged during the past decade by two general kinds of arguments, both invalid in my judgement. I shall cite a specific example in each category.

(1) *Direct denial of causal hierarchy.* Dawkins (1976, 1982) has presented the strongest argument for reduction even beyond the traditional Darwinian level of individuals to the gene itself as an ultimate unit of selection. But I regard the argument as deeply though interestingly fallacious, for what I regard as two logical errors in its formulation (not to mention a host of empirical difficulties, see Wright, 1980).

First, Dawkins confuses bookkeeping with causality. He notes correctly that any evolutionary change involves an alteration in the frequencies of genes. He then assumes that changes at the lowest level imply a causal reduction to objects of that level, as in the following comment upon group selection models (1982, p. 115). 'The end result of the selection discussed is a change in gene frequencies, for example an increase of "altruistic genes" at the expense of "selfish genes". It is still genes that are regarded as the replicators which actually survive (or fail to survive) as a consequence of the (vehicle) selection process.'

But it is a property of any hierarchy that causal intervention at one level results in a sorting of objects at all lower levels. The existence of such a sorting simply does not imply that objects sorted are causal agents. Genes do have more intergenerational stability than bodies, and we may therefore choose, for convenience in bookkeeping, to *record* evolutionary change by the sorting that occurs at this lowest level. If species selection exists, it must sort genes within a clade, because some subclades diversify and others disappear, and monophyletic groups share genes absent from other groups. But the existence of sorting does not deny a direct and irreducible causal role to species. Bookkeeping is not causality.

Second, Dawkins presents a false induction by arguing that legitimate cases of gene selection imply an exclusive causal basis for the phenomenon. Gene selection surely exists; the notion of hierarchy defends its presence with all the force that Dawkins' view musters – but not its exclusivity or its preferential nature. Dawkins argues, for example (1982, p. 158). 'The theory of selfish DNA is in a way revolutionary. But once we deeply imbibe the fundamental truth that an organism is a tool of DNA, rather than the other way around, the idea of "selfish DNA" becomes compelling, even obvious.' I only point out that selfish DNA is equally 'compelling, even obvious' under the concept of hierarchy and its postulate of a legitimate causal level (among many) for genes (Gould, 1981).

So committed is Dawkins to his own exclusively gene-based view of selfish DNA that he even falls into the absurdity of describing its absence as a gene, in order to get round the part of the hypothesis that depends upon hierarchy: stabilization by negative interaction between gene-level selection to augment copies, and organism-level selection to reduce them as a consequence of their energetic costs. Dawkins would describe the whole process – both the increase and the stabilization – as results of gene-level selection; and he is forced to describe the absence of some copies as a gene, and therefore an evolutionary agent in its own right, battling with actual copies for survival at its own level. 'Here we are recognizing that the deletion itself, the absence of the selfish DNA, is itself a replicating entity (a replicating absence!), which can be favored by selection (Dawkins, 1982, p. 164).' I take it, from his exclamation point, that Dawkins himself has some inkling of the difficulty and mystification of such an argument.

(2) *Failure to consider hierarchical alternatives* through the venerable 'consistency argument' used to halt rather than to encourage further thought.

When supporters of punctuated equilibrium began to emphasize the puzzling phenomenon of stasis, some orthodox neo-Darwinians responded by simply searching their catalogue of processes for an item that could, in principle, render the result; and, having found one (stabilizing selection), essentially declaring the problem solved and devoid of further interest. Charlesworth (1982, p. 135) for example, writes:

The most likely explanation of stasis is that the character in question is sufficiently well adjusted to its functional requirements that stabilizing selection is operating ...Under stabilizing selection, we could expect a population to evolve rapidly to a state in which the optimum value for a character coincides with the mean, and then to remain constant until the optimum changes.

Now I don't doubt that stabilizing selection acts to maintain stasis in phyletic lineages. How could it not? But is there nothing else to the stasis that often endures ten million years or more in fossil invertebrates living through all the vicissitudes of environment that must characterize any period of such length? Indeed, for an instant, Charlesworth seems to grasp the problem, but then he dismisses it again, this time citing presumed unanswerability rather than lack of potential interest:

It is, of course, a surprising fact that some characteristics of some organisms should apparently have an optimal value that has remained nearly constant over very long periods of time. It is important to bear in mind, however, that we are usually so ignorant of the genetic structure of fossil populations and of the relations between environment, fitness and morphology that we cannot provide explanations for any particular historical pattern of evolution (1982, p. 135).

My complaint is not so much with the invocation of stabilizing selection as with the viewpoint hidden within the phrase: 'should apparently have an optimal value that has remained nearly constant...' And, with all deference to Darwin's reminder that 'the old saying of vox populi, vox Dei...cannot be trusted in science' (1872, p. 134), I want to invoke something akin to common sense.

Shells of the common quahog *Mercenaria* are common along the beaches of Scientists' Cliffs, Maryland. But they are Miocene fossils, some fifteen million years old, weathered out of the cliffs and deposited on the beach along with modern shells. They may be a bit thicker on average, but are otherwise indistinguishable from modern quahogs. Is it reasonable to believe that these shallow-burrowing clams have not changed because their progenitors attained such adaptive perfection that all impetus for selective alteration has disappeared? Species of shallow-burrowing clams come in hundreds of shapes and sizes, and I simply cannot believe that each has attained an optimum for some finely constituted microniche. I cannot believe, in short, that the morphospace of shallow burrowers is a map of ideal engineering solutions to slightly differing circumstances. Thus I advance the idea, simple-minded though it may seem, that stable form may be constrained by genetic and developmental coherences that do not only reflect best adaptive solutions – and that half a million species of beetles do not represent preserved peaks of best design. Why should we close off the study of 'internal' sources of stability, just because our ignorance about them is so profound at present, and because a standard 'externalist' explanation can be stretched to account consistently for our observations?

I think that much of our disinclination to consider internal coherences (perhaps not constructed by natural selection and not easily modified by it) arises from an exaggerated fear of 'essentialism' – the proclaimed source of all that is evil in biological thinking. Olson (1960, p. 526) expressed a similar frustration in noting that strict neo-Darwinians often dismissed reasonable opposition by arguing that 'those who do not agree with the synthetic theory do not understand evolution and are incapable of so doing, in most cases because they think typologically'.

Let me pay my dues and state the obvious. The downfall of old-fashioned typology was a prerequisite to the rise of evolutionary thinking. The antiessentialism of modern thought has a host of salutary implications and consequences. Essentialists did reason in ways that barred modern understanding. Quetelet's essentialistic statistical search for *l'homme moyen* is antithetical in concept to a statistics that regards variation as fundamental rather than accidental.

Nonetheless, we may have gone too far in rejecting the idea that species have 'essences' – defined in a useful modern way as genetic and developmental coherences that resist selective pressures of the moment, and impose a higher level, or macroevolutionary, constraint upon change within local populations. The radical anti-essentialism of modern evolutionary biology leads us to ignore or deemphasize the interplay of internal design and external selection.

This overextended anti-essentialism pervades our approach to all evolutionary issues, often constraining our vision of reasonable possibilities. I have no doubt, for example, that it subtly underlies the preference most neo-Darwinians feel for stabilizing selection as a sole explanation of stasis. For if variation is fundamental and species have no meaningful essence, then what can hold a species in one region of morphology except an external force that culls and regulates its defining variation? Developmental programs and genetic systems are 'essential' properties 'distributed' to individuals by virtue of their membership in a species (and subject to important variation among these individuals of course). The admission of such essential properties might force evolutionists once again to consider the role of 'internal' factors in evolutionary change and stability.

This perspective both looks back into history and forward to our expanding knowledge of genetics and development. Looking back, it affirms the validity of a central postulate in the 'laws of form' tradition – that internal features of design can constrain and channel a course of change. Looking forward, it encourages us to view mechanisms of internal structure and construction as more than a source of random variation for the attention of natural selection.

Selection and drift are often cited as the only important, sometimes as the only conceivable, modes of evolutionary change. But perhaps it is time to consider a third major way – the rarely recognized category (these days) that includes most non-Darwinian proposals of the past century – internally generated changes of the genome sufficiently rapid in occurrence and great in extent to present fundamentally new features as *faits accomplis* to forces of selection (that may then accept or reject) or drift (that may then fix or eliminate). This category includes all the wreckage of rejected and invalid theories (classical saltationism) and phenomena valid in theory but deemed unimportant in practice (meiotic drive). But our exploding knowledge of molecular and genomic organization may provide new and reasonable proposals. Dover's theory of molecular drive as a mechanism of speciation (Dover *et al.*, 1982; Dover, 1982) strikes me as a first promising entry.

Evolutionary theory at the third centennial is bursting with new life and excitement. We are experiencing none of the anarchy of 1909, since Darwinism is now secure as a centrepiece. But we are also moving away from the smugness of 1959 and considering a flood of useful expansions and additions. The central Darwinian metaphor of 'trial and error' (R. C. Lewontin, personal communication), with direction supplied by external selection and internal factors reduced to sources of undirected raw material, may eventually yield to another metaphor based on a more equalized interaction of external selection and internal production and constraint.

Darwinians should welcome this fruitful ferment, not fear that its outcome might compromise or dilute the contributions of history's greatest biologist, whom we honour again at his third centennial. William Bateson, that old 'laws of form' man, did Darwin his greatest honour in 1909 (though he was among the least Darwinian of all participants) in writing

(1909, p. 85): 'We shall honour most in him not the rounded merit of finite accomplishment, but the creative power by which he inaugurated a line of discovery endless in variety and extension.'

References

Arnold, A. & Fristrup, K. (1982). The theory of evolution by natural selection: a hierarchical expansion. *Paleobiology*, **8**, 113–29.

Bateson, W. (1894). *Materials for the Study of Variation*. London: John Murray.

Bateson, W. (1909). Heredity and variation in modern lights. In *Darwin and Modern Science*, ed. A. C. Seward, pp. 85–101. Cambridge: Cambridge University Press.

Chambers, R. (1844). *Vestiges of the Natural History of Creation*. London: John Churchill.

Charlesworth, B. (1982). Neo-Darwinism – the plain truth. *New Scientist*, **94**, 133–37.

Darwin, C. (1859). *On the Origin of Species*, 1st edn. London: John Murray.

Darwin, C. (1872). *The Origin of Species*, 6th edn. London: John Murray.

Dawkins, R. (1976). *The Selfish Gene*. Oxford University Press.

Dawkins, R. (1982). *The Extended Phenotype*. San Francisco: W. H. Freeman.

de Vries, H. (1909). Variation. In *Darwin and Modern Science*, ed. A. C. Seward, pp. 66–84. Cambridge: Cambridge University Press.

Dobzhansky, T. (1937). *Genetics and the Origin of Species*. New York: Columbia University Press.

Dobzhansky, T. (1951). *Genetics and the Origin of Species*, 3rd edn. New York: Columbia University Press.

Dover, G. A. (1982). A molecular drive through evolution. *Bioscience*, **32**, 526–33.

Dover, G. A. *et al.* (1982). The dynamics of genome evolution and species differentiation. In *Genome Evolution*, ed. G. A. Dover & R. B. Flavell, pp. 343–72. *Systematics Association Special Volume No. 20*, March.

Eldredge, N. & Cracraft, J. (1980). *Phylogenetic Patterns and the Evolutionary Process*. New York: Columbia University Press.

Gilinsky, N. L. (1981). Stabilizing species selection in the Archaeogastropoda. *Paleobiology*, **7**, 316–31.

Goldschmidt, R. (1940). *The Material Basis of Evolution*. New Haven: Yale University Press.

Gould, S. J. (1971). D'Arcy Thompson and the science of form. *New Literary History*, **II** (2), 229–58.

Gould, S. J. (1980). *The Panda's Thumb*. New York: W. W. Norton.

Gould, S. J. (1981). The ultimate parasite. *Natural History*, **90** (11), 7–14.

Gould, S. J. (1982a). The meaning of punctuated equilibrium and its role in validating a hierarchical approach to macroevolution. In *Perspectives on Evolution*, ed. R. Milkman, pp. 83–104. Sunderland, MA: Sinauer Associates.

Gould, S. J. (1982b). Darwinism and the expansion of evolutionary theory. *Science*, **216**, 380–7.

Gould, S. J. (1982c). Introduction to T. Dobzhansky, Genetics and the Origin of Species. In *The Columbia Classics in Evolution Series*, ed. N. Eldredge & S. J. Gould, pp. xvii–xli. New York: Columbia University Press.

Gould, S. J. (1982d). The Uses of Heresy: An Introduction to Richard Goldschmidt, *The Material Basis of Evolution*, pp. xiii–xlii. New Haven and London: Yale University Press.

Gould, S. J. (1982e). The hardening of the modern synthesis. In *Conference Volume for Wenner-Reimer Stiftung*, ed. M. Grene.

Gould, S. J. et al. (1977). The shape of evolution: a comparison of real and random clades. *Paleobiology*, 3 (1), 23–40.

Hansen, T. (1978). Larval dispersal and species longevity in Lower Tertiary gastropods. *Science*, 199, 885–7.

Jepsen, G. L., Mayr, E. & Simpson, G. G. (1949). *Genetics, Paleontology and Evolution*. Princeton, N.J.: Princeton University Press.

Lamarck, J. B. (1809). *Philosophie zoologique*. Paris.

Mayr, E. (1942). *Systematics and the Origin of Species*. New York: Columbia University Press.

Mayr, E. (1963). *Animal Species and Evolution*. Cambridge, MA: Belknap Press of Harvard University.

Olson, E. C. (1960). Morphology, paleontology, and evolution. In *Evolution after Darwin*, vol. 1, ed. S. Tax, pp. 523–45. Chicago: University of Chicago Press.

Osborn, H. F. (1922). Orthogenesis as observed from paleontological evidence beginning in the year 1889. *American Naturalist*, 56, 134–43.

Provine, W. (1983). Adaptation and mechanisms of evolution after Darwin: a study in persistent controversies. In *Symposium on Persistent controversies in evolutionary theory*. University of Chicago.

Raup, D. M. et al. (1973). Stochastic models of phylogeny and the evolution of diversity. *Journal of Geology*, 81, 525–42.

Raup, D. M. & Gould, S. J. (1974). Stochastic simulation and evolution of morphology – towards a nomothetic paleontology. *Systematic Zoology*, 23, 305–22.

Russell, E. S. (1916). *Form and Function*. London: John Murray.

Scott, W. B. (1909). The palaeontological record. I. Animals. In *Darwin and Modern Science*, ed. A. C. Seward, pp. 185–99. Cambridge: Cambridge University Press.

Sepkoski, J. J., Jr. (1978). A kinetic model of Phanerozoic taxonomic diversity I. Analysis of marine orders. *Paleobiology*, 4, 223–51.

Sepkoski, J. J., Jr. (1979). A kinetic model of Phanerozoic taxonomic diversity. II. Early Phanerozoic families and multiple equilibria. *Paleobiology*, 5, 222–51.

Seward, A. C. (ed.) (1909). *Darwin and Modern Science*. Cambridge: Cambridge University Press.

Simpson, G. G. (1944). *Tempo and Mode in Evolution*. New York: Columbia University Press.

Simpson, G. G. (1953). *The Major Features of Evolution*. New York: Columbia University Press.

Simpson, G. G. (1965). *The Geography of Evolution*. Philadelphia: Chilton Company. (Republication of 1950 essay Evolution and the History of Life, pp. 1–14.)

Stanley, S. M. (1979). *Macroevolution*. San Francisco: W. H. Freeman.

Stebbins, G. L. & Ayala, F. J. (1981). Is a new evolutionary synthesis necessary. *Science*, 213, 967–71.

Tax, S. (ed.) (1960). *Evolution after Darwin*, 3 vols. Chicago: University of Chicago Press.

Thompson, D'Arcy W. (1917). *On Growth and Form*. Cambridge: Cambridge University Press.

Van Valen, L. (1973). A new evolutionary law. *Evolutionary Theory* 1, 1–30.
Vrba, E. S. (1980). Evolution, species and fossils. How does life evolve? *South African Journal of Science*, **76**, 61–84.
Wallace, A. R. (1910). *The World of Life*. London: Chapman and Hall.
Watson, D. M. S. (1949). The evidence afforded by fossil vertebrates on the nature of evolution. In *Genetics, Paleontology and Evolution*, ed. G. L. Jepsen, E. Mayr & G. G. Simpson, pp. 45–63. Princeton: Princeton University Press.
Wright, S. (1980). Genic and organismic selection. *Evolution*, **34**, 825–43.

18

Plate tectonics and evolution

A. HALLAM

In his famous concluding paragraph of *The Origin of Species* Darwin expressed wonder that the diversity of the organic world can have been produced from one or a few ancestors by the operation of several natural laws – 'there is grandeur in this view of life...'. The preferential replication of genes by means of natural selection may well be a necessary condition for evolution to take place, but it is hardly a sufficient explanation for how the enormous diversity of life in space and time has come about. That there are, for instance, kangaroos in Australia and lemurs in Madagascar, and that antelopes rather than dinosaurs currently roam the plains of Africa, is among other things consequent upon a whole series of historical contingencies. The study of palaeontology in conjunction with geology ought to throw light on the interaction of historical events and the evolution of a substantial sample of the organisms that have inhabited this planet.

Darwin viewed biotic interactions – the struggle for existence – as being the major promoter of evolution. This is clearly indicated in the following passage from *The Origin of Species*. 'As species are produced and exterminated by slowly acting causes, and not by miraculous acts of creation; and as the most important of all causes of organic change is one which is almost independent of altered and perhaps suddenly altered physical conditions, namely, the mutual relation of organism to organism – the improvement of one organism entailing the improvement or the extermination of others'.

Such a view would imply, for instance, that the mammals progressively outcompeted the dinosaurs in the late Mesozoic to become the dominant terrestrial vertebrates in the early Tertiary. We have known for some time that this cannot have been the case. For many millions of years, through the Jurassic and Cretaceous periods, primitive mammals coexisted with dinosaurs, but remained low in diversity and small in size. Not until after the dinosaurs finally became extinct at the end of the Cretaceous did the mammals radiate explosively into a great diversity of forms such as we see today, to occupy an even wider range of ecological niches than those vacated by the dinosaurs. It is quite likely that the early mammals were nocturnal in habit and thereby avoided direct competition with the smaller dinosaurs. With hindsight we can envisage them as biding their time, as it were, until their reptilian competitors disappeared.

This pattern of change is by no means exceptional in the fossil record (see, for example, Gould & Calloway, 1980). Indeed it seems to be rather characteristic, as is the close coincidence in time of episodes of mass extinction and radiation of a wide variety of animal and plant groups, both terrestrial and marine. There is a clear implication that such significant evolutionary episodes may be the consequence of major physical events in earth history. Following the earth sciences revolution within the last couple of decades (Hallam, 1973) it is natural to investigate what relationship, if any, exists between major biogeographic, radiation and extinction episodes and plate tectonics, which appear to have controlled first order events in the physical environment for at least as long as a diverse metazoan fauna became established in the Cambrian, nearly six hundred million years ago. Because of my limited space, I can do no more, of course, than outline some of the more significant features as they are currently understood.

Physical effects of plate movements

The theory of plate tectonics states that the outer layer of the earth, the *lithosphere* ($\sim$ 100 km thick) comprises a small number of (relatively) rigid *plates* which are separated by narrow zones along which most tectonic, seismic and volcanic activity is concentrated. These *plate margins* are of three types: (1) *divergent*, where crustal material moves apart, under the oceans by a process known as sea-floor spreading; (2) *convergent*, where one plate plunges down into the underlying mantle (also known as *subduction zones*); (3) *transform faults*, where one plate slides laterally with respect to its neighbour, crust being here neither created (1) nor destroyed (2) (Fig. 18.1).

The most obvious effect of plate tectonics is that continents can be split and their components driven apart if a divergent plate margin becomes established beneath them, and can be caused to collide with each other along the lines of subduction zones, where mountain belts such as the

Himalayas may thereby be generated. Other mountain belts such as the Andes are also produced by subduction but at the boundary of continent and ocean floor, which have very different geological character. Plate tectonics is not the same as continental drift. Continents are carried passively on moving sectors of plates which also embrace ocean floor; they do not 'drift' across the latter (Fig. 18.2).

Migrating continents have obvious implications for biogeography and evolution, but there are other consequences of plate movements which may alter the physical environment in such a way as to affect the biosphere just as profoundly.

Firstly, it is widely accepted, though admittedly not conclusively established, that major ice ages may result from the siting of large continental masses in the polar regions, because only in such circumstances can extensive ice sheets become established, with significant consequences for world climate (Frakes, 1979). In Phanerozoic time (the time which has elapsed since the beginning of the Cambrian) there have been three such ice ages, separated by intervals of a few hundred million years when the world enjoyed a more equable climate and lacked extensive polar ice caps. Such a gross cyclicity is not, of course, to be confused with the much shorter-phase climatic cycles such as within the most recent, Pleistocene, ice age, which appear to be the consequence of an interaction of several astronomical variables (Imbrie & Imbrie, 1979).

The short-term climatic fluctuations of the Pleistocene have resulted in rapid world-wide (or *eustatic*) falls and rises of sea level as polar ice has alternately frozen and melted. The stratigraphic record indicates, however,

Fig. 18.1. Block diagram schematically illustrating plate tectonics. *A*, divergent plate boundary; *B*, convergent plate boundary; *C*, transform fault.

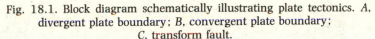

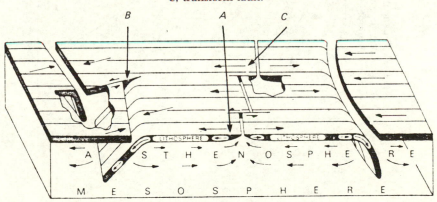

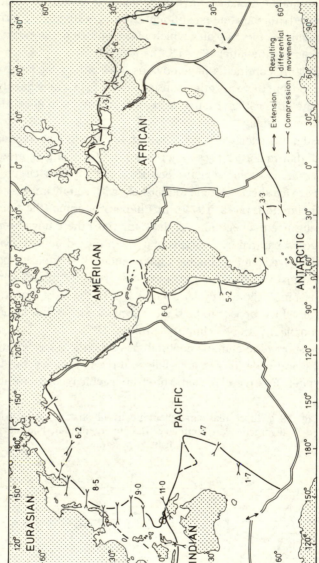

Fig. 18.2. The earth's surface divided into six major plates. Double lines, constructive margins with known rates of opening; single continuous lines, destructive margins with computed rates of lithosphere consumption recorded in cm year^{-1}; dashed lines, boundaries of possible minor plates.

that eustatic changes have also occurred during the long equable periods when the available evidence suggests that polar ice was absent. These sea-level fluctuations must therefore have been produced by changes in the cubic capacity of the ocean basins. By plotting the areal distribution of marine strata and making reasonable inferences from facies distributions about the location of former shorelines, it has been established that for long periods of time in the Phanerozoic, especially in the early Palaeozoic and late Cretaceous, the continents have been inundated by up to two thirds of the present area by *epicontinental seas*. The present shelf, or neritic zone, is in comparison a mere pericontinental fringe (Hallam, 1981*c*).

By far the most effective way of changing the cubic capacity of the ocean basins is to vary the volume of the mid-ocean ridges; increase in volume will cause a displacement of sea water on to the continents and *vice versa*. One popular hypothesis has it that variations in rates of sea-floor spreading are the controlling factor (Hays & Pitman, 1973). Since ocean-floor basalt subsides as it cools while migrating away from the spreading axis, a faster-spreading ridge will be hotter and more buoyant over a larger area and hence will cause more displacement of sea water over the continents. The rate of eustatic change producable by this process has been estimated to be about three orders of magnitude slower than that produced by the climatically induced fluctuations of the Pleistocene (Pitman, 1978).

Alternatively, eustatic changes in non-glacial periods can be produced by changes in the cumulative length of active spreading ridges, which will obviously relate to changes in plate patterns (Hallam, 1977).

Epicontinental seas have been biologically significant for several reasons. They were the site of habitation of a very large proportion of all adult aquatic organisms, and, because of their overall shallowness – generally less than (and often much less than) 200 m – quite modest eustatic and other physical changes may have had disproportionately large environmental consequences compared with the open ocean. Furthermore, extensive spreads of such seas can as effectively isolate pieces of emergent continent as spreading ocean floor, thereby creating barriers to migration of terrestrial organisms, and should also promote equability of the continental climate. It has recently been proposed that variations in albedo with respect to latitude (with considerable consequences for global climate) are a result of both the changing distribution of continents and sea-level oscillations. The latter, causing a change in land–sea proportions, is apparently the more important (Barron, Sloan & Harrison, 1980).

The organic response

For the purposes of a general survey, changes in various biological groups through time are most usefully analysed in terms of *diversity* (or, more strictly, taxonomic *richness*), and changes in spatial distribution by the amount of provinciality or *endemism*. Obviously, significant radiations will appear as marked increases, and extinctions as decreases of diversity, while the degree to which free migration is inhibited by geographic barriers, thereby promoting genetic isolation according to the classic model (Mayr, 1963), should be expressed by the ratio of endemic to pandemic organisms. Pioneer attempts to relate temporal diversity changes to plate tectonics were undertaken by Valentine & Moores (1972) and Flessa & Imbrie (1973), while the relationship of changing patterns of endemism through time in relation to plate movements was outlined by Hallam (1974). For more up-to-date palaeobiogeographic reviews see Gray & Boucot (1979) and Hallam (1981*a*, *b*).

The organic response to the changes in the physical environment induced by plate tectonics can be considered under three headings.

Migrating continents

I have proposed four simple distributional patterns for both marine and terrestrial animals which involve changes in time. *Convergence* (not to be confused with phyletic convergence) refers to the degree of resemblance of faunas in different regions increasing from an earlier to a later period, and *divergence* refers to the reverse phenomenon (Fig. 18.3). *Disjunct endemism* refers to a type of regionally restricted distribution of a fossil taxon in which two or more component parts are separated by a major physical barrier and hence is not readily explicable in terms of present-day geography. *Complementarity* in the distributional changes of contiguous marine and terrestrial animals is recognisable when one group exhibits convergence and the other divergence (Fig. 18.4). This happens, for example, when a land connection is created between two hitherto isolated areas of continent, so allowing convergence of the terrestrial faunas to take place, while severing of a once-continuous landmass gives rise to divergence as a result of genetic isolation.

Pliocene uplift of the isthmus linking North and South America, apparently related to movement of the Cocos Plate in the East Pacific, is a classic example of complementarity and is also significant from a Darwinian point of view. Substantial cross-migration of terrestrial mammal faunas coincided with mass extinction of the endemic South

American mammals which had been isolated by sea through Tertiary times. This has generally been assumed to have been the result of competition from the adaptively superior North American mammals, but Marshall (1981) has recently argued that at least some of the extinction was caused by changes in the physical environment.

In contrast, the Australian marsupial fauna has remained isolated by sea and not until late Tertiary times, when Australia–New Guinea had moved close to the Indonesian islands, was even limited and chance colonisation possible of Asian placentals across small water barriers (Whitmore, 1981).

More generally, the substantial diversity increase of faunas through the

Fig. 18.3. Schematic representation of convergence of neritic faunas of different continents, signified by vertical lines. As the continents approach each other as a result of seafloor spreading, the faunas (symbolized by A, B, C and D) progressively merge.

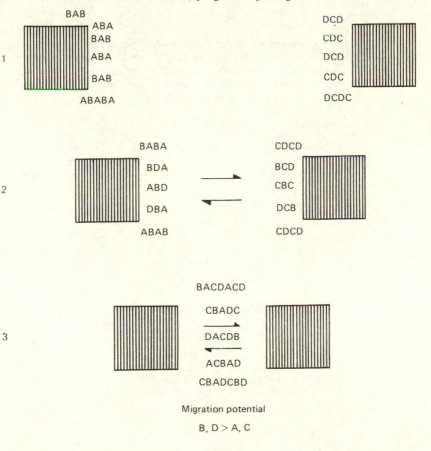

Migration potential

B, D > A, C

Mesozoic and Cainozoic, accelerating from mid-Cretaceous times onwards, which has recently been confirmed statistically (Sepkoski, Bambach, Raup & Valentine, 1981), is evidently related in substantial part to the progressive break-up of the late Palaeozoic–early Mesozoic supercontinent known as Pangaea, with the consequent increase in endemism of both terrestrial and neritic groups (Valentine, 1973).

Sea-level changes

Among the multiplicity of causes proposed to account for mass extinction events, both terrestrial and extra-terrestrial, strong supporting evidence for a succession of such events is available for only one, related to eustatic changes of sea level (Fig. 18.5). Whether the extinctions among neritic organisms were the consequence of regression of epicontinental seas (Newell, 1967) or the widespread bottom-water anoxia characteristic of the initial phase of subsequent transgression (Hallam, 1981c) a significant

Fig. 18.4. Complementarity in the distribution of terrestrial and marine faunas as two continents approach each other and eventually become sutured together. The terrestrial faunas exhibit convergence because of the creation of a corridor, while the marine organisms diverge from ancestors X to faunas Y and Z.

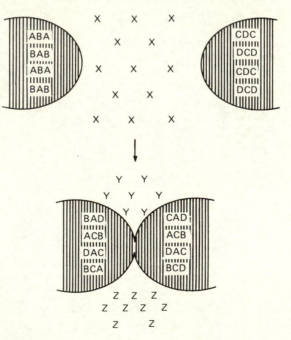

reduction of habitat area and hence deterioration of the environment would have been produced by either phenomenon.

By far the most important episode of mass extinction took place at the end of the Permian period (Sepkoski *et al.*, 1981), when it is estimated that as many as 96% of all marine species died out (Raup, 1979). Schopf (1974) and Simberloff (1974) have demonstrated that diversity changes of marine invertebrates across the Permian–Triassic boundary correlate closely with areas of epicontinental sea, in fact seem to obey rather well the ecologists' well known 'species-area' relationship.

The even more familiar end-Cretaceous mass extinction event is also associated with substantial regression (Fig. 18.5), as are the equally striking events in the marine realm at or near the end of the Ordovician and Triassic, and to a lesser extent the Devonian (Raup & Sepkoski, 1982,

Fig. 18.5. Ammonite appearances and extinctions in relation to the area of continents covered by sea, which is related to the relative height of sea level. After Kennedy (1977).

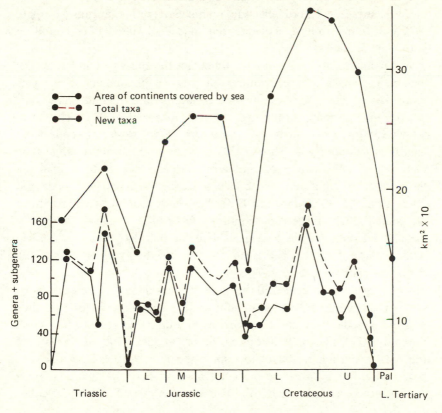

and Fig. 18.6). It is by no means clear, however, with what particular plate tectonic events they may have been associated, either directly or indirectly. The end-Ordovician regression is indeed widely thought to be the result of the growth of a Saharan ice sheet.

The question may be raised: to what extent were the extinctions selective or random? Computer modelling exercises have persuaded Raup (1981) that for the important group of trilobites it was more a matter of 'bad genes' than 'bad luck'. On the other hand, a species extinction rate for the end of the Permian as high as 96% implies that the role of chance may in extreme cases have been dominant (Raup, 1979). Generally speaking, organisms in warm, shallow seas that either build or are closely associated with reefs have been relatively vulnerable to extinction, as have planktonic foraminifera and ammonites, which have undergone a succession of 'boom and bust' cycles (Vermeij, 1978; Hallam, 1981c). Among terrestrial vertebrates (as well as at least some marine invertebrates) large organisms have been more vulnerable (Bakker, 1977).

With regard to phases of spectacular radiation, the two most important in the Phanerozoic record, affecting a large variety of organisms, correlate closely with major physical events that appear ultimately to be bound up with plate tectonics.

The radiation of marine faunas following the massive end-Palaeozoic extinctions was a long-continuing, progressive phenomenon, but was marked by a pulse of acceleration and replacement in the mid-Cretaceous, with diversity increase continuing into the Cainozoic (Fig. 18.6). The better known end-Cretaceous extinction event can in many respects be considered as a mere temporary setback in this very dramatic faunal change. The more-or-less coincident, spectacular mid-Cretaceous radiations included teleost fish, infaunal veneroid bivalves, carnivorous neogastropods and crabs (Vermeij, 1977; Stanley, 1977). There were contemporary radiations of plankton, including coccolithophores, foraminifera, diatoms and dino-flagellates (Lipps, 1970) and deep-sea ichnofauna (Frey & Seilacher, 1980), while on land the angiosperms rapidly replaced the gymnosperms as the dominant plants (Doyle, 1977).

It can hardly be coincidental that these remarkable evolutionary events, taking place within only a few million years, correspond so closely in time with an episode of exceptional igneous and orogenic activity (Larsen & Pitman, 1972), the rapid disintegration of Pangaea (Hallam, 1980) and the biggest marine transgression since the mid-Palaeozoic, apparently produced either by a phase of accelerated sea-floor spreading or by a dramatic increase in the length of the ocean ridge system. In particular, plankton evolution could perhaps have been stimulated by a drastic change in ocean current systems consequent upon Pangaea breakup, but the sea-level rise might have been just as significant.

Thus Hart (1980) has put forward a factually well-supported model relating the explosive diversification of late Cretaceous planktonic foraminifera to increasing depth of epicontinental seas. This is a surprising conclusion with intriguing implications, bearing in mind the problem of near total extinction of this group at the end of the Cretaceous, which has appeared to be an even greater enigma than the contemporary extinction of the dinosaurs. The latter might well have suffered from an increase in continentality of climate following regression, but it has not unreasonably been assumed by most palaeontologists that a planktonic group such as the globigerinid foraminifera should have been indifferent to what was happening to epicontinental seas.

Darwin was troubled by the sudden appearance of a wide diversity of fossils in the Cambrian, and tentatively suggested that there might have

Fig. 18.6. Diversity plot through Phanerozoic time of the three 'evolutionary faunas' of marine fossils distinguished by factor analysis. The first, Cambrian-Ordovician, fauna is trilobite-dominated, the second, in the later Palaeozoic, is brachiopod-dominated and the third, in the Mesozoic and Cainozoic, is mollusc-dominated. Note the sharp diversity falls, marking notable extinction phases, at the end of the Ordovician, Permian, Triassic and Cretaceous. Cam. = Cambrian, O = Ordovician, S = Silurian, D = Devonian, Carb. = Carboniferous, P = Permian, Tr = Triassic, J = Jurassic, Cr = Cretaceous, T = Tertiary. Modified from Sepkoski (1981).

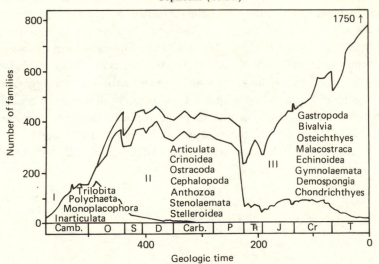

been a long interval prior to this period when no stratal record was preserved on the continents. For a variety of reasons this type of explanation must be rejected today and the explosive diversification of Metazoa across the Precambrian–Cambrian boundary, as recorded in the strata, is now generally accepted as being a true reflection of what actually happened (Stanley, 1976; Seilacher, 1977).

This diversification has been shown to correspond closely to a simple exponential growth model. As the number of taxa increased the rate of diversification seems to have become diversity-dependent (Sepkoski, 1978). The close correspondence of diversity increase with rise of sea level in the Cambrian suggests the possibility of a causal correlation (Brasier, 1979). We lack the information necessary to relate it with confidence to specific plate tectonic events, as is possible for the Cretaceous, but the Cambrian sea-level rise might well have been a consequence of opening of the Iapetus Ocean, with the growth of a spreading ridge (Anderton, 1980, 1982).

Climatic change

The pronounced increase in latitudinal temperature zonation through the course of the Cainozoic, as the world altered progressively from its Mesozoic condition of equability, must have had the effect of creating a great number of ecological niches. This is seen as a major factor contributing towards the marked increase in faunal diversity (Valentine, 1973). Increase in annual temperature range on the continents as a consequence of regression of epicontinental seas might well have played a significant role in the mass extinctions of large reptiles at the end of the Palaeozoic and Mesozoic. Pangaea at the end of the Palaeozoic must have experienced a climate of extreme continentality, not only because of its coherence (Valentine & Moores, 1972) but because of the high albedo of extensive low-latitude deserts (Barron et al., 1980). The extinction of many marine foraminiferal and ostracode species at or close to the Eocene–Oligocene boundary could be bound up with the establishment of the layer of cold, deep water in the oceans known as the psychrosphere. This in turn relates to the establishment of the Circum-Antarctic current as first Australia and then South America broke away from that continent, and to the formation of south polar sea ice (Hallam, 1981c).

Discussion

I should now like to broaden the scope of this essay by discussing the relative importance of stochastic and deterministic, as well as biotic and physical, factors as promoters of evolution.

In recent years there has been an increasing tendency in some circles to investigate the evolutionary record of fossils in terms of general rules and processes without regard to specific causes operating on specific taxa. Thus Van Valen (1973) applied the survivorship curve technique of population biologists to the study of extinction rates for numerous fossil taxa, and claimed to demonstrate a general approximation to linearity in his curves, which are cumulative frequency distributions of taxonomic durations with logarithmic ordinates. This led him to propose a new evolutionary 'law', which, in brief, states that within a relatively homogeneous higher taxon, subtaxa tend to become extinct at a stochastically constant rate.

In explanation Van Valen put forward what he termed the Red Queen's hypothesis, named after the Lewis Carroll character who found it took all the running one can do to keep in the same place. It is thoroughly Darwinian in its stress on the paramount importance of biotic interactions. All species within a given adaptive zone compete intensively. A successful adaptive response by one species is assumed to occur at the expense of other species, which must either adapt by themselves speciating or become extinct, as the 'quality' of their environment is reduced. This phenomenon leads to an endless chain of adaptive responses and in the long run means that fitness and rate of extinction remain constant.

The high rate of diversification and evolutionary turnover in mammals is thought likely to be the result of a variety of factors, such as strong competitive interactions leading to specialisation in feeding methods, limitations on food supply, high mobility and energy use, interspecific aggression and territoriality. Such factors will conspire to lower the 'resource threshold' needed to prevent extinction, compared with other animals. Epistandard rates of evolution are required to make up the loss through extinction.

Though the Red Queen model could conceivably apply to mammals, there are doubts about its more general validity. Thus Stanley (1973) analyses the effect of competition on evolutionary rates by comparing mammals with bivalve molluscs. In sharp contrast to mammals, bivalves are nearly all benthic suspension feeders which appear to mind their own business, being characterized by weak interactions with other species, primitive inflexible behaviour, uncrowded, largely sedentary mode of life and generalised feeding habits. Limits on bivalve populations are imposed more by predation and fluctuations in the physical environment than by food resources, and biotic competition is minimal. As Stanley remarks laconically, 'Interspecific aggression is not characteristic of bivalve behaviour'. What is true of bivalves is without much doubt true of the majority of benthic invertebrates.

More general criticisms of the Red Queen hypothesis have been voiced by Foin, Valentine & Ayala (1975), Raup (1975), Salthe (1975) and Sepkoski (1975). In a nutshell, it is argued that either Van Valen's results show linearity with time, which is held to be biologically without significance, or most do not, in which case Van Valen's 'law' breaks down.

Other stochastic models have been explored by the computer generation of phylogenetic diagrams (cladograms), with termination and branching events being controlled using random numbers (Raup *et al.*, 1973; Gould *et al.*, 1977). Part of the input dealt with the required establishment of an equilibrium diversity. The cladograms so produced displayed a variety of patterns, many of which appear to simulate temporal diversity changes in well-studied fossil groups.

It does not, of course, follow that the radiation and extinction of the monophyletic units known as clades do not have deterministic explanations. Stanley, Signor, Lidgard & Karr (1981) present some cogent criticism of the work of Raup and his associates and establish that chance factors have not played a dominant role in producing dramatic changes in diversity. Hoffman (1981) gives a sophisticated critique both of stochastic modelling and the application of equilibrium theory in palaeontology.

The most powerful case against randomness is that afforded by mass extinction and radiation events, whereby a wide variety of taxonomic groups with different modes of life, and effective biological independence, have experienced synchronised diversity reduction or increase. This strongly implies the operation of some form of control by the physical environment, and some likely examples of an ultimate link with plate tectonics have been presented above.

The interesting question arises: were the several widely acknowledged major episodes of mass extinction in the Phanerozoic caused by exceptional events different in kind from what occurred in the much lengthier intervening periods, or are they merely the spectacular end-members of a whole series? Those who have recourse to the *deus ex machina* of lethal rays or thunderbolts from outer space may incline to the former view, but only at the expense of disregarding the abundant evidence in the stratigraphic record of a correlation between mass extinctions and physical events on earth (Hallam, 1981c).

Even in the one case where independent evidence is claimed in the form of an abnormal enrichment in thin sedimentary layers of iridium and other platinum group metals, namely the much-publicized asteroid impact hypothesis for the end-Cretaceous extinctions (Alvarez, Alvarez, Asaro & Michel, 1980) serious doubts have been raised by a careful analysis of some key palaeontological evidence (Clemens, Archibald & Hickey, 1981).

There is in fact evidence to indicate that regressions at intermediate and small scales correlate with coordinated extinction events in particular fossil groups (e.g. Hallam, 1978; Williamson, 1981) but much more work is required to establish how generally such a relationship holds.

If it does turn out to be of general significance at a variety of scales it will imply that the increased environmental stress associated in some way with regressions promotes an increase in extinction rate and corresponding vacation of ecological niches, with a consequent opportunity for new species to establish themselves. Alternatively, or in association, it could be that the pronounced environmental changes associated with the regression events act to destabilize intraspecific selection pressures and hence promote speciation. The comparatively long time intervals between such environmental vicissitudes may be characterized by stasis in ecosystems as well as the component species.

A brief consideration of the well-documented Pleistocene sea-level oscillations caused by glaciation and deglaciation suggests an apparent problem, because these events do not, by and large, correlate with episodes of pronounced extinction or speciation. Indeed, the characteristic response of both terrestrial and marine organisms to the pronounced climatic changes of the Pleistocene has been to migrate to ecological refuges, in effect to track their environment. Without such a phenomenon, in fact, stratigraphic correlation would operate under a crippling handicap. Why therefore did not organisms respond in a similar way to the much slower changes of sea level in the lengthy periods of climatic equability?

At least two possibilities readily suggest themselves. Perhaps the Pleistocene sea-level falls, though dramatically rapid in geological terms, were too short-lived to have the kind of environmental impact required to cause extinction. Or perhaps the increasingly unstable environments of the late Cainozoic associated with climatic deterioration caused a selection for eurytopic organisms well adapted to withstand environmental instability. In contrast, the comparative stability of, for instance, Mesozoic environments might have allowed the establishment of complex ecosystems characterized by comparatively stenotopic organisms, which would have been vulnerable to even modest environmental vicissitudes.

Such ideas need to be tested, and further studies made on the relative susceptibility of different fossil groups to extinction as a result of particular environmental events. Furthermore, we need to enquire further into the factors governing the response of organisms to given disturbances of their environment, whether it involves migration to refuges or extinction and speciation. There is clearly a rich field for future investigation into matters of such major importance.

Though I believe that Darwin laid too much stress on biotic interactions as a promoter of evolution and extinction, to the extent of substantially dismissing changes in the physical environment, there is no justification for going to the other extreme, as several examples will illustrate.

That the South American mammals might have gone extinct in the early Pleistocene as a consequence of competition from the North American invaders has already been noted, though we should take due account of the caveats of Marshall (1981).

With regard to marine benthic communities, the expansion of mobile, infaunal deposit- and suspension-feeding populations in the Mesozoic correlates with a decline of immobile suspension feeders on soft substrates (Thayer, 1979). Some of the groups in decline, such as stalked crinoids, cidaroid echinoids (Kier, 1974) and brachiopods, apparently sought refuge in the deep sea. The late Mesozoic radiation of predatory crabs, neogastropods and teleosts correlates with an increase in resistance to destruction of the shells of their molluscan prey (Vermeij, 1977; Ward, 1981).

Further likely examples of coevolution concern the rise of the angiosperms. From the late Cretaceous onwards the new sea-grass communities transformed the shallow neritic environment and seem to have promoted the contemporary radiation of deposit-feeding and epiphytic gastropods and miliolid foraminifera (Brasier, 1975). Although the fossil record is much poorer, coevolution with the terrestrial angiosperms probably accounts for a major component of the radiation of insects and birds. In all these cases, however, the initial trigger to change might have been physical rather than biotic.

A final intriguing example involves a subtle interplay between biotic and physical factors, and concerns the common phyletic trend towards size increase known as Cope's Rule, which I suspect may be the only important gradualistic exception to punctuated equilibria (Hallam, 1978). There is no shortage of adaptive explanations for size increase, but Darwin is one of the few to have pointed out that large organisms such as mastodonts and dinosaurs would have been more vulnerable to extinction because of the limitations of food resources. I have proposed that, on the reasonable assumption that resources remained more or less constant for the time in question, the price exacted for (phyletically) growing larger was to become rarer, thereby increasing the probability of extinction (Hallam, 1975). Whether one applies the older notion of a trend from less to more specialized or the newer concept of a trend from an r- to a K-selected adaptive strategy (ecologists seem to have a love–hate relationship towards r and K selection (Dawkins, 1981)), the phyletically younger organisms would have become progressively more vulnerable to environmental disturbance.

The stratigraphic record has been compared to the traditional life of a soldier – long periods of boredom interrupted by moments of terror (Ager, 1973). The fossil record suggests that the largest members of a phyletic series usually had the most reason to be apprehensive of the future.

Further reading

Hallam, A. (1973). *A revolution in the earth sciences: from continental drift to plate tectonics*. Oxford: Oxford University Press.

Hallam, A. (1981). *Facies interpretation and the stratigraphic record*. San Francisco: Freeman.

Oxburgh, E. R. (1974). The plain man's guide to plate tectonics. *Proceedings of the Geological Association, London*, **85**, 299–357.

References

Ager, D. V. (1973). *The Nature of the Stratigraphic Record*. London: Macmillan.

Alvarez, L. W., Alvarez, W., Asaro, F. & Michel, H. V. (1980). Extraterrestrial cause for the Cretaceous-Tertiary extinction: experiment and theory. *Science*, **208**, 1095–108.

Anderton, R. (1980). Did Iapetus start to open during the Cambrian? *Nature*, **286**, 706–8.

Anderton, R. (1982). Dalradian deposition and the late Precambrian-Cambrian history of the N. Atlantic region: a review of the early history of the Iapetus Ocean. *Journal of the Geological Society, London*, **139**, 423–34.

Bakker, R. T. (1977). Tetrapod mass extinctions – a model of the regulation of speciation rates and immigration by cycles of topographic diversity. In *Patterns of Evolution as illustrated by the Fossil Record*, ed. A. Hallam, pp. 439–68. Amsterdam: Elsevier.

Barron, E. J., Sloan, J. L. & Harrison, C. G. A. (1980). Potential significance of land-sea distribution and surface albedo variations as a climatic factor: 180 m.y. to the present. *Palaeogeography, Palaeoclimatology, Palaeoecology*, **30**, 17–40.

Brasier, M. D. (1975). An outline history of seagrass communities. *Palaeontology*, **18**, 681–702.

Brasier, M. D. (1979). The Cambrian radiation event. In *The Origin of Major Invertebrate Groups*, ed. M. R. House, pp. 103–59. London and New York: Academic Press.

Clemens, W. A., Archibald, J. D. & Hickey, L. J. (1981). Out with a whimper not a bang. *Paleobiology*, **7**, 293–8.

Dawkins, R. (1981). *The Extended Phenotype*. Oxford and San Francisco: Freeman.

Doyle, J. A. (1977). Patterns of evolution in early angiosperms. In *Patterns of Evolution as illustrated by the Fossil Record*, ed. A. Hallam, pp. 501–46. Amsterdam: Elsevier.

Flessa, K. W. & Imbrie, J. (1973). Evolutionary pulsations: evidence from Phanerozoic diversity patterns. In *Implications of Continental Drift to the Earth Sciences*, vol. 1, ed. D. H. Tarling & S. K. Runcorn, pp. 245–84. London and New York: Academic Press.

Foin, T. C., Valentine, J. W. & Ayala, F. J. (1975). Extinction of taxa and Van Valen's law. *Nature*, **257**, 514–5.

Frakes, L. A. (1979). *Climates through geologic time*. Amsterdam: Elsevier.

Frey, R. W. & Seilacher, A. (1980). Uniformity in marine invertebrate ichnology. *Lethaia*, **13**, 183–207.

Gould, S. J. & Calloway, C. B. (1980). Clams and brachiopods – ships that pass in the night. *Paleobiology*, **6**, 383–6.

Gould, S. J., Raup, D. M., Sepkoski, J. J., Schopf, T. J. M. & Simberloff, D. S. (1977). The shape of evolution: a comparison of real and random clades. *Paleobiology*, **3**, 23–40.

Gray, J. & Boucot, A. J. (eds) (1979). *Historical biogeography, plate tectonics and the changing environment*. Corvallis: Oregon State University Press.

Hallam, A. (1973). *A revolution in the earth sciences: from continental drift to plate tectonics*. Oxford: University Press.

Hallam, A. (1974). Changing patterns of provinciality and diversity of fossil animals in relation to plate tectonics. *Journal of Biogeography*, **1**, 213–25.

Hallam, A. (1975). Evolutionary size increase and longevity in Jurassic bivalves and ammonites. *Nature*, **258**, 193–6.

Hallam, A. (1977). Secular changes in marine inundation of USSR and North America through the Phanerozoic. *Nature*, **269**, 769–72.

Hallam, A. (1978). How rare is phyletic gradualism? Evidence from Jurassic bivalves. *Paleobiology*, **4**, 16–25.

Hallam, A. (1980). A reassessment of the fit of Pangaea components and the time of their initial breakup. In *The Continental Crust and its Mineral Deposits*, ed. D. W. Strangway. *Geological Association of Canada Special Paper 20*, pp. 375–87.

Hallam, A. (1981a). Relative importance of plate movements, eustasy and climate in controlling major biogeographic changes since the early Mesozoic. In *Vicariance Biogeography: a Critique*, ed. G. Nelson & D. E. Rosen, pp. 303–40. New York: Columbia University Press.

Hallam, A. (1981b). Biogeographic relations between the northern and southern continents during the Mesozoic and Cenozoic. *Geologische Rundschau*, **70**, 583–95.

Hallam, A. (1981c). *Facies Interpretation and the Stratigraphic Record*. Oxford and San Francisco: Freeman.

Hart, M. B. (1980). A water depth model for the evolution of the planktonic Foraminiferida. *Nature*, **286**, 252–4.

Hays, J. D. & Pitman, W. C. (1973). Lithospheric plate motion, sea level changes and climatic and ecological consequences. *Nature*, **246**, 16–22.

Hoffman, A. (1981). Stochastic versus deterministic approach to paleontology: the question of scaling or metaphysics? *Neues Jahrbuch für Geologie und Paläontologie*, **162**, 80–96.

Imbrie, J. & Imbrie, K. P. (1979). *Ice Ages: Solving the Mystery*. London: Macmillan.

Kennedy, W. J. (1977). Ammonite evolution. In *Patterns of Evolution as illustrated by the Fossil Record*, ed. A. Hallam, pp. 251–304. Amsterdam: Elsevier.

Kier, P. M. (1974). Evolutionary trends and their functional significance in the post-Paleozoic echinoids. *Palaentological Society Memoirs*, **5**.

Larsen, R. L. & Pitman, W. C. (1972). World-wide correlation of Mesozoic magnetic anomalies and its implications. *Bulletin of the Geological Society of America*, **83**, 3645–62.

Lipps, J. H. (1970). Plankton evolution. *Evolution*, **24**, 1–22.

Marshall, L. G. (1981). The great American interchange – an invasion-induced crisis for South American mammals. In *Biotic Crises in ecological and evolutionary time*, ed. M. H. Nitecki, pp. 133–229. New York: Academic Press.

Mayr, E. (1963). *Animal Species and Evolution*. Cambridge, Mass.: Harvard University Press.

Newell, N. J. (1967). Revolutions in the history of life. *Special Papers of the Geological Society of America* no. 89, 63–91.

Pitman, W. C. (1978). Relationship between eustasy and stratigraphic sequences of passive margins. *Bulletin of the Geological Society of America*, **89**, 1389–403.

Raup, D. M. (1975). Taxonomic survivorship curves and Van Valen's Law. *Paleobiology*, **1**, 82–96.

Raup, D. M. (1979). Size of the Permo-Triassic bottleneck and its evolutionary implications. *Science*, **206**, 217–8.

Raup, D. M. (1981). Extinction: bad genes or bad luck? *Acta Geologica Hispanica*, **16**, 25–33.

Raup, D. M., Gould, S. J., Schopf, T. J. M. & Simberloff, D. S. (1973). Stochastic models of phylogeny and the evolution of diversity. *Journal of Geology*, **81**, 525–42.

Raup, D. M. & Sepkoski, J. J. (1982). Mass extinctions in the marine fossil record. *Science*, **215**, 1501–3.

Salthe, S. N. (1975). Some comments on Van Valen's law of extinction. *Paleobiology*, **1**, 356–8.

Schopf, T. J. M. (1974). Permo-Triassic extinctions: relation to sea floor spreading. *Journal of Geology*, **82**, 129–43.

Seilacher, A. (1977). Evolution of trace fossil communities. In *Patterns of Evolution as illustrated by the Fossil Record*, ed. A. Hallam, pp. 359–376. Amsterdam: Elsevier.

Sepkoski, J. J. (1975). Stratigraphic biases in the analysis of taxonomic survivorship. *Paleobiology*, **1**, 343–55.

Sepkoski, J. J. (1978). A kinetic model of Phanerozoic taxonomic diversity. 1. Analysis of marine orders. *Paleobiology*, **4**, 223–51.

Sepkoski, J. J. (1981). A factor analytic description of the Phanerozoic marine fossil record. *Paleobiology*, **7**, 36–53.

Sepkoski, J. J., Bambach, R. K., Raup, D. M. & Valentine, J. W. (1981). Phanerozoic marine diversity and the fossil record. *Nature*, **293**, 435–7.

Simberloff, D. S. (1974). Permo-Triassic extinctions: effects of an area on biotic equilibrium. *Journal of Geology*, **82**, 267–74.

Stanley, S. M. (1973). Effects of competition on rates of evolution, with special reference to bivalve molluscs and mammals. *Systematic Zoology*, **22**, 486–506.

Stanley, S. M. (1976). Ideas on the timing of metazoan diversification. *Paleobiology*, **2**, 209–19.

Stanley, S. M. (1977). Trends, rates and patterns of evolution in the Bivalvia. In *Patterns of Evolution as illustrated by the Fossil Record*, ed. A. Hallam, pp. 209–250.

Stanley, S. M., Signor, P. W., Lidgard, S. & Karr, A. F. (1981). Natural clades differ from 'random' clades: simulations and analyses. *Paleobiology*, **7**, 115–27.

Thayer, C. W. (1979). Biological bulldozers and the evolution of marine benthic communities. *Science* **203**, 458–61.

Valentine, J. W. (1973). *Evolutionary Paleoecology of the Marine Biosphere*. New Jersey: Prentice-Hall.

Valentine, J. W. & Moores, E. M. (1972). Global tectonics and the fossil record. *Journal of Geology*, **80**, 167–84.

Van Valen, L. (1973). A new evolutionary law. *Evolutionary Theory*, **1**, 1–30.

Vermeij, G. J. (1977). The Mesozoic marine revolution: evidence from snails, predators and grazers. *Paleobiology*, **3**, 245–58.

Vermeij, G. J. (1978). *Biogeography and Adaptation: Patterns of Marine Life*. Cambridge: Harvard University Press.

Ward, P. (1981). Shell structure as a defensive adaptation to ammonoids. *Paleobiology*, **7**, 96–100.

Whitmore, T. C. (ed.) (1981). *Wallace's Line and Plate Tectonics*. Oxford: Oxford University Press.

Williamson, P. G. (1981). Palaeontological documentation of speciation in Cenozoic molluscs from Turkana Basin. *Nature*, **293**, 437–43.

19

Microevolution and macroevolution

FRANCISCO J. AYALA

In his widely known book, *Chance and Necessity*, Jacques Monod has written that 'the elementary mechanisms of evolution have been not only understood in principle but identified with precision...the problem has been resolved and evolution now lies well to this side of the frontier of knowledge' (Monod, 1972, p. 139).

Monod's unbridled optimism is unwarranted. The causes of evolution and the patterning of the processes that bring it about are far from completely understood (Dobzhansky, Ayala, Stebbins & Valentine, 1977). But there is no justification either for the condemnation of the modern theory of evolution voiced by a few palaeontologists. S. J. Gould, for example, has written that 'The modern synthesis, as an exclusive proposition, has broken down on both of its fundamental claims: extrapolationism (gradual allelic substitution as a mode for all evolutionary change) and nearly exclusive reliance on selection leading to adaptation' (Gould, 1980, p. 119); and, further: 'the synthetic theory...is effectively dead, despite its persistence as textbook orthodoxy' (p. 120).

Gould's critique of the modern theory of evolution is grounded on a distorted version of the modern synthesis and it has been adequately refuted (see, e.g., Levinton & Simon, 1980; Stebbins & Ayala, 1981; Charlesworth, Lande & Slatkin, 1982). After the publication of these rebuttals, and possibly because of them, Gould has had second thoughts. He now explains that 'Nothing about microevolutionary population genetics, or any other aspect of microevolutionary theory, is wrong or inadequate at its level...But it is not everything' (Gould, 1982a, p. 104). The criticisms, he has now qualified, propose 'much less than a revolution...The modern synthesis is incomplete, not incorrect' (Gould, 1982b, p. 382). That microevolutionary theory 'is not everything' and that 'the modern synthesis is incomplete' are tame propositions with which one can only agree.

Punctualism versus gradualism

Gould, S. M. Stanley (1979, 1982), E. S. Vrba (1980) and a few other palaeontologists have criticized neo-Darwinism in order to set the stage for a positive proposition; namely, that macroevolution – the evolution of species, genera, and higher taxa – is an autonomous field of study, independent of microevolutionary theory (and the intellectual turf of palaeontologists). This claim for autonomy has been expressed as a 'decoupling' of macroevolution from microevolution (e.g., Stanley, 1979, pp. x, 187, 193) or as a rejection of the notion that microevolutionary mechanisms can be extrapolated to explain macroevolutionary processes (e.g. Gould, 1980, p. 383).

It is precisely this question 'whether the mechanisms underlying microevolution can be extrapolated to explain macroevolution' (Lewin, 1980, p. 883) that I propose here to examine. The palaeontologists who argue for the autonomy of macroevolution base their claim on the notion that large-scale evolution is 'punctuated', rather than 'gradual'. The model of punctuated equilibrium proposes that morphological evolution happens in bursts, with most phenotypic change occurring during speciation events, so that new species are morphologically quite distinct from their ancestors, but do not thereafter change substantially in phenotype over a lifetime that may encompass many millions of years. The punctuational model is contrasted with the gradualistic model, which sees morphological change as a more or less gradual process, not strongly associated with speciation events (Fig. 19.1).

Lest anybody doubt that punctualists predicate the autonomy of macroevolution on the alleged punctuational nature of large-scale evolution, I shall offer two quotations. 'If rapidly divergent speciation interposes discontinuities between rather stable entities (lineages), and if there is a strong random element in the origin of these discontinuities (in speciation), then phyletic trends are essentially *decoupled* from phyletic trends within lineages. Macroevolution is decoupled from microevolution' (Stanley, 1979, p. 187, italics in the original). 'Punctuated equilibrium is crucial to the independence of macroevolution – for it embodies the claim that species are legitimate individuals, and therefore capable of displaying irreducible properties' (Gould, 1982a, p. 94).

Whether phenotypic change in macroevolution occurs in bursts or is more or less gradual, is a question to be decided empirically. Examples of rapid phenotypic evolution followed by long periods of morphological stasis are known in the fossil record. But there are instances as well in which phenotypic evolution appears to occur gradually within a lineage. The question is the relative frequency of one or the other mode; and palaeontologists disagree in their interpretation of the fossil record (Eldredge (1971), Eldredge & Gould (1972), Hallam (1978), Raup (1978), Stanley (1979), Gould (1980) and Vrba (1980) are among those who favour punctualism; whereas Kellogg (1975), Gingerich (1976), Levinton & Simon (1980), Schopf (1979, 1981), Cronin, Boaz, Stringer & Rak (1981) and Douglas & Avise (1982) favour phyletic gradualism). Whatever the palaeontological record may show about the frequency of smooth, relative to jerky, evolutionary patterns, there is one fundamental reservation that must be raised against the theory of punctuated equilibrium. This evolutionary model argues not only that most morphological change

Fig. 19.1. Simplified representation of two models of phenotypic evolution: punctuated equilibrium (left) and phyletic gradualism (right). According to the punctuated model, most morphological evolution in the history of life is associated with speciation events, which are geologically instantaneous. After their origin, established species generally do not change substantially in phenotype over a lifetime that may encompass many million years. According to the gradualist model, morphological evolution occurs during the lifetime of a species, with rapidly divergent speciation playing a lesser role. The figures are extreme versions of the models. Punctualism does not imply that phenotypic change never occurs between speciation events. Gradualism does not imply that phenotypic change is occurring continuously at a more or less constant rate throughout the life of a lineage, or that some acceleration does not take place during speciation, but rather that phenotypic change may occur at any time throughout the lifetime of a species.

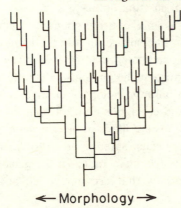

←Morphology→   ←Morphology→

occurs in rapid bursts followed by long periods of phenotypic stability, but also that the bursts of change occur during the origin of new species. Stanley (1979, 1982), Gould (1982a, b), and other punctualists have made it clear that what is distinctive of the theory of punctuated equilibrium is this association between phenotypic change and speciation. One quotation should suffice: 'Punctuated equilibrium is a specific claim about speciation and its deployment in geological time; *it should not be used as a synonym for any theory of rapid evolutionary change at any scale*...Punctuated equilibrium holds that accumulated speciation is the root of most major evolutionary change, and that what we have called anagenesis is usually no more than repeated cladogenesis (branching) filtered through the net of differential success at the species level' (Gould, 1982a, pp. 84–5; italics added).

Species are groups of interbreeding natural populations that are reproductively isolated from any other such groups (Mayr, 1963; Dobzhansky *et al.*, 1977). Speciation involves, by definition, the development of reproductive isolation between populations previously sharing in a common gene pool. But it is no way apparent how the fossil record could provide evidence of the development of reproductive isolation. Palaeontologists recognize species by their different morphologies as preserved in the fossil record. New species that are morphologically indistinguishable from their ancestors (or from contemporary species) go totally unrecognized. Sibling species are common in many groups of insects, in rodents, and in other well studied organisms (Mayr, 1963; Dobzhansky, 1970; Nevo & Shaw, 1972; Dobzhansky *et al.*, 1977; White, 1978; Benado, Aguilera, Reig & Ayala, 1979). Moreover, morphological discontinuities in a time series of fossils are usually interpreted by palaeontologists as speciation events, even though they may represent phyletic evolution within an established lineage, without splitting of lineages.

Thus, when palaeontologists use evidence of rapid phenotypic change in favour of the punctuational model, they are committing a definitional fallacy. Speciation as seen by the palaeontologist always involves substantial morphological change because palaeontologists identify new species by the eventuation of substantial morphological change. Stanley (1979, p. 144) has argued that 'rapid change is concentrated in small populations and...that such populations are likely to be associated with speciation and unlikely to be formed by constriction of an entire lineage'. But the two points he makes are arguable. First, rapid (in the geological scale) change may occur in populations that are not small. Second, bottle-necks in population size are not necessarily rare (again, in the geological scale) within a given lineage.

Punctualists speak of evolutionary change 'concentrated in geologically instantaneous events of branching speciation' (Gould, 1982*b*, p. 383). But events that appear instantaneous in the geological time scale may involve thousands, even millions of generations. Gould (1982*a*, p. 84), for example, has made operational the fuzzy expression 'geologically instantaneous' by suggesting that 'it be defined as 1 percent or less of later existence in stasis. This permits up to 100 000 years for the origin of a species with a subsequent life span of 10 million years'. But 100 000 years encompasses one million generations of an insect such as *Drosophila*, and tens or hundreds of thousands of generations of fish, birds, or mammals. Speciation events or morphological changes deployed during thousands of generations may occur by the slow processes of allelic substitution that are familiar to the population biologist. Hence, the problem faced by microevolutionary theory is not how to account for rapid palaeontological change, because there is ample time for it, but why lineages persist for millions of years without apparent morphological change. Although other explanations have been proposed, it seems that stabilizing selection may be the process most often responsible for the morphological stasis of lineages (Stebbins & Ayala, 1981; Charlesworth *et al.*, 1982). Whether microevolutionary theory is sufficient to explain punctuated as well as gradual evolution is, however, a different question, to which I now turn.

The issue of reductionism

According to the proponents of punctuated equilibrium, phyletic evolution proceeds at two levels. First, there is change within a population that is continuous through time. This consists largely of allelic substitutions prompted by natural selection, mutation, genetic drift, and the other processes familiar to the population geneticist, operating at the level of the individual organism. Thus, most of evolution within established lineages rarely, if ever, yields any substantial morphological change. Second, there is the process of origination and extinction of species. Most morphological change is associated with the origin of new species. Evolutionary trends result from the patterns of origination and extinction of species, rather than from evolution within established lineages. Hence, the relevant unit of macroevolutionary study is the species rather than the individual organism. It follows from this argument that the study of microevolutionary processes provides little, if any, information about macroevolutionary patterns, the tempo and mode of large-scale evolution. Thus, macroevolution is autonomous relative to microevolution, much in the same way as biology is autonomous relative to physics. Gould (1982*b*, p. 384) has summarized the argument: 'Individuation of higher-level units is enough to invalidate the reductionism of traditional Darwinism – for pattern and style of evolution depend critically on the disposition of higher-level individuals [i.e., species]'.

The question raised is the general issue of reduction of one branch of science to another. But as so often happens with questions of reductionism, the issue of whether microevolutionary mechanisms can account for macroevolutionary processes is muddled by confusion of separate issues. Identification of the issues involved is necessary in order to resolve them and to avoid misunderstanding, exaggerated claims, or unwarranted fears.

The issue 'whether the mechanisms underlying microevolution can be extrapolated' to macroevolution involves, at least, three separate questions. (1) Whether microevolutionary processes *operate* (and have operated in the past) throughout the organisms which make up the taxa in which macroevolutionary phenomena are observed. (2) Whether the microevolutionary processes identified by population geneticists (mutation, random drift, natural selection) are sufficient to *account for* the morphological changes and other macroevolutionary phenomena observed in higher taxa, or whether additional microevolutionary processes need to be postulated. (3) Whether theories concerning evolutionary trends and other macroevolutionary patterns can be *derived* from knowledge of microevolutionary processes.

The distinctions that I have made may perhaps become clearer if I state them as they might be formulated by a biologist concerned with the question whether the laws of physics and chemistry can be extrapolated to biology. The first question would be whether the laws of physics and chemistry apply to the atoms and molecules present in living organisms. The second question would be whether interactions between atoms and molecules according to the laws known to physics and chemistry are sufficient to account for biological phenomena, or whether the workings of organisms require additional kinds of interactions between atoms and molecules. The third question would be whether biological theories can be derived from the laws and theories of physics and chemistry.

The first issue raised can easily be resolved. It is unlikely that any biologist would seriously argue that the laws of physics and chemistry do not apply to the atoms and molecules that make up living things. Similarly, it seems unlikely that any palaeontologist or macroevolutionist would claim that mutation, drift, natural selection, and other microevolutionary processes do not apply to the organisms and populations which make up the higher taxa studied in macroevolution. There is, of course, an added snarl – macroevolution is largely concerned with phenomena of the past. Direct observation of microevolutionary processes in populations of long-extinct organisms is not possible. But there is no reason to doubt that the genetic structures of populations living in the past were in any fundamental way different from the genetic structures of living populations. Nor is there any reason to believe that the processes of mutation, random drift and natural selection, or the nature of the interactions between organisms and the environment would have been different in nature for, say, Palaeozoic trilobites or Mesozoic ammonites than for modern molluscs or fishes. Extinct and living populations – like different living populations – may have experienced quantitative differences in the relative importance of one or another process, but the processes could have hardly been different in kind. Not only are there reasons to the contrary lacking, but the study of biochemical evolution reveals a remarkable continuity and gradual change of informational macromolecules (DNA and proteins) over the most diverse organisms, which advocates that the current processes of population change have persisted over evolutionary history (Dobzhansky *et al.*, 1977).

Microevolutionary processes and the tempo of evolution

The second question raised above is considerably more interesting than the first: Can the microevolutionary processes studied by population geneticists account for macroevolutionary phenomena or do we need to postulate new kinds of genetic processes? The large morphological (phenotypic) changes observed in evolutionary history, and the rapidity with which they appear in the geological record, is one major matter of concern. Another issue is 'stasis', the apparent persistence of species, with little or no morphological change, for hundreds of thousands or millions of years. The dilemma is that microevolutionary processes apparently yield small but continuous changes, while macroevolution as seen by punctualists occurs by large and rapid bursts of change followed by long periods without change.

Goldschmidt (1940, p. 183) argued long ago that the incompatibility is real: 'The decisive step in evolution, the first step towards macroevolution, the step from one species to another, requires another evolutionary method than that of sheer accumulation of micromutations.' Goldschmidt's solution was to postulate 'systematic mutations', yielding 'hopeful monsters' that, on occasion, would find a new niche or way of life for which they would be eminently preadapted. The progressive understanding of the nature and organization of the genetic material acquired during the last forty years excludes the 'systemic mutations' postulated by Goldschmidt, which would involve transformations of the genome as a whole.

Single-gene or chromosome mutations are known that have large effects on the phenotype because they act early in the embryo and their effects become magnified through development. Examples of such 'macro-mutations' carefully analysed in *Drosophila* are 'bithorax' and the homeotic mutants that transform one body structure, e.g., antennae, into another, e.g., legs. Whether the kinds of morphological differences that characterize different taxa are due to such 'macromutations' or to the accumulation of several mutations with small effect, has been examined particularly in plants where fertile interspecific, and even intergeneric, hybrids can be obtained. The results do not support the hypothesis that the establishment of macromutations is necessary for divergence at the macroevolutionary level (Stebbins, 1950; Clausen, 1951; Grant, 1971; see Stebbins & Ayala, 1981). Moreover, Lande (1981 and references therein; see also Charles-worth *et al.*, 1982) has convincingly shown that major morphological changes, such as in the number of digits or limbs, can occur in a geologically rapid fashion through the accumulation of mutations each with a small effect. The analysis of progenies from crosses between races or species that differ greatly (by as much as 30 phenotypic standard deviations) in a quantitative trait indicates that these extreme differences can be caused by the cumulative effects of no more than 5 to 10 independently segregating genes.

The punctualists' claim that mutations with large phenotypic effects must have been largely responsible for macroevolutionary change is based on the rapidity with which morphological discontinuities appear in the fossil record (Stanley, 1979; Gould, 1980). But the alleged evidence does not necessarily support the claim. Microevolutionists and macroevolutionists use different time scales. As pointed out earlier, the 'geological instants' during which speciation and morphological shifts occur may involve expands of the order of 100000 years. There is little doubt that the gradual accumulation of small mutations may yield sizeable morphological changes during periods of that length.

Anderson's (1973) study of body size in *Drosophila pseudoobscura* provides an estimate of the rates of gradual morphological change produced by natural selection. Large populations, derived from a single set of parents, were set up at different temperatures and allowed to evolve on their own.

A gradual, genetically determined, change in body size ensued, with flies kept at lower temperature becoming, as expected, larger than those kept at higher temperatures. After 12 years, the mean size of the flies from the population kept at 16 °C had become, when tested under standard conditions, approximately 10% greater than the size of the flies kept at 27 °C; the change of mean value being greater than the standard deviation in size at the time when the tests were made. Assuming 10 generations per year, the populations diverged at a rate of 8×10^{-4} of the mean value per generation.

Palaeontologists have emphasized the 'extraordinary high *net* rate of evolution that is the hallmark of human phylogeny' (Stanley, 1979). Interpreted in terms of the punctualist hypothesis, human phylogeny would have occurred as a succession of jumps, or geologically instantaneous saltations, interspersed by long periods without morphological change. Could these bursts of phenotypic evolution be due to the gradual accumulation of small changes? Consider cranial capacity, the character undergoing the greatest relative amount of change. The fastest rate of net change occurred between 500 000 years ago, when our ancestors were represented by *Homo erectus* and 75 000 years ago, when Neanderthal man had acquired a cranial capacity similar to that of modern humans. In the intervening 425 000 years, cranial capacity evolved from about 900 cc in Peking man to about 1400 cc in Neanderthal people. Let us assume that the increase in brain size occurred in a single burst at the rate observed in *D. pseudoobscura* of 8×10^{-4} of the mean value per generation. The change from 900 cc to 1400 cc could have taken place in 540 generations or, assuming generously 25 years per generation, in 13 500 years. Thirteen thousand years are, of course, a geological instant. Yet, this evolutionary 'burst' could have taken place by gradual accumulation of mutations with small effects at rates compatible with those observed in microevolutionary studies.

The known processes of microevolution can, then, account for macroevolutionary change, even when this occurs according to the punctualist model – i.e., at fast rates concentrated on geologically brief time intervals. But what about the problem of stasis? The theory of punctuated equilibrium argues that after the initial burst of morphological change associated with their origin, 'species generally do not change substantially in phenotype over a lifetime that may encompass many million years' (Gould, 1982*b*, p. 383). Is it necessary to postulate new processes, yet unknown to population genetics, in order to account for the long persistence of lineages with apparent phenotypic change? The answer is 'no'.

The geological persistence of lineages without morphological change was already known to Darwin, who wrote in the last edition of *The Origin of Species* (1872, p. 375). 'Many species once formed never undergo any further change...; and the periods, during which species have undergone modification, though long as measured by years, have probably been short in comparison with the periods during which they retain the same form.' A successful morphology may remain unchanged for extremely long periods of time, even through successive speciation events – as manifested, e.g., by the existence of sibling species, which in many known instances have persisted for millions of years (Stebbins & Ayala, 1981).

Evolutionists have long been aware of the problem of palaeontological stasis and have explored a number of alternative hypotheses consistent with microevolutionary principles and sufficient to account for the phenomenon. Although the issue is far from definitely settled, the weight of the evidence favours stabilizing selection as the primary process responsible for morphological stasis of lineages through geological time (Stebbins & Ayala, 1981; Charlesworth *et al.*, 1982).

The autonomy of macroevolution

Macroevolution and microevolution are *not* decoupled in the two senses so far expounded: identity at the level of events and compatibility of theories. First, the populations in which macroevolutionary patterns are observed are the same populations that evolve at the microevolutionary level. Second, macroevolutionary phenomena can be accounted for as the result of known microevolutionary processes. That is, the theory of punctuated equilibrium (as well as the theory of phyletic gradualism) is consistent with the theory of population genetics. Indeed, any theory of macroevolution that is correct must be compatible with the theory of population genetics, to the extent that this is a well established theory.

Now, I pose the third question raised earlier: can macroevolutionary theory be derived from microevolutionary knowledge? The answer can only be 'no'. If macroevolutionary theory were deducible from micro-evolutionary principles, it would be possible to decide between competing macroevolutionary models simply by examining the logical implications of microevolutionary theory. But the theory of population genetics is compatible with both, punctualism and gradualism; and, hence, logically it entails neither. Whether the tempo and mode of evolution occur predominantly according to the model of punctuated equilibria or according to the model of phyletic gradualism is an issue to be decided by studying macroevolutionary patterns, not by inference from microevolutionary processes. In other words, macroevolutionary theories are not reducible (at least at the present state of knowledge) to microevolution. Hence, macroevolution and microevolution are decoupled in the sense (which is epistemologically most important) that macroevolution is an autonomous field of study that must develop and test its own theories.

Punctualists have claimed autonomy for macroevolution because species – the units studied in macroevolution – are higher in the hierarchy of organization of the living world than individual organisms. Species, they argue, have therefore 'emergent' properties, not exhibited by, nor predictable from, lower-level entities. In Gould's (1980, p. 121) words, the study of evolution embodies 'a concept of hierarchy – a world constructed not as a smooth and seamless continuum, permitting simple extrapolation from the lowest level to the highest, but as a series of ascending levels, each bound to the one below it in some ways and independent in others... "emergent" features not implicit in the operation of processes at lower levels, may control events at higher levels'. Although I agree with the thesis that macroevolutionary theories are not reducible to microevolutionary principles, I shall argue that it is a mistake to ground this autonomy on the hierarchical organization of life, or on purported emergent properties exhibited by higher-level units.

The world of life is hierarchically structured. There is a hierarchy of levels that go from atoms, through molecules, organelles, cells, tissues, organs, multicellular individuals and populations, to communities. Time adds another dimension of the evolutionary hierarchy, with the interesting consequence that transitions from one level to another occur: as time proceeds the descendants of a single species may include separate species, genera, families, etc. But hierarchical differentiation of subject matter is neither necessary nor sufficient for the autonomy of scientific disciplines. It is not necessary, because entities of a given hierarchical level can be the subject of diversified disciplines: cells are appropriate subject of study for cytology, genetics, immunology, and so on. Even a single *event* can be the subject matter of several disciplines. My writing of this paragraph can be studied by a physiologist interested in the workings of muscles and nerves, by a psychologist concerned with thought processes, by a philosopher interested in the epistemological question at issue, and so on. Nor is the hierarchical differentiation of subject matter a sufficient condition for the autonomy of scientific disciplines: relativity theory obtains all the way from subatomic particles to planetary motions and genetic laws apply to multicellular organisms as well as to cellular and even subcellular entities.

One alleged reason for the theoretical independence of levels within a hierarchy is the appearance of 'emergent' properties, which are 'not implicit in the operation of lesser levels, [but] may control events at higher levels'. The question of emergence is an old one, particularly in discussions on the reducibility of biology to the physical sciences. The issue is, for example, whether the functional properties of the kidney are simply the properties of the chemical constituents of that organ. In the context of macroevolution, the question is: do species exhibit properties different from those of the individual organisms of which they consist? I have argued elsewhere (Dobzhansky *et al.*, 1977, ch. 16), that questions about the emergence of properties are ill-formed, or at least unproductive, because they can only be solved by definition. The proper way of formulating questions about the relationship between complex systems and their component parts is by asking whether the properties of complex systems can be *inferred* from knowledge of the properties that their components have in isolation. The issue of emergence cannot be settled by discussion about the 'nature' of things or their properties, but it is resolvable by reference to our *knowledge* of those objects.

Consider the following question. Are the properties of common salt, sodium chloride, simply the properties of sodium and chlorine when they are associated according to the formula NaCl? If among the properties of sodium and chlorine I include their association into table salt and the properties of the latter, the answer is 'yes'; otherwise, the answer is 'no'. But the solution, then, is simply a matter of definition; and resolving the issue by a definitional manoeuvre contributes little to understanding the relationships between complex systems and their parts.

Is there a rule by which one could decide whether the properties of complex systems should be listed among the properties of their component parts? Assume that by studying the components in isolation we can infer the properties they will have when combined with other component parts in certain ways. In such a case, it would seem reasonable to include the 'emergent' properties of the whole among the properties of the component parts. (Notice that this solution to the problem implies that a feature that may seem emergent at a certain time, might not appear as emergent any longer at a more advanced state of knowledge.) Often, no matter how exhaustively an object is studied in isolation, there is no way to ascertain the properties it will have in association with other objects. We cannot infer the properties of ethyl alcohol, proteins, or human beings from the study of hydrogen, and thus it makes no good sense to list their properties among those of hydrogen. The important point, however, is that the issue of emergent properties is spurious and that it needs to be reformulated in terms of propositions expressing our knowledge. It is a legitimate question to ask whether the *statements* concerning the properties of organisms (but not the properties themselves) can be logically deduced from statements concerning the properties of their physical components.

The question of the autonomy of macroevolution, like other questions of reduction, can only be settled by empirical investigation of the logical consequences of propositions, and not by discussions about the 'nature' of things or their properties. What is at issue is not whether the living world is hierarchically organized – it is; or whether higher level entities have emergent properties – which is a spurious question. The issue is whether in a particular case, a set of *propositions* formulated in a defined field of knowledge (e.g., macroevolution) can be derived from another set of propositions (e.g., microevolutionary theory). Scientific theories consist, indeed, of propositions about the natural world. Only the investigation of the logical relations between propositions can establish whether or not one theory or branch of science is reducible to some other theory or branch of science. This implies that a discipline which is autonomous at a given stage of knowledge may become reducible to another discipline at a later time. The reduction of thermodynamics to statistical mechanics became possible only after it was discovered that the temperature of a gas bears a simple relationship to the mean kinetic energy of its molecules. The reduction of genetics to chemistry could not take place before the discovery of the chemical nature of the hereditary material (I am not, of course, intimating that genetics can now be fully reduced to chemistry, but only that a partial reduction may be possible now, whereas it was not before the discovery of the structure and mode of replication of DNA).

Nagel (1961; see also Ayala, 1968) has formulated the two conditions that are necessary and jointly sufficient to effect the reduction of one theory or branch of science to another. These are the condition of derivability and the condition of connectability.

The *condition of derivability* requires that the laws and theories of the branch of science to be reduced be derived as logical consequences from the laws and theories of some other branch of science. The *condition of connectability* requires that the distinctive terms of the branch of science to be reduced be redefined in the language of the branch of science to which it is reduced – this redefinition of terms is, of course, necessary in order to analyse the logical connections between the theories of the two branches of science.

Microevolutionary processes, as presently known, are compatible with the two models of macroevolution – punctualism and gradualism. From microevolutionary knowledge, we cannot infer which one of those two macroevolutionary patterns prevails, nor can we deduce answers for many other distinctive macroevolutionary issues, such as rates of morphological evolution, patterns of species extinctions, and historical factors regulating taxonomic diversity. The condition of derivability is not satisfied: the theories, models, and laws of macroevolution cannot be logically derived, at least at the present state of knowledge, from the theories and laws of population biology.

In conclusion, then, macroevolutionary processes are underlain by microevolutionary phenomena and are compatible with microevolutionary theories, but macroevolutionary studies require the formulation of autonomous hypotheses and models (which must be tested using macroevolutionary evidence). In this (epistemologically) very important sense, macroevolution is decoupled from microevolution: macroevolution is an autonomous field of evolutionary study.

Further reading

Charlesworth, B., Lande, R. & Slatkin, M. (1982). A neo-Darwinian commentary on macroevolution. *Evolution*, **36**, 474–98.
Gould, S. J. (1982). Darwinism and the expansion of evolutionary theory. *Science*, **216**, 380–7.
Stanley, S. M. (1979). *Macroevolution: Pattern and Process*. San Francisco: W. H. Freeman.
Stebbins, G. L. & Ayala, F. J. (1981). Is a new evolutionary synthesis necessary? *Science*, **213**, 967–71.

References

Anderson, W. W. (1973). Genetic divergence in body size among experimental populations of *Drosophila pseudoobscura* kept at different temperatures. *Evolution*, **27**, 278–84.
Ayala, F. J. (1968). Biology as an autonomous science. *American Scientist*, **56**, 207–21.
Benado, M., Aguilera, M., Reig, D. A. & Ayala, F. J. (1979). Biochemical genetics of Venezuelan spiny rats of the *Proechimys guainae* and *Proechimys trinitatis* superspecies. *Genetica*, **50**, 89–97.
Charlesworth, B., Lande, R. & Slatkin, M. (1982). A neo-Darwinian commentary on macroevolution. *Evolution*, **36**, 474–98.
Clausen, J. (1951). *Stages in the Evolution of Plant Species*. Ithaca: Cornell University.
Cronin, J. E., Boaz, N. T., Stringer, C. B. & Rak, Y. (1981). Tempo and mode in hominid evolution. *Nature*, **292**, 113–22.
Darwin, C. R. (1872). *The Origin of Species*. 6th edn, London: John Murray. (Reprinted 1936. New York: Random House).

Dobzhansky, Th. (1970). *Genetics of the Evolutionary Process*. New York: Columbia University Press.

Dobzhansky, Th., Ayala, F. J., Stebbins, G. L. & Valentine, J. W. (1977). *Evolution*. San Francisco: W. H. Freeman & Co.

Douglas, M. E. & Avise, J. C. (1982). Speciation rates and morphological divergence in fishes: Tests of gradual versus rectangular modes of evolutionary change. *Evolution*, **36**, 224–32.

Eldredge, N. (1971). The allopatric model and phylogeny in Paleozoic invertebrates. *Evolution*, **25**, 156–67.

Eldredge, N. & Gould, S. J. (1972). Punctuated equilibria: an alternative to phyletic gradualism. In *Models in Paleobiology*, ed. T. J. M. Schopf, pp. 82–115. Freeman, Cooper Co.

Gingerich, P. D. (1976). Paleontology and phylogeny: patterns of evolution at the species level in early Tertiary mammals. *American Journal of Science*, **276**, 1–28.

Goldschmidt, R. B. (1940). *The Material Basis of Evolution*. New Haven, Conn.: Yale University Press.

Gould, S. J. (1980). Is a new general theory of evolution emerging? *Paleobiology*, **6**, 119–30.

Gould, S. J. (1982a). The meaning of punctuated equilibrium and its role in validating a hierarchical approach to macroevolution. In *Perspectives in Evolution*, ed. R. Milkman. Sunderland, Mass.: Sinauer.

Gould, S. J. (1982b). Darwinism and the expansion of evolutionary theory. *Science*, **216**, 380–7.

Grant, V. (1971). *Plant Speciation*. New York: Columbia University.

Hallam, A. (1978). How rare is phyletic gradualism and what is its evolutionary significance? Evidence from Jurassic bivalves. *Paleobiology*, **4**, 16–25.

Kellogg, D. E. (1975). The role of phyletic change in the evolution of *Pseudocubus vema* (Radiolaria). *Paleobiology*, **1**, 359–70.

Lande, R. (1981). The minimum number of genes contributing to quantitative variation between and within populations. *Genetics*, **99**, 541–53.

Levinton, J. S. & Simon, C. M. (1980). A critique of the punctuated equilibria model and implications for the detection of speciation in the fossil record. *Systematic Zoology*, **29**, 130–42.

Lewin, R. (1980). Evolution theory under fire. *Science*, **210**, 883–7.

Mayr, E. (1963). *Animal Species and Evolution*. Cambridge, Mass.: Harvard University Press.

Monod, J. (1972). *Chance and Necessity*. New York: Vintage Books.

Nagel, E. (1961). *The Structure of Science*. New York: Harcourt, Brace and World, Inc.

Nevo, E. & Shaw, C. R. (1972). Genetic variation in a subterranean mammal, *Spalax ehrenbergi*. *Biochemical Genetics*, **7**, 235–41.

Raup, D. M. (1978). Cohort analysis of generic survivorship. *Paleobiology*, **4**, 1–15.

Schopf, T. J. M. (1979). Evolving paleontological views on deterministic and stochastic approaches. *Paleobiology*, **5**, 337–52.

Schopf, T. J. M. (1981). Punctuated equilibrium and evolutionary stasis. *Paleobiology*, **7**, 156–66.

Stanley, S. M. (1979). *Macroevolution: Pattern and Process*. San Francisco: W. H. Freeman.

Stanley, S. M. (1982). Macroevolution and the fossil record. *Evolution*, **36**, 460–73.

Stebbins, G. L. (1950). *Variation and Evolution in Plants*. New York: Columbia University.

Stebbins, G. L. & Ayala, F. J. (1981). Is a new evolutionary synthesis necessary? *Science*, **213**, 967–71.

Vrba, E. S. (1980). Evolution, species, and fossils: how does life evolve? *South African Journal Science*, **76**, 61–84.

White, M. J. D. (1978). *Modes of Speciation*. San Francisco: W. H. Freeman.

20

Universal Darwinism

RICHARD DAWKINS

It is widely believed on statistical grounds that life has arisen many times all around the universe (Asimov, 1979; Billingham, 1981). However varied in detail alien forms of life may be, there will probably be certain principles that are fundamental to all life, everywhere. I suggest that prominent among these will be the principles of Darwinism. Darwin's theory of evolution by natural selection is more than a local theory to account for the existence and form of life on Earth. It is probably the only theory that *can* adequately account for the phenomena that we associate with life.

My concern is not with the details of other planets. I shall not speculate about alien biochemistries based on silicon chains, or alien neurophysiologies based on silicon chips. The universal perspective is my way of dramatizing the importance of Darwinism for our own biology here on Earth, and my examples will be mostly taken from Earthly biology. I do, however, also think that 'exobiologists' speculating about extraterrestrial life should make more use of evolutionary reasoning. Their writings have been rich in speculation about how extraterrestrial life might work, but poor in discussion about how it might *evolve*. This essay should, therefore, be seen firstly as an argument for the general importance of Darwin's theory of natural selection; secondly as a preliminary contribution to a new discipline of 'evolutionary exobiology'.

The 'growth of biological thought' (Mayr, 1982) is largely the story of Darwinism's triumph over alternative explanations of existence. The chief weapon of this triumph is usually portrayed as *evidence*. The thing that is said to be wrong with Lamarck's theory is that its assumptions are factually wrong. In Mayr's words: 'Accepting his premises, Lamarck's theory was as legitimate a theory of adaptation as that of Darwin. Unfortunately, these premises turned out to be invalid.' But I think we can say something stronger: *even accepting his premises*, Lamarck's theory is *not* as legitimate a theory of adaptation as that of Darwin because, unlike Darwin's, it is *in principle* incapable of doing the job we ask of it – explaining the evolution of organized, adaptive complexity. I believe this is so for all theories that have ever been suggested for the mechanism of evolution except Darwinian natural selection, in which case Darwinism rests on a securer pedestal than that provided by facts alone.

Now, I have made reference to theories of evolution 'doing the job we ask of them'. Everything turns on the question of what that job is. The answer may be different for different people. Some biologists, for instance, get excited about 'the species problem', while I have never mustered much enthusiasm for it as a 'mystery of mysteries'. For some, the main thing that any theory of evolution has to explain is the diversity of life – cladogenesis. Others may require of their theory an explanation of the observed changes in the molecular constitution of the genome. I would not presume to try to convert any of these people to my point of view. All I can do is to make my point of view clear, so that the rest of my argument is clear.

I agree with Maynard Smith (1969) that 'The main task of any theory of evolution is to explain adaptive complexity, i.e. to explain the same set of facts which Paley used as evidence of a Creator'. I suppose people like me might be labelled neo-Paleyists, or perhaps 'transformed Paleyists'. We concur with Paley that adaptive complexity demands a very special kind of explanation: either a Designer as Paley taught, or something such as natural selection that does the job of a designer. Indeed, adaptive complexity is probably the best diagnostic of the presence of life itself.

Adaptive complexity as a diagnostic character of life

If you find something, anywhere in the universe, whose structure is complex and gives the strong appearance of having been designed for a purpose, then that something either is alive, or was once alive, or is an artefact created by something alive. It is fair to include fossils and artefacts since their discovery on any planet would certainly be taken as evidence for life there.

Complexity is a statistical concept (Pringle, 1951). A complex thing is a statistically improbable thing, something with a very low *a priori* likelihood of coming into being. The number of possible ways of arranging the 10^{27} atoms of a human body is obviously inconceivably large. Of these possible ways, only very few would be recognized as a human body. But this is not, by itself, the point. Any existing configuration of atoms is, *a posteriori*, unique, as 'improbable', with hindsight, as any other. The point is that, of all possible ways of arranging those 10^{27} atoms, only a tiny minority would constitute anything remotely resembling a machine that worked to keep itself in being, and to reproduce its kind. Living things are not just statistically improbable in the trivial sense of hindsight: their statistical improbability is limited by the *a priori* constraints of design. They are *adaptively* complex.

The term 'adaptationist' has been coined as a pejorative name for one who assumes 'without further proof that all aspects of the morphology, physiology and behavior of organisms are adaptive optimal solutions to problems' (Lewontin, 1979, and this volume). I have responded to this elsewhere (Dawkins, 1982*a*, Chapter 3). Here, I shall be an adaptationist in the much weaker sense that I shall only be *concerned* with those aspects of the morphology, physiology and behaviour of organisms that are undisputedly adaptive solutions to problems. In the same way, a zoologist may specialize on vertebrates without denying the existence of inverte-brates. I shall be preoccupied with undisputed adaptations because I have defined them as my working diagnostic characteristic of all life, anywhere in the universe, in the same way as the vertebrate zoologist might be preoccupied with backbones because backbones are the diagnostic character of all vertebrates. From time to time I shall need an example of an undisputed adaptation, and the time-honoured eye will serve the purpose as well as ever (Paley, 1828; Darwin, 1859; any fundamentalist tract). 'As far as the examination of the instrument goes, there is precisely the same proof that the eye was made for vision, as there is that the telescope was made for assisting it. They are made upon the same principles; both being adjusted to the laws by which the transmission and refraction of rays of light are regulated' (Paley 1828, V. 1, p. 17).

If a similar instrument were found upon another planet, some special explanation would be called for. Either there is a God, or, if we are going to explain the universe in terms of blind physical forces, those blind physical forces are going to have to be deployed in a very peculiar way. The same is not true of non-living objects, such as the moon or the solar system (see below). Paley's instincts here were right.

My opinion of Astronomy has always been, that it is *not* the best medium through which to prove the agency of an intelligent Creator...The very simplicity of [the heavenly bodies'] appearance is against them...Now we deduce design from relation, aptitude, and correspondence of *parts*. Some degree therefore of *complexity* is necessary to render a subject fit for this species of argument. But the heavenly bodies do not, except perhaps in the instance of Saturn's ring, present themselves to our observation as compounded of parts at all (1828, Vol. 2, pp. 146–7).

A transparent pebble, polished by the sea, might act as a lens, focussing a real image. The fact that it is an efficient optical device is not particularly interesting because, unlike an eye or a telescope, it is too simple. We do not feel the need to invoke anything remotely resembling the concept of design. The eye and the telescope have many parts, all coadapted and working together to achieve the same functional end. The polished pebble has far fewer coadapted features: the coincidence of transparency, high refractive index and mechanical forces that polish the surface in a curved shape. The odds against such a threefold coincidence are not particularly great. No special explanation is called for.

Compare how a statistician decides what P value to accept as evidence for an effect in an experiment. It is a matter of judgment and dispute, almost of taste, exactly when a coincidence becomes too great to stomach. But, no matter whether you are a cautious statistician or a daring statistician, there are some complex adaptations whose 'P value', whose coincidence rating, is so impressive that nobody would hesitate to diagnose life (or an artefact designed by a living thing). My definition of living complexity is, in effect, 'that complexity which is too great to have come about through a single coincidence'. For the purposes of this paper, the problem that any theory of evolution has to solve is how living adaptive complexity comes about.

In the book referred to above, Mayr (1982) helpfully lists what he sees as the six clearly distinct theories of evolution that have ever been proposed in the history of biology. I shall use this list to provide me with my main headings in this paper. For each of the six, instead of asking what the evidence is, for or against, I shall ask whether the theory is *in principle* capable of doing the job of explaining the existence of adaptive complexity. I shall take the six theories in order, and will conclude that only Theory 6, Darwinian selection, matches up to the task.

Theory 1. Built-in capacity for, or drive toward, increasing perfection
To the modern mind this is not really a theory at all, and I shall not bother to discuss it. It is obviously mystical, and does not explain anything that it does not assume to start with.

Theory 2. Use and disuse plus inheritance of acquired characters
It is convenient to discuss this in two parts.

Use and disuse

It is an observed fact that on this planet living bodies sometimes become better adapted as a result of use. Muscles that are exercised tend to grow bigger. Necks that reach eagerly towards the treetops may lengthen in all their parts. Conceivably, if on some planet such acquired improvements could be incorporated into the hereditary information, adaptive evolution could result. This is the theory often associated with Lamarck, although there was more to what Lamarck said. Crick (1982, p. 59) says of the idea: 'As far as I know, no one has given *general* theoretical reasons why such a mechanism must be less efficient than natural selection...' In this section and the next I shall give two general theoretical objections to Lamarckism of the sort which, I suspect, Crick was calling for. I have discussed both before (Dawkins, 1982*b*), so will be brief here. First the shortcomings of the principle of use and disuse.

The problem is the crudity and imprecision of the adaptation that the principle of use and disuse is capable of providing. Consider the evolutionary improvements that must have occurred during the evolution of an organ such as an eye, and ask which of them could conceivaby have come about through use and disuse. Does 'use' increase the transparency of a lens? No, photons do not wash it clean as they pour through it. The lens and other optical parts must have reduced, over evolutionary time, their spherical and chromatic aberration; could this come about through increased use? Surely not. Exercise might have strengthened the muscles of the iris, but it could not have built up the fine feedback control system which controls those muscles. The mere bombardment of a retina with coloured light cannot call colour-sensitive cones into existence, nor connect up their outputs so as to provide colour vision.

Darwinian types of theory, of course, have no trouble in explaining all these improvements. Any improvement in visual accuracy could significantly affect survival. Any tiny reduction in spherical aberration may save a fast flying bird from fatally misjudging the position of an obstacle. Any minute improvement in an eye's resolution of acute coloured detail may crucially improve its detection of camouflaged prey. The genetic basis of any improvement, however slight, will come to predominate in the gene pool. The relationship between selection and adaptation is a direct and close-coupled one. The Lamarckian theory, on the other hand, relies on a much cruder coupling: the rule that the more an animal uses a certain bit of itself, the bigger that bit ought to be. The rule occasionally might have some validity but not generally, and, as a sculptor of adaptation it is a blunt hatchet in comparison to the fine chisels of natural selection.

This point is universal. It does not depend on detailed facts about life on this particular planet. The same goes for my misgivings about the inheritance of acquired characters.

Inheritance of acquired characters

The problem here is that acquired characters are not always improvements. There is no reason why they should be, and indeed the vast majority of them are injuries. This is not just a fact about life on earth. It has a universal rationale. If you have a complex and reasonably well-adapted system, the number of things you can do to it that will make it perform less well is vastly greater than the number of things you can do to it that will improve it (Fisher, 1958). Lamarckian evolution will move in adaptive directions only if some mechanism – selection – exists for distinguishing those acquired characters that are improvements from those that are not. Only the improvements should be imprinted into the germ line.

Although he was not talking about Lamarckism, Lorenz (1966) emphasized a related point for the case of learned behaviour, which is perhaps the most important kind of acquired adaptation. An animal learns to be a better animal during its own lifetime. It learns to eat sweet foods, say, thereby increasing its survival chances. But there is nothing inherently nutritious about a sweet taste. Something, presumably natural selection, has to have built into the nervous system the arbitrary rule: 'treat sweet taste as reward', and this works because saccharine does not occur in nature whereas sugar does.

Similarly, most animals learn to avoid situations that have, in the past, led to pain. The stimuli that animals treat as painful tend, in nature, to be associated with injury and increased chance of death. But again the connection must ultimately be built into the nervous system by natural selection, for it is not an obvious, necessary connection (M. Dawkins, 1980). It is easy to imagine artificially selecting a breed of animals that enjoyed being injured, and felt pain whenever their physiological welfare was being improved. If learning is adaptive *improvement*, there has to be, in Lorenz's phrase, an innate teaching mechanism, or 'innate schoolmarm'. The principle holds even where the reinforcers are 'secondary', learned by association with primary reinforcers (P. P. G. Bateson, this volume).

It holds, too, for morphological characters. Feet that are subjected to wear and tear grow tougher and more thick-skinned. The thickening of the skin is an acquired adaptation, but it is not obvious why the change went in this direction. In man-made machines, parts that are subjected to wear get thinner not thicker, for obvious reasons. Why does the skin on the feet do the opposite? Because, fundamentally, natural selection has worked in the past to ensure an adaptive rather than a maladaptive response to wear and tear.

The relevance of this for would-be Lamarckian evolution is that there has to be a deep Darwinian underpinning even if there is a Lamarckian surface structure: a Darwinian choice of which potentially acquirable characters shall in fact be acquired and inherited. As I have argued before (Dawkins, 1982a, pp. 164–77), this is true of a recent, highly publicized immunological theory of Lamarckian adaptation (Steele, 1979). Lamarckian mechanisms cannot be fundamentally responsible for adaptive evolution. Even if acquired characters are inherited on some planet, evolution there will still rely on a Darwinian guide for its adaptive direction.

Theory 3. Direct induction by the environment

Adaptation, as we have seen, is a fit between organism and environment. The set of conceivable organisms is wider than the actual set. And there is a set of conceivable environments wider than the actual set. These two subsets match each other to some extent, and the matching is adaptation. We can re-express the point by saying that information from the environment is present in the organism. In a few cases this is vividly literal – a frog carries a picture of its environment around on its back. Such information is usually carried by an animal in the less literal sense that a trained observer, dissecting a new animal, can reconstruct many details of its natural environment.

Now, how could the information get from the environment into the animal? Lorenz (1966) argues that there are two ways, natural selection and reinforcement learning, but that these are both *selective* processes in the broad sense (Pringle, 1951). There is, in theory, an alternative method for the environment to imprint its information on the organism, and that is by direct 'instruction' (Danchin, 1979). Some theories of how the immune system works are 'instructive': antibody molecules are thought to be shaped directly by moulding themselves around antigen molecules. The currently favoured theory is, by contrast, selective (Burnet, 1969). I take 'instruction' to be synonymous with the 'direct induction by the environment' of Mayr's Theory 3. It is not always clearly distinct from Theory 2.

Instruction is the process whereby information flows directly from its environment into an animal. A case could be made for treating imitation learning, latent learning and imprinting (Thorpe, 1963) as instructive, but for clarity it is safer to use a hypothetical example. Think of an animal on some planet, deriving camouflage from its tiger-like stripes. It lives in long dry 'grass', and its stripes closely match the typical thickness and spacing of local grass blades. On our own planet such adaptation would come about through the selection of random genetic variation, but on the imaginary planet it comes about through direct instruction. The animals go brown except where their skin is shaded from the 'sun' by blades of grass. Their stripes are therefore adapted with great precision, not just to any old habitat, but to the precise habitat in which they have sunbathed, and it is this same habitat in which they are going to have to survive. Local populations are automatically camouflaged against local grasses. Information about the habitat, in this case about the spacing patterns of the grass blades, has flowed into the animals, and is embodied in the spacing pattern of their skin pigment.

Instructive adaptation demands the inheritance of acquired characters if it is to give rise to permanent or progressive evolutionary change. 'Instruction' received in one generation must be 'remembered' in the genetic (or equivalent) information. This process is in principle cumulative and progressive. However, if the genetic store is not to become overloaded by the accumulations of generations, some mechanism must exist for discarding unwanted 'instructions', and retaining desirable ones. I suspect that this must lead us, once again, to the need for some kind of selective process.

Imagine, for instance, a form of mammal-like life in which a stout 'umbilical nerve' enabled a mother to 'dump' the entire contents of her memory in the brain of her foetus. The technology is available even to our nervous systems: the corpus callosum can shunt large quantities of information from right hemisphere to left. An umbilical nerve could make the experience and wisdom of each generation automatically available to the next, and this might seem very desirable. But without a selective filter, it would take few generations for the load of information to become unmanageably large. Once again we come up against the need for a selective underpinning. I will leave this now, and make one more point about instructive adaptation (which applies equally to all Lamarckian types of theory).

The point is that there is a logical link-up between the two major theories of adaptive evolution – selection and instruction – and the two major theories of embryonic development – epigenesis and preformationism. Instructive evolution can work only if embryology is preformationistic. If embryology is epigenetic, as it is on our planet, instructive evolution cannot work. I have expounded the argument before (Dawkins, 1982*a*, pp. 174–6), so I will abbreviate it here.

If acquired characters are to be inherited, embryonic processes must be reversible: phenotypic change has to be read back into the genes (or equivalent). If embryology is preformationistic – the genes are a true blueprint – then it may indeed be reversible. You can translate a house back into its blueprint. But if embryonic development is epigenetic: if, as on this planet, the genetic information is more like a recipe for a cake (Bateson, 1976) than a blueprint for a house, it is irreversible. There is no one-to-one mapping between bits of genome and bits of phenotype, any more than there is mapping between crumbs of cake and words of recipe. The recipe is not a blueprint that can be reconstructed from the cake. The transformation of recipe into cake cannot be put into reverse, and nor can the process of making a body. Therefore acquired adaptations cannot be read back into the 'genes', on any planet where embryology is epigenetic.

This is not to say that there could not, on some planet, be a form of life whose embryology was preformationistic. That is a separate question. How likely is it? The form of life would have to be very different from ours, so much so that it is hard to visualize how it might work. As for reversible embryology itself, it is even harder to visualize. Some mechanism would have to scan the detailed form of the adult body, carefully noting down, for instance, the exact location of brown pigment in a sun-striped skin, perhaps turning it into a linear stream of code numbers, as in a television camera. Embryonic development would read the scan out again, like a television receiver. I have an intuitive hunch that there is an objection in principle to this kind of embryology, but I cannot at present formulate it clearly. All I am saying here is that, if planets are divided into those where embryology is preformationistic and those, like Earth, where embryology is epigenetic, Darwinian evolution could be supported on both kinds of planet, but Lamarckian evolution, even if there were no other reasons for doubting its existence, could be supported only on the preformationistic planets – if there are any.

The close theoretical link that I have demonstrated between Lamarckian evolution and preformationistic embryology gives rise to a mildly entertaining irony. Those with ideological reasons for hankering after a neo-Lamarckian view of evolution are often especially militant partisans of epigenetic, 'interactionist', ideas of development, possibly – and here is the irony – for the very same ideological reasons (Koestler, 1967; Ho & Saunders, 1982).

Theory 4. Saltationism

The great virtue of the idea of evolution is that it explains, in terms of blind physical forces, the existence of undisputed adaptations whose statistical improbability is enormous, without recourse to the supernatural or the mystical. Since we *define* an undisputed adaptation as an adaptation that is too complex to have come about by chance, how is it possible for a theory to invoke only blind physical forces in explanation? The answer – Darwin's answer – is astonishingly simple when we consider how self-evident Paley's Divine Watchmaker must have seemed to his contemporaries. The key is that the coadapted parts do not have to be assembled *all at once*. They can be put together in small stages. But they really do have to be *small* stages. Otherwise we are back again with the problem we started with: the creation by chance of complexity that is too great to have been created by chance!

Take the eye again, as an example of an organ that contains a large number of independent coadapted parts, say N. The *a priori* probability of any one of these N features coming into existence by chance is low, but not incredibly low. It is comparable to the chance of a crystal pebble being washed by the sea so that it acts as a lens. Any one adaptation on its own could, plausibly, have come into existence through blind physical forces. If each of the N coadapted features confers some slight advantage on its own, then the whole many-parted organ can be put together over a long period of time. This is particularly plausible for the eye – ironically in view of that organ's niche of honour in the creationist pantheon. The eye is, *par excellence*, a case where a fraction of an organ is better than no organ at all; an eye without a lens or even a pupil, for instance, could still detect the looming shadow of a predator.

To repeat, the key to the Darwinian explanation of adaptive complexity is the replacement of instantaneous, coincidental, multi-dimensional luck, by gradual, inch by inch, smeared-out luck. Luck is involved, to be sure. But a theory that bunches the luck up into major steps is more incredible than a theory that spreads the luck out in small stages. This leads to the following general principle of universal biology. Wherever in the universe adaptive complexity shall be found, it will have come into being gradually through a series of small alterations, never through large and sudden increments in adaptive complexity. We must reject Mayr's 4th theory, saltationism, as a candidate for explanation of the evolution of complexity.

It is almost impossible to dispute this rejection. It is implicit in the definition of adaptive complexity that the only alternative to gradualistic evolution is supernatural magic. This is not to say that the argument in favour of gradualism is a worthless tautology, an unfalsifiable dogma of the sort that creationists and philosophers are so fond of jumping about on. It is not *logically* impossible for a full-fashioned eye to spring *de novo* from virgin bare skin. It is just that the possibility is statistically negligible.

Now it has recently been widely and repeatedly publicized that some modern evolutionists reject 'gradualism', and espouse what Turner (1982) has called theories of evolution by jerks. Since these are reasonable people without mystical leanings, they must be gradualists in the sense in which I am here using the term: the 'gradualism' that they oppose must be defined differently. There are actually two confusions of language here, and I intend to clear them up in turn. The first is the common confusion between 'punctuated equilibrium' (Eldredge & Gould, 1972) and true saltationism. The second is a confusion between two theoretically distinct kinds of saltation.

Punctuated equilibrium is not macromutation, not saltation at all in the traditional sense of the term. It is, however, necessary to discuss it here, because it is popularly regarded as a theory of saltation, and its partisans quote, with approval, Huxley's criticism of Darwin for upholding the principle of *Natura non facit saltum* (Gould, 1980). The punctuationist theory is portrayed as radical and revolutionary and at variance with the 'gradualistic' assumptions of both Darwin and the neo-Darwinian synthesis (e.g. Lewin, 1980). Punctuated equilibrium, however, was originally conceived as what the orthodox neo-Darwinian synthetic theory should truly predict, on a palaeontological timescale, if we take its embedded ideas of allopatric speciation seriously (Eldredge & Gould, 1972). It derives its 'jerks' by taking the 'stately unfolding' of the neo-Darwinian synthesis, and *inserting* long periods of stasis separating brief bursts of gradual, albeit rapid, evolution.

The plausibility of such 'rapid gradualism' is dramatized by a thought experiment of Stebbins (1982). He imagines a species of mouse, evolving larger body size at such an imperceptibly slow rate that the differences between the means of successive generations would be utterly swamped by sampling error. Yet even at this slow rate Stebbins's mouse lineage would attain the body size of a large elephant in about 60000 years, a time-span so short that it would be regarded as instantaneous by palaeontologists. Evolutionary change too *slow* to be detected by microevolutionists can nevertheless be too *fast* to be detected by macroevolutionists. What a palaeontologist sees as a 'saltation' can in fact be a smooth and gradual change so slow as to be undetectable to the microevolutionist. This kind of palaeontological 'saltation' has nothing to do with the one-generation macromutations that, I suspect, Huxley and Darwin had in mind when they debated *Natura non facit saltum*. Confusion has arisen here, possibly because some individual champions of punctuated equilibrium have also, incidentally, championed macromutation (Gould, 1982). Other 'punctuationists' have either confused their theory with macromutationism, or have explicitly invoked macromutation as one of the mechanisms of punctuation (e.g. Stanley, 1981).

Turning to macromutation, or true saltation itself, the second confusion that I want to clear up is between two kinds of macromutation that we might conceive of. I could name them, unmemorably, saltation (1) and saltation (2), but instead I shall pursue an earlier fancy for airliners as metaphors, and label them 'Boeing 747' and 'Stretched DC-8' saltation. 747 saltation is the inconceivable kind. It gets its name from Sir Fred Hoyle's much quoted metaphor for his own cosmic misunderstanding of Darwinism (Hoyle & Wickramasinghe, 1981). Hoyle compared Darwinian selection to a tornado, blowing through a junkyard and assembling a Boeing 747 (what he overlooked, of course, was the point about luck being 'smeared-out' in small steps – see above). Stretched DC-8 saltation is quite different. It is not in principle hard to believe in at all. It refers to large and sudden changes in *magnitude* of some biological measure, without an accompanying large increase in adaptive information. It is named after an airliner that was made by elongating the fuselage of an existing design, not adding significant new complexity. The change from DC-8 to Stretched DC-8 is a big change in magnitude – a saltation not a gradualistic series of tiny changes. But, unlike the change from junk-heap to 747, it is not a big increase in information content or complexity, and that is the point I am emphasizing by the analogy.

An example of DC-8 saltation would be the following. Suppose the giraffe's neck shot out in one spectacular mutational step. Two parents had necks of standard antelope length. They had a freak child with a neck of modern giraffe length, and all giraffes are descended from this freak. This is unlikely to be true on Earth, but something like it may happen elsewhere in the universe. There is no objection to it in principle, in the sense that there is a profound objection to the (747) idea that a complex organ like an eye could arise from bare skin by a single mutation. The crucial difference is one of complexity.

I am assuming that the change from short antelope's neck to long giraffe's neck is *not* an increase in complexity. To be sure, both necks are exceedingly complex structures. You couldn't go from *no*-neck to either kind of neck in one step: that would be 747 saltation. But once the complex organization of the antelope's neck already exists, the step to giraffe's neck is just an elongation: various things have to grow faster at some stage in embryonic development; existing complexity is preserved. In practice, of course, such a drastic change in magnitude would be highly likely to have deleterious repercussions which would render the macromutant unlikely to survive. The existing antelope heart probably could not pump the blood up to the newly elevated giraffe head. Such practical objections to evolution by 'DC-8 saltation' can only help my case in favour of gradualism, but I still want to make a separate, and more universal, case against 747 saltation.

It may be argued that the distinction between 747 and DC-8 saltation is impossible to draw in practice. After all, DC-8 saltations, such as the proposed macromutational elongation of the giraffe's neck, may appear very complex: myotomes, vertebrae, nerves, blood vessels, all have to elongate together. Why does this not make it a 747 saltation, and therefore rule it out? But although this type of 'coadaptation' has indeed often been thought of as a problem for any evolutionary theory, not just macro-mutational ones (see Ridley, 1982, for a history), it is so only if we take an impoverished view of developmental mechanisms. We know that single mutations can orchestrate changes in growth rates of many diverse parts of organs, and, when we think about developmental processes, it is not in the least surprising that this should be so. When a single mutation causes a *Drosophila* to grow a leg where an antenna ought to be, the leg grows in all its formidable complexity. But this is not mysterious or surprising, not a 747 saltation, because the organization of a leg is already present in the body before the mutation. Wherever, as in embryogenesis, we have a hierarchically branching tree of causal relationships, a small alteration at a senior node of the tree can have large and complex ramified effects on the tips of the twigs. But although the change may be large in magnitude, there can be no large and sudden increments in adaptive information. If you think you have found a particular example of a large and sudden increment in adaptively complex information in practice, you can be certain the adaptive information was already there, even if it is an atavistic 'throwback' to an earlier ancestor.

There is not, then, any objection in principle to theories of evolution by jerks, even the theory of hopeful monsters (Goldschmidt, 1940), provided that it is DC-8 saltation, not 747 saltation that is meant. Gould (1982) would clearly agree: 'I regard forms of macromutation which include the sudden origin of new species with all their multifarious adaptations intact *ab initio*, as illegitimate'. No educated biologist actually believes in 747 saltation, but not all have been sufficiently explicit about the distinction between DC-8 and 747 saltation. An unfortunate consequence is that creationists and their journalistic fellow-travellers have been able to exploit saltationist-sounding statements of respected biologists. The biologist's intended meaning may have been what I am calling DC-8 saltation, or even non-saltatory punctuation; but the creationist *assumes* saltation in the sense that I have dubbed 747, and 747 saltation would, indeed, be a blessed miracle.

I also wonder whether an injustice is not being done to Darwin, owing to this same failure to come to grips with the distinction between DC-8 and 747 saltation. It is frequently alleged that Darwin was wedded to gradualism, and therefore that, if some form of evolution by jerks is proved, Darwin will have been shown wrong. This is undoubtedly the reason for the ballyhoo and publicity that has attended the theory of punctuated equilibrium. But was Darwin really opposed to all jerks? Or was he, as I suspect, strongly opposed only to 747 saltation?

As we have already seen, punctuated equilibrium has nothing to do with saltation, but anyway I think it is not at all clear that, as is often alleged, Darwin would have been discomfited by punctuationist interpretations of the fossil record. The following passage, from later editions of the *Origin*, sounds like something from a current issue of *Paleobiology*: 'the periods during which species have been undergoing modification, though very long as measured by years, have probably been short in comparison with the periods during which these same species remained without undergoing any change'.

Gould (1982) shrugs this off as somehow anomalous and away from the mainstream of Darwin's thought. As he correctly says: 'You cannot do history by selective quotation and search for qualifying footnotes. General tenor and historical impact are the proper criteria. Did his contemporaries or descendants ever read Darwin as a saltationist?' Certainly nobody ever accused Darwin of being a saltationist. But to most people saltation means macromutation, and, as Gould himself stresses, 'Punctuated equilibrium is not a theory of macromutation'. More importantly, I believe we can reach a better understanding of Darwin's general gradualistic bias if we invoke the distinction between 747 and DC-8 saltation.

Perhaps part of the problem is that Darwin himself did not have the distinction. In some anti-saltation passages it seems to be DC-8 saltation that he has in mind. But on those occasions he does not seem to feel very strongly about it: 'About sudden jumps', he wrote in a letter in 1860, 'I have no objection to them – they would aid me in some cases. All I can say is, that I went into the subject and found no evidence to make me believe in jumps [as a source of new species] and a good deal pointing in the other direction' (quoted in Gillespie, 1979). This does not sound like a man fervently opposed, in principle, to sudden jumps. And of course there is no reason why he *should* have been fervently opposed, if he only had DC-8 saltations in mind.

But at other times he really is pretty fervent, and on those occasions, I suggest, he is thinking of 747 saltation: '...it is impossible to imagine so many co-adaptations being formed all by a chance blow' (quoted in Ridley, 1982). As the historian Neal Gillespie puts it: 'For Darwin, monstrous births, a doctrine favored by Chambers, Owen, Argyll, Mivart, and others, from clear theological as well as scientific motives, as an explanation of how new species, or even higher taxa, had developed, was no better than a miracle: "it leaves the case of the co-adaptation of organic beings to each other and to their physical conditions of life, untouched and unexplained". It was "no explanation" at all, of no more scientific value than creation "from the dust of the earth"' (Gillespie, 1979, p. 118).

As Ridley (1982) says of the 'religious tradition of idealist thinkers [who] were committed to the explanation of complex adaptive contrivances by intelligent design', 'The greatest concession they could make to Darwin was that the Designer operated by tinkering with the generation of diversity, designing the variation'. Darwin's response was: 'If I were convinced that I required such additions to the theory of natural selection, I would reject it as rubbish...I would give nothing for the theory of Natural selection, if it requires miraculous additions at any one stage of descent'.

Darwin's hostility to monstrous saltation, then, makes sense if we assume that he was thinking in terms of 747 saltation – the sudden invention of new adaptive complexity. It is highly likely that that is what he was thinking of, because that is exactly what many of his opponents had in mind. Saltationists such as the Duke of Argyll (though presumably not Huxley!) wanted to believe in 747 saltation, precisely because it *did* demand supernatural intervention. Darwin did not believe in it, for exactly the same reason. To quote Gillespie again (p. 120): '...for Darwin, designed evolution, whether manifested in saltation, monstrous births, or manipulated variations, was but a disguised form of special creation'.

I think this approach provides us with the only sensible reading of Darwin's well known remark that 'If it could be demonstrated that any complex organ existed, which could not possibly have been formed by numerous, successive, slight modifications, my theory would absolutely break down'. That is not a plea for gradualism, as a modern palaeobiologist uses the term. Darwin's theory is falsifiable, but he was much too wise to make his theory *that* easy to falsify! Why on earth *should* Darwin have committed himself to such an arbitrarily restrictive version of evolution, a version that positively invites falsification? I think it is clear that he didn't. His use of the term 'complex' seems to me to be clinching. Gould (1982) describes this passage from Darwin as 'clearly invalid'. So it is invalid if the alternative to slight modifications is seen as DC-8 saltation. But if the alternative is seen as 747 saltation, Darwin's remark is valid and very wise. Notwithstanding those whom Miller (1982) has unkindly called Darwin's more foolish critics, his theory is indeed falsifiable, and in the passage quoted he puts his finger on one way in which it might be falsified.

There are two kinds of imaginable saltation, then, DC-8 saltation and 747 saltation. DC-8 saltation is perfectly possible, undoubtedly happens in the laboratory and the farmyard, and may have made important contributions to evolution. 747 saltation is statistically ruled out unless there is supernatural intervention. In Darwin's own time, proponents and opponents of saltation often had 747 saltation in mind, because they believed in – or were arguing against – divine intervention. Darwin was hostile to (747) saltation, because he correctly saw natural selection as an *alternative* to the miraculous as an explanation for adaptive complexity. Nowadays saltation either means punctuation (which isn't saltation at all) or DC-8 saltation, neither of which Darwin would have had strong objections to in principle, merely doubts about the facts. In the modern context, therefore, I do not think Darwin should be labelled a strong gradualist. In the modern context, I suspect that he would be rather open-minded.

It is in the anti-747 sense that Darwin was a passionate gradualist, and it is in the same sense that we must all be gradualists, not just with respect to life on earth, but with respect to life all over the universe. Gradualism in this sense is essentially synonymous with evolution. The sense in which we may be non-gradualists is a much less radical, although still quite interesting, sense. The theory of evolution by jerks has been hailed on television and elsewhere as radical and revolutionary, a paradigm shift. There is, indeed, an interpretation of it which is revolutionary, but that interpretation (the 747 macromutation version) is certainly wrong, and is apparently not held by its original proponents. The sense in which the theory might be right is not particularly revolutionary. In this field you may choose your jerks so as to be revolutionary, *or* so as to be correct, but not both.

Theory 5. Random evolution

Various members of this family of theories have been in vogue at various times. The 'mutationists' of the early part of this century – De Vries, W. Bateson and their colleagues – believed that selection served only to weed out deleterious freaks, and that the real driving force in evolution was mutation pressure. Unless you believe mutations are directed by some mysterious life force, it is sufficiently obvious that you can be a mutationist only if you forget about adaptive complexity – forget, in other words, most of the consequences of evolution that are of any interest! For historians there remains the baffling enigma of how such distinguished biologists as De Vries, W. Bateson and T. H. Morgan could rest satisfied with such a crassly inadequate theory. It is not enough to say that De Vries's view was blinkered by his working only on the evening primrose. He only had to look at the adaptive complexity in his own body to see that 'mutationism' was not just a wrong theory: it was an obvious non-starter.

These post-Darwinian mutationists were also saltationists and anti-gradualists, and Mayr treats them under that heading, but the aspect of their view that I am criticizing here is more fundamental. It appears that they actually thought that mutation, on its own without selection, was sufficient to explain evolution. This *could* not be so on any non-mystical view of mutation, whether gradualist or saltationist. If mutation is undirected, it is clearly unable to explain the adaptive directions of evolution. If mutation is directed in adaptive ways we are entitled to ask how this comes about. At least Lamarck's principle of use and disuse makes a valiant attempt at explaining how variation might be directed. The 'mutationists' didn't even seem to see that there was a problem, possibly because they under-rated the importance of adaptation – and they were not the last to do so. The irony with which we must now read W. Bateson's dismissal of Darwin is almost painful: 'the transformation of masses of populations by imperceptible steps guided by selection is, as most of us now see, so inapplicable to the fact that we can only marvel...at the want of penetration displayed by the advocates of such a proposition...' (1913, quoted in Mayr, 1982).

Nowadays some population geneticists describe themselves as supporters of 'non-Darwinian evolution'. They believe that a substantial number of the gene replacements that occur in evolution are non-adaptive substitutions of alleles whose effects are indifferent relative to one another (Kimura, 1968). This may well be true, if not in Israel (Nevo, this volume) maybe somewhere in the Universe. But it obviously has nothing whatever to contribute to solving the problem of the evolution of adaptive complexity.

Modern advocates of neutralism admit that their theory cannot account for adaptation, but that doesn't seem to stop them regarding the theory as interesting. Different people are interested in different things.

The phrase 'random genetic drift' is often associated with the name of Sewall Wright, but Wright's conception of the relationship between random drift and adaptation is altogether subtler than the others I have mentioned (Wright, 1980). Wright does not belong in Mayr's fifth category, for he clearly sees selection as the driving force of adaptive evolution. Random drift may make it easier for selection to do its job by assisting the escape from local optima (Dawkins, 1982a, p. 40), but it is still selection that is determining the rise of adaptive complexity.

Recently palaeontologists have come up with fascinating results when they perform computer simulations of 'random phylogenies' (e.g. Raup, 1977). These random walks through evolutionary time produce trends that look uncannily like real ones, and it is disquietingly easy, and tempting, to read into the random phylogenies apparently adaptive trends which, however, are not there. But this does not mean that we can admit random drift as an explanation of real adaptive trends. What it might mean is that some of us have been too facile and gullible in what we think are adaptive trends. That does not alter the fact that there are some trends that really *are* adaptive – even if we don't always identify them correctly in practice – and those real adaptive trends can't be produced by random drift. They must be produced by some non-random force, presumably selection.

So, finally, we arrive at the sixth of Mayr's theories of evolution.

Theory 6. Direction (order) imposed on random variation by natural selection
Darwinism – the non-random selection of randomly varying replicating entities by reason of their 'phenotypic' effects – is the only force I know that can, in principle, guide evolution in the direction of adaptive complexity. It works on this planet. It doesn't suffer from any of the drawbacks that beset the other five classes of theory, and there is no reason to doubt its efficacy throughout the universe.

The ingredients in a general recipe for Darwinian evolution are replicating entities of some kind, exerting phenotypic 'power' of some kind over their replication success. I have referred to these necessary entities as 'active germ-line replicators' or 'optimons' (Dawkins, 1982a, Chapter 5). It is important to keep their replication conceptually separate from their phenotypic effects, even though, on some planets, there may be a blurring in practice. Phenotypic adaptations can be seen as tools of replicator propagation.

Gould (this volume) disparages the replicator's-eye view of evolution as preoccupied with 'book-keeping'. The metaphor is a superficially happy one: it is easy to see the genetic changes that accompany evolution as book-keeping entries, mere accountant's records of the really interesting phenotypic events going on in the outside world. Deeper consideration, however, shows that the truth is almost the exact opposite. It is central and essential to Darwinian (as opposed to Lamarckian) evolution that there shall be causal arrows flowing from genotype to phenotype, but not in the reverse direction. Changes in gene frequencies are not passive book-keeping records of phenotypic changes: it is precisely because (and to the extent that) they actively *cause* phenotypic changes that evolution of the phenotype can occur. Serious errors flow, both from a failure to understand the importance of this one-way flow (Dawkins, 1982*a*, Chapter 6), and from an over-interpretation of it as inflexible and undeviating 'genetic determinism' (Dawkins, 1982*a*, Chapter 2).

The universal perspective leads me to emphasize a distinction between what may be called 'one-off selection' and 'cumulative selection'. Order in the non-living world may result from processes that can be portrayed as a rudimentary kind of selection. The pebbles on a seashore become sorted by the waves, so that larger pebbles come to lie in layers separate from smaller ones. We can regard this as an example of the selection of a stable configuration out of initially more random disorder. The same can be said of the 'harmonious' orbital patterns of planets around stars, and electrons around nuclei, of the shapes of crystals, bubbles and droplets, even, perhaps, of the dimensionality of the universe in which we find ourselves (Atkins, 1981). But this is all one-off selection. It does not give rise to progressive evolution because there is no replication, no succession of generations. Complex adaptation requires many generations of cumulative selection, each generation's change building upon what has gone before. In one-off selection, a stable state develops and is then maintained. It does not multiply, does not have offspring.

In life the selection that goes on *in any one generation* is one-off selection, analogous to the sorting of pebbles on a beach. The peculiar feature of life is that successive generations of such selection build up, progressively and cumulatively, structures that are eventually complex enough to foster the strong illusion of design. One-off selection is a commonplace of physics and cannot give rise to adaptive complexity. Cumulative selection is the hallmark of biology and is, I believe, the force underlying all adaptive complexity.

Other topics for a future science of Universal Darwinism

Active germ-line replicators together with their phenotypic consequences, then, constitute the general recipe for life, but the form of the system may vary greatly from planet to planet, both with respect to the replicating entities themselves, and with respect to the 'phenotypic' means by which they ensure their survival. Indeed the very distinction between 'genotype' and 'phenotype' may be blurred (L. Orgel, personal communication). The replicating entities do not have to be DNA or RNA. They do not have to be organic molecules at all. Even on this planet it is possible that DNA itself is a late usurper of the role, taking over from some earlier, inorganic crystalline replicator (Cairns-Smith, 1982). It is also arguable that today selection operates on several levels, for instance the levels of the gene and the species or lineage, and perhaps some unit of cultural transmission (Lewontin, 1970).

A full science of Universal Darwinism might consider aspects of replicators transcending their detailed nature and the time-scale over which they are copied. For instance, the extent to which they are 'particulate' as opposed to 'blending' probably has a more important bearing on evolution than their detailed molecular or physical nature. Similarly, a universe-wide classification of replicators might make more reference to their dimensionality and coding principles than to their size and structure. DNA is a digitally coded one-dimensional array. A 'genetic' code in the form of a two-dimensional matrix is conceivable. Even a three-dimensional code is imaginable, although students of Universal Darwinism will probably worry about how such a code could be 'read'. (DNA is, of course, a molecule whose 3-dimensional structure determines how it is replicated and transcribed, but that doesn't make it a 3-dimensional code. DNA's meaning depends upon the 1-dimensional sequential arrangement of its symbols, not upon their 3-dimensional position relative to one another in the cell.) There might also be theoretical problems with analogue, as opposed to digital codes, similar to the theoretical problems that would be raised by a purely analogue nervous system (Rushton, 1961).

As for the phenotypic levers of power by which replicators influence their survival, we are so used to their being bound up into discrete organisms or 'vehicles' that we forget the possibility of a more diffuse extra-corporeal or 'extended' phenotype. Even on this Earth a large amount of interesting adaptation can be interpreted as part of the extended phenotype (Dawkins, 1982a, Chapters 11, 12 and 13). There is, however, a general theoretical case that can be made in favour of the discrete organismal body, with its own recurrent life cycle, as a necessity in any process of evolution of advanced adaptive complexity (Dawkins, 1982a, Chapter 14), and this topic might have a place in a full account of Universal Darwinism.

Another candidate for full discussion might be what I shall call divergence, and convergence or recombination of replicator lineages. In the case of Earthbound DNA, 'convergence' is provided by sex and related processes. Here the DNA 'converges' within the species after having very recently 'diverged'. But suggestions are now being made that a different kind of convergence can occur among lineages that originally diverged an exceedingly long time ago. For instance there is evidence of gene transfer between fish and bacteria (Jacob, this volume). The replicating lineages on other planets may permit very varied kinds of recombination, on very different time-scales. On Earth the rivers of phylogeny are almost entirely divergent: if main tributaries ever recontact each other after branching apart it is only through the tiniest of trickling cross-streamlets, as in the fish/bacteria case. There is, of course, a richly anastomosing delta of divergence and convergence due to sexual recombination *within* the species, but only within the species. There may be planets on which the 'genetic' system permits much more cross-talk at all levels of the branching hierarchy, one huge fertile delta.

I have not thought enough about the fantasies of the previous paragraphs to evaluate their plausibility. My general point is that there is one limiting constraint upon all speculations about life in the universe. If a life-form displays adaptive complexity, it must possess an evolutionary mechanism capable of generating adaptive complexity. However diverse evolutionary mechanisms may be, if there is no other generalization that can be made about life all around the Universe, I am betting it will always be recognizable as Darwinian life. The Darwinian Law (Eigen, this volume) may be as universal as the great laws of physics.

As usual I have benefited from discussions with many people, including especially Mark Ridley, who also criticized the manuscript, and Alan Grafen. Dr F. J. Ayala called attention to an important error in the original spoken version of the paper.

References

Asimov, I. (1979). *Extraterrestrial civilizations*. London: Pan.
Atkins, P. W. (1981). *The creation*. Oxford: W. H. Freeman.
Bateson, P. P. G. (1976). Specificity and the origins of behavior. *Advances in the study of behavior*, **6**, 1–20.
Billingham, J. (1981). *Life in the universe*. Cambridge, Mass.: MIT Press.
Burnet, F. M. (1969). *Cellular immunology*. Melbourne: Melbourne University Press.

Cairns-Smith, A. G. (1982). *Genetic takeover*. Cambridge: Cambridge University Press.

Crick, F. H. C. (1982). *Life itself*. London: Macdonald.

Danchin, A. (1979). Themes de la biologie: theories instructives et theories selectives. *Revue des Questions Scientifiques*, **150**, 151–64.

Darwin, C. R. (1859). *The origin of species*. 1st edition, reprinted (1968), London: Penguin.

Dawkins, M. (1980). *Animal Suffering: the science of animal welfare*. London: Chapman & Hall.

Dawkins, R. (1982*a*). *The Extended Phenotype*. Oxford: W. H. Freeman.

Dawkins, R. (1982*b*). The necessity of Darwinism. *New Scientist*, **94**, 130–32, reprinted in *Darwin up to date*, ed. J. Cherfas. London: New Scientist.

Eldredge, N. & Gould, S. J. (1972). Punctuated equilibria: an alternative to phyletic gradualism. In *Models in Paleobiology*, ed. T. J. M. Schopf. San Francisco: Freeman Cooper.

Fisher, R. A. (1958). *The genetical theory of natural selection*. New York: Dover.

Gillespie, N. C. (1979). *Charles Darwin and the problem of creation*. Chicago: University of Chicago Press.

Goldschmidt, R. (1940). *The material basis of evolution*. New Haven: Yale University Press.

Gould, S. J. (1980). *The Panda's Thumb*. New York: W. W. Norton.

Gould, S. J. (1982). The meaning of punctuated equilibrium and its role in validating a hierarchical approach to macroevolution. In *Perspectives on evolution*, ed. R. Milkman, pp. 83–104. Sunderland, Mass.: Sinauer.

Ho, M-W. & Saunders, P. T. (1982). Adaptation and natural selection: mechanism and teleology. In *Towards a liberatory biology* (Dialectics of Biology Group, general editor S. Rose), 85–102.

Hoyle, F. & Wickramasinghe, N. C. (1981). *Evolution from space*. London: J. M. Dent.

Kimura, M. (1968). Evolutionary rate at the molecular level. *Nature*, **217**, 624–6.

Koestler, A. (1967). *The ghost in the machine*. London: Hutchinson.

Lewin, R. (1980). Evolutionary theory under fire. *Science*, **210**, 883–7.

Lewontin, R. C. (1970). The units of selection. *Annual Review of Ecology and Systematics*, **1**, 1–18.

Lewontin, R. C. (1979). Sociobiology as an adaptationist program. *Behavioral Science*, **24**, 5–14.

Lorenz, K. (1966). *Evolution and modification of behavior*. London: Methuen.

Maynard Smith, J. (1969). The status of neo-Darwinism. In *Towards a Theoretical Biology*, ed. C. H. Waddington. Edinburgh: University Press.

Mayr, E. (1982). *The growth of biological thought*. Cambridge Mass.: Harvard University Press.

Miller, J. (1982). *Darwin for beginners*. London: Writers and Readers.

Paley, W. (1828). *Natural Theology*, 2nd edition. Oxford: J. Vincent.

Pringle, J. W. S. (1951). On the parallel between learning and evolution. *Behaviour*, **3**, 90–110.

Raup, D. M. (1977). Stochastic models in evolutionary palaeontology. In *Patterns of evolution*, ed. A. Hallam. Amsterdam: Elsevier.

Ridley, M. (1982). Coadaptation and the inadequacy of natural selection. *British Journal for the History of Science*, **15**, 45–68.

Rushton, W. A. H. (1961). Peripheral coding in the nervous system. In *Sensory communication*, ed. W. A. Rosenblith. Cambridge, Mass.: MIT Press.

Stanley, S. M. (1981). *The new evolutionary timetable*. New York: Basic Books.
Stebbins, G. L. (1982). *Darwin to DNA, molecules to humanity*. San Francisco: W. H. Freeman.
Steele, E. J. (1979). *Somatic selection and adaptive evolution*. Toronto: Williams and Wallace.
Thorpe, W. H. (1963). *Learning and instinct in animals*, 2nd edition. London: Methuen.
Turner, J. R. G. (1982). Review of R. J. Berry, *Neo-Darwinism. New Scientist*, **94**, 160–2.
Wright, S. (1980). Genic and organismic selection. *Evolution*, **34**, 825–43.

EVOLUTION OF SOCIAL BEHAVIOUR

21

The development of an evolutionary ethology

RICHARD W. BURKHARDT, JR

Of all the individuals who took up the study of animal behaviour in the generation after Charles Darwin's death, the most irascible was undoubtedly the English birdwatcher Edmund Selous, a man who combined a deep love of nature and respect for Darwin on the one hand with a vigorous antipathy toward the scientific establishment of his day and society in general on the other. Typical of Selous's comments was the statement: 'Darwin, as it seems to me, has never been properly assimilated...Instead of reading new books upon evolution it would be better – more profitable – to read him over and over again' (Selous, 1927, p. 227).

Today, a whole century after Darwin's death, Selous's counsel is not as appropriate as it was in Selous's own day. This is not to say, however, that there is nothing left to learn from Darwin's works. The reality of evolution and the central role of natural selection in the evolutionary process – the two major premises of Darwin's book *The Origin of Species* (1859) – have by now been incorporated in biological thinking for a relatively long time. The same does not hold, though, for some of Darwin's other ideas, particularly ideas that he developed in writings other than *The Origin of Species*. For example, sexual selection, one of the central concepts of Darwin's book, *The Descent of Man* (1871), attracted a good deal of attention up to the early twentieth century, but then it limped along in semi-obscurity until being rejuvenated only recently. The issue of individual versus group selection was also confronted by Darwin in *The Descent of Man*. He did not devote as much attention to this subject as he did to sexual selection. None the less, in discussing the origins of altruistic behaviour in humans, he showed an appreciation of the problems involved in group selection that appears remarkably astute when one considers how little was done with this subject until V. C. Wynne-Edwards' 1962 book on animal dispersion caused attention to be focussed on it again. With respect to current discussions of the extent to which the different characters of a given species can be considered adaptive, it is interesting to find that Darwin commented on this subject too. In 1871, in *The Descent of Man*, he acknowledged that in the early editions of *The Origin of Species*, he had 'perhaps attributed too much to the action of natural selection...' As he put it 'I had not formerly sufficiently considered the existence of many structures which appear to be, as far as we can judge, neither beneficial nor injurious; and this I believe to be one of the greatest oversights as yet detected in my work' (1: 152).

I have not cited these comments by Darwin to endorse a particular view of evolution. Darwin himself continued to be of two minds on the subject of the adaptiveness of characters. The reason I have cited Darwin's comments is to call attention to the *historical* insight embodied in Darwin's explanation of his 'oversight'. In writing the *Origin*, Darwin said, he had been primarily concerned '...firstly, to shew that species had not been separately created, and secondly, that natural selection had been the chief agent of change, though largely aided by the inherited effects of habit, and slightly by the direct action of the surrounding conditions'. (1: 152–3). His excuse for his oversight was:

Nevertheless, I was not able to annul the influence of my former belief, then widely prevalent, that each species had been purposely created; and this led to my tacitly assuming that every detail of structure, except rudiments, was of some special, though unrecognized, service. Any one with this assumption in his mind would naturally extend the action of natural selection, either during past or present times, too far.' (1: 153).

Darwin perceived, in other words, that in the way he had structured his response to the creationist position, he had unconsciously led himself to overlook a potentially important idea. What I would like to suggest here is that the study of behavioural evolution from Darwin's time to the present has involved a number of cases similar to the one just mentioned. The past century of animal-behaviour study has witnessed numerous sharp debates and the aggressive formation of new disciplines and subdisciplines. It has witnessed investigators of new generations seeking to distinguish themselves from previous generations. These activities have resulted in a host of impressive advances in the understanding of animal behaviour. At the same time, they have also provided occasions for certain important ideas to be overlooked, either briefly or for substantial periods of time. This is one of the reasons why students of animal behaviour are just now rediscovering the interest of some of Darwin's ideas. To put it another way, this is one of the reasons why it has taken some time for the development of an *evolutionary* ethology.

It can be argued, of course, that from its very beginning ethology has been essentially an evolutionary science. This is what is suggested by Konrad Lorenz in his latest book, *Foundations of Ethology* (1981). According to Lorenz, 'Ethology, the comparative study of behaviour, is easy to define: it is the discipline which applies to the behaviour of animals and humans all those questions asked and methodologies used as a matter of course in all other branches of biology since Charles Darwin's time' (p. 1).

Lorenz's definition of ethology is an appealing one. It is also characteristic of Lorenz's assertive style of expression. From a historical standpoint, however, the definition is incomplete. It is incomplete not merely because Lorenz neglected to mention his own pre-eminent role in establishing ethology as a discipline. It is also incomplete because the very activity of constructing ethology as a new scientific field shaped the field's early cognitive dimensions, and shaped them in such a way that there was not a perfect fit between the interpretation of behaviour in Darwinian terms and the foundation of ethology as a scientific discipline. Significantly, as late as 1963, by which time ethology was well-established as a discipline, Ernst Mayr was able to write 'There are vast areas of modern biology, for instance...the study of behaviour, in which the application of evolutionary principles is still in the most elementary stage' (Mayr, 1963, p. 7).

To say this is not to belittle the enormous significance of the achievement of Lorenz and those other biologists who, in the second third of this century, made ethology into an exciting, viable, and scientifically respected field. It is to say, however, that the relations between ethology and evolutionary theory have not been as straightforward as Lorenz's definition of ethology would seem to imply. Here only a few aspects of the story can be mentioned. My focus will be on Darwin's work and the work of the major founders of ethology in the 1930s and 1940s, Konrad Lorenz and Niko Tinbergen. In a detailed study of ethology's development, the ideas of many other individuals would also have to be analysed.

Darwin's own work illustrates that the interpretation of behaviour in Darwinian terms did not have to await ethology's founding: throughout his career Darwin himself was profoundly interested in behaviour – behaviour of animals, behaviour of humans, and behaviour of plants (Burkhardt, 1983). When he embarked on his *Beagle* voyage, he appreciated fully that his role as voyager-naturalist involved observing not only the physical but also the behavioural characteristics of the different organisms –

and peoples – he would encounter. During his travels he was indeed greatly struck by the habits of native animals and native peoples. Once he returned to England, and the idea of organic mutability had taken shape in his mind, behavioural phenomena he had witnessed on his voyage played a conspicuous role in his thinking about species change. He studied behaviour further at the London Zoo, and in the summer of 1838 he wrote in his *Notebook C*: 'Let man visit Ourang-outang in domestication, hear expressive whine, see its intelligence when spoken [to], as if it understood every word said – see its affection to those it knows, – see its passion & rage, sulkiness & very extreme of despair; let him look at savage, roasting his parent, naked, artless, not improving, yet improvable and then let him dare to boast of his proud eminence.' (de Beer, 1960–61, p. 91). Though Darwin did not declare himself publicly on the subject for another third of a century, he had little doubt that the mental faculties of animals and humans were continuous and that even the higher moral faculties of humans could be studied effectively from the standpoint of the naturalist.

Behaviour, it must be stressed, was not something to which Darwin simply *applied* his evolutionary views. His attention to behavioural phenomena contributed to his theorizing in several important ways, as his manuscripts and books indicate. In his initial evolutionary speculations, Darwin attributed considerable importance to habits and instincts as agents of organic change. Later, in *The Origin of Species*, in dealing with the instincts (and structures) of neuter castes of insects, he presented an unexcelled example of the explanatory power of natural selection as contrasted to the inheritance of acquired characters (Burkhardt, 1981a; Richards, 1981). Later still, in his book *The Descent of Man*, in elaborating his idea of sexual selection, he underlined the role of behaviour in reproductive success, and he showed at the same time that he did not regard reproductive success and adaptiveness to be necessarily the same thing (Darwin, 1871; Mayr, 1972). A year afterwards, in discussing the expression of the emotions in man and animals, Darwin provided his most extended argument that not all characters of animals need be understood in terms of their adaptiveness (Darwin, 1872; Ghiselin, 1969). Since this last case provides another instance where the pursuit of a good cause evidently facilitated one insight on Darwin's part while making another insight *less* accessible to him, it may be useful here to say a bit more about Darwin's thoughts on emotional expression.

In light of Darwin's acknowledgment that he may have tended to attribute too much to the action of natural selection, it seems anomalous that Darwin attached virtually no importance to selection as an agent in the evolution of emotional expression. This seems even more anomalous when one considers that in the 1930s and afterwards, Lorenz and Tinbergen argued that the most distinctive forms of the expression of the emotions in animals were distinctive precisely because they had been selected – selected for their role in intraspecific communication (Lorenz, 1935, 1937a; Tinbergen, 1940, 1948, 1952). What is crucial for understanding this 'oversight' on Darwin's part is that Darwin's work on the expression of the emotions was conceived in response to a treatise by Sir Charles Bell (1844).

Bell had maintained that emotional expression in humans was on a wholly different level from emotional expression in animals. In humans, he said, there was a unique set of facial muscles designed by the Creator for the specific purpose of man's nonverbal communication. Darwin saw Bell's treatise as a threat to the idea that man had evolved from a more primitive form. In distinguishing his views from Bell's, Darwin argued not only that there was a continuity in the emotional expression of animals and humans, but also that emotional expression could be explained without ascribing any communicative function to it at all (Darwin, 1872).

In his treatment of the evolution of emotional expression, Darwin showed a clear appreciation that merely imagining how a character might be useful by no means constituted a satisfactory explanation of how that character in fact originated. This was a critical insight. It is an insight that apparently still needs repeating today. The additional point to be emphasized here, nonetheless, is that having constructed his argument against the idea that special facial muscles in man had been providentially designed for man's nonverbal communication, Darwin left himself ill-prepared to develop the idea that certain expressive actions, having arisen independently of any communicative function, might thereafter have been further developed by natural selection. Darwin was capable of making this kind of argument. Indeed, he had already offered such an argument a decade earlier in discussing the fertilization of orchids (Darwin, 1862). In the context of reacting to Bell, however, he did not apply this kind of argument to the explanation of emotional expression.

Considering the wealth of behavioural evidence Darwin cited in support of the general idea of evolution, and the variety of ways in which behavioural phenomena were related to key features of his evolutionary theorizing, and his strong claims about the continuity between the mental faculties of the higher animals and man, one might expect that Darwin's work would have launched a vigorous new field devoted to the evolutionary interpretation of behaviour. This is not, however, what happened. The reasons for this are complicated, and evidently include the methodological difficulties posed by the subject of animal behaviour, the reaction occasioned by the anecdotal and anthropomorphic writings of figures such as G. J. Romanes, and the way certain biological fields and styles of research flourished, at the expense of others, late in Darwin's lifetime and after his death. As E. B. Poulton (1890) remarked on the last of these factors, Darwin himself had been primarily interested in questions concerning 'the living animal as a whole and its relations to the organic world'. The 'great impetus given to biological inquiry' by Darwin's work, however, had 'chiefly manifested itself in the domain of Comparative Anatomy, and especially in that of Embryology...' Said Poulton '...There are comparatively few true naturalists – men who would devote much time and the closest study to watching living animals in their natural surroundings, and who would value a fresh observation more than a beautiful dissection of a rare specimen.' (pp. 286–7).

If the existence of 'true naturalists' was necessary for the emergence of an evolutionary science of behaviour, it was not, however, sufficient. An effort had to be put not only into studying living animals in their natural surroundings but also into carving out a niche for a new field of investigation. This kind of dual effort was not really evidenced at the beginning of the twentieth century in the activities of such capable amateurs as George and Elizabeth Peckham, Edmund Selous, or H. Eliot Howard (Burkhardt, 1981b). Nor was it even evidenced in the work of Julian Huxley, a professional biologist. It was evidenced in the 1930s, however, in the endeavours of the young Austrian biologist Konrad Lorenz. Lorenz's commitment is perhaps best conveyed in his correspondence with Oscar Heinroth, the distinguished ornithologist who was at one and the same time Lorenz's clearest role model and greatest supporter. In 1931, for example, Lorenz wrote excitedly to Heinroth saying 'Are you aware, Herr Doktor, that you are actually the founder of a science, namely animal psychology as a branch of biology? That that is the great value of [your book (Heinroth & Heinroth, 1924–33)] "the Birds of Europe"? That it is a question of an approach and a method of research that in fact must be extended to the "animals of the world"?' (K. Heinroth, 1971, p. 155).

Lorenz kept in close contact with Heinroth throughout the 1930s. After the *Deutsche Gesellschaft für Tierpsychologie* was founded in 1936, Lorenz urged Heinroth to collaborate with him in a campaign to ensure that 'the phylogenetic view' had an influence on the new society and the developing field of animal psychology. It was up to the two of them to do this, Lorenz said, since the only other scientists capable of the task were Jan Verwey in Holland, who was occupied with other matters, and Wallace Craig in the United States, who was old and no longer publishing anything. 'Our way of inquiry', Lorenz told Heinroth, 'could possibly be adopted if we only proceed skilfully' (K. Heinroth, 1971, p. 160).

History has confirmed how astute Lorenz was in believing that a skilful campaign could gain ethology an important place in the world of science. It was not Heinroth, however, despite his exceptional knowledge of the particulars of animal behaviour and his encouragement of Lorenz's efforts, who proved to be Lorenz's greatest ally in the development of Lorenz's campaign. Shortly after Lorenz wrote to Heinroth about what the two of them had to do, Lorenz travelled to Leiden to participate in a conference on the subject of instinct. There he met Niko Tinbergen, a naturalist three years his junior, who had developed at the University of Leiden a research programme characterized by the observational and experimental study of animals in their natural environments. It was primarily through the combined efforts of Lorenz and Tinbergen, bolstered by important contributions from a number of other co-workers, that ethology emerged in the next decade and a half with a clear identity of its own. By the early 1950s ethology could be recognized by its subject matter; its special methods, explanatory models, and terminology; its own journals, societies, and congresses; and its institutionalization at universities and research centres (Thorpe, 1979). It had also gained sufficient distinction to have attracted attacks by representatives of other traditions of behaviour study (especially Lehrman, 1953). Lorenz was ethology's father figure. Tinbergen, with his book, *The Study of Instinct* (1951), provided the new discipline with its classic text and gave the rest of the scientific world the clearest single statement of what ethology was all about.

Let us now examine more closely some of ethology's cognitive dimensions and some of the ways in which the very activity of establishing ethology as a new science may have influenced these dimensions.

In the fall of 1936 Lorenz delivered to the new *Deutsche Gesellschaft für Tierpsychologie* a major address calling for the rethinking of the study of animal behaviour. He entitled the address 'The Framing of Biological Questions in Animal Psychology' (1937b). All the natural sciences, Lorenz told his audience, shared a common concern for causal analysis. What was unique about biology, he said, was that it had to address other kinds of questions as well. These additional questions, he maintained, were not being taken into consideration by animal psychologists, and the field of animal psychology was suffering as a consequence. Contemporary animal psychology, he complained, was dominated by a debate between mechanists and vitalists, and the way these two groups opposed each other precluded the framing of the kinds of biological questions the subject of animal behaviour really demanded. If one wanted to understand animal behaviour, Lorenz said, one not only had to subject it to causal analysis, one also had to appreciate three other things about it: first, its purposefulness, that is, its species-preserving function; second, its relation to the whole pattern or *Gestalt* of the animal's natural activities; and third, its evolutionary history.

Lorenz was quite astute in pointing out how, in their opposition to each other, neither the mechanists nor the vitalists were able to deal satisfactorily with phenomena such as the 'purposefulness' of behaviour. What must be considered here are the ways in which the ethologists' opposition to the traditions in contemporary animal psychology structured their own work on animal behaviour.

Interestingly enough, in their early efforts to secure a place for the science of ethology, what the ethologists concentrated upon more than anything else were not the questions that Lorenz identified as being unique to biology, but rather the question of causation. In 1942 Tinbergen described ethology as being represented by '...a whole school of students of animal behaviour, whose principal characteristic is the faith in the preeminent value of causal analysis for a better understanding of behaviour' (p. 39). What the new 'objectivistic ethology' was doing, he said, was 'applying physiological methods to the objects of animal Psychology' (p. 40). Even while indicating that ethologists were also interested in the function, or survival value, of behaviour, Tinbergen stressed the primacy of causal analysis: 'The statement that an activity has a goal or a function, does not satisfy the ethologist; it challenges him to find out, by which causal relations it attains that goal' (p. 92).

This concern with physiological causation can be seen in Lorenz's 'hydro-mechanical' model of the accumulation and consumption of 'action-specific energy' (1950) and in Tinbergen's diagrams of the hierarchical organization of instinctive behaviour (1942, 1950, 1951). Significantly, of the eight chapters of Tinbergen's book, *The Study of Instinct*, the first five were devoted to the subject of causation. What Tinbergen called 'the more or less neglected fields of our science' – the fields dealing with the ontogeny, function, and evolution of behaviour – received only one chapter each.

The ethologists' early emphasis on physiological causation can be explained partly in terms of their feeling that such phenomena as threshold reduction, vacuum activities, displacement activities, and so on simply cried out for causal analysis. Erich von Holst's discovery in the 1930s of the

endogenous production and central coordination of motor impulses provided particular encouragement that instinctive behaviour patterns could
be explained in physiological terms. In addition, though, focussing on
causation served a broad, strategic purpose for the ethologists. It enabled
them to claim to the scientific world that theirs was a perfectly legitimate
science, involving the same kind of objective, experimental, and analytical
methods as the hard sciences. More specifically, it enabled them to
distinguish themselves from such prominent vitalists as the Dutch animal
psychologist J. A. Bierens de Haan (1935, 1940) who had maintained that
instincts were, in the last analysis, inexplicable.

As for the ethologists' work on the function and evolution of behaviour,
the most important early studies were Lorenz's famous 'Kumpan' paper
(1935) and his comparative analysis of the motor patterns of the *Anatinae*
(1941). In his 'Kumpan' paper he discussed – among many other things –
the way in which special structures and behaviour patterns had evolved
in the service of intraspecific communication. In his paper on the *Anatinae*
he showed how motor patterns could be used in reconstructing phylogenies.
While Lorenz promoted the study of behavioural evolution in these ways,
however, it nonetheless appears that not all the lessons of the Darwinian
revolution had been fully impressed upon him. For one thing, his statements
about species revealed virtually no engagement on his part with one of the
key concepts of Darwin's thought, the concept of intraspecific variation.
For another thing, while insisting upon the importance of studying animals
in nature, Lorenz was disinclined to relate the signalling behaviour
of animals to the particular ecological niches the animals occupied. Each
of these attitudes on Lorenz's part appears to have been related to his
acceptance of comparative anatomy as the guiding discipline for evolutionary studies. Each of these attitudes was further reinforced by Lorenz's
desire to distinguish his work on animal behaviour from the work of other
scientists of his day.

Consider first Lorenz's commitment to the comparative anatomical
approach to evolutionary studies. Lorenz undeniably qualifies as a 'true
naturalist' in Poulton's terms, that is, a man willing to 'devote much time
and the closest study to watching animals in their natural surroundings'.
Nonetheless, Lorenz's understanding of what it meant for a science to be
evolutionary was not formed by a direct reading of Darwin or by a direct
engagement with all the kinds of problems that Darwin confronted.
Instead, it was mediated by that turn-of-the-century emphasis on comparative anatomy about which Poulton had complained. Indeed, Lorenz has
stated on numerous occasions that ethology essentially began when

C. O. Whitman (1898) and Oscar Heinroth (1911) discovered that the principles of comparative anatomy could be applied to animal behaviour, that is, that behaviour patterns could be used like animal structures to reconstruct phylogenies.

Historically, comparative anatomy seems to have lent itself rather readily to thinking of species as ideal types rather than as populations of individuals. Whether Lorenz's own strongly typological thinking derived from his training in comparative anatomy at the University of Vienna is, however, impossible to say, for typological thinking was a commonplace in his broader intellectual milieu. It is the case, at any rate, that while the theme of the preservation of the species appears again and again throughout Lorenz's writings, the theme of individual advantages among differently endowed members of the same species does not, even though the existence of heritable variations in populations was an issue to which a number of investigators of animal behaviour at the turn of the century were especially sensitive (Peckham & Peckham, 1898). Lorenz's lack of engagement with the concept of intraspecific variation might be explained in part as a function of the relatively low phenotypic variability of species-specific signals, on which there is a selective pressure to be unmistakable (Mayr, 1974). Or it might simply be dismissed as an incidental oversight compared to the important things Lorenz was studying. What I would like to suggest here is that Lorenz's lack of engagement with the concept of intraspecific variation was in fact functionally related to the special claims he wanted to make about the scientific significance of what he was studying. To appreciate how this worked, it is crucial to observe that Lorenz defined his research not so much by comparing it with what was going on in biology as by comparing it with what was going on in animal psychology.

The animal psychologists of Lorenz's day dealt primarily with learning – learning displayed by domesticated animals under laboratory conditions. Lorenz, in contrast, studied the instinctive behaviour of wild animals living freely. With regard to the respective advantages of his programme and that of the animal psychologists, Lorenz was quick to point out that the animal psychologists were hindered on each count by the *variability* of what they studied: learned behaviour, as opposed to instinctive behaviour, was characterized by its variability; domesticated animals – in addition to being aesthetically repulsive – '[varied] quite unpredictably in their performance of instinctive behaviour patterns'; and as for animals in captivity, the traumatic nature of captivity made them act less uniformly than did animals living freely (1935 [1970, 1, 114–15, 220–21]). The idea of intraspecific variation was not logically precluded by any of these contrasts Lorenz drew, and Lorenz did not deny its existence. Nonetheless, his general orientation was unmistakable: variation was a somewhat unpalatable phenomenon that he associated with the subject matter of the schools of behaviour from which he sought to distinguish himself.

As for Lorenz's lack of attention to ecological influences on the signalling behaviour of animals, this was related to a claim he wished to make about the special value of instincts in reconstructing phylogenies. His claim was that the investigator of instinctive behaviour was 'often able to determine genetic relationships with a degree of accuracy seldom available to the comparative morphologist' (1935 [1970, 1, 249]). What made this possible, he said, was that releasing actions were 'pure conventions' that were 'largely independent of environmental influences' (1937a, p. 254). They therefore could be treated quite reliably as homologies.

Lorenz's emphasis on the study of the instinctive behaviour of free-living, wild animals truly revolutionized animal behaviour studies in the twentieth century. His promotion of Whitman's and Heinroth's idea that instincts could be used for phylogenetic purposes was also very valuable. In each instance, however, he seems to have predisposed himself against an important scientific insight. In contrasting his subject matter with that of the animal psychologists, he stressed the invariability of instinctive behaviour and failed to attend to the phenomenon of intraspecific variation. In emphasizing the value of instincts for studying phylogenies, he aligned himself against the idea that intraspecific signals might be related to the particular selection pressures operating upon species in particular ecological niches.

The ethologist who led the way in the 1950s and 1960s to a broader appreciation of the multiform ways in which animal behaviour is related to environmental pressures was Tinbergen. Unlike Lorenz, Tinbergen did not imbibe the paradigm of comparative morphology in his early career, though he was exposed to it. Instead, he remained first and foremost a field naturalist. When he took up comparative studies in the 1950s, he had already developed a special appreciation of 'the animal in its world' (1972–3), and he was both temperamentally and intellectually prepared to identify and analyse further the particular cases of ecological adaptiveness that he and his students (e.g. Cullen, 1957) discovered. In his studies of adaptive radiation in gulls (1959, 1967), he proceeded to identify environmental influences upon species-specific signals in a way that Lorenz had not. By the early 1960s, Tinbergen was able to report that through their study of 'entire adaptive systems', his group was coming to appreciate

'...how manifold are the selection pressures to which a species is subjected, how conflicting their demands can be, and how the animal as a whole has, under the influence of these selection pressures, developed innumerable compromises.' (1962, p. 5).

In considering the ways in which the cognitive dimensions of ethology corresponded to the social activity of fostering ethology's development as a new science, it is interesting to find that Tinbergen perceived his studies on survival value in animals to be related to the survival of ethology. Indeed, he couched his analysis of ethology's needs in terms very much like those he used in describing the needs of the animal in its world. When he left Leiden in 1949 to take up a new post at Oxford, he found himself with an opportunity for a kind of adaptive radiation of his own. As he later recalled it, in thinking about how to ensure the viability of his new Animal Behaviour Research Group at Oxford, he and his colleagues concluded that it would be better for the group to concentrate in depth on some selected issues than to spread its efforts too thin. They were

...further convinced that ethology would be doomed if it did not attempt to enter certain "no-man's lands" between it and sister disciplines. Of these, however, there are so many...that a small group such as ours would have to choose. We considered that some work on neurophysiological aspects of behaviour ought to be done in any such group, and that for the rest our choice should be guided by opportunity, such as the possibility of close contact with specialists. For this reason we decided to pay special attention to problems of ecology, genetics and evolution, fields in which Oxford can boast to have flourishing schools under Dr A. J. Cain, C. S. Elton, Dr E. B. Ford, Professor A. C. Hardy and Dr D. Lack.

It was also necessary, Tinbergen said, 'to be economical with our resources, such as space and money', and this had some influence on the choice of animals studied. 'These considerations', in Tinbergen's words, 'determined the character of the research programme, which consequently is a compromise between broadness of approach and penetration in depth' (1963, p. 206).

The parallel between Tinbergen's comments on the adaptive compromises displayed by animal species and the adaptive compromises involved in his own group's research programme is an appealing one. It supports the kind of ecological and evolutionary view of scientific activity presented in this paper. Ethology has not been portrayed here as a simple intellectual entity. Indeed, instead of viewing the ethologists as having simply moved into a previously unoccupied niche in the domain of scientific knowledge, it appears more instructive to think of them as having actually *constructed* a niche, and shaped their own ideas in the process.

I have suggested here that focussing on the socially interactive nature of science may help account for a number of cases in the history of the study of behavioural evolution that otherwise appear difficult to explain. Darwin's failure to develop the idea of the communicative function of the expression of the emotions in animals seems strange in light of his customary emphasis on the adaptiveness of characters and his specific appreciation of the function of displays in mating. Seen in terms of the particular challenge to which his book on the expression of the emotions was a response, however, this oversight makes more historical sense. The typological character of Lorenz's comments about species seems out of place for someone who endorsed the idea that natural selection is the primary agent of behavioural evolution. However, given Lorenz's commitment to developing a 'comparative morphology of behaviour', and given the claims he wished to make about the special virtues of studying the innate behaviour of free-living, wild animals, one can see that his inattentiveness to intraspecific variation was consistent with the main orientation of his agenda. As for Lorenz's position on the degree to which species-specific signals are independent of environmental influences, this position appears to have been not just a function of the evidence with which he was most familiar but also a function of his desire to invest the study of instinct with a special role in phylogenetic studies. Tinbergen's efforts in the 1930s and 1940s were directed more toward the study of behavioural causation than toward the study of the function and evolution of behaviour. When he came to the comparative study of species-specific signals in the 1950s, he not only had a background somewhat different from Lorenz's, he had a different view of ethology's needs at that particular time, and he ended up developing an insight that Lorenz's programme had virtually precluded. The very activity of constructing a scientific niche for ethology, it would seem, had conceptual consequences, especially with respect to when, and in what ways, the founders of ethology took up evolutionary questions.

As indicated at the beginning of this paper, Charles Darwin appreciated an instance in his own thinking where the act of distinguishing his views from previously-held views influenced his perception of natural phenomena more than he at first appreciated. Biologists have done well to follow up other insights of Darwin's. Historians of science would do well to follow up this one. To have biologists and historians alike invoking Darwin as a patron saint would be fitting testimony, one hundred years after Darwin's death, to the continuing value of Darwin's insights – and to the continuing need of subsequent investigators to legitimate their own researches. What is more, in contemplating Darwin's insight about how his pursuit of a good cause carried with it a constraint on his thinking, biologists and historians might be inspired to scrutinize their own good causes, and consider the ways in which these might be attended by intellectual constraints.

Further reading

Burkhardt, R. W. (1981). On the emergence of ethology as a scientific discipline. *Conspectus of History*, 1 (no. 7), 62–81.

Burkhardt, R. W. (1983). Darwin on animal behavior and evolution. In *The Darwinian Heritage*, ed. D. Kohn. Princeton, N.J.: Princeton University Press.

Durant, J. R. (1981). Innate character in animals and man: a perspective on the origins of ethology. In *Biology, Medicine and Society 1840–1940*, ed. C. Webster, pp. 157–92. Cambridge: Cambridge University Press.

Mayr, E. (1974). Behavior programs and evolutionary strategies. *American Scientist*, 62, 650–59.

Tinbergen, N. (1963). On aims and methods of ethology. *Zeitschrift für Tierpsychologie*, 20, 410–33.

Tinbergen, N. (1969). Ethology. In *Scientific Thought 1900–1960*, ed. R. Harré, pp. 238–68. Oxford: Clarendon Press.

References

Bell, Sir C. (1844). *The Anatomy and Philosophy of Expression, as Connected with the Fine Arts*. Third edition. London: John Murray.

Bierens de Haan, J. A. (1935). Probleme des tierischen Instinktes. *Die Naturwissenschaften*, 23, 711–17, 733–37.

Bierens de Haan, J. A. (1940). *Die tierischen Instinkte und ihr Umbau durch Erfahrung*. Leiden: E. J. Brill.

Burkhardt, R. W. (1981a). Lamarck's understanding of animal behavior. In *Lamarck et son temps, Lamarck et notre temps*, Centre de Recherche sur l'Histoire des Idées de l'Université de Picardie, pp. 11–28. Paris: Librairie Philosophique J. Vrin.

Burkhardt, R. W. (1981b). On the emergence of ethology as a scientific discipline. *Conspectus of History*, 1 (no. 7), 62–81.

Burkhardt, R. W. (1983). Darwin on animal behaviour and evolution. In *The Darwinian Heritage*, ed. D. Kohn. Princeton, N. J.: Princeton University Press.

Cullen, E. (1957). Adaptations to cliff-nesting in the kittiwake. *Ibis*, 99, 275–302.

Darwin, C. (1859). *On the Origin of Species by Means of Natural Selection*. London: John Murray.

Darwin, C. (1862). *On the Various Contrivances by which British and Foreign Orchids are Fertilized by Insects, and on the Good Effects of Intercrossing*. London: John Murray.

Darwin, C. (1871). *The Descent of Man, and Selection in Relation to Sex*. 2 vols. London: John Murray.

Darwin, C. (1872). *The Expression of the Emotions in Man and Animals*. London: John Murray.

Ghiselin, M. T. (1969). *The Triumph of the Darwinian Method*. Berkeley and Los Angeles: University of California Press.

Heinroth, K. (1971). *Oskar Heinroth. Vater der Verhaltensforschung, 1871–1945*. Stuttgart: Wissenschaftliche Verlagsgesellschaft.

Heinroth, O. (1911). Beiträge zur Biologie, namentlich Ethologie und Psychologie der Anatiden. *Verhandlungen des V. Internationalen Ornithologischen-Kongresses, Berlin 1910*, pp. 589–702.

Heinroth, O. & Heinroth, M. (1924–33). *Die Vögel Mitteleuropas in allen Lebens- und Entwicklungsstufen photographisch aufgenommen und in ihrem Seelenleben bei der Aufzucht vom Ei ab beobachtet*. 4 vols. Berlin, Lichterfelde: H. Bermühler.

Lehrman, D. S. (1953). A critique of Konrad Lorenz's theory of instinctive behaviour. *Quarterly Review of Biology*, **28**, 337–63.

Lorenz, K. Z. (1935). Der Kumpan in der Umwelt des Vogels. *Journal für Ornithologie*, **83**, 137–213, 289–413.

Lorenz, K. Z. (1937a). The companion in the bird's world. *The Auk*, **54**, 245–73.

Lorenz, K. Z. (1937b). Biologische Fragestellung in der Tierpsychologie. *Zeitschrift für Tierpsychologie*, **1**, 24–32.

Lorenz, K. Z. (1941). Vergleichende Bewegungsstudien an Anatinen. *Journal für Ornithologie*, **89** (Ergänzungsband III), 194–293.

Lorenz, K. Z. (1970–71). *Studies in Animal and Human Behaviour*. 2 vols. Cambridge, Mass.: Harvard University Press.

Lorenz, K. Z. (1981). *Foundations of Ethology*. New York: Springer Verlag.

Mayr, E. (1963). *Animal Species and Evolution*. Cambridge, Mass.: Harvard University Press.

Mayr, E. (1972). Sexual selection and natural selection. In *Sexual Selection and the Descent of Man 1871–1971*, ed. B. G. Campbell, pp. 87–104. Chicago: Aldine Publishing Company.

Mayr, E. (1974). Behavior programs and evolutionary strategies. *American Scientist*, **62**, 650–9.

Peckham, G. W. & Peckham, E. G. (1898). On the instincts and habits of the solitary wasps. *Bulletin of the Wisconsin Geological and Natural History Survey*, **2**, 1–245.

Poulton, E. B. (1890). *The Colours of Animals*. New York: D. Appleton and Company.

Richards, R. J. (1981). Instinct and intelligence in British natural theology: some contributions to Darwin's theory of the evolution of behavior. *Journal of the History of Biology*, **14**, 193–230.

Selous, E. (1927). *Realities of Bird Life. Being Extracts from the Diaries of a Life-Loving Naturalist*. London: Constable.

Thorpe, W. H. (1979). *The Origins and Rise of Ethology*. London: Heinemann.

Tinbergen, N. (1940). Die Uebersprungbewegung. *Zeitschrift für Tierpsychologie*, **4**, 1–40.

Tinbergen, N. (1942). The objectivistic study of the innate behaviour of animals. *Bibliotheca biotheoretica*, **1**, 39–98.

Tinbergen, N. (1948). Social releasers and the experimental method required for their study. *Wilson Bulletin*, **60**, 6–52.

Tinbergen, N. (1950). The hierarchical organisation of nervous mechanisms underlying instinctive behaviour. *Symposia of the Society for Experimental Biology*, **4**, 305–12.

Tinbergen, N. (1951). *The Study of Instinct*. Oxford: Clarendon Press.

Tinbergen, N. (1952). "Derived" activities: their causation, biological significance, origin, and emancipation during evolution. *Quarterly Review of Biology*, **27**, 1–32.

Tinbergen, N. (1959). Comparative studies of the behaviour of gulls (Laridae): a progress report. *Behaviour*, **15**, 1–70.

Tinbergen, N. (1962). The evolution of animal communication – a critical examination of methods. *Symposia of the Zoological Society of London*, **8**, 1–8.

Tinbergen, N. (1963). The work of the Animal Behaviour Research Group in the Department of Zoology, University of Oxford. *Animal Behaviour*, **11**, 206–9.

Tinbergen, N. (1967). Adaptive features of the Black-headed Gull *Larus ridibundus* L. *Proceedings of the XIV International Orinithological Congress*, 43–59.

Tinbergen, N. (1972–73). *The Animal in its World. Explorations of an Ethologist 1932–1972*. 2 vols. London: George Allen & Unwin.

Whitman, C. O. (1898). Animal Behavior. *Biological Lectures for the Marine Biological Laboratory, Woods Hole*, pp. 285–338.

Wynne-Edwards, V. C. (1962). *Animal Dispersion in Relation to Social Behaviour*. Edinburgh and London: Oliver and Boyd.

22

Game theory and the evolution of cooperation

J. MAYNARD SMITH

Evolutionary game theory and conventional behaviour

Maynard Smith and Price (1973) introduced the concept of an 'Evolutionarily Stable Strategy', or ESS, in a paper analysing the evolution of animal conflicts. In this paper, I use the same concept to analyse the evolution of cooperation, and ask whether analogous models can be useful for thinking about the stability of societies. First, I introduce the basic ideas of evolutionary game theory; readers already familiar with these ideas should skip to the next section.

In the simplest model, we imagine a population of individuals pairing off at random, and playing a 'game' in which each has a choice of two strategies, H and D (Hawk and Dove); H is an aggressive and risky strategy, and D an unaggressive and safe one. The results of the game can be summarized in a 'payoff matrix' (Table 22.1). By a 'payoff' is meant the change in Darwinian fitness of an individual resulting from the contest. The actual values are arbitrary, but the argument to follow depends on the inequalities being as in the matrix, as would probably be the case for an aggressive and a safe strategy.

Each animal adopts either a 'pure' strategy, H or D, or a 'mixed' strategy,

Table 22.1. *Payoff matrix for the Hawk–Dove game*

The values in the matrix are the payoffs to an individual adopting the strategy on the left, given that his opponent adopts the strategy above.

	H	D
H	−1	2
D	0	1

'play H with probability P; play D with probability $1-P$'. After one or a number of contests against random opponents, a new generation is produced. Each strategy type breeds true, and each individual produces a number of offspring proportional to its accumulated payoff. We ask how such a population will evolve. In particular, we seek an ESS, which has the property that, if almost all members of the population adopt it, no alternative, mutant, strategy can invade. The mathematical criteria for uninvadability are discussed in Maynard Smith (1982).

For the matrix of Table 22.1, it is easy to see that H is not an ESS, because a population of Hawks can be invaded by D; similarly, D is not an ESS. The only ESS of the matrix is 'Play H with $P = \frac{1}{2}$; play D with $P = \frac{1}{2}$'; that is, it is a mixed strategy. For our present purpose, however, the interesting question is what happens if each contest has an asymmetry perceivable by the contestants. Suppose, for example, all contests are between 'owners' and 'intruders'. For simplicity, assume that ownership does not affect payoffs, or the likelihood of winning an escalated contest. Assume also that animals adopting different strategies are equally likely to be owner or intruder at the start of a contest.

Given the asymmetry, new strategies dependent on it become possible. One such is the strategy B (Bourgeois), 'play H if owner; play D if intruder'. The new payoff matrix is shown in Table 22.2. The only ESS of this matrix is B, which cannot be invaded either by H or D. Hence, even in the absence of learning or rational behaviour, asymmetric cues can be used to settle contests conventionally. It is a consequence of the conventional behaviour that the average payoff per contest is higher than it would be at the mixed ESS in the absence of an asymmetry. It is important, however, that an ESS is a consequence of selection acting at the individual level; individuals do not calculate or attempt to maximize the group payoff.

Table 22.2. *Payoff matrix for the asymmetric Hawk–Dove game*

	H	D	B
H	-1	2	$\frac{1}{2}$
D	0	1	$\frac{1}{2}$
B	$-\frac{1}{2}$	$1\frac{1}{2}$	1

The evolution of cooperation – basic mechanisms

Animals sometimes act cooperatively; i.e., they do things which favour the fitness (survival and reproduction) of others as well as, or even at the expense of, their own fitness. A number of selective mechanisms have been suggested for the evolution of such behaviour, but two – kin selection and mutualism – seem likely to have been most important.

An appreciation of the importance of kin selection we owe primarily to Hamilton (1964 and later). Essentially, the idea is that a gene whose presence reduces the fitness of the individual in which it finds itself can nevertheless increase in frequency in the population if it so influences the behaviour of its carrier that relatives of that carrier become more likely to survive and/or reproduce. I shall not be concerned with kin selection further here, and will make only two comments. First, the 'genes-eye view' of the matter expressed above seems to me the easiest way of understanding what is happening (see, particularly, Dawkins (1976); for a contrary view, see Cavalli-Sforza & Feldman (1978)). Second, the main reason for thinking that kin selection has been an important mechanism in the evolution of cooperation is that most animal societies are in fact composed of relatives.

The role of mutualism in the evolution of social behaviour has, in recent years, been emphasized particularly by West-Eberhard (1975). I want here to present a somewhat more formal model of what is required. Consider first a case in which individuals interact in pairs. During an interaction, an individual can cooperate, C, or defect, D. The effects on fitness are shown in Table 22.3.

Table 22.3. *Additive and synergistic interactions*

Types of pair	C+C		C+D		D+D	
Individual fitnesses						
additive	3	3	1	4	2	2
synergistic	5	5	1	4	2	2

	Additive			Synergistic	
	C	D		C	D
C	3	1	C	5	1
D	4	2	D	4	2

It is supposed that a pair of defectors each acquires a payoff of 2. If one of the pair is a cooperator, it loses one unit, and benefits its partner by two units. If both cooperate, two assumptions can be made. On the 'additive' assumption, each cooperator loses one unit and benefits its partner by two units, just as if its partner were D. On the synergistic assumption there is a more-than-additive advantage of cooperation. It is not hard to imagine situations in which this would be true; for example, two male lions may be able to take over and hold a pride, but one cannot.

The payoffs can be shown in matrix form. The additive case corresponds to the classical 'prisoner's dilemma' game. The only ESS is to defect; cooperation is not stable. In the synergistic case both pure D *and* pure C are ESSs. Cooperation is stable if once it can evolve. However, as pointed out by Charlesworth (1979), it is hard to get cooperation started without kin selection; C cannot invade D.

The argument can readily be extended to interacting groups of more than two. If we want to know the stability of a population of C individuals (and not the dynamics of mixtures of C and D), all we need are the payoffs to a C individual, and to a D individual, in a population consisting otherwise entirely of C. If the former is larger, then cooperation is stable. Common sense suggests that cooperation is more likely to be stable in small groups; a beneficial synergistic effect requiring the cooperation of all the members of a group is more plausible for a small group.

Cooperation in repeated interactions

In the model of mutualism just discussed, it was assumed that an individual plays either C or D, and cannot modify its behaviour in the light of what its partner does. Suppose, in contrast, that the strategy of 'conditional cooperation', C/C or 'cooperate if partner cooperates, otherwise defect', is possible. The payoff matrices are shown in Table 22.4.

In both cases, conditional cooperation is the only ESS. Of course, one

Table 22.4. *Payoff matrices when conditional cooperation, C/C, is possible*

	Additive			Synergistic	
	C/C	D		C/C	D
C/C	3	2	C/C	5	2
D	2	2	D	2	2

animal cannot make a conditional choice until the other has revealed its intentions. Hence it is better to model the interaction as a series of games between the same two animals. This, in effect, is what Trivers (1971) did in his paper on 'reciprocal altruism'. Acknowledging a suggestion of Hamilton's, he pointed out that the problem could be treated as follows. In a single game the payoffs are as in Table 22.5a. Then, in a series of n games, the payoffs are as in Table 22.5b. C/C is a strategy which cooperates in the first game, but defects if its partner defects. Provided n is large, C/C is an ESS. However, D is also an ESS, so we have not escaped the problem of how to get mutualism started.

More recently, Axelrod & Hamilton (1981) have pursued the repeated Prisoner's Dilemma further. They suppose that after each game there is a constant probability w of a further game. This avoids the complications that arise if players know they are playing the last game of the series. They show that the strategy TIT FOR TAT (similar to C/C above; i.e., cooperate in first game, and in subsequent games do what your opponent did in the last game) is an ESS of the game, provided w is large enough. They show this for all possible alternative strategies, not just for the alternatives 'always C' and 'always D'.

Trivers thought of reciprocal altruism evolving in species with the intelligence to remember the behaviour of particular partners. Clearly, the cooperative ESS does not exist in a series of games against different opponents. Axelrod & Hamilton point out that cooperation will evolve if an animal interacts with only one other during its life, whether or not the species can learn and remember individuals. Thus cooperative interactions should be commoner in sessile species.

Reciprocal altruism as envisaged by Trivers requires that an animal which interacts with a number of others should be able to treat each individual differently. Packer's (1977) study of coalitions in olive baboons shows that some animals have this ability. Jon Seger has pointed out to me that cooperation can be stable for the Prisoner's Dilemma game played

Table 22.5. *Payoffs for the repeated Prisoner's Dilemma*

		(a) single games		(b) n games against the same opponent	
	C	D		C/C	D
C	R	S	C/C	nR	$S+(n-1)P$
D	T	P	D	$T+(n-1)P$	nP

where $S < P < R < T$ defines the Prisoner's dilemma game.

against a succession of different opponents, provided that it occurs in a social context, so that the strategy 'cooperate only with those opponents who have been seen to cooperate with others' is a possibility. Such a strategy may not be beyond the capacity of non-human primates: in man, it is the reason why 'he that filches from me my good name, robs me of that which not enriches him, and leaves me poor indeed'.

The stability of queues

So far I have been writing about animals. I turn now to games played by people. Instead of seeking evolutionarily stable states, we have to seek states which are stable if behaviour is determined by perceived self-interest, subject to custom and to legal constraints. I start by discussing the queuing game, which has the advantage of being played by both men and animals. This will illustrate the complexity even of a very simple game.

I then turn to what I call the 'social contract' and 'class war' games. The choice of names should make it clear that I do not claim to have invented these games. My hope is that, by presenting very simple formal models, it may be possible to reveal something of the logical structure of the real world, in the same way that the Hawk–Dove game, although enormously simpler than any actual animal contest, does help us to think about actual contests.

In some countries, and in some circumstances, queues are stable; in England, queues at bus stops are usually stable but queues never form at bars. A particular form of queue, which I will call an age queue, is known to exist in some social animals, and may be commoner than we have realized. To give an example (Wiley, 1981), striped wrens live in groups containing males and females. Only one pair breeds. Young males usually stay in the group into which they were born; females transfer to another group if an opportunity arises to become a breeding female. The males form an age queue, with the oldest at the head and the youngest at the tail; the bird at the head of the queue will take over as breeding male if an opportunity arises. Occasionally a male will leave a group, pair with a female, and the two birds will start breeding as a pair; the success of such pairs is low, mainly because of predation on the young. Age queues are rather stable, and it is rare for the order to be disturbed.

What mechanisms might maintain the stability of a queue? I can think of at least five possibilities.

(1) The order corresponds to fighting ability, so that challenges would not pay. This is not true for bus queues, and probably is not true of age queues in birds.

(2) The risks of injury are large compared to the benefits of winning. Hierarchical position in bachelor herds of red deer out of the rutting season is a poor predictor of success in fighting during the rut (Gibson, 1978), and is probably stable because it is not worth fighting about. The case is similar to the 'Bourgeois' strategy discussed earlier; conventional acceptance of a position in a queue resembles conventional acceptance of ownership, and is stable only if the cost of fighting is greater than the value of winning.

(3) If any individual challenges another this is collectively resisted by other members of the queue, because it is individually advantageous for them to do so. For example, imagine a queue of three individuals, in the order $\alpha - \beta - \gamma$, and suppose that γ challenges α. It would pay β to help resist the challenge, because if he does not the order might become $\gamma - \alpha - \beta$, and β has dropped to the bottom of the queue. It is less clear that all possible challenges should be resisted in this way. However, analysis suggests that collective resistance might be individually advantageous, particularly if it is assumed that a complete break-up of the queue would be disadvantageous to all. Despite the apparent plausibility of this mechanism, I know of no evidence that age queues in animals are maintained in this way.

(4) Any attempt to jump the queue is collectively resisted, because of the existence of a 'social contract' to do so. I discuss the stability of social contracts in the next section; I would not expect to find such behaviour in animals.

(5) The queue is stabilized by factors external to itself. This also is plausible only for human queues, when the stabilizing factor could be legal enforcement or regard for reputation.

In the present context, the main interest of queuing is that it is a game played by both men and animals.

The social contract game

Imagine a small population of n individuals. Each individual is free to 'cooperate', C, or to 'defect', D. If we are interested only in the stability or otherwise of particular behaviour, we need to know the payoffs to an individual doing C or D, depending on whether the remaining $n-1$ individuals are doing C or D. Suppose the payoffs are as in Table 22.6a.

I have chosen payoffs corresponding to the classic Prisoner's Dilemma. An animal population playing such a game would evolve to D, defect (I am ignoring the possibility of repeated contests). In the absence of further social controls, rational individuals would also adopt D.

Suppose now that each of n rational individuals agree to the following social contract: 'I will play C; if any other individual plays D, I will join in punishing him'. Suppose further that there is a cost of -20 for being punished, and of -1 for helping to punish. Let the act of punishing be symbolized by P. Then the payoffs to various individual strategies, in a population adopting both C and P, are given in Table 22.6*b*.

It will be apparent that 'C and P' is not stable. There is nothing to prevent individuals cheating by taking the benefits of cooperation, but not paying the (small) cost of punishing; this is the classic problem of the 'free-rider'. To make the system stable, an additional clause must be added to the contract, so that it becomes 'I will play C; if any other individual plays D, I will join in punishing him; if any other individual fails to join in punishing, I will treat this as equivalent to D'.

I have assumed that the contract is reached by rational means, requiring symbolic communication. If so, the solution is not available to animals other than man. It is conceivable that, for some cooperative act C, an innate behaviour corresponding to the contract might exist, but it seems implausible, and I can think of no examples. Social contracts clearly do exist between people, but it is rare for them to be enforced solely by legalistic means. Socially acceptable norms of behaviour are enforced by elaborate systems of myth and ritual, and 'punishment' more often takes the form of social disapproval than of physical restraint.

Before turning to a more realistic model of human cooperation, let me clarify the structure of the social contract game. It has three components.

(1) A 'strategy set' (C, D, P), corresponding to the phenotype set in

Table 22.6. *Payoffs in the social contract game*

(*a*) Payoffs to an individual in a group

| | | Policy of rest of group | |
		C	D
Payoff to	C	20	0
individual	D	30	10

(*b*) Payoffs to an individual in a group adopting the policy, 'cooperate and punish'

		Payoff
Individual	C and P	$20 - 1 = 19$
strategy	D	$30 - 20 = 10$
	C, not P	20

evolutionary game theory. The model says nothing about possible additional strategies – for example, appear to cooperate, but actually defect. That is, the strategy set is an assumption of the model, not a conclusion from it.

(2) A set of payoffs. Thus it is assumed that a group of cooperators do better than a group of defectors, and that 'punishment' is cheap for each of a group of punishers, but expensive for the punished. These payoffs are outside the control of the players, and will change if the techniques available to the society change.

(3) A 'contract'. In a sense, this is not a logically necessary part of the model. So long as all members of a group adopt 'C and P' then the strategy is stable, regardless of how they came to adopt it. In man, however, adoption of such a strategy will involve law, custom, reasoning, and so on.

The class war game

The weakness of the preceding game as a model of human society lies in the absence of any division of the population into social classes. For the present, and in the context of the model, social classes can be thought of as groups with different available strategy sets. For example, men who own land or factories have a different strategy set from men who do not, and groups of men with weapons have different sets from unarmed men.

In constructing a model of a society containing more than one group, one must allow for the possibility that the members of a particular group may acquire a common strategy, perceived to be in their common interest, and enforced by custom and/or a process of 'policing' internal to the group; such a strategy would be the analogue of the social contract strategy considered in the last section.

The simplest possible game of this type – with two groups, and two strategies available to members of each – is shown in Table 22.7. To make it easier to grasp, I present it as a game between factory owners and workers, but the basic logic is more general. Individual owners can pay high wages (PH) or low wages (PL); individual workers can accept low wages (AL), or accept only high wages (AH). Owners who can find no workers, or workers who can find no jobs, get zero payoff; owners who get workers for low wages get 4, and for high wages get 3; workers who get jobs for low wages get 1, and for high wages get 2.

To find the stable states of such a game, it may be sufficient to know the payoffs to workers when all owners behave alike, and to owners when all workers behave alike. The relevant payoffs are shown in the matrix. The game has two stable states. (1) All owners pay low wages and all workers accept them. (Given this state of affairs, it would not pay individual workers to insist on high wages, or individual owners to pay them.) (2) All owners pay high wages and all workers accept them.

Clearly, workers prefer state 2, and owners state 1. If either group can exercise greater political power (by the acquisition of the appropriate common strategy), it can ensure that the stable state reached will be the one favourable to itself. Notice also that, whichever state the society is in, owners do better than workers. Hence it would pay individual workers to become owners, if that is a possible strategy.

Hence, if we want to understand the stability of such a system, we have to ask which group(s) benefit from the status quo, how they exercise power, and how group membership is controlled. To understand what leads to change, we must ask which groups benefit from change, and whether there are technical developments which lead to changes in payoffs and hence destabilize a previously stable system. We also have to ask how group membership is determined.

If we ask how such a model relates to the real world, the critical question is: what constitutes a group? I started by supposing that a social class was a group with an available strategy set different from other groups. But this will not quite do. Thus a group of armed men would, on this definition, be a social class. Sometimes groups of armed men do act collectively in the political arena in their own interests: the praetorian guard once auctioned the Roman empire to the highest bidder. But it is not the case that all

Table 22.7. *The class war game*

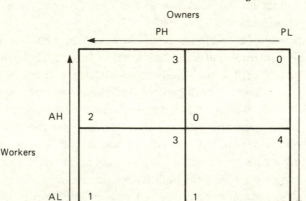

societies are military dictatorships, as too literal an interpretation of the model suggests they should be.

What are the criteria whereby a group comes to recognize itself as capable of independent political activity? A Marxist might insist that such a group must have a specific relationship to the means of production. This is often – perhaps usually – the case, but it is not necessary. In so far as Catholics and Protestants in Northern Ireland have different economic circumstances, this is a consequence and not a cause of their recognizing themselves as political groups. Still more relevant to the Marxist argument, the communist party in Russia was not, at the time of its origin, defined by any special role in production, yet it is becoming a group which acts in its own interests, acquiring special economic privileges, and, significantly, providing educational and career advantages to its children.

A point which inevitably strikes a biologist in this context is that, almost always, groups which play an independent political role are also breeding groups: that is, children tend to inherit the group membership of their parents. There are two obvious reasons for this correspondence. First, people's political aspirations are often aspirations for their children; second, children acquire their value systems partly from their parents. The correspondence has some interesting consequences. It may be one reason why women have found it hard to act as a political unit, despite their manifest common interests. It has also provided a means whereby particular groups, whose political or military power might lead them to act in their own interests at the expense of others, can be inhibited from doing so. A group without children is less of a political threat. Hence the role of eunuchs in Byzantium, and of the Janisseries under the Turks. The significance of celibacy in the Catholic church is more subtle, but may be related to the same point.

My remarks about human societies are not intended to be taken too seriously. What is intended seriously is the idea that simple models have done much to illuminate evolutionary biology, and may do the same in the human sciences. Game theory was borrowed by biologists from sociology; it would be nice if we could return it, somewhat sharpened up.

Further reading

Axelrod, R. & Hamilton, W. D. (1981). The evolution of cooperation. *Science*, **211**, 1390–6.

Maynard Smith, J. (1982). *Evolution and the Theory of Games*. Cambridge: Cambridge University Press.

References

Axelrod, R. & Hamilton, W. D. (1981). The evolution of cooperation. *Science,* **211,** 1390–6.

Cavalli-Sforza, L. L. & Feldman, M. W. (1978). Darwinian selection and altruism. *Theoretical Population Biology,* **14,** 268–80.

Charlesworth, B. (1979). A note on the evolution of altruism in structured demes. *American Naturalist,* **113,** 601–5.

Dawkins, R. (1976). *The Selfish Gene.* Oxford: Oxford University Press.

Gibson, R. M. (1978). Behavioural factors affecting reproductive success in red deer stags. Ph.D. thesis. University of Sussex.

Hamilton, W. D. (1964). The genetical evolution of social behaviour. *Journal of Theoretical Biology,* **7,** 1–32.

Maynard Smith, J. (1982). *Evolution and the Theory of Games.* Cambridge: Cambridge University Press.

Maynard Smith, J. & Price, G. R. (1973). The logic of animal conflict. *Nature,* **246,** 15–18.

Packer, C. (1977). Reciprocal altruism in *Papio anubis. Nature,* **265,** 441–3.

Trivers, R. L. (1971). The evolution of reciprocal altruism. *Quarterly Review of Biology,* **46,** 35–57.

West-Eberhard, M. J. (1975). The evolution of social behaviour by kin selection. *Quarterly Review of Biology,* **50,** 1–33.

Wiley, R. H. (1981). Social structure and individual ontogenies: problems of description, mechanism, and evolution. In *Perspectives in Ethology,* **4,** ed. P. P. G. Bateson & P. H. Klopfer. New York: Plenum.

23

Selection in relation to sex

T. H. CLUTTON-BROCK

In many animal species, differences between the sexes are pronounced. In fact, there are relatively few morphological, physiological or behavioural characteristics that do not differ to some extent between males and females (see Glucksman, 1974) and the extent of these differences varies widely between species. For example, sexual differences in body size range from species, like worms of the genus *Bonnellia*, where females can be over 25 times the length of males (Barnes, 1974) to species like the southern elephant seal, *Mirounga leonina*, where mature males average eight times the weight of females (Bryden, 1969). Other sex differences are less apparent – such as those in fat deposition, in haemoglobin levels and in metabolic rate among mammals (see Glucksman, 1974) or those in auditory apparatus among frogs (Narins & Capranica, 1976, 1978). And many have only recently been explored: for example, recent research shows that in some species there are pronounced sex differences in feeding ecology (Gautier-Hion, 1980) as well as in the effects of starvation on survival (Widdowson, 1976).

It was to provide an explanation for the evolution of sex differences that Charles Darwin formulated the theory of sexual selection, described first in *The Origin of Species* (1859) and later, in greater depth, in *The Descent of Man* (1871). In this chapter, I briefly trace the development of the theory and describe recent attempts to measure variation in reproductive success in males and females, as well as some of the practical problems involved. However, as I argue in the final section, the extent of sexual dimorphism will depend not on the extent to which reproductive success varies in the two sexes but on the comparative effects of particular phenotypic traits on the breeding success of males and females.

A brief history of sexual selection

The origin of the theory of sexual selection can be traced to a peacock – or, rather, a peahen owned by Lady Tynte. This bird, born around 1762, lived in such comfortable circumstances that it had already reared eight broods when, to the consternation of its noble owner, it suddenly developed the plumage and spurs of a male and thereafter refused to lay another egg. No possible confusion of identities could have occurred since Lady Tynte was able to recognize her favourite by the nobs on its toes, which were unaffected by its change in appearance (Hunter, 1837).

It is not known whether Lady Tynte investigated just what had happened to the bird's genitalia but subsequent studies of female pheasants showing similar transvestite tendencies revealed that only sexual differences developing at or after puberty were affected and that the reproductive organs themselves remained unaltered (Yarrell, 1827). This finally led John Hunter, the eminent surgeon, anatomist and classifier of monsters to produce a seminal paper (*An account of an extraordinary pheasant*, Hunter, 1837) in which he proposed that differences between the sexes were of two kinds: those involving the sexual organs themselves, which were evident from birth and did not change during an individual's lifetime; and those that did not develop until the animal approached breeding age, such as differences in body size, plumage and in the tendency to be fat which he termed 'secondary' marks or characters of sex (Hunter, 1837, 1861). Some of these, like the plumage of gallinaceous birds, could change during an individual's lifetime but these differences were not evident at birth.

Hunter realized both that 'secondary' sexual characters were functionally related to fighting or display and that their extent varied with ecology.

The males of almost every class of animals are probably disposed to fight, being, as I have observed, stronger than the females; and in many of these there are parts destined solely for that purpose, as the spurs of the cock, and the horns of the bull...

One of the most general marks (of sex) is the superior strength of make in the male; and another circumstance, perhaps equally so, is this strength being directed to one part more than another, which parts is that most immediately employed in fighting. This difference in external form is more particularly remarkable in the animals whose females are of a peaceable nature, as are the greatest number of those which feed on vegetables, and the marks to discriminate the sexes are in them very numerous. (Hunter, 1837)

Hunter's distinction between primary and secondary sexual characters was adopted by Charles Darwin (1871) with a subtle distinction. Darwin was aware that many 'secondary' sex differences were evident at birth (or hatching) and distinguished between the two categories on functional rather than on ontogenetic grounds. Darwin's primary sexual characters were those connected with the act of reproduction itself while secondary sexual characters were used in acquiring mating partners. To these two categories, Darwin added a third: sex differences 'related to different habits of life, and not at all, or only indirectly, to the reproductive functions', among which he included structures associated with sex differences in feeding behaviour (Darwin, 1871).

The theory of sexual selection was intended to provide an explanation only of secondary sexual characteristics. Darwin realized that many secondary sexual differences were a consequence of the greater intensity of competition between males for access to mates and that many traits were more highly developed in males either because they conferred an advantage in fights or because they rendered their possessor more attractive to females.

...when the females and males of any animal have the same general habits of life, but differ in structure, colour, or ornament, such differences have been mainly caused by sexual selection: that is by individual males having had, in successive generations, some slight advantage over other males, in their weapons, means of defence, or charms which they have transmitted to their male offspring alone.

He distinguished sexual selection from natural selection on two grounds: first, that it was a consequence of competition between members of the same sex rather than between members of different sexes or species; and, second, that it depended on variation in reproductive success rather than survival. '... This form of selection depends, not on a struggle for existence in relation to other organic beings or to external conditions, but on a struggle between the individuals of one sex, generally the males, for the possession of the other sex. The result is not death to the unsuccessful competitor but few or no offspring.'

Darwin's theory of sexual selection was less readily accepted by scientists than the theory of natural selection. Wallace (1889) agreed that combat between males was an important source of selection pressures leading to sexual dimorphism but regarded this as a form of *natural* selection on the grounds that it increased 'the vigour and fighting power of the male animal, since, in every case, the weaker are either killed, wounded or driven away'. He regarded Darwin's second mode of sexual selection – female choice of particular males – as unimportant on the grounds that any consequences which female choice might have would be annulled by natural selection – unless females selected the fittest males, in which case the results of sexual and natural selection would be inseparable. He also pointed to the lack of evidence of consistent female choice for mates carrying particular characteristics. Some fifty years later, the same points were reaffirmed in two influential papers by Huxley (1938a, b).

Wallace's objection that sexual selection is a form of natural selection is semantically correct – after all, Darwin originally coined the term 'natural selection' in order to mark its relation to *man's* power of selection, and the opposite of natural selection is not sexual but artificial selection (see Brown, 1975; Halliday, 1978). However, his insistence that the process of sexual selection described by Darwin could only increase the average reproductive success or survival of males is clearly wrong (Lande, 1980). Especially in polygynous species, the costs of combat are frequently high (Geist, 1971; Clutton-Brock, Albon, Gibson & Guinness, 1979) and so, too are the costs of many sexually dimorphic characters associated with combat, such as increased male body size and weapon development (Clutton-Brock, Guinness & Albon, 1982): in species where males are substantially larger than females, both growing and adult males are often more likely to die than females (Robinette, Gashwiler, Low & Jones, 1957; Grubb, 1974; Howe, 1977) and in one reindeer population which crashed from 6000 to 42, only one of the remaining adults was a male (Klein, 1968). Sexual selection on males may also reduce the average fitness of females (Lande, 1980): in species where adult males are substantially larger than females, producing sons appears to depress the mother's subsequent reproductive success more than producing daughters (Clutton-Brock, Albon & Guinness, 1981).

Wallace's theoretical objections to the importance of female choice as a source of sexual selection on males can also be discounted. R. A. Fisher (1930) demonstrated that female choice for particular male characteristics (such as tail size) can cause them to develop to a point at which they reduce the average fitness of males. Subsequent treatments have confirmed Fisher's conclusions (O'Donald, 1980; Lande, 1981) and shown that the process need not depend on the initial female preference favouring more viable males (Kirkpatrick, 1982). However, while there is extensive evidence of assortative mating (O'Donald, 1980), of the importance of plumage characteristics in courtship (Williams, 1982) and of female preference for males who can defend superior breeding territories (Pleszczynska, 1978) only very recently has it been demonstrated that consistent female choice for any continuous morphological character in males is an important source of variation in male reproductive success. By experimental manipulation of tail length in widow birds of the African genus *Euplectes*, Andersson has been able to alter both the extent to which males are favoured by females and their mating success (Andersson, 1982). This scarcity of evidence does not mean that the evolution of secondary sexual characters through female choice is uncommon, for mating preferences are usually difficult to demonstrate, particularly where inter-male competition is also involved. Nevertheless, the possibility remains that, as Wallace argued, many of the sex differences in plumage and coloration ascribed by Darwin to the action of female choice may have evolved because they help the sexes to recognize or locate each other or because they improve male success in competitive interactions.

Measures of sexual selection

Darwin was not specific as to why males should typically compete more strongly for access to breeding partners than females and it was left to biologists of this century to provide the answer (Fisher, 1930; Bateman, 1948; Trivers, 1972). The reason why males usually compete more intensely is most easily understood by considering the energetic costs of reproduction to each sex. In most animals, the energetic costs of fertilization to the male are minimal whereas the costs of reproduction to the female are substantial. Consequently, males are capable of fathering more progeny than females can bear and rear: in current terminology (Trivers, 1972) they invest less heavily in their offspring than females. In species where successful males can monopolize breeding access to large numbers of females but similar numbers of males and females are recruited, direct competition between males is likely to be intense, aggressive interactions may be frequent and the selective advantages of possessing traits that affect success in combat may be higher among males than among females.

However, neither Trivers' explanation of the prevalence of increased competition among males nor Darwin's description of the theory of sexual selection provide an operational definition of the intensity of sexual selection (Wade, 1979). Many different definitions of sexual selection have been proposed but two kinds are in common use. First, some workers argue that the intensity of sexual selection will depend on the *relative* variability of reproductive success among males and females. For example, Ralls (1977) argues that 'the intensity of intrasexual selection in a species should be proportional to the ratio of the lifetime number of offspring sired by a highly successful male compared to the number born by a highly successful female in her lifetime' while Payne (1979) suggests that the extent to which variance in breeding success differs between the sexes is important.

This position is also sometimes mistakenly attributed to Bateman (see Wade & Arnold, 1980).

In contrast, other workers use measures of the extent to which male breeding success deviates from mean *male* success as estimates of the intensity of sexual selection on males (see Wade, 1979; Howard, 1979; Wade & Arnold, 1980). Similarly, the intensity of sexual selection in females can be estimated by measuring the extent to which female breeding success deviates from mean female success. Several statistical measures are in use, including the coefficient of variation, Pielou's (1969) index of evenness and variance in breeding success divided by the square of mean breeding success – a measure derived from Crow's (1958) index of the intensity of selection, $I = V/\overline{W}^2$ where V is variance in fitness and $\overline{W}$ is the mean fitness of the population. This can either be calculated using total variance in breeding success or using only variance attributable to differences in the number of mates per individual (see Wade & Arnold, 1980).

Ratios of variation in male and female success can be used to estimate the relative rate at which male and female characteristics can change. However, since the extent of variation among females differs widely between species (see below), they are of limited value as measures of the comparative intensity of sexual selection (see Wade & Arnold, 1980).

The three measures of the extent to which reproductive success varies within each sex are fundamentally similar though the last is the most convenient since it offers a measure of the potential change in fitness between generations, relative to the average (see Crow, 1958).

It is still necessary to decide whether to use total variation in breeding success in any calculation or just variation due to differences in mate number. Where the aim of the calculations is to estimate the potential rate of genetic change, total variance in breeding success is clearly the relevant measure to use. In contrast, where calculations are carried out to assess the comparative importance of differences in reproductive success versus survival for each sex, only variation in mate number may be included (see Wade & Arnold, 1980) – though measures of the intensity of sexual selection as conceived by Darwin should also include variation due to differences in mate quality, which may be an important cause of variation in lifetime breeding success both among males and females in monogamous species.

Lastly, where the aim is to investigate the functional significance of particular phenotypic sex differences, it may be necessary to calculate selection intensities for specific episodes of selection, such as mating success at particular ages (see Arnold & Wade, 1983) for the effect of particular traits on lifetime reproductive success may often be obscured by the influence of other variables or by random variation.

One general point concerning all indices of sexual selection must be emphasized. Since many of the traits affecting breeding success are strongly influenced by rearing conditions (see Clutton-Brock *et al.*, 1982), these measures reflect the *potential* rate of genetic change and may provide little indication of actual rates of change in established populations.

Practical problems

Although it is relatively easy to decide how the intensity of sexual selection should be estimated, collecting the relevant data poses a variety of logistic problems (Ralls, 1977; Davies, 1982). In particular, in many polygynous species a proportion of males avoid competing directly with larger or older animals and adopt a policy of surreptitious fertilization or kleptogamy (e.g. Clutton-Brock *et al.*, 1977; Wirtz, 1982). As a result, it is often difficult to be sure that breeding males fertilize all the females in the groups that they guard: for example, even if territorial male red-winged blackbirds (*Agelaius phoeniceus*) are vasectomized, their females sometimes lay fertile eggs (Bray, Kennelly & Guarino, 1975).

Moreover, all indices of the intensity of sexual selection beg a problem of fundamental importance: over what period should reproductive success be measured? It is widely agreed that the lifetime reproductive success of individuals is the most satisfactory measure of fitness that it is usually possible to collect (see Falconer, 1960; Cavalli-Sforza & Bodmer, 1961; Maynard Smith, 1969; Grafen, 1982). This may not always be the case – for example, in rapidly expanding populations, selection may favour reproductive rate at the expense of lifetime reproductive success (see Lewontin, 1965; Elliot, 1975) – but exceptions are probably rare, especially in long-lived species where variation in longevity greatly exceeds variation in age at first breeding.

While many studies of sexual selection pay lip service to the importance of basing calculations on measures of lifetime success, most use estimates of success calculated over a part of the animal's lifespan. The time dimension used varies widely: some studies use measures of instantaneous reproductive success (IRS) calculated across different classes of males at a particular point in time (e.g. Mason, 1964; Scheiring, 1977; McCauley & Wade, 1980), while others measure daily reproductive success (DRS) or seasonal reproductive success (SRS) (see Howard, 1979; Payne, 1979). Very recently, estimates of individual variation in lifetime reproductive success (LRS) have become available from field studies of a small number of species, including a territorial and promiscuous invertebrate, the dragonfly *Erythemis simplicicolis* (McVey, 1981), two monogamous birds, the great tit, *Parus major* (McGregor, Krebs & Perrins, 1981) and the kittiwake *Rissa tridactyla* (J. Coulson & C. Thomas, personal communication), and one polygynous mammal, the red deer, *Cervus elaphus* (Clutton-Brock *et al.*, 1982) while in several other species, including sage grouse (*Centrocercus urophasianus*: Wiley, 1973) estimates of individual success are available which span several breeding seasons. These make it possible, for the first time, to test the extent to which variation in instantaneous, daily and seasonal reproductive success reflect variation in lifetime success.

There appears to be no consistent relationship between variation in IRS, DRS, SRS and LRS (see Table 23.1). There are several reasons why variation in daily reproductive success may not reflect variation in either seasonal or lifetime success. In many polygynous species reproductive success varies from hour to hour and day to day and this may cause variation in IRS or DRS to overestimate variation in SRS and LRS. For example, red deer stags which hold a large harem in a sheltered site on one day may have few or no hinds the next day if the wind changes (see Clutton-Brock *et al.*, 1982). As a result, measures of variation in DRS overestimate SRS and LRS in red deer (see Table 23.2). Estimates of variation in DRS and SRS in sage grouse (*Centrocercus urophasianus*) show the same trend though estimates of variation in DRS and LRS were similar in McVey's (1981) study of dragonflies.

In addition, variation in SRS will not reflect variation in LRS (see Gadgil, 1972) if individuals that breed successfully show reduced success in future or are likely to cease breeding or die earlier than less successful breeders. There is evidence that successful reproduction reduces the future reproductive capabilities or survival of females in several species (e.g. Wooller & Coulson, 1977; Altmann, Altmann & Hausfater, 1978; Guinness, Albon & Clutton-Brock, 1978; Bryant, 1979) and in some species successful males are likely to die before unsuccessful ones (Geist, 1971). However, several other studies have shown that successful breeders live *longer* (*Drosophila melanogaster*: Partridge & Farquhar, 1981). For example, in red deer stags, where harem size is one of the principal determinants of reproductive success (Clutton-Brock *et al.*, 1979, 1982), not only do stags that hold large harems hold them for longer within particular breeding

Table 23.1. *Measures of variation in breeding success ($\sigma^2/\bar{X}^2$) in different species, calculated from the available literature*

	Daily	Seasonal	Lifetime
Dragonfly: *Erythemis simplicicolis*[a] (McVey, 1981)			
Male[a]	1.74	(1.85)	1.85
Female[b]	0.36	—	—
Sage grouse: *Centrocercus urophasianus* (Wiley, 1973)			
Male[c]	4.35	2.29	—
Kittiwake: *Rissa tridactyla* (J. Coulson & C. Thomas, personal communication)			
Male[d]	—	0.687	0.826
Female[e]	—	0.687	0.691
Red deer: *Cervus elaphus* (Clutton-Brock, Guinness & Albon, unpublished)			
Male[f]	2.82	2.23	1.202
Female[g]	—	2.049	0.354

[a] Calculated for 82 adults. Estimates measured in terms of eggs fertilized per lifetime. Adults live for a single season.

[b] Calculated for 12 adults over 90 days. Estimates measured in terms of number of eggs laid.

[c] Calculated for 8 adults visiting mating centre no. 3 lek during the 1969 season. Estimates measured in terms of the number of copulations.

[d] Seasonal estimates calculated for 85, 79, 70 and 63 pairs (over 3 years old) breeding in 1972, 1973, 1974 and 1975 respectively. Lifetime estimates based on 220 individuals reaching breeding age. Estimates measured in terms of the number of young fledged.

[e] Seasonal estimates calculated as for males. Lifetime estimates based on 250 individuals reaching breeding age. Estimates measured in terms of the number of young fledged.

[f] Daily estimates based on number of hinds held by different stags of $\geqslant 5$ years old. Seasonal estimates based on number of surviving calves sired by 32, 33, 31 and 32 males holding harems in 1977, 1978, 1979 and 1980 respectively. Lifetime estimates measured in terms of the number of surviving calves sired by 10 males reaching breeding age who completed their reproductive lifespan during our study. A calf was counted as surviving if it reached 1 year old.

[g] Seasonal estimates based on number of surviving calves produced by 103 females $\geqslant 3$ years old in 1980. Lifetime estimates based on number of surviving calves produced by 30 females reaching breeding age. Relative to variation in lifetime success, variation in seasonal success was high for females since they commonly breed in alternate years.

seasons than those which hold smaller harems (Fig. 23.1*a*), but individuals that are consistently successful in securing large harems throughout their lives tend to live longer than their less successful competitors (Fig. 23.1*b*). Trends of this kind will produce an opposite bias and may cause variation in SRS to *underestimate* variation in LRS. For example, the fighting ability and reproductive success of red deer stags shows a pronounced peak between the ages of seven and ten years (see Fig. 23.2). If variation in breeding success is calculated for stags of above a year old (the age of sexual maturity), it greatly exceeds more realistic measures of variation in breeding success such as variation in seasonal success within cohorts or lifetime success (see Table 23.3). This effect will also tend to overestimate variation in male success relative to variation in female success (see Fig. 23.2).

A final problem is that unless breeding success is measured across the lifetime, a biased set of males may be sampled. Especially in polygynous species, males that fail to win a breeding territory spend their time on the fringes of the breeding population and often show high mortality. The field observer approaching a breeding colony is likely to see mostly (if not only) breeding males. If he calculates the extent to which reproductive success varies across adults that have managed to gain a breeding territory, he will underestimate the real variance since he will exclude non-breeders. Dividing by the mean success of males in this sample will accentuate this error since the mean success of the males sampled will be higher than that of the male population as a whole. Table 23.4 shows how the restriction of the sample to harem holding males in red deer reduces estimates of daily variation in the number of females held. Estimates that are probably biased in this way are already widespread in the literature (see Trivers, 1976; Arnold & Wade, 1983).

Table 23.2. *Variation in number of hinds held by all red deer stags* ≥ 1 *year old*

	October					
	6	11	17	23	29	Season[a]
$\bar{X}$	0.74	1.30	1.03	0.82	0.80	24.02
$\sigma^2/\bar{X}^2$	12.31	6.68	6.03	7.08	7.48	3.49
N	61	71	64	68	65	68

[a] Calculations based on number of hind/days held by different individuals over the whole breeding season.

Fig. 23.1. Duration of breeding plotted against mean success in red deer. (*a*) Number of days in a single breeding season (1980) on which individual males defended harems plotted against the mean size of their harems. (*b*) Age at death plotted against mean harem size during the years when different stags held a harem during the rut.

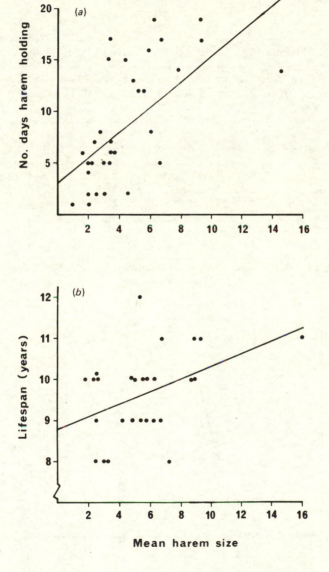

Table 23.3. *Variation in number of calves sired*

(*a*) Calculated across all male red deer ⩾ 1 year old present in each of seven breeding seasons and (*b*) across stags of the same age within breeding seasons. These estimates do not take into account calves dying in their first year of life. Variation in lifetime success calculated across all individuals reaching breeding age (5 years) is also shown (LRS).

(*a*)

	Year						
	1974	1975	1976	1977	1978	1979	1980
$\bar{X}$	0.875	1.130	0.884	0.654	0.725	0.825	0.725
$\sigma^2/\bar{X}^2$	3.090	2.108	4.762	6.126	8.072	3.624	3.293
N	46	56	69	78	91	61	68

(*b*)

	1–4	5	6	7	8	9	10	11	12	13+	LRS
$\bar{X}$	0	0.32	0.72	1.65	2.07	1.95	2.17	1.35	1.43	0.33	10.15
$\sigma^2/\bar{X}^2$	—	7.21	2.17	1.61	1.02	1.22	1.35	0.71	2.26	2.25	1.22
N		28	29	41	28	19	12	10	7	8	28

Fig. 23.2. Mean number of calves sired/borne per year by red deer stags and hinds of different ages.

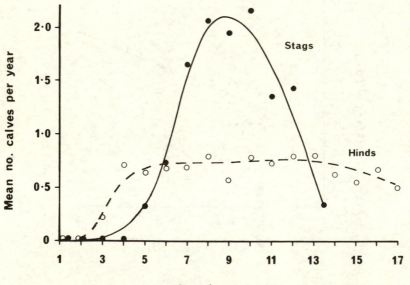

None of these four biases are likely to be consistent either across the two sexes or across different species: all four are more likely to affect estimates of variation in male success than estimates of variation in female success in polygynous species, though in contrary directions. Ignoring short term variation and age effects will tend to overestimate variation in male success relative to variation in female success, which is usually less strongly age-dependent in polygynous species (see Fig. 23.2) and less likely to vary widely from day to day. Conversely, positive correlations between seasonal success and longevity or sampling only successful males will lead to underestimates of variation in male success relative to variation in female success. That the two sets of biases oppose each other is no guarantee that short term measures of reproductive success will provide reliable estimates of variation in lifetime success for the comparative strengths of the different biases probably vary widely too.

Moreover, sex differences in the length of effective reproductive lifespans and in the influence of age on breeding success are likely to be more pronounced in polygynous species than in monogamous ones. As a result, short-term estimates of variation in male breeding success are likely to overestimate the extent to which male success varies in polygynous species relative to similar measures for monogamous ones.

Biased estimates of variation in reproductive success may also cause the effects of particular phenotypic traits on reproductive success to be overestimated. This applies particularly to traits, such as body size, which are themselves related to age. For example, where the body size and reproductive success of males both increase with age, the effects of size on breeding success may be grossly exaggerated if age differences in size are ignored. This effect is again likely to be stronger in males than females.

Table 23.4. *Variation in harem size*

Calculated (*a*) across all red deer stags $\geq$ 1 year old on different days in the October rut, (*b*) across only those holding harems.

October	6	11	17	23	29	Season
			(a)			
$\bar{X}$	0.738	1.296	1.032	0.824	0.800	24.015
$\sigma^2/\bar{X}^2$	12.305	6.677	6.029	7.084	7.480	3.493
N	61	71	64	68	65	68
			(b)			
$\bar{X}$	4.09	7.08	5.00	4.31	3.71	50.875
$\sigma^2/\bar{X}^2$	1.499	0.421	0.467	0.569	0.863	1.138
N	11	13	13	13	14	32

Table 23.5. *Variation in lifetime reproductive success for red deer and kittiwakes measured in terms of the offspring surviving to one year (red deer) or fledging kittiwakes*

(a) Figures calculated for all individuals born/hatched. (b) Figures calculated for all individuals reaching breeding age. Sample sizes in brackets.

	Male	Female
	(a)	
Red deer	1.681 (12)	1.065 (42)
Kittiwake	1.976 (367)	1.896 (417)
	(b)	
Red deer	1.202 (10)	0.354 (30)
Kittiwake	0.826 (220)	0.691 (250)

Fig. 23.3. Distribution of lifetime reproductive success for male and female red deer (from Clutton-Brock, Guinness & Albon, 1982). The histograms show the proportions of all individuals born that produced different numbers of offspring during their lifetime. Only calves surviving to one year old were included.

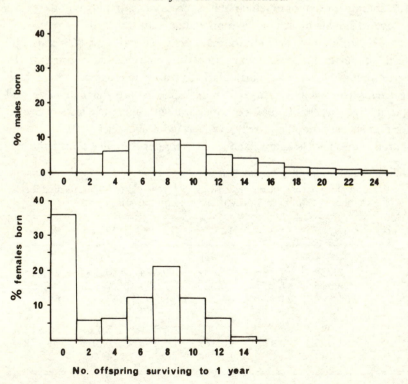

The most obvious conclusion to be drawn is the need for studies of the extent to which lifetime success varies and of the factors which affect it. Where this is impossible, an alternative approach is to concentrate on studies of variation in breeding success within cohorts but, even here, problems will arise if breeding success is consistently related to longevity.

Polygyny, reproductive success and sexual dimorphism

The theory of sexual selection has led to four common predictions about the relationship between breeding success and sexual dimorphism.

Variation in reproductive success should be greater in males than females in polygynous species but similar in the two sexes in monogamous ones.
Both in red deer and in *Erythemis*, variation in lifetime breeding success is substantially greater among males than females whereas, in kittiwakes, variation is similar in both sexes (Figs. 23.3, 23.4). When the red deer

Fig. 23.4. Distribution of lifetime reproductive success in (*a*) male and (*b*) female kittiwakes (from J. Coulson & C. Thomas, personal communication). The histograms show the percentage of all individuals hatched that produced different numbers of fledged young during their lifetime. Numbers dying before reaching one year old calculated from known mortality figures.

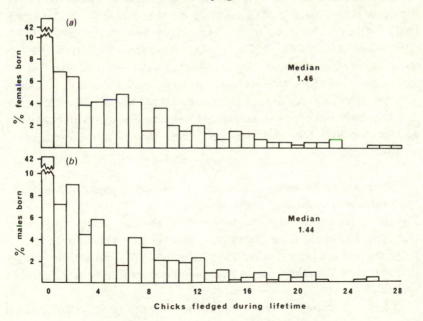

samples are restricted to animals that reach breeding age, sex differences in the extent to which breeding success varies are accentuated (Table 23.5).

Considering that male red deer can hold harems of over thirty hinds, it is, perhaps, surprising that male success does not vary more widely. This is partly because only a proportion of hinds conceive in a given year and individual stags rarely hold harems throughout the whole breeding season, and partly because few stags breed successfully for more than four years.

In contrast, the range of breeding success among hinds is greater than might be expected because their potential breeding lifespans are long (over 12 years) and individuals tend to be either consistently successful or consistently unsuccessful breeders.

Variation in reproductive success should be greater among males of polygynous species than among males of monogamous ones.
One surprising result of the comparison between red deer and kittiwakes is that variation in lifetime breeding success is little greater in red deer stags than in male kittiwakes (see Table 23.5): in fact if all individuals born/hatched are included, it is slightly (though not significantly) greater in male kittiwakes.

The comparison is an unsatisfactory one since there are important differences in the life histories of the two species: male kittiwakes can breed for many more seasons, adult mortality is not so strongly age-dependent and females can fledge up to three young per year (see Coulson, 1966, 1968; Coulson & Wooller, 1976; Wooller & Coulson, 1977). However, this example serves to emphasize how misleading it can be to assume that the breeding sex ratio necessarily reflects the extent to which male reproductive success varies for, even among closely related species, it is likely to be the case that males have substantially longer breeding lifespans in monogamous species than in polygynous ones (see Wiley, 1974; Clutton-Brock *et al.*, 1982).

Direct competition for mates will be more intense among males of polygynous species than among males of monogamous ones.
Fights between males can be common and dangerous in polygynous species (see Geist, 1971). However, competition between males can also be intense in monogamous species (Lack, 1954; Kleiman, 1977) and data are not yet available which would permit a meaningful comparison between the two groups of species.

In fact, it is unsafe to assume that the intensity of direct competition between males should necessarily be reduced in monogamous species. Though variation in male success may be caused principally by differences in mate or territory quality (whereas, in polygynous species, differences in mate number are the main cause of differences in success: Bateman, 1948; Wade, 1979; Clutton-Brock *et al.*, 1982), monogamous males might be expected to compete as intensely for the best mates or territories as do polygynous males for the biggest harems. Where this is not the case, it may be because males cannot identify the breeding potential of young females or because female choice pre-empts male competition rather than because variation in success is slight among males.

Sexual dimorphism will be most developed among strongly polygynous species and least developed among monogamous ones.
In many different groups of animals there is an association between polygyny and sexual dimorphism and Darwin himself was well aware of the relationship.

The practice of polygamy leads to the same results as would follow from an actual inequality in the number of sexes; for if each male secures two or more females, many males cannot pair, and the latter assuredly will be the weaker or less attractive individuals...That some relation exists between polygamy and the development of secondary sexual characters, appears nearly certain.

More recently, a variety of studies have demonstrated statistical relationships between the degree of polygyny and the development of sexual dimorphism, though the relationship is not always a close one (Ralls, 1977). Compared to monogamous species, polygynous ones usually show greater sexual dimorphism in body size (Clutton-Brock, Harvey & Rudder, 1977; Shine, 1979; Alexander *et al.*, 1979) while weapons used in intraspecific combat, such as the canines of primates and the antlers of deer, are also more developed in the males of polygynous species (Harvey, Kavanagh & Clutton-Brock, 1978; Clutton-Brock, Albon & Harvey, 1980).

However, there are many exceptions (Ralls, 1977). In some groups of animals, the relationship between the extent of polygyny and the degree of sexual dimorphism is not a close one (Clutton-Brock *et al.*, 1977). In addition, some polygynous species, like Burchell's zebra (*Equus burchelli*), show little or no size dimorphism while others, like the spotted hyena (*Crocuta crocuta*) and the Weddell seal (*Leptonychotes weddelli*), even show reversed dimorphism (Klingel, 1972; Kruuk, 1972; Stirling, 1969).

While it is possible that some of these exceptions result from the differing energetic requirements of males and females and attendant selection pressures affecting the relative size of the two sexes (Selander, 1972;

Downhower, 1976), the common association between sex differences in size and the development of male weaponry suggests that selection pressures associated with breeding competition are frequently involved. To interpret exceptions to the general rule that sexual dimorphism increases with the degree of polygyny, we need to remember that it is the comparative effects of phenotypic traits on reproductive success in males and females that will determine the degree of dimorphism and not the amount of variation in reproductive success *per se* (see Price, 1970; Lande, 1980). For example, while sexual dimorphism in size is likely to evolve where variation in male success is greater than female success and a given increment in body size has the same effect on breeding success in both sexes, it will also evolve if variation in reproductive success is similar in both sexes but size has a greater influence on success in males or even if variation in success is greater in females but the effects of size are greater among males. Conversely, sexual dimorphism in size is unlikely to evolve in circumstances where variation in reproductive success is greater among males but the effects of size on reproductive success are slight in both sexes.

In the simplest of all possible worlds sexual dimorphism should, perhaps, be predicted by the relative slope of lifetime reproductive success on body size in males and females. However, there is no reason to suppose that relationships between size and reproductive success will be linear or that they will follow a similar pattern in both sexes. Indeed, where size differences are heritable and stabilizing selection is operating, there is every reason to suppose that relationships between size and reproductive success will *not* be linear. In addition, even a knowledge of the relationship between body size and lifetime reproductive success in the two sexes will not answer whether the association occurs for reasons connected with breeding competition or because the two sexes differ in their energy requirements (see Downhower, 1976). To sort out these questions, it will be necessary to identify the particular episodes of selection during which size influences breeding success in males and females (see Arnold & Wade, 1983).

This argument raises the question of why it is that sexual dimorphism and polygyny are related at all. The most likely explanation is that the factors determining breeding success in males and females tend to be most similar in monogamous species and most different in highly polygynous ones. The factors affecting breeding success in males and females certainly differ widely in polygynous species. For example, in red deer, longevity, offspring survival and home range quality have a greater effect on the reproductive success of hinds than on that of stags. In contrast, fighting ability, body size and (because of its effect on adult size) early growth have a more important effect on the reproductive success of stags than hinds (Clutton-Brock *et al.*, 1982). The factors affecting lifetime success in males and females of monogamous species have yet to be described. While it is clear that they will not be identical (see McGregor *et al.*, 1981), it is reasonable to suppose that, especially among species that pair for life, they are likely to be more similar than in polygynous species.

Selection pressures on males and females

Emphasis on the importance of understanding the factors affecting breeding success in males and females has the advantage that it forces us to ask specific comparative questions concerning the functional significance of particular sex differences. For example, does body size have a less important effect on the reproductive success of male zebras than male bovids because zebras fight with their teeth and hooves rather than by pushing? Similarly, does size have a greater effect on reproductive success in female Weddell seals compared to land or pack-ice breeding species because they breed on fast ice and defend access to water holes? Conversely, is body size less important in male Weddells compared to other species because they defend underwater territories where success depends on manoeuvrability and because females are widely dispersed?

Thinking in these terms may help us to understand the distribution of many other sex differences. For example, among hermaphroditic reef fish, some species begin life as females and a proportion of individuals later become males (protogyny) (Warner, Robertson & Leigh, 1975; Robertson & Hoffman, 1977). But in a few species, such as the clown fishes (*Amphiprion*), individuals start life as males and a proportion later become females (protandry) (Fricke & Fricke, 1977). The females of most protogynous species spawn on the edge of the reef, releasing their eggs into the plankton, and seldom compete for spawning sites. In contrast, female clownfish lay their eggs around sea anemones which they subsequently help to defend. Does resource defence by females again increase the benefits of large body size to female clownfish while reducing the benefits of size to males as a consequence of female dispersion?

Understanding the comparative effects of size may even help to explain variation in birth sex ratios. It has been suggested that in species where reproductive success varies more widely among males than females and is influenced by parental investment, parents who can afford to invest heavily in their offspring should produce sons while those that cannot do so should tend to produce daughters (Trivers & Willard, 1973). In apparent contradiction to the theory, dominant female baboons and macaques produce more daughters than sons while subordinates produce more sons than daughters (Altmann, 1980; Simpson & Simpson, 1982). One possible explanation is that even though reproductive success probably varies more widely among males in these species, maternal rank has a stronger *effect* on the success of daughters since females remain in their mother's troop and inherit her rank while sons disperse to other groups and may be unable to benefit substantially from their mother's rank (Altmann, 1980).

Perhaps most importantly, emphasis on examining the effects of particular traits on the breeding success of males and females should encourage us to investigate the adaptive significance of sex differences whose functions are not immediately obvious. For example, in many mammals growing males lay down less body fat than females (Glucksman, 1974) and suffer heavier mortality during periods of food shortage as a consequence (Clutton-Brock *et al.*, 1982). One possible explanation is that because early growth exerts a greater effect on reproductive success in males than females (see above), selection favours increased investment by young males in growth at the expense of laying down body fat to assure survival during periods of food shortage. This prompts the question as to whether young *females* lay down less fat in species showing reversed size dimorphism. Recent research shows that this is the case in at least one species belonging to this category (the European sparrow hawk: I. Newton, personal communication).

These explanations are cautiously worded – and necessarily so, for our current knowledge of the factors affecting reproductive success in males and females is rudimentary. Nevertheless, the way ahead is clear. If we wish to fulfil Charles Darwin's ambition of understanding the reasons for the distribution of differences between the sexes, we shall need to examine the causes of variation in lifetime breeding success among males and females in natural populations.

I am very grateful to Michael Reiss, Derek Bendall, Paul Harvey, John Coulson, Callum Thomas and John Maynard Smith for their comments; Anthony Arak, Dafila Scott, Jon Seeger, Steve Albon, Joanne Reiter, Meg McVey, Steve Arnold, Russ Lande and Burney Le Boeuf for helpful discussion; to John Coulson, Callum Thomas and Meg McVey for permission to quote unpublished results; and to Fiona Guinness, Steve Albon, Glenn Iason and Callan Duck, my colleagues on the Rhum deer project, for their respective parts in collecting and analysing estimates of lifetime breeding success in red deer.

Further reading

Darwin, C. (1871). *The descent of man and selection in relation to sex*. London: John Murray.

Trivers, R. L. (1972). Parental investment and sexual selection. Pp. 136–179 in: *Sexual selection and the descent of man 1871–1971* ed. B. Campbell. Chicago: Aldine-Atherton.

Wade, M. J. & Arnold, S. J. (1980). The intensity of sexual selection in relation to male sexual behaviour, female choice, and sperm precedence. *Animal Behaviour*, **28**, 446–61.

Clutton-Brock, T. H., Guinness, F. E. & Albon, S. D. (1982). *Red deer: behavior and ecology of two sexes*. Chicago: University of Chicago Press.

References

Alexander, R. D., Hoogland, J. L., Howard, R. D., Noonan, M. & Sherman, P. W. (1979). Sexual dimorphisms, and breeding systems in pinnipeds, ungulates, primates and humans. In *Evolutionary Biology and Human Social Behaviour, and Anthropological Perspective*, ed. N. A. Chagnon & W. Irons, pp. 402–604. Massachusetts: Duxbury Press.

Altmann, J. (1980). *Baboon Mothers and Infants*. Cambridge, Mass.: Harvard University Press.

Altmann, J., Altmann, S. A. & Hausfater, G. (1978). Primate infant's effects on mother's future reproduction. *Science*, **201**, 1028–30.

Andersson, M. (1982). Female choice selects for extreme tail length in a widdowbird. *Nature*, **299**, 818–20.

Arnold, J. J. & Wade, M. J. (1983). On the measurement of natural and sexual selection in field and laboratory populations. *Evolution*.

Barnes, R. D. (1974). *Invertebrate Zoology*. Philadelphia: Saunders.

Bateman, A. J. (1948). Intrasexual selection in *Drosophila*. *Heredity*, **2**, 349–68.

Bray, O. E., Kennelly, J. S. & Guarino, J. J. (1975). Fertility of eggs produced on territories of vasectomized red-winged blackbirds. *Wilson Bulletin*, **87**, 187–95.

Brown, J. L. (1975). *The Evolution of Behavior*. New York: Norton.

Bryant, D. M. (1979). Reproductive costs in the house martin. *Journal of Animal Ecology*, **48**, 655–75.

Bryden, M. M. (1969). Growth of the southern elephant seal, *Mirounga leonina* (Linn.). *Growth*, **33**, 531–6.

Cavalli-Sforza, L. L. & Bodmer, W. F. (1961). *The Genetics of Human Populations*. San Francisco: Freeman.

Clutton-Brock, T. H., Albon, S. D., Gibson, R. M. & Guinness, F. E. (1979). The logical stag: adaptive aspects of fighting in red deer (*Cervus elaphus* L.). *Animal Behaviour*, **27**, 211–25.

Clutton-Brock, T. H., Albon, S. D. & Guinness, F. E. (1981). Parental investment in males and female offspring in polygynous mammals. *Nature*, **289**, 487–9.

Clutton-Brock, T. H., Albon, S. D. & Harvey, P. H. (1980). Antlers, body size and breeding systems in the Cervidae. *Nature*, **285**, 565–7.

Clutton-Brock, T. H., Guinness, F. E. & Albon, S. D. (1982). *Red deer: the behavior and ecology of two sexes*. Chicago: University of Chicago Press.

Clutton-Brock, T. H., Harvey, P. H. & Rudder, B. (1977). Sexual dimorphism, socionomic sex ratio and body weight in primates. *Nature*, **269**, 797–800.

Coulson, J. C. (1966). The influence of the pair bond and age on the breeding biology of the kittiwake gull. *Journal of Animal Ecology*, **35**, 269–79.

Coulson, J. C. (1968). Differences in the quality of birds nesting in the centre and on the edge of a colony. *Nature*, **217**, 478–99.

Coulson, J. C. & Wooller, R. D. (1976). Differential survival rates among breeding kittiwake gulls *Rissa tridactyla* L. *Journal of Animal Ecology*, **45**, 205–14.

Crow, J. F. (1958). Some possibilities for measuring selection intensities in man. *Human Biology*, **30**, 1–13.

Darwin, C. (1859). *On the Origin of Species*. London: John Murray.

Darwin, C. (1871). *The Descent of Man and Selection in Relation to Sex*. London: John Murray.

Davies, N. B. (1982). Behaviour and competition for scarce resources. In *Current Problems in Sociobiology*, ed. King's College Sociobiology Group, pp. 363–80. Cambridge: Cambridge University Press.

Downhower, J. F. (1976). Darwin's finches and the evolution of sexual dimorphism in body size. *Nature*, **263**, 558–63.

Elliot, P. F. (1975). Longevity and the evolution of polygyny. *American Naturalist*, **109**, 281–7.

Falconer, D. S. (1960). *Introduction to Quantitative Genetics*. New York: Ronald Press.

Fisher, R. A. (1930). *The Genetical Theory of Natural Selection*. Oxford: Oxford University Press.

Fricke, H. & Fricke, S. (1977). Monogamy and sex change by aggressive dominance in coral reef fish. *Nature*, **266**, 830–2.

Gadgil, M. (1972). Male dimorphism as a consequence of sexual selection. *American Naturalist*, **106**, 574–80.

Gautier-Hion, A. (1980). Seasonal variations of diet related to species and sex in a community of *Cercopithecus* monkeys. *Journal of Animal Ecology*, **49**, 237–69.

Geist, V. (1971). *Mountain Sheep: a Study in Behavior and Evolution*. Chicago: University of Chicago Press.

Glucksman, A. (1974). Sexual dimorphism in mammals. *Biological Reviews*, **49**, 423–75.

Grubb, P. (1974). Population dynamics of the Soay sheep. In *Island Survivors: the Ecology of the Soay Sheep of St. Kilda*, ed. P. A. Jewell, C. Milner & J. M. Boyd, pp. 242–72. London: Athlone Press.

Guinness, F. E., Albon, S. D. & Clutton-Brock, T. H. (1978). Factors affecting reproduction in red deer (*Cervus elaphus* L.). *Journal of Reproduction and Fertility*, **54**, 325–34.

Grafen, A. (1982). How not to measure inclusive fitness. *Nature*, **298**, 419–20.

Halliday, T. R. (1978). Sexual selection and mate choice. In *Behavioural Ecology: an Evolutionary Approach*, ed. J. R. Krebs & N. B. Davies, pp. 180–213. Oxford: Blackwell.

Harvey, P. H., Kavanagh, M. J. & Clutton-Brock, T. H. (1978). Sexual dimorphism in primate teeth. *Journal of Zoology*, **186**, 475–85.

Howard, R. D. (1979). Estimating reproductive success in natural populations. *American Naturalist*, **114**, 221–31.

Howe, H. (1977). Nestling sex ratio adjustment among common grackles. *Science*, **198**, 744–6.

Hunter, J. (1837). An account of an extraordinary pheasant. In *Observations on Certain Parts of the Animal Oeconomy* with notes by Richard Owen. London: Longman, Orme, Brown, Green and Longmans.

Hunter, J. (1861). Observations on generation. In *Essays and Observations on Natural History, Anatomy, Physiology, Psychology and Geology* Vol. I. London: John van Voorst.

Huxley, J. S. (1938*a*). The present standing of the theory of sexual selection. In *Evolution: Essays on Aspects of Evolutionary Biology*, ed. G. R. de Beer, pp. 11–42. Oxford: Clarendon Press.

Huxley, J. S. (1938*b*). Darwin's theory of sexual selection and the data subsumed by it, in the light of recent research. *American Naturalist*, 72, 416–33.

Kirkpatrick, M. (1982). Sexual selection and the evolution of female choice. *Evolution*, 36, 1–12.

Kleiman, D. G. (1977). Monogamy in mammals. *Quarterly Review of Biology*, 52, 39–69.

Klein, D. R. (1968). The introduction, increase and crash of reindeer on St. Mathew Island. *Journal of Wildlife Management*, 32, 350–67.

Klingel, H. (1972). Social behaviour of African Equidae. *Zoologica Africana*. 7, 175–85.

Kruuk, H. (1972). *The Spotted Hyena: a Study of Predation and Social Behavior*. Chicago: University of Chicago Press.

Lack, D. (1954). *Population Studies of Birds*. Oxford: Clarendon Press.

Lande, R. (1980). Sexual dimorphism, sexual selection, and adaptation in polygenic characters. *Evolution*, 34, 292–305.

Lande, R. (1981). Models of operation of sexual selection on polygenic traits. *Proceedings of the National Academy of Sciences, USA*, 78, 3721–5.

Lewontin, R. C. (1965). Selection for colonizing ability. In *The Genetics of Colonizing Species*, ed. H. G. Baker & G. L. Stebbins, pp. 79–94. New York: Academic Press.

McCauley, D. E. & Wade, M. J. (1980). Female choice and the mating structure of a natural population of the soldier beetle *Chaulignathus pennsylvanicus*. *Evolution*, 32, 771–5.

McGregor, P. K., Krebs, J. R. & Perrins, C. M. (1981). Song repertoires and lifetime reproductive success in the great tit, *Parus major*. *American Naturalist*, 118, 149–59.

McVey, M. (1981). Lifetime reproductive practice in a territorial dragonfly *Erythemis simplicicolis*. Ph.D. thesis, Rockefeller University, New York.

Mason, L. G. (1964). Stabilizing selection for mating fitness in natural populations of *Tetraopes*. *Evolution*, 18, 492–7.

Maynard Smith, J. (1969). The status of neo-Darwinism. In *Towards a Theoretical Biology*, ed. C. H. Waddington, pp. 82–9. Edinburgh: Edinburgh University Press.

Narins, P. & Capranica, R. R. (1976). Sexual differences in the auditory system of the tree frog *Eleutherodactylus coqui*. *Science*, 192, 378–80.

Narins, P. M. & Capranica, R. R. (1978). Communicative significance of the two-note call of the treefrog *Eleutherodactylus coqui*. *Journal of Comparative Physiology*, 127, 1–9.

O'Donald, P. (1980). *Genetic Models of Sexual Selection*. Cambridge: Cambridge University Press.

Partridge, L. & Farquhar, M. (1981). Sexual activity reduces lifespan of male fruitflies. *Nature*, 294, 580–2.

Payne, R. B. (1979). Sexual selection and intersexual differences in variance of breeding success. *American Naturalist*, **114**, 447–52.

Pielou, E. C. (1969). *An Introduction to Mathematical Ecology*. New York: Wiley.

Pleszczynska, W. K. (1978). Microgeographic prediction of polygyny in the lark bunting. *Science*, **201**, 935–7.

Price, G. R. (1970). Selection and covariance. *Nature*, **227**, 520–1.

Ralls, K. (1977). Sexual dimorphism in mammals: avian models and unanswered questions. *American Naturalist*, **111**, 917–38.

Robertson, D. R. & Hoffman, S. G. (1977). The roles of female mate choice and predation in the mating systems of some tropical labroid fishes. *Zeitschrift für Tierpsychologie*, **45**, 298–320.

Robinette, W. L., Gashwiler, J. S., Low, J. B. & Jones, D. A. (1957). Differential mortality by sex and age among mule deer. *Journal of Wildlife Management*, **21**, 1–16.

Scheiring, J. E. (1977). Stabilizing selection for size as related to mating fitness in *Tetraopes*. *Evolution*, **31**, 447–9.

Selander, R. K. (1972). Sexual selection and dimorphism in birds. In *Sexual selection and the descent of man*, ed. B. C. Campbell, pp. 180–220. Chicago: Aldine-Atherton.

Shine, R. (1979). Sexual selection and sexual dimorphism in the Amphibia. *Copeia*, 1979, 297–306.

Simpson, M. J. & Simpson, A. E. (1982). Birth sex ratios and social rank in rhesus monkey mothers. *Nature*, **300**, 440–1.

Stirling, I. (1969). Ecology of the Weddell seal in McMurdo Sound, Antarctica. *Ecology*, **50**, 573–86.

Trivers, R. L. (1972). Parental investment and sexual selection. In *Sexual Selection and the Descent of Man 1871–1971*, pp. 136–79. Chicago: Aldine-Atherton.

Trivers, R. L. (1976). Sexual selection and resource-accruing abilities in *Anolis garmani*. *Evolution*, **30**, 253–69.

Trivers, R. L. & Willard, D. E. (1973). Natural selection of parental ability to vary the sex ratio of offspring. *Science*, **179**, 90–2.

Wade, M. J. (1979). Sexual selection and variance in reproductive success. *American Naturalist*, **114**, 742–6.

Wade, M. J. & Arnold, S. J. (1980). The intensity of sexual selection in relation to male sexual behaviour, female choice, and sperm precedence. *Animal Behaviour*, **28**, 446–61.

Wallace, A. R. (1889). *Darwinism, an Exposition of the Theory of Natural Selection*. London: Macmillan.

Warner, R. R. D., Robertson, R. & Leigh, E. G. Jnr. (1975). Sex change and sexual selection. *Science*, **190**, 633–8.

Widdowson, E. M. (1976). The response of the sexes to nutritional stress. *Proceedings of the Nutrition Society*, **35**, 1175–80.

Wiley, R. H. (1973). Territoriality and non-random mating in sage grouse, *Centrocercus urophasianus*. *Animal Behaviour Monographs*, **6**, 87–169.

Wiley, R. H. (1974). Evolution of social organisation and life history patterns among grouse. *Quarterly Review of Biology*, **49**, 201–27.

Williams, D. M. (1982). Agonistic behaviour and male selection in the mallard, *Anos platyrhinchos*. Ph.D thesis, University of Leicester.

Wirtz, P. (1982). Territory holders, satellite males and bachelor males in a high density population of water buck (*Kobus ellipsiprymnus*) and their associations with conspecifics. *Zeitschrift für Tierpsychologie*, **58**, 277–300.

Wooller, R. D. & Coulson, J. C. (1977). Factors affecting the age at first breeding of the kittiwake *Rissa tridactyla*. *Ibis*, **119**, 339–49.

Yarrell, W. (1827). On the change in the plumage of some hen pheasants. *Philosophical Transactions of the Royal Society*, 1827.

24

Rules for changing the rules

PATRICK BATESON

Few biologists have not at one time or another marvelled at the exquisite fit that can be found between the characteristics of an organism and the characteristics of its environment. Darwin was not the first to marvel but he made the notion of such adaptedness scientifically respectable by providing an explanation of how it might have come about. Nowadays, most discussion about adaptation assumes that when it is found, a fit between organism and environment is the result of evolutionary selection pressures and nothing else (e.g. Lewontin, 1978). Furthermore, each adaptation is supposedly transmitted from one generation to the next by genetic means alone. However, these assumptions are manifestly false when applied to behaviour – particularly the behaviour of complex animals (e.g. Lorenz, 1965; Hinde, 1968).

Quite obviously a great many animals are able to tune their behaviour to their environments by learning and by other developmental mechanisms which rely on external triggering. When faced with an example of adaptedness in behaviour, at least three explanations can be offered for the process of adaptation. Consider an actual case, the way in which a long-tailed tit makes a strong, cryptic and well-insulated place in which to lay eggs.

Once a site in a bush or tree has been selected, the pair of long-tailed tits search for moss and bring it back to the site. When some moss has stuck, each tit collects spiders' webs and stretches them across the moss. More moss is collected and then more spiders' webs until a platform has been formed. The bird that is building can now place moss and webs around itself – building up the sides of the nest. When the nest-cup is well formed the bird fetches lichen and weaves this on to the outside of the nest. Building up the sides of the nest is resumed but is periodically interrupted so that more lichen can be added to the outside. Eventually, the bird builds the walls up and over itself to form a dome, but leaves a neat entrance hole at the side. Finally, the nest is lined with a large number of feathers (Tinbergen, 1953, cited in Thorpe, 1956).

In principle, the behaviour of the tits could be adapted to the job of building a safe, warm nest for offspring in three separate ways. First, birds performing the appropriate actions could have had more surviving offspring than those making less good nests; consequently, in the course of time, genes necessary for the expression of the appropriate actions spread through the long-tailed tit population. Secondly, the bird could copy what another more experienced bird had done; the process of selecting the actions best adapted to the environment had gone on in previous generations and been transmitted socially. Thirdly, by experimenting on its own with different materials and different actions, each bird could assemble the appropriate repertoire for building nests. The three processes of adaptation and the three sources of adaptedness for an individual are shown in Table 24.1.

All three processes could contribute to the adaptedness of the nest-building. We should not expect three classes of behaviour corresponding to the three processes of adaptation. Furthermore, we should not be surprised by the extent of learning and imitation that can be found, particularly in complex animals. Cultural transmission of adapted behaviour is by no means confined to humans (see Galef, 1976; Bonner, 1980). One of the first examples to be discovered in animals was the opening of milk bottles in Britain by great tits, blue tits and coal tits. The spread of the habit from a few scattered locations before 1930 to a great many in 1947 was well documented (Fisher & Hinde, 1949; Hinde & Fisher, 1951).

The animal which is learning does not operate like an idiot photographer

Table 24.1. *Three sources on which an individual's adapted behaviour may depend and the related processes of adaptation*

Source of adaptedness for individual	Process of adaptation
Genes	Natural selection during evolution
Social companions	Learning by previous generations
Environment	Learning and facultative responses by individual

attempting to take a snap of everything. What an animal learns is highly selective and highly ordered. The instruments for changing behaviour show an adapted regularity which suggests that they themselves have been subject to natural selection during the course of evolution. With precisely this point in mind, Konrad Lorenz (1965) referred to the 'innate school marm' who, he imagined, was busy directing the course of learning. It is worth noting that what Lorenz meant by the memorable phrase was not that there were unlearned instructions for learning, but the instructions were adapted for their present use by natural selection. The importance of this distinction will become apparent later.

It is more usual nowadays to use computer metaphors and refer to the programming of learning (e.g. Pulliam & Dunford, 1980). I mildly distrust such metaphors. It can be easily assumed that because we know how computers work we therefore know how learning is programmed. My distrust turns into hostility when phrases like 'genetically programmed' are used instead. Such phrases confuse the way in which coded information is transmitted from one generation to another with the regularities of a nervous system, which itself is the outcome of an ordered developmental process. For these reasons I prefer to use the phrase 'rules for changing the rules' (Bateson, 1976).

A necessary preliminary is to clarify the meaning of the term 'rule'. Behaviour is very far from being disorderly. It may be complex but it is certainly not chaotic. Something provides direction and keeps it in order. Something is responsible for the regularity. 'Rules' refer to these consistencies of behaviour. In more complex animals, consistencies are not easily found on the surface and many of us feel the need to postulate structural regularity beneath the surface if we are to make sense of what we see. A famous example of this theoretical approach is Chomsky's (1965) analysis of language in terms of underlying grammatical rules. Not everybody likes this style and, among psychologists, Skinner (1959) in particular has derided the use of concepts based on inferred structures and processes. It must be admitted that concepts referring to unseen processes tend to acquire additional meanings that are not suggested by the evidence they are intended to explain (see MacCorquodale & Meehl, 1948). Muddle ensues when the rules are treated as though they are tangible and can be observed directly. But the confusion can be avoided if we treat unseen high-order rules for what they are, namely, as explanatory devices. When that is done the thinking can be creative and rewarding. The usage of 'rule' by biologists is clearly different from that employed by social scientists when they talk about verbally transmitted instructions for what humans may and may not do. It would obviously avoid punning and confusion if different terms were used. However, it is unlikely that biologists will be misunderstood when they apply the term to animals.

In this chapter I first consider the underlying rules for associative learning and suggest that some useful general principles have already been uncovered. I go on to argue that, despite the underlying regularities, the behaviour of an individual animal is only predictable when a lot is known about the conditions in which the animal has grown up. In order to emphasize the importance of this point, I devote the rest of the chapter to a discussion of the developmental rules that may influence mate choice in humans. A growing understanding of such rules has tempted biologists into making exaggerated claims about the invariance of human mating preferences and also about the origins of incest taboos. I argue that even if only one process were involved, the outcome would depend on conditions and, since conditions vary, so must the behavioural outcome. For all that, I conclude that the postulated underlying rules for development may usefully account for *some* of the variation in human sexual behaviour and possibly even the variation in marriage laws.

Rules for associative learning

Any animal with even a rudimentary nervous system will be better placed if it can compute the arrival of impending danger or the location of valuable resources such as food or mates. Its nervous system does not have to be modifiable in order to work with reasonable efficiency in this way. Even so, the power to predict and control the environment is enormously enhanced by a capacity to associate neutral events with those that already have some importance for the animal. With such capacity, initially meaningless cues and initially haphazard or exploratory acts can acquire causal significance. What could be the rules for the necessary associative learning processes? One very obvious possibility would be a time-window preceding the important event. If a neutral event occurs within this time-window then it loses its neutrality.

To give a text-book example, college students were trained in a situation in which a buzzer was sounded before, after or together with the delivery of a mild electric shock to the finger (Spooner & Kellogg, 1947). Periodically, the students' responses to the buzz alone were tested. The students who, in the training trials, heard the buzz half a second before the shock, jerked their finger back more consistently than those who had had a longer gap between the buzz and the shock during training, and they did so much more markedly than those who heard the buzz at the time of the shock or after it (see Fig. 24.1).

The brief interval allowed for the establishment of a link between neutral and significant events was at one time elevated into a general law of associative learning. Equipped with this rule and with knowledge of what are important events, the animal seems to be well set up to acquire the ability to use initially meaningless environmental cues as predictors of what will happen, and initially haphazard acts as instruments for controlling the environment. While it might seem to make good inituitive sense that a time-window should be small, substantial delay in detectable effect can sometimes follow the performance of an activity. If you eat some contaminated food, you will not necessarily feel the ill-effects immediately. Indeed, it is well known now that many mammals and birds can develop aversions to novel foods that were followed by ill-effects hours after ingestion (reviewed by Domjan, 1980). Experiments involve a spurious association between the novel food and the illness which is usually induced chemically or by X-rays. Nevertheless, the animals subsequently avoid the novel food. They do not avoid familiar food which has similarly been followed by illness, and only certain cues such as smells and tastes associated with the novel food are attended to. Others, such as noise, are treated as being irrelevant (see Revusky, 1971; LoLordo, 1979). The implication is that the

Fig. 24.1. The strength of response of college students to a buzz previously associated with shock as measured by the mean percentage of occasions when their fingers were withdrawn from an electrode. Different groups of students were trained with different intervals between buzz and shock or shock and buzz. (Data from Spooner & Kellogg, 1947.)

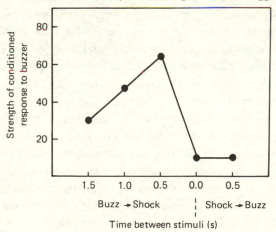

animal is able to classify neutral events prior to learning and has a rule for what classes are relevant to particular outcomes.

The phenomenon of modifiable taste aversion is often taken as one of the prime pieces of evidence for doubting general principles of associative learning. It has led some people to argue that the only sensible way to study learning is by examining it in the ecological conditions to which it is adapted (Johnston, 1981). However, common features can be found, whether a rat is learning to avoid poison or shock. If two neutral events of the same class are used, the second one interferes with learning about the first. For instance, rats were given novel saccharin solution and 15 min later were given novel vinegar solution, and finally they were made ill with lithium chloride. Subsequently, they were much less likely to avoid the saccharin solution than rats which had been given water instead of vinegar (Revusky, 1971). The vinegar had over-shadowed the saccharin.

The time-window idea probably has to be retained in a watered-down form because if an animal is given novel food followed by weeks of familiar food and finally made sick, it is unlikely that it will avoid novel food. Nonetheless, the notion of a time-window is not sufficient to account for what is found. The rule would seem to need a triple condition attached to it. *If* a neutral event has occurred within a certain time of an already important event, *if* it was of a certain category and *if* another of the same category had not been interposed between it and the important event, *then* that event itself acquires significance for the animal.

At this point it would be fair to ask: what has all the work on the avoidance of shock and poisons got to do with social behaviour? The answer is that associative aspects of learning enable the individual to cope not only with its physical environment but also with its social environment. Humphrey (1976) has argued convincingly that animals are in many ways over-equipped for the inanimate environment, but the environment provided by other animals (particularly clever ones) is especially complex, difficult to predict and difficult to control. Predators and prey have to be coped with one way or the other, but strong pressures also come from social companions. Social groups are clearly not just bands of competitors and individuals often need each other for survival. Nevertheless, group members also have to compete with each other for many necessary resources. In such competitions, it is not simply the one who is strongest who wins. It can frequently be the one with the best abilities to make complex calculations about what the others are up to. In terms of predicting and controlling the social environment, high technology can quite clearly be every bit as important as brute force. It follows that in complex animals in particular, the rules for learning the rules can be as much to do with social interaction as anything else.

Development of rules

If we are right in our inferences about the rules for associative learning, they clearly do seem to have adapted qualities. They fit the animal's information-gathering equipment to particular problems and, presumably, they have been subject to natural selection during evolution. Therefore, they must be transmitted in some way, usually genetically, from one generation to the next. At this point, though, we should not forget the hard-learned lesson that an evolutionary argument is not the same as a developmental one. The rules for modifying behaviour do not spring fully armed out of the genome. They themselves have to develop and, clearly, they represent the workings of an already functional nervous system and body. The extent to which their development involves various kinds of experience raises an entirely separate issue. As a matter of fact we know that, at least in complicated animals, many features of the rules are profoundly modified by experience.

I discovered this painfully myself when I went as a visitor to a beautifully equipped laboratory to work on learning in rhesus monkeys. The laboratory had some elegant computer-controlled apparatus for teaching the monkeys to discriminate between visual forms such as letters. If the monkey pressed the correct letter it was rewarded with a peanut by a mechanical dispenser which was specially designed for this kind of food. Everything was perfect except that when I came to train experimentally naive monkeys, I discovered that they did not like peanuts. The monkeys had to be deprived of their regular food and accustomed to the peanuts for weeks before they would take them with any readiness, let alone treat the nuts as rewards for appropriate behaviour. In this case, which is not exceptional (see Weiskrantz & Cowey, 1963), experience expanded what the monkeys regarded as acceptable food, and at an earlier stage in development experience had narrowed the range. It could be argued that in such instances an unlearnt program could still be detected at work behind the scenes since the general category of food, and its effectiveness as a reward, was in some sense built in. In other cases though, it becomes more difficult to pinpoint what might or might not act as a reward without very extensive knowledge of the animal's previous experience. For instance, the conditions in which it becomes possible for an animal to perform an act that would bring it food becomes rewarding in itself. So the animal will work in order to provide itself with those conditions. In this way lengthy chains of behaviour can be developed with any one event providing the terminating condition for one action and the enabling condition for the next (see Kelleher, 1966). This is the basis for many complex circus acts performed by animals.

Beyond this, the knowledge of the ways in which initially neutral cues are treated as potentially relevant or ignored is growing, and suggests once again that the rules for learning can be influenced by the nature of prior experience (e.g. Dickinson, 1980). Presumably, if the rules for learning are to have any universality in natural conditions, the experience which affects them must be a common feature of all the animals having the rules – in other words, the variance due to the environment is normally small in the environment to which the animal is adapted. Also, when considering development, it must be stressed that we do not have to depend on an infinite regress. Quite clearly learning does not have to be involved in the development of the rules for learning. For instance, even very young rats selectively associate taste with poison and texture with electric shock (Gemberling, Domjan & Amsel, 1980; Domjan, 1980). So, in this case, it looks as though the rules for forming some associations, but not others, are not dependent on learning for their development. But that should not make anybody complacent about the developmental processes. The classic mistake has been to confuse experiences involved in learning with all other kinds of experience which the animal can have during its development (see Lehrman, 1970). All sorts of environmental conditions can have non-specific but profound influences on behavioural development without involving learning (see Bateson, 1981). Change the social or physical conditions in which the animal is growing up and you may find it ends up with a different set of rules for learning.

It will seem obvious, I hope, that a rule for learning or for any other kind of developmental process, is not a gene written large. We have no reason to suppose that there is any simple correspondence between gene and rule for changing behaviour any more than there is an isomorphic relationship between gene and behaviour. So why make an issue out of it? The reason is that some of the influential popularizers of modern evolutionary theory still manage to confuse the developmental issue with the evolutionary one. Because gene frequencies are generally presumed to change in phylogeny, then it is suggested that genes must be doing the *real* work in ontogeny. As a result of this category mistake, total non-communication has occurred between the sociobiologists and their critics (see Bateson, 1982b). The distinct issues are relatively easy to sort out in the case of associative learning but, as will become apparent in the remainder of this chapter, they are all too easily muddled when functional approaches are brought to bear on development problems.

A functional approach to morality?

I am going to consider now a famous case in which the supposed regularities of human morality are attributed to the workings of adaptive rules, so providing an evolutionary explanation for part of human culture. The evolutionary costs in this case are those due to inbreeding, and the cultural outcome is the incest taboo. Two quite distinct arguments are mounted. Since these are sometimes confusingly conflated, it is helpful to keep them separate even though they are not mutually exclusive (see Fig. 24.2). The first argument is the classical one and runs as follows. Human beings, being observant and intelligent, spot the consequences of matings between close relatives and make safety laws about them. The incest taboo is equivalent to a legal requirement to wear seat belts or crash helmets. It is sometimes claimed that people in many cultures are aware of the ill-effects of inbreeding (Lindzey, 1967), but nobody, as far as I know, has claimed that such knowledge is universal. Certainly in modern statistical studies, differences between children of inbred and outbred marriages can be scarcely detectable, particularly when the inbred marriages occur in communities where spouses are traditionally at least first-cousins (e.g. Rao & Inbaraj, 1977). Even in outbred communities it would be very difficult to detect inbreeding costs when infant mortality was high and its causes and those of deformities in offspring were numerous and varied. The second argument about the origins of the incest taboo is the one that relates to the major theme of this chapter and I shall consider it at some length. I should emphasize first, that the two possibilities shown in Fig. 24.2 are not the only explanations for the origins of the incest taboo. Most anthropologists would prefer to look elsewhere, partly I suspect because of the way the biological arguments have been overstated.

The biological part of the second argument shown in Fig. 24.2 is well-trodden ground and has been reviewed in numerous places recently (e.g. Alexander, 1980; Fox, 1980; Thiessen & Gregg, 1980; Bixler, 1981). Westermarck (1891) believed that satisfying sexual relationships are not formed between people who have spent their childhood together. This view is supported by the behaviour of members of Israeli kibbutzim who very rarely marry the people they have grown up with (Spiro, 1958; Talmon, 1964; Shepher, 1971). Other evidence in support of the Westermarck hypothesis comes from the work of Wolf and his colleagues, summarized in a recent book (Wolf & Huang, 1980). They have analysed a form of arranged marriage which was practised in Taiwan. The wife-to-be was adopted into the family of the husband-to-be when she was a young girl. The marriage was formalized and consummated when the partners were adolescent. This form of arranged marriage, the 'minor marriage', could be compared with a more common form of arranged marriage, the 'major marriage', in which the partners met each other for the first time when they were adolescents. In a great many respects the minor marriages were less successful than the major marriages. They generated fewer children, the rates of infidelity were higher, and so forth. For instance, 15% of the 1117 minor marriages ended in divorce whereas only 6% of the 1651 major marriages did so. On a note of caution, it should be pointed out that the people involved in minor marriages were considerably younger at the time of the formalization of the marriage than those involved in the major marriages. Also a minor marriage was a much cheaper option for the parents than a major marriage and was, therefore, considered to be socially disgraceful. These factors may have contributed markedly to the relative lack of success of the minor marriages.

Another independent piece of evidence corroborating the Westermarck

Fig. 24.2. Two arguments used to relate the ill-effects of inbreeding to culturally transmitted incest taboos.

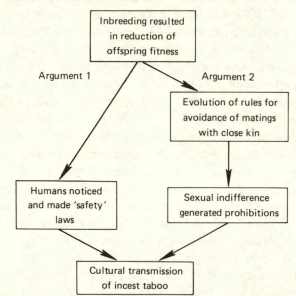

hypothesis is provided by stable incestuous relationships (Weinberg, 1956). Weinberg found that whereas most incestuous relationships were unstable and short-lived, six that he examined involved strong and lasting attachments between the partners. In each case the siblings concerned had been separated from each other when they were babies.

Three criticisms are commonly directed at the evidence for the Westermarck hypothesis. First, it is pointed out that a most preferred sexual partner is not necessarily a spouse (e.g. Solomon, 1978); however, it is difficult to see how this perfectly valid point is relevant to the evidence given above. Secondly, overt sexuality is found between siblings (Finkelhor, 1980) and among kibbutz members of the same age (Spiro, 1958; Kaffmann, 1977). Finally, despite supposed indifference to familiar members of the opposite sex, incest does occur quite frequently (Livingstone, 1980). I shall not ignore these criticisms, but at this point I think it is helpful to consider the data from animals. Apart from strengthening the view that reduced sexual responsiveness to familiar members of the opposite sex is quite widespread in birds and mammals, the animal studies also suggest ways of dealing with criticisms of the human evidence. They also point to other sourcs of variation in mate choice.

Imprinting and the discrepancy hypothesis

The effects of early experience on the mating preferences of birds and mammals has been known for a long time. The effects of imprinting on the sexual preferences of birds was made famous many years ago by Konrad Lorenz (1935). Numerous quantitative studies have been done on both birds and mammals in the last twenty years (reviews in Immelmann, 1972; Bateson, 1978a) and have shown that early experience can have profound and lasting effects on sexual preferences. Admittedly, imprinting was usually thought of as the process by which animals normally learn about the characteristics of their species, even though a reference was occasionally made to 'asexual imprinting' (Aberle *et al.*, 1963) and a number of studies were done on the reduced sexual responsiveness to familiar members of the opposite sex in rodents (reviews in Dewsbury, 1982; D'Udine & Alleva, 1983). In general, though, thinking was retarded by the dichotomous classifications that were in use at the time. Assortative mating was either positive or negative. Animals, like humans, were either endogamous or exogamous, they preferred the familiar or they preferred the novel.

It is possible to break out of the straight-jacket by applying an idea that had been used for many years in thinking about the psychology of classification and aesthetics (McClelland & Clark, 1953; Berlyne, 1960). This is known as the 'discrepancy hypothesis'. In general, what people find most stimulating and most attractive is a bit different but not too different from what they know already. As is shown in Fig. 24.3, it was a simple step to translate this into a hypothesis about the effects of early experience on mating preferences (Bischof, 1972; Bateson, 1978*a*). Experimental studies on both birds and mammals followed quickly (Bateson, 1978*b*; Gilder & Slater, 1978; McGregor & Krebs, 1982). However, it was not all plain sailing. When they were given a choice between a familiar and a novel member of the opposite sex, birds might actually choose the familiar even when the novel really did not look so very different to our eyes (Miller, 1979; Bateson, 1980; Slater & Clements, 1981). The results suggested that the birds might have sharply tuned preferences only slightly displaced away from siblings when normally reared.

The difficulty is that, if the most preferred mate is slightly different from a familiar member of the opposite sex and if we do not know how to measure the difference, we can unwittingly present the animal with a novel object which is less attractive than the familiar. It might seem as though the hypothesis is so slippery that it cannot be falsified. Indeed, this point

Fig. 24.3. The postulated relationship between the sexual response to an individual and its degree of novelty relative to familiar objects which would be kin in natural conditions (Bateson, 1978*b*). A positive response means sexual approach, and a negative response could mean escape.

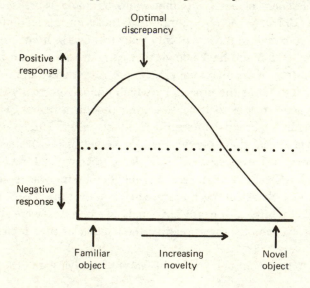

has been used as a general criticism of the discrepancy hypothesis (Thomas, 1971). However, the issue is settled by positive evidence not by ingenious explanation of the failure to confirm the idea. If members of the opposite sex were graded along a continuum from familiar to very novel and the animals were allowed to choose between all the possibilities, then progress can be made. While we do not yet know what cues the animals might use, we can exploit the likelihood that, when other qualities such as physical well-being are equal, an optimal choice of mate is likely to be one that minimizes the costs of both inbreeding and outbreeding (Bateson, 1980, 1983). In other words, the genetic relatedness of the partner is likely to be important.

It is a relatively easy matter, when we know the pedigrees of animals, to arrange choices between members of the opposite sex of different degrees of relatedness. I have done this with Japanese quail. Birds that had been reared with siblings were tested in apparatus that allowed them to be given up to six alternatives (Bateson, 1982a). In one experiment birds were given choices between members of the opposite sex that were either familiar siblings, novel siblings, novel first-cousins, novel third-cousins, or novel unrelated individuals. Fig. 24.4 shows the mean percentage durations

Fig. 24.4. The mean percentage time spent by adult Japanese quail near members of the opposite sex that were either familiar siblings, novel siblings, novel first-cousins, novel third-cousins, or novel unrelated individuals. The males (N = 22) are shown as triangles and the females (N = 13) as circles (Bateson, 1982a).

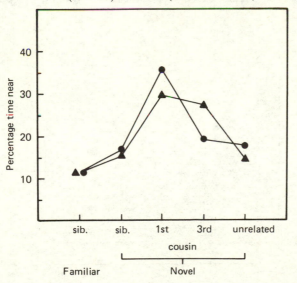

spent in front of each category of stimulus bird by both adult males and females. The time spent near novel first-cousins was significantly greater than the time spent near both the familiar and novel siblings and novel unrelated individuals. Despite the clear overall preference for first-cousins, the data were highly variable. The variability was expected as the degree of relationship between two individuals only indicates the probability that the two share heritable characters. Knowledge by a bird of a sibling's appearance must necessarily be an imperfect guide to what a first-cousin will look like.

Of course, remaining near a member of the opposite sex is not the same as mating with it. However, other experiments have shown that, in adult male Japanese quail, the time spent near a female in a choice test is strongly linked to the copulation preference (Bateson, 1978b). Furthermore, the males are observed to court the females in the choice tests. Finally, I had found in other experiments that birds show no consistent preferences for members of the same sex.

The results with quail provide direct support for the discrepancy hypothesis as applied to sexual preferences and indirect support for the

Fig. 24.5. A possible way in which a preference for a mate, differing slightly from familiar members of the opposite sex, might be generated. In the idealized case shown here, all the animals along a graded continuum initially have the same value. Imprinting reduced responsiveness to the novel and habituation reduced responsiveness to the familiar. The shaded part represents the area under the resulting preference curve.

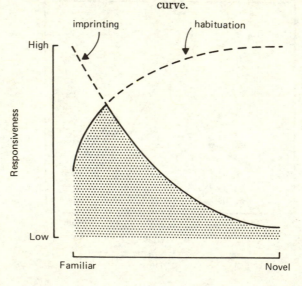

notion of optimal outbreeding. I have written elsewhere about the evolutionary pressures which might generate a balance between inbreeding and outbreeding (Bateson, 1983). However, it is worth emphasizing here that at least four costs have been proposed for inbreeding, and at least seven for outbreeding. The costs of outbreeding may include the risks of infections from pathogens carried by the partner and the breaking up in the offspring of co-adapted complexes of genes found in the parents.

Not all costs can apply to all species and some naturally outbreeding species may use other mechanisms for avoiding the costs of inbreeding, such as dispersal away from the natal area by one sex (reviewed by Greenwood, 1980). Even in those species that choose a mate that is a bit different but not too different from close kin, other factors are also important. Qualities such as the physical condition of the member of the opposite sex, the resources it holds and the extent to which it bears characters that have been subject to sexual selection can all affect whether or not it is chosen (see Halliday, 1983). It would be quite wrong to suggest that the *only* influence on mate choice is relative familiarity.

Despite the variety and the complexity, the animal work indicates first and foremost that mate choice can be profoundly influenced by early experience. Secondly, in some species the choice is remarkably finely tuned so that under certain circumstances familiarity may be preferred over novelty. The fine tuning might be achieved by employing two well-known mechanisms as shown in Fig. 24.5. Filial imprinting is known to restrict preferences to the familiar (see Bateson, 1979), and sexual imprinting could operate in exactly the same way. Habituation, by contrast, reduces responsiveness to the familiar. The net effect of superimposing habituation on imprinting would be to displace the preference away from the familiar. The combination of the two learning processes could produce a sharply peaked preference for something a bit different from the familiar when other things are equal.

Mating preferences in humans

The animal evidence enriches the discussion of human mating preferences in several important ways. First, by emphasizing that preferences are displaced away somewhat from the familiar, it is possible to explain two facets of the data from humans that would otherwise have seemed incompatible. Apart from the evidence of reduced sexual interest in familiar members of the opposite sex, which I have already mentioned, the great mass of data shows that freely chosen human spouses are more like each other than would be expected on a chance basis. Similarities are not only social and psychological, but also found in measures of body dimensions such as length of earlobe (e.g. Eckland, 1968; Lewis, 1975; Thiessen & Gregg, 1980).

A second point is that the method of testing choices draws attention to the relative nature of a measured preference. It will rarely be the case that either an animal or a human will be provided with the opportunity to mate with an absolutely ideal member of the opposite sex. Furthermore, the best available mating may be with a sibling or an offspring on certain occasions. Sexual responsiveness to a familiar member of the opposite sex may not be zero – particularly when the time allotted to searching for an alternative has run out. The conclusion is, therefore, that when an individual has no choice or an impoverished set of choices, he or she may inbreed.

Finally, the precocious sexual behaviour, which is often observed between siblings and was the basis for Freud's (1950) thinking about the development of sexual preferences, may play a role. If habituation is involved in displacing preferences away from the familiar to individuals that are slightly different, then the learning process may be facilitated by the performance of precocious sexual behaviour which is common enough in humans (Finkelhor, 1980) as it is in other animals. I am not convinced that overt sexual behaviour is essential for the development of indifference, even though it may help. But I think it is highly misleading to suggest, as Shepher (1971) has done, that development of sexual preferences is *complete* by the age of six in humans. He based this conclusion on a very few individuals who married within their peer group in the kibbutz and were found to have entered the kibbutz during their childhood and usually after the age of six. To demonstrate a sensitive period of the type he was proposing, it would be necessary to show that adults who had left a kibbutz at the age of six were not sexually attracted by members of the opposite sex whom they had been reared with while still in the kibbutz. As things stand, the existing evidence has been wildly over-interpreted both by Shepher and by others who have uncritically accepted his conclusions (e.g. Lumsden & Wilson, 1981; van den Berghe, 1982).

If we reject the naive application of the sensitive period concept and accept that familiarity of a certain kind does reduce sexual attractiveness, then it may be possible to reconcile the thinking of Freud with that of Westermarck. The sexual attraction of their siblings and parents, which people under psychoanalysis reported they felt, may have created the conditions for developing subsequent indifference. This line of thought might also be applied rewardingly to explain one striking feature of divorce statistics. For instance, in British women who married before the age of 20, the proportion of marriages that ended in divorce has been approximately double that of the marriages of women who married between 20 and 24 (Office of Population Censuses and Surveys, 1978). This has been true at any time between four and 25 years after marriage. Many factors, such as differences between social classes in attitudes to marriage, could explain or contribute to explaining the difference. Clever research design could sort out some of the confounded variables, so I shall add to the possibilities a speculation arising from my point about habituation. Early marriages may involve a great deal of intimacy but relatively little sexual satisfaction. Indeed, people often report that their early sex lives were relatively unrewarding. If the effects of habituation are not powerfully offset by rewarding sexual experience, the partner may lose his or her attractiveness and become the equivalent of a sibling.

While a great deal is still unknown about the development of sexual preferences in both animals and humans, the similarities are quite striking. In my view a good case has been made for the view that the learning processes involved in the formation of mating preference of humans have been subject to natural selection during the course of evolution. However, acceptance of this point has to be tempered by an awareness that mate choice is influenced by many qualities. Early experience with particular individuals is not the only source of variation in adults' mating preferences. So even if Westermarck was right, as I believe he was, it would be extremely surprising if his hypothesis explained all of what humans do.

Incest taboos and marriage laws

In a society in which spouses are freely chosen, it is easy to confuse the influences on sexual preference with those on marriage. The anthropologists have to point again and again to the great many societies in which spouses are arranged and not freely chosen. A biologist who was so minded could counter by arguing that, if by and large the marriages generate the children, then marriage laws could still have been influenced by evolutionary pressures. While that argument would beg an important question, on the face of it the marriage laws do seem to promote the function of optimal outbreeding. Marriage with close kin is generally forbidden in most societies and so, commonly, is marriage with people of dissimilar culture. A general bias in favour of spouses who come from nearby – both spatially and socially – has often been noted by anthropologists (Fortes, 1962). Charles Darwin married a cousin and, indeed, such marriages were quite common in nineteenth-century Europe. It must be said that the apparently convincing evidence has been processed into non-significance by elegant mathematical analysis (Hajnal, 1963). This seems to have been done in the interests of retaining the theoreticians' assumption of panmixia (Charlesworth, 1980). But maybe Darwin knew better, since the assumption of random mating could hardly apply in those numerous societies that actually *favour* first-cousin marriages (Murdock, 1967). Is there some functional similarity with mate choice in quail? Two important and unresolved problems are raised by this line of thought. How does an inhibition get translated into a prohibition? And how much of the variation in the prohibition is explained by the character of the inhibitions?

Incest taboos take many different shapes and forms. They sometimes include certain types of cousin and people related by marriage only. Prohibitions on sexual relations with parents, siblings or children, are nearly always universal – but not quite. Hopkins (1980) has analysed the marriage records of Roman Egypt in which the census data were especially complete. He confirmed the view that among perfectly ordinary people, who were neither Pharoahs nor priests, full brother–sister marriages occurred in a minimum of nine out of the 113 marriages he analysed. If the less certain cases are also included along with marriages between half-siblings, the proportion of incestuous marriages was of the order of 20%. Whatever view one takes of the origins of the incest taboos, it would be intellectually shoddy simply to ignore these data.

Can we find a common underlying principle that explains the variation? Levi-Strauss (1969) proposed that, women being the most important resource that men have, a system for exchanging women always underlies the social control of marriage. His arguments have something of the character of the Ptolemaic theory of the universe – brilliant, logical, grand, but despite all these things, extremely complex. His theory is in stark contrast to the biologists' attempts to find a relationship between the prohibitions on certain types of marriage partner, and the inhibitions about having sexual contact with such classes of people. These ideas are simple. So simple, indeed, that usually the argument is not stated at all. The correlation between prohibition and inhibition is offered as though it explained everything, or at best, the prohibition is held to have arisen by 'myth making' or 'ritualisation' (Bischof, 1972). As Williams (1978) pointed out, such statements are nothing more than promissory notes. They do not provide an explanation of how one turns into the other.

In a recent book, Lumsden & Wilson (1981) attempted to deal with the issue mathematically. However, their efforts have not provided the explanation that we need. They merely assumed that inhibition generates the incest taboo without pointing to any behavioural mechanism that could translate one into the other. In lieu of anything better I shall make a suggestion. Prohibitions may have arisen from the social pressure directed against unorthodox behaviour. What is normal behaviour is itself influenced by the pattern of early experiences which are common to that society. The implication is, therefore, that there will be some correlation between child-rearing practices and taboos (see Fig. 24.6). People often strongly disapprove of others who behave in unusual ways. The most obvious example is the moral repugnance that many people show for homosexuality between consenting adults. Why should they mind? They are not harmed by the homosexuality. But the conventional response is nonetheless a violent one – in some societies homosexuality may be punished by death. If fear of nonconformity and the unusual has driven the cultural evolution of incest taboos, then a comparable argument should apply to taboos on marriages with strangers or members of other castes and races. Levi-Strauss (1969) noted that such taboos certainly exist and a notorious modern example of it is found in the immorality laws of South Africa which forbid sexual relations between blacks and whites.

If this approach is anywhere near correct, we should expect a correlation between the class of people who are prohibited as sexual partners and the likelihood that they will be familiar or extremely novel (see Fig. 24.6). Obviously many things might muddy the correlation between social structure and prohibited partners, but one piece of existing evidence points in the right direction. While first-cousins are often favoured as marriage partners, a distinction is very often made between parallel and cross-cousins. The sibling parents of parallel cousins are the same sex and the sibling parents of cross-cousins are of the opposite sex. Sometimes parallel cousins

Fig. 24.6. Possible links between particular types of early experience and the particular prohibitions against unusual behaviour.

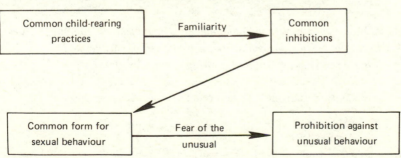

are forbidden as spouses and cross-cousins are favoured. Alexander (1980) has gone through Murdock's (1967) ethnographic atlas and found that this asymmetrical treatment of cousins is strongly associated with the type of marriage common in that culture. The results are shown in Table 24.2. Alexander argues that because brothers may share wives in polygynous societies, parallel cousins may in fact be half-siblings. He seems, therefore, to be using the evidence as an ingenious updating of the classical safety law argument. However, parallel cousins who may in reality be half-siblings are also likely to have lived in the same household. Once again, genetic relatedness is confounded with familarity. So, another possible explanation is that in a polygynous society, the parallel cousins will be much more likely to grow up together than the cross-cousins. In effect, the parallel cousins are as familiar as siblings.

It does not follow from the argument I have mounted here that the conformism generating prohibitions is an adaptive response that evolved in the service of maintaining optimal outbreeding. The conformism might have arisen for quite different reasons and among its other consequences happened incidentally to amplify the beneficial effects of the inhibitions. Certainly, it would be difficult to argue that *all* the variation in human marriage laws could be explained in terms of their evolutionary benefits. The brother–sister marriages of Roman Egypt probably had a great deal to do with the preservation of property and nothing to do with the preservation of genes (Hopkins, 1980). And similar explanations can account for a lot of the variation found in human societies (Goody, 1976).

It would be absurd to adopt a rigidly determinist view of what has gone on in the formation of culturally transmitted marriage laws. To say the least, it is unfortunate that a multiply influenced process with many stages in it should be thought of by sociobiological proponents and their critics alike (e.g. Solomon, 1978) as having an invariant outcome and a single explanation. It is a bit like arguing that if the hypothesis that smoking

Table 24.2. *The numbers of polygynous and monogamous human societies that have symmetrical or asymmetrical treatment of parallel and cross-cousins in respect of marriage (from Alexander, 1980)*

Cousin treatment	Polygynous	Monogamous
Symmetrical	4	66
Asymmetrical	75	35

caused lung cancer is to be believed, *everybody* who smokes must get cancer; what is more *nothing* else (such as asbestos) can be admitted as having the same effect. The opposed parties in such disputes, evidently believing that causality has a chain-like character to it, have an impoverished notion of how things actually work. In most complex systems the sources of variation are likely to be numerous. It follows that alternative explanations do not have to be mutually exclusive. The best that can be said about marriage laws is that *some* of the variation may be explained along the lines proposed here. Even within this explanatory framework, little useful understanding will be obtained without first studying the problem at many different levels. What is needed, therefore, is constructive collaboration between biologists and social scientists and a proper respect for the insights that the different disciplines can provide.

Conclusion

The arguments presented in this chapter do not lead to comfortable conclusions that can be instantly assimilated. However, they are not, I trust, obscurantist and a positive point does emerge. Regularities *can* be found in the way that behaviour is tuned to the environment during development. To be effective, though, the rules have a conditional character to them which means, of course, that they generate variation in behaviour in response to variable environments. It is not inexplicable variation, but it is variation nonetheless.

The old reductionist's vision was that one day when we knew enough about genes we would be able to predict every detail of every adult's behaviour. It always was a pipe-dream. Imagine civil servants working in a capital city and trying to make *all* the decisions required for running a large country. They simply would not have the flexibility or the speed of reaction to cope with the complexities of everyday life. And so it is with the genes, the natural bureaucrats. On their own they are too clumsy in their form of regulation to provide the necessary adaptations, especially those required for social living. The genes had to delegate control.

When we examine animals with nervous systems that were built with conditional rules for dealing with the external environment, the business of predicting how they will respond on the basis of knowing how they were made becomes impossible. It is like trying to predict the outcome of a game of chess before anyone has made a move. What we *can* do is attempt to get hold of the rules of the game so that we can make sense of a game as it is played. At that stage I concede happily that we may be able to predict what a clever animal will do in a particular set of circumstances. In the meantime, we should expect to be surprised very often.

I have greatly benefited from the comments of friends in a number of different disciplines. I am particularly grateful to the social scientists who went to a lot of trouble to read and comment on a draft of this chapter while remaining sceptical about the relevance of biological thought to their subject. I should like to thank the following: Tony Dickinson, Meyer Fortes, Nick Humphrey, Robert Hinde, Edmund Leach, Alan Macfarlane, Steven Rose, Joan Stevenson-Hinde and Bernard Williams.

Further reading

For a discussion of the distinctiveness and interplay of ideas about the evolution of behaviour and the development of behaviour, see Bateson, P. (1982), Behavioural development and evolutionary processes, in: *Current Problems in Sociobiology*, ed. King's College Sociobiology Group, pp. 133–51. Cambridge: Cambridge University Press. Argument for and against a biological approach to human marriage customs can be found in Stent, G. S. (ed.) (1978) *Morality as a biological phenomenon*. Berlin: Dahlem Konferenzen and also in the article and commentary following van den Berghe, P. L. (1983) Human inbreeding avoidance: Culture in nature. *Behavioral and Brain Sciences*.

References

Aberle, D. F., Bronfenbrenner, J., Hess, E. H., Miller, D. R., Schneider, D. M. & Spuhler, J. N. (1963). The incest taboo and the mating patterns of animals. *American Anthropologist*, **65**, 253–65.

Alexander, R. D. (1980). *Darwinism and human affairs*. Pitman: London.

Bateson, P. P. G. (1976). Rules and reciprocity in behavioural development. In *Growing Points in Ethology*. ed. P. P. G. Bateson & R. A. Hinde, pp. 401–21. Cambridge: Cambridge University Press.

Bateson, P. P. G. (1978a). Early experience and sexual preferences. In *Biological Determinants of Sexual Behaviour*. ed. J. B. Hutchison, pp. 29–53. Chichester: Wiley.

Bateson, P. [P. G.] (1978b). Sexual imprinting and optimal outbreeding. *Nature*, **273**, 659–60.

Bateson, P. [P. G.] (1979). How do sensitive periods arise and what are they for? *Animal Behaviour*, **27**, 470–86.

Bateson, P. [P. G.] (1980). Optimal outbreeding and the development of sexual preferences in Japanese quail. *Zeitschrift für Tierpsychologie*, **53**, 231–44.

Bateson, P. [P. G.] (1981). Ontogeny of behaviour. *British Medical Bulletin*, **37**, 159–64.

Bateson, P. [P. G.] (1982a). Preferences for cousins in Japanese quail. *Nature*, **295**, 236–37.

Bateson, P. [P. G.] (1982b). Behavioural development and evolutionary processes. In *Current Problems in Sociobiology*. ed. King's College Sociobiology Group, pp. 133–51. Cambridge: Cambridge University Press.

Bateson, P. [P. G.] (1983). Optimal outbreeding. In *Mate Choice*, ed. P. Bateson, pp. 257–77. Cambridge: Cambridge University Press.

Berlyne, D. E. (1960). *Conflict, arousal and curiosity.* New York: McGraw Hill.

Bischof, N. (1972). The biological foundations of the incest taboo. *Social sciences information*, 11, 7–36.

Bixler, R. H. (1981). The incest controversy. *Psychological Reports*, 49, 267–83.

Bonner, J. T. (1980). *The Evolution of Culture in Animals.* Princeton: Princeton University Press.

Charlesworth, B. (1980). *Evolution in age-structured populations.* Cambridge: Cambridge University Press.

Chomsky, N. (1965). *Aspects of the Theory of Syntax.* Cambridge, Mass.: MIT Press.

Dewsbury, D. A. (1982). Avoidance of incestuous breeding in two species of *Peromyscus* mice. *Biology of Behavior*, 7, 157–69.

Dickinson, A. (1980). *Contemporary Animal Learning Theory.* Cambridge: Cambridge University Press.

Domjan, M. (1980). Ingestional aversion learning: unique and general processes. In *Advances in the Study of Behavior*, vol. 11, ed. J. S. Rosenblatt, R. A. Hinde, C. Beer & M.-C. Busnel, pp. 275–336. New York: Academic Press.

D'Udine, B. & Alleva, E. (1983). Sexual preferences in rodents. In *Mate Choice*, ed. P. Bateson, pp. 311–27. Cambridge: Cambridge University Press.

Eckland, B. K. (1968). Theories of mate selection. *Eugenics Quarterly*, 15, 71–84.

Finkelhor, D. (1980). Sex among siblings: A survey on prevalence, variety, and effects. *Archives of Sexual Behavior*, 9, 171–94.

Fisher, J. & Hinde, R. A. (1949). The opening of milk bottles by birds. *British Birds*, 42, 347–57.

Fortes, M. (1962). Introduction. In *Marriages in Tribal Societies*, ed. M. Fortes. Cambridge: Cambridge University Press.

Fox, R. (1980). *The Red Lamp of Incest.* London: Hutchison.

Freud, S. (1950). *Totem and Taboo.* London: Routledge & Kegan Paul.

Galef, B. G. (1976). Social transmission of acquired behavior: a discussion of tradition and social learning in vertebrates. In *Advances in the Study of Behavior*, vol. 6, ed. J. S. Rosenblatt, R. A. Hinde, E. Shaw & C. Beer, pp. 77–100. New York: Academic Press.

Gemberling, G. A., Domjan, M. & Amsel, A. (1980). Aversion learning in 5-day-old rats: Taste-toxicosis and texture-shock associations. *Journal of Comparative and Physiological Psychology*, 94, 734–000.

Gilder, P. M. & Slater, P. J. B. (1978). Interest of mice in conspecific male odours is influenced by degree of kinship. *Nature*, 274, 364–5.

Goody, J. (1976). *Production and Reproduction.* Cambridge: Cambridge University Press.

Greenwood, P. J. (1980). Mating systems, philopatry and dispersal in birds and mammals. *Animal Behaviour*, 28, 1140–62.

Hajnal, J. (1963). Concepts of random mating and the frequency of consanguineous marriages. *Proceedings of the Royal Society of London*, B, 159, 125–77.

Halliday, T. R. (1983). The study of mate choice. In *Mate Choice*, ed. P. Bateson, pp. 3–32. Cambridge: Cambridge University Press.

Hinde, R. A. (1968). Dichotomies in the study of development. In *Genetic and Environmental Influences on Behavior*, ed. J. M. Thoday & A. S. Parkes, pp. 3–14. Edinburgh: Oliver & Boyd.

Hinde, R. A. & Fisher, J. (1951). Further observations on the opening of milk bottles by birds. *British Birds*, **44**, 393–96.

Hopkins, K. (1980). Brother-sister marriage in Roman Egypt. *Comparative Studies in Society and History*, **22**, 303–54.

Humphrey, N. K. (1976). The social function of intellect. In *Growing Points in Ethology*, ed. P. P. G. Bateson & R. A. Hinde, pp. 303–21. Cambridge: Cambridge University Press.

Immelmann, K. (1972). Sexual and other long-term aspects of imprinting in birds and other species. In *Advances in the Study of Behavior*, vol. 4, ed. D. S. Lehrman, J. S. Rosenblatt, R. A. Hinde & E. Shaw, pp. 147–74. New York: Academic Press.

Johnston, T. D. (1981). Contrasting approaches to a theory of learning. *Behavioural and Brain Sciences*, **4**, 125–73.

Kaffmann, M. (1977). Sexual standards and behavior of kibbutz adolescents. *American Journal of Orthopsychiatry*, **47**, 207–17.

Kelleher, R. T. (1966). Chaining and conditioned reinforcement. In *Operant Behavior*. ed. W. K. Honig, pp. 160–212. New York: Appleton-Century-Crofts.

Lehrman, D. S. (1970). Semantic and conceptual issues in the nature-nurture problem. In *Development and Evolution of Behavior*, ed. L. E. Aronson, E. Tobach, D. S. Lehrman & J. S. Rosenblatt, pp. 17–52. San Francisco: Freeman.

Levi-Strauss, C. (1969). *The Elementary Structures of Kinship*. Boston: Beacon Press.

Lewis, R. A. (1975). Social influences on marital choice. In *Adolescence in the Life Cycle*, ed. S. E. Dragastin & G. H. Elder, pp. 211–24. New York: John Wiley.

Lewontin, R. C. (1978). Adaptation. *Scientific American*, **239** (3), 157–69.

Lindzey, G. (1967). Some remarks concerning incest. The incest taboo and psychoanalytic theory. *American Psychologist*, **22**, 1051–9.

Livingstone, F. B. (1980). Cultural causes of genetic change. In *Sociobiology: beyond nature/nurture?* ed. G. W. Barlow & J. Silverberg, pp. 307–29. Boulder, Colorado: Westview.

LoLordo, V. M. (1979). Selective associations. In *Mechanisms of Learning and Motivation: A memorial volume to Jerzy Konorski*, ed. A. Dickinson & R. A. Boakes, pp. 367–98. Hillside, New Jersey: Erlbaum.

Lorenz, K. (1935). Der Kumpan in der Umvelt des Vogels. *Journal für Ornithologie*, **83**, 137–213, 289–413.

Lorenz, K. (1965). *Evolution and modification of behavior*. Chicago, Illinois: University of Chicago Press.

Lumsden, C. J. & Wilson, E. O. (1981). *Genes, Mind, and Culture*. Cambridge, Mass.: Harvard University Press.

MacCorquodale, K. & Meehl, P. E. (1948). On a distinction between hypothetical constructs and intervening variables. *Psychological Review*, **55**, 95–107.

McClelland, D. C. & Clark, R. A. (1953). Discrepancy hypothesis. In *The Achievement Motive*, ed. D. C. McClelland, J. W. Atkinson, R. A. Clark & E. L. Lowell, pp. 42–66. New York: Appleton-Century-Crofts.

McGregor, P. K. & Krebs, J. R. (1982). Mating and song types in the great tit. *Nature*, **297**, 60–1.

Miller, D. B. (1979). Long-term recognition of father's song by female zebra finches. *Nature*, **280**, 389–91.

Murdock, G. P. (1967). *Ethnographic atlas*. Pittsburgh: University of Pittsburgh Press.

Office of Population Censuses and Surveys (1978). *Demographic Review*. London: Her Majesty's Stationery Office.

Pulliam, H. R. & Dunford, C. (1980). *Programmed to learn*. New York: Columbia University Press.

Rao, P. S. S. & Inbaraj, S. G. (1977). Inbreeding effects on human reproduction in Tamil Nadu of South India. *Annals of Human Genetics*, **41**, 87–98.

Revusky, S. (1971). The role of interference in association over a delay. In *Animal Memory*, ed. W. K. Honig & P. H. R. James, pp. 155–213. New York: Academic Press.

Shepher, J. (1971). Mate selection among second generation kibbutz adolescents and adults: Incest avoidance and negative imprinting. *Archives of Sexual Behavior*, **1**, 293–307.

Skinner, B. F. (1959). A case history in scientific method. In *Psychology: A study of a Science* Vol. 2, ed. S. Koch, pp. 359–79. New York: McGraw Hill.

Slater, P. J. B. & Clements, F. A. (1981). Incestuous mating in zebra finches. *Zeitschrift für Tierpsychologie*, **57**, 201–8.

Solomon, R. C. (1978). Sociobiology, morality, and culture group report. In *Morality as a biological phenomenon*, ed. G. S. Stent, pp. 283–308. Berlin: Dahlem Konferenzen.

Spiro, M. E. (1958). *Children of the Kibbutz*. Cambridge, Mass.: Harvard University Press.

Spooner, A. & Kellogg, W. N. (1947). The backward conditioning curve. *American Journal of Psychology*, **60**, 321–34.

Talmon, Y. (1964). Mate selection in collective settlements. *American Sociological Review*, **29**, 491–508.

Thiessen, F. & Gregg, B. (1980). Human assortative mating and genetic equilibrium: An evolutionary perspectives. *Ethology & Sociobiology*, **1**, 111–40.

Thomas, H. (1971). Discrepancy hypothesis: methodological and theoretical considerations. *Psychological Reviews*, **78**, 249–59.

Thorpe, W. H. (1956). *Learning and Instinct in Animals*. London: Methuen.

Tinbergen, N. (1953). Specialists in nest-building. *Country Life*, 30 January, 270–71.

van den Berghe, P. L. (1983). Human inbreeding avoidance: Culture in nature. *Behavioral and Brain Sciences*.

Weinberg, S. K. (1956). *Incest Behavior*. Secaucus, New Jersey: Citadel Press.

Weiskrantz, L. & Cowey, A. (1963). The aetiology of food reward in monkeys. *Animal Behaviour*, **11**, 225–34.

Westermarck, E. (1891). *The History of Human Marriage*. London: Macmillan.

Williams, B. A. O. (1978). Conclusion. In *Morality as a biological phenomenon*, ed. G. S. Stent, pp. 309–20. Berlin: Dahlem Konferenzen.

Wolf, A. P. & Huang, C. (1980). *Marriage and adoption in China, 1845–1945*. Stanford, California: Stanford University Press.

25

Aspects of human evolution

GLYNN LL. ISAAC

Understanding the literature on human evolution calls for the recognition of special problems that confront scientists who report on this topic. Regardless of how the scientists present them, accounts of human origins are read as replacement materials for genesis. They fulfil needs that are reflected in the fact that all societies have in their culture some form of origin beliefs, that is, some narrative or configurational notion of how the world and humanity began. Usually, these beliefs do more than cope with curiosity, they have allegorical content, and they convey values, ethics and attitudes. The Adam and Eve creation story of the Bible is simply one of a wide variety of such poetic formulations.

We are conscious of a great change in all this, starting in the eighteenth and nineteenth centuries. The scientific movement which culminated in Darwin's compelling formulation of evolution as a mode of origin, seemed to sweep away earlier beliefs and relegate them to the realm of myth and legend. Following on from this, it is often supposed that the myths have been replaced by something quite different, which we call 'science'. However, this is only partly true; scientific theories and information about human origins have been slotted into the same old places in our minds and our cultures that used to be occupied by the myths. The information component has then inevitably been expanded to fill the same needs. Our new origin beliefs are in fact surrogate myths, that are themselves part science, part myths.

It is also true that the study of human evolution is a meeting ground of science and humanism. We can and should seek to be rigorous in our testing of propositions and hypotheses so that we achieve an expanding corpus of secure information and orderly knowledge – but as I have already indicated, the meaning that people attach to these findings will surely be affected in subtle and complex ways by variations in individual experience of humanity and by the ethos of the times. Just as historians expect to have to rewrite continuously the comprehension of history, so will consumers of human evolution evidence want to re-evaluate the meaning of their 'facts'.

Clear examples of myth-making extensions of scientific information include the embellishment of the man-the-hunter theme. Archaeology does provide strong indications of early hominid involvement in the acquisition and consumption of meat from large animals. However, romantic and symbolic meanings that go far beyond the empirical information have commonly been attached to the evidence. In fact, themes that are common in folklore, mythology and the scriptures are unconsciously attached to the archaeological findings (cf. Morgan, 1972; Perper & Schrire, 1977). Similarly, in recent years, in order to redress the imbalances of years of unconscious male bias in versions of the story of human evolution, women writers have set forth female-gathering hypotheses as rivals. Various of these are perfectly plausible and deserve to be tested, but meanwhile they clearly have the same social function as legitimizing myths. Further evidence of the emotional charge that attaches to the form and content of interpretations of human evolution can be seen in the reception given to sociobiology. People clearly do want to be free to choose their evolutionary origin stories. Bear this in mind as you read this and other accounts of human evolution.

As a starting point, I am going to take a question: What has science learned about human evolution that was not known to Charles Darwin when in 1871 he wrote *The Descent of Man*? I shall then go on to look briefly at some points of current debate, and at lines of enquiry that are now getting under way.

In 1871 the Neanderthal and Gibraltar skulls were the only significant human fossils known. In his 1863 essay, *Man's Place in Nature*, Huxley had already shown that the Neanderthal form was effectively a variant of the human type rather than an evolutionary link, so that Darwin's concept of human evolution was of necessity 'fossil free' (cf. Pilbeam, 1980). Darwin based his ideas on comparative anatomy plus what little was known of the natural history of apes, plus a general knowledge of human behaviour patterns (augmented by his own ethnographic observations during the voyage of the *Beagle*). Darwin's notions were configurational rather than fully narrative (Fig. 25.1). He envisaged a series of interconnected ingredients which promoted or had been promoted by natural selection to produce humanity. The key elements were non-arboreal habitat plus upright bipedal stance with the hands free. Darwin perceived this condition to be helpless and defenceless, and he envisaged two concurrent lines of evolutionary solution – the use of the hands to make and wield tools and weapons plus the development of '...social qualities which lead him to give and receive aid from his fellow men'. To this mix was added 'the natural selection arising from the competition of tribe with tribe...together with the inherited effects of habit,...which...would under favourable conditions have sufficed to raise man to his present high position in the organic scale'. Darwin wrote 'The small strength and speed of man, his want of natural weapons are more than counterbalanced...by his intellectual powers, through which he has formed for himself weapons, tools etc'. One can thus read Darwin's writing as arguing both that the brain led the way in evolutionary change and also that it followed. For this reason the elements are best viewed as a configuration rather than a

Fig. 25.1. Darwin's concept of human evolution involved an initial event, 'leaving the trees', followed by the adoption of a bipedal stance with hands free. Natural selection then acted on a system of adaptation that involved tools, weapons, social cooperation and tribal warfare...finally delivering an animal with an enlarged brain and a strong moral sense.

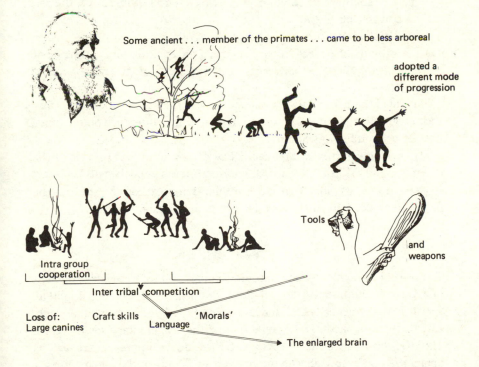

Some ancient ... member of the primates ... came to be less arboreal

adopted a different mode of progression

Tools

and weapons

Intra group cooperation

Inter tribal competition

Loss of: Large canines

Craft skills

Language

'Morals'

The enlarged brain

narrative. However, there is one storyline, namely that the process began with leaving the trees and adopting an upright stance.

So, in *The Descent of Man*, there is a first approximation that brought a whole series of potentially important topics and issues up for consideration: habitat shifts, bipedalism, tool use and social systems involving reciprocal altruism...to say nothing of group selection. As I shall show, with the exception of group selection, these elements are still part of discourse; they contend for pride of place in explanation, they are all subjects of active investigation and we have by no means yet succeeded in evaluating their interaction, or their relative importance. It might well be asked, what in that case is new? In reply one can best respond that since Darwin, there have been surges of growth in several major fields of enquiry that relate closely to understanding human evolution:

(1) recovery of fossil hominoids and hominids;
(2) excavation of archaeological evidence covering two or three million years of co-evolution of brain and culture;
(3) field studies of primate behaviour and ecology;
(4) analytic studies of the ecology and social systems of human hunter-gatherers;
(5) effective enquiry into climatic and environmental changes over the past several million years;
(6) biochemical measurement of degrees of relatedness;
(7) the growth of explicit theory concerning evolutionary ecology and concerning relations between social systems, ecology and population genetics.

Several of these lines have developed only in the last 10 or 20 years and their eventual implications for our understanding of human evolution have not yet been fully worked out.

My own research is on archaeological evidence for early stages in the differentiation of human-like behaviour patterns, and because a short review cannot be comprehensive, I am going to focus on the study of the past, rather than on the implications of neontological studies.

Biochemical evidence

The rise of biochemistry and molecular biology has led to the development of rigorous, quantitative information on degrees of relationship (Fig. 25.2). This in turn calls both for revisions to the family tree and for some careful thinking about interpretations of the fragmentary fossil record. There has been heated debate over both these issues, but rearguard actions apart, the battle now appears to be over, in favour of the biochemistry.

Fig. 25.2 summarizes the evidence that the human lineage diverged from a common ancestor with African apes no more than four to six, or seven million years ago. Chimpanzees and gorillas emerge as sibling species with humans.

If rather than molecules, the physiology and behaviour of humans is compared with that of their close living relatives, then particularly important contrasts can be seen in three, or maybe four, systems.

(1) The *locomotor system* with modifications especially of the foot, pelvic complex, the back and the hands.

(2) The *socio-reproductive system*, with human females losing conspicuous oestrus and concealing ovulation, and with the males investing in the feeding of offspring and mates.

(3) The *brain–speech–technology–culture system*, involving an enlarged modified brain, prolonged infancy with an extended learning period, during which young humans must assimilate language, the making and using of tools and complex bodies of customs, rules and information.

(4) The *food choice–masticatory and digestive system*. Modern humans tend to eat a rather different diet from apes with strong biases towards including more animal tissue and more starch. Relative to apes, humans have small front teeth and thick-enamelled back teeth. This was even more pronounced at an early stage in hominoid divergence.

Accepting these contrasts, students of human evolution are then confronted by a number of questions. What was the common ancestral condition? When did each of these systems start to become modified? Under what circumstances did change occur? What have been the selection patterns favouring these trends of change? We are thus facing a challenge first to determine the sequence of changes, which I shall term the *narrative* and second to enquire into the mechanisms, which can be called the *dynamics*. It is the first of these that has commanded most explicit attention during the century since Darwin, and indeed this is often treated as being synonymous with the study of human evolution. However, in recent years, curiosity about mechanisms has become steadily more conspicuous and in the last part of this chapter I shall argue that it is here that some of the most exciting growing points in research are to be discerned.

Narrative

For the past 100 years, most of the specific effort devoted to this branch of science has been in pursuit of missing chapters in the story. I will attempt, with the aid of diagrams, to provide a simplified summary of the sequence

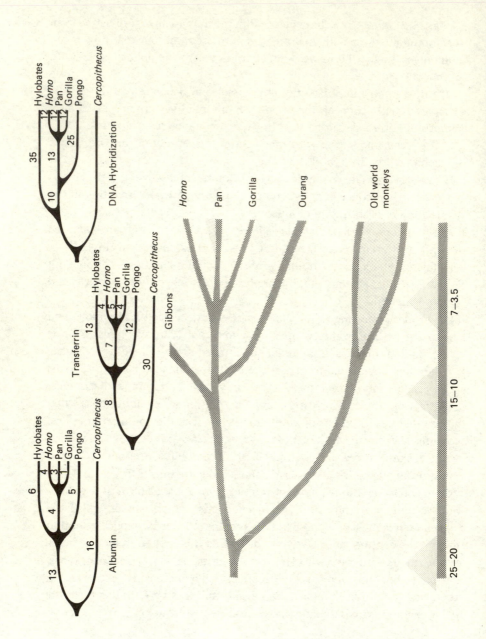

of changes as they are now known. However, before doing so, it is fun to be able to draw attention to a recent analysis of what goes on when scientists deal with origins.

Misia Landau (1981) has done a careful analysis of a series of accounts of human evolution starting with Darwin's own writings. She points out two main things: (1) the same elements tend to recur in the accounts though they may be arranged in different order. These elements or episodes are 'terrestriality' (coming to the ground), 'bipedalism', 'encephalization' (brain and intellect enlargement), 'civilization' (technology, custom, tradition, social morals, etc.); (2) although the specifics and the order may vary, the accounts tend to a common structure, which under scrutiny emerges as the structure of folkloric hero tales (cf. Propp, 1968). One version of this analysis is playfully suggested in Fig. 25.3. Misia Landau has called the literary genre of the narratives the 'anthropogenic'.

When I first read Landau's work, I became worried. Was there any way to present a sequential account that did not involve the hero-story structure? If not, did that disqualify study of the narrative of human evolution from being science?

I have got used to the idea now – and would counter-argue that provided the fit between the stories and empirical evidence is improvable through testing and falsification, then this is indeed science. (If any of the rest of the scientific community is inclined to snigger at the embarrassment of palaeoanthropologists over all this, pause and reflect. I bet that the same basic findings would apply to accounts of the origin of mammals, or of flowering plants, or of life... or even the big bang and the cosmos.)

One of the major developments since the time of Darwin has been the

---

Fig. 25.2. Biochemical evidence concerning degrees of relatedness among the apes and Old World monkeys. Along the top are examples of a series of molecule types for which comparative matrices have been determined. From these matrices trees can be drawn showing branching sequence and quantitative estimates of differences attributed to each internode. Taken together, these and other biochemical determinations provide unambiguous evidence of the branching sequence shown below. Given strong evidence of stochastic consistency in the rates of cumulative change, estimates of divergence times can be offered. The time scale is a 'rubber ruler', which can be proportionately stretched or shrunk, but if this portion of the overall vertebrate phylogenetic tree were to be stretched beyond the limits shown, one would have great difficulty accommodating the estimates for divergence times in the rest of the Mammalia and Vertebrata. (Based on Sarich & Cronin, 1976; see also Wilson, Carlson & White, 1977 for a review and this volume for new confirmatory evidence.)

recovery of substantial numbers of hominoid and hominid fossils from all over the Old World. Our grasp of several phases of human evolution need no longer be fossil-free.

Fig. 25.4 provides histograms that give some idea of the distribution of finds shown in relation to sequence of discovery and to time and geography. This summary chart deals with the romance of exploration and discovery as far as I intend to take it in this review. The legitimate excitement which scientists and lay folk alike feel over the finding of missing links, is thoroughly familiar from newspapers and magazines – as is the existence of still another involvement of palaeoanthropology with hero mythologies. (As an aside, it is also apparent from this popular literature that bits of old hominid bone arouse excitements quite beyond their information content. Our field has unwittingly got mixed up in latter-day sacred-ancestor fetishism!)

Fig. 25.5 shows that for the last four million years a useful if still somewhat patchy fossil record of members of the family Hominidae has

Fig. 25.3. Accounts of human origins are almost invariably cast as narratives that follow the format of hero tales in folklore (Landau, 1981). Here one common version is shown involving leaving the forest, struggling in the harsh savannah, being granted aid by a donor (natural selection) which promotes solutions in the form of brain enlargement, tools, hunting and social living. Final triumph takes the form of a spread across the globe and the development of art and civilization.

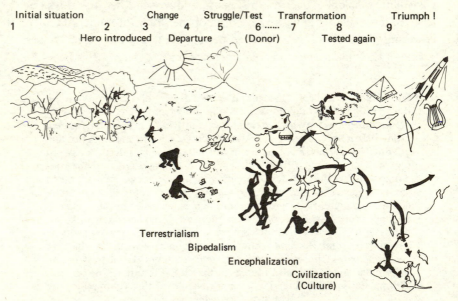

Fig. 25.4. The romance of discovery expressed as a histogram. Bars represent numbers of specimens recovered in each continent in each decade since 1840 (based on Oakley, Campbell & Mollison, 1971–77). The upper frame shows all specimens, the lower shows only crania, maxillae, mandibles and ± whole skeletons. Major discoveries are indicated by place and person. The sequence is shown in which the various taxa were recovered. In general, the most human-like taxa were found first and this has affected interpretation (see Reader, 1980). Records for 1970–80 are incomplete.

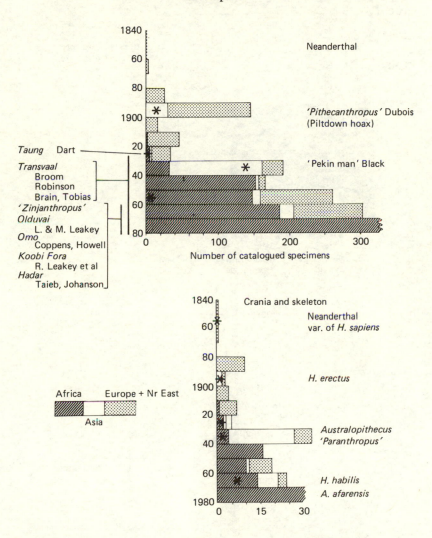

Fig. 25.5. As for Fig. 25.4 but with specimen numbers shown relative to variable divisions of a chronometric time scale. The time ranges of taxa in common usage are also shown (bottom right). *Note*: Before about 1–2 mya all material comes from Africa while Europe dominates the Upper Pleistocene record because of the large number of caves which have been excavated there. These are mainly 'neanderthal' fossils.

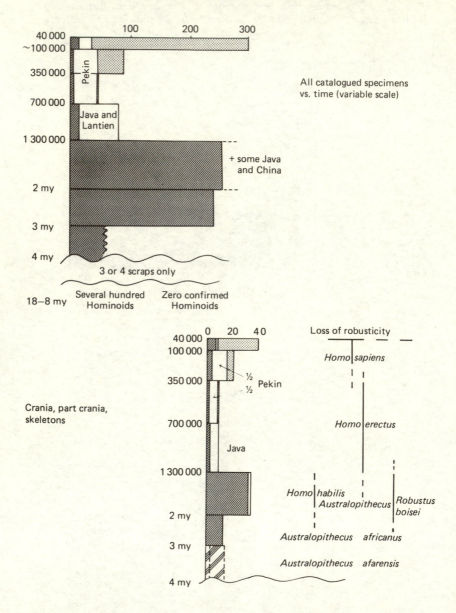

.been recovered. We need more, but what we have is a handy start. A similar useful but patchy series of hominoid fossils has been recovered from the time range from about eighteen to eight million years ago. In spite of loudly enunciated early claims to the contrary, a concensus is now emerging that none of these earlier hominoids can be classified as members of the family Hominidae.

Between the Miocene and the Pliocene to Pleistocene fossil samples, there is a four to five million year gap, a period for which we have as yet virtually no hominoid fossils of any kind. This is the period during which the biochemical evidence would indicate that hominids, chimps and gorillas separated. Although we know more than Darwin did about the range of skeletal organization patterns that existed before divergence, our interpretation of the divergence itself is still obliged to be fossil free.

Perhaps 95% of the hominoid palaeontology literature deals directly or indirectly with taxonomy and naming. This is a necessary, but boring topic, and I propose to deal with it in diagrams (Fig. 25.6a, b). The genera represented here are clustered by Pilbeam into four families. Each of these can be thought of as a small-scale adaptive radiation. Drawing phyletic lines is much more speculative. Fig. 25.6b shows three schemes for deriving the later families (radiations) from the earlier ones. It should be noted that Scheme a, which has for a long time been confidently advocated by palaeontologists, is contradicted by the steadily strengthening corpus of biochemical evidence.

Note that while there is a four-million year record of fossil hominidae there is no record at all for the chimp or gorilla. Pilbeam (1980) has rightly pointed out that securing such a fossil record would be a major contribution to understanding human evolution.

Fig. 25.7 presents a highly simplified summary of successive anatomical shifts that can be detected from the samples of hominid fossils which have so far been recovered. These span the time range four million years to the present. As I understand the record, it divides in two parts with the features of the first (4–2 mya) part being:

(1) The earliest known hominids were *fully bipedal* (although their feet, hands and shoulders may still have been more adapted for tree-climbing than are modern human limbs and extremities). All subsequent hominids were fully bipedal (Lovejoy, 1978; McHenry & Temerin, 1979).

(2) The earliest hominids had *large cheek teeth with thick enamel* plus relatively small anterior teeth. Canines were reduced relative to both Dryopithecine apes and modern pongids, but in the very earliest sample

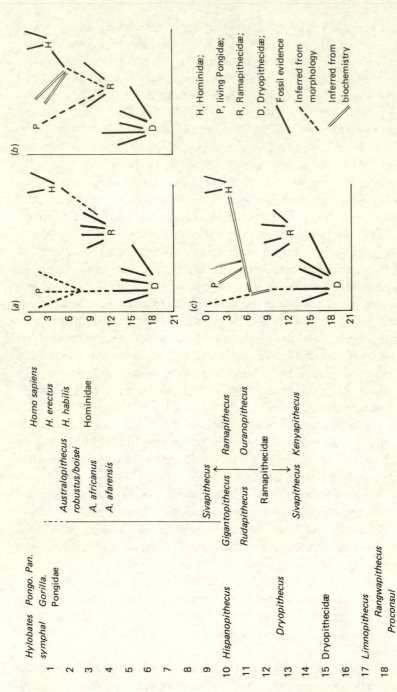

1 *Hylobates Pongo. Pan.*
symphal Gorilla.
Pongidae

2 *Australopithecus* *Homo sapiens*
robustus/boisei *H. erectus*
3 *A. africanus* *H. habilis*
4 *A. afarensis* Hominidae
5
6
7
8
9 *Sivapithecus*
10 *Hispanopithecus* *Gigantopithecus* *Ramapithecus*
11 *Rudapithecus* *Ouranopithecus*
12 *Dryopithecus* Ramapithecidæ
Sivapithecus *Kenyapithecus*
13
14
15 Dryopithecidæ
16
17 *Limnopithecus*
18 *Rangwapithecus*
19 *Proconsul*
20

(a) (b) (c)

H, Hominidæ;
P, living Pongidæ;
R, Ramapithecidæ;
D, Dryopithecidæ;
—— Fossil evidence
---- Inferred from morphology
══ Inferred from biochemistry

series, *Australopithecus afarensis* these projected further than those of subsequent hominids (Wolpoff, 1973, 1975; Johanson & White, 1979; Wood, 1981).

(3) The earliest hominids had brains the size of modern pongids in bodies that were probably a little smaller than modern ape bodies (Pilbeam & Gould, 1974; Holloway, 1981; Tobias, 1981).

(4) Many workers feel that no known fossiliferous formation that dates to between 2.5 and 4 mya contains more than one sympatric species of hominid, though if this is true of the Hadar, then the early forms were as sexually dimorphic as gorillas. (For a contrary view see R. Leakey, 1981, p. 70.)

At about two million years ago, the second part begins and the situation became more complex in interesting ways. Samples from all richly fossiliferous localities start to be differentiated into two (or perhaps more?) sympatric species. One of these forms shows a tendency to cheek tooth enlargement and to increase in body size (*Australopithecus robustus* in South Africa and *A. boisei* in East Africa). Others show some reduction in tooth size and a marked increase in cranial capacity (*Homo habilis*).

Between two and one million years ago in Africa, specimens of both species are often found in the same layers, but thereafter the ultra megadont form disappears from the record. The discovery of the existence of two species of hominid in the time range 2.2 to 1.2 mya is one of the exciting, original contributions of palaeontology. It could not have been predicted from any other class of evidence.

Dating from two million years to the present, fossil specimens have been found which are usually classified into three successive species of the genus *Homo*. If cranial capacities are plotted against time they show a tendency to increase until a levelling off occurs in the last few hundred thousand years. If cheek-tooth size is plotted in the same way, it shows a decrease until in the recent past a size range equivalent to that of both Dryopithecine and chimpanzee cheek teeth is reached.

A final anatomical shift occurs which is much less well known but which

---

Fig. 25.6. Left: Described genera of fossil and living hominoids shown against a time scale with family groupings indicated (based on Pilbeam, 1980, with modifications). Right: Three alternative phylogenetic schemes of which (*a*) is incompatible with the biochemical evidence. The hominid species are widely but not universally accepted (see Tobias, 1980, White *et al.*, 1981; Leakey, 1981). *Note*: from biochemistry the living Pongidae emerge as a polyphyletic group.

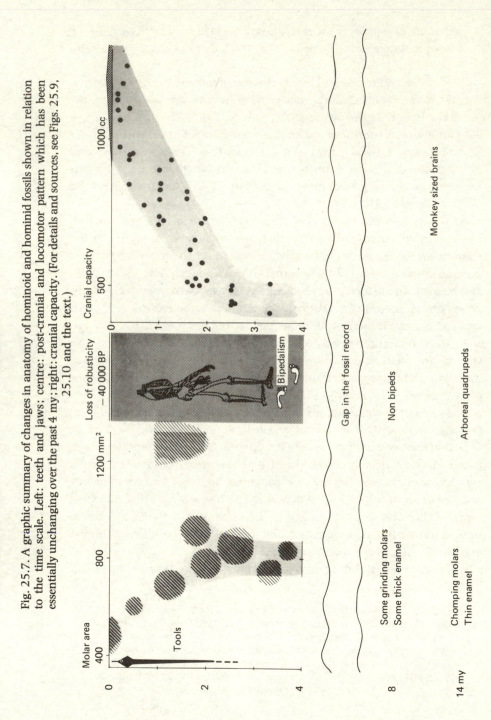

Fig. 25.7. A graphic summary of changes in anatomy of hominoid and hominid fossils shown in relation to the time scale. Left: teeth and jaws; centre: post-cranial and locomotor pattern which has been essentially unchanging over the past 4 my; right: cranial capacity. (For details and sources, see Figs. 25.9, 25.10 and the text.)

may be of fairly profound importance. Between about 50000 or so and 30000 years ago, with precise timing varying from region to region, all surviving human populations show a marked reduction in skeletal robusticity. Virtually all previous hominid fossils show a thickness of bone, plus muscular ridging that is outside the range that can be induced in modern humans even by extreme muscular training and stresses. There are also some subtle changes in skull architecture and pelvic form (Trinkaus & Howells, 1979). These are the major contrasts separating modern humans from neanderthals, neanderthaloids and 'archaic *Homo sapiens*'. The biological meaning of this loss of robusticity is as yet poorly understood (J. D. Clark, 1982).

One of the specific characteristics of the human evolutionary lineage has been the propensity to make tools – and to discard them. This has created a trail of litter that can be traced back some two to two and a half million years. Archaeological study of this trail of refuse represents a major contribution to our knowledge of what has happened during the final two million years or so of the co-evolution of the brain and culture (Fig. 25.8).

Stone tools comprise the most widespread and persistent element of this record. The oldest known sets from sites such as Olduvai, Omo, Koobi Fora, Hadar, Melka Kunture and Swartkrans are all in East and South Africa. They are simple in terms of technology and design. Rocks were broken by conchoidal fracture so as to generate a varied set of sharp-edged forms. Experiment shows that these forms can be used effectively to cut off branches and to sharpen them as digging sticks or spears, or to cut up animal carcasses, ranging in size from gazelles to elephants. Newly developed techniques for determining use patterns from microscopically detectable polishes on the edges join other lines of evidence to show that some early examples were indeed used for cutting up carcasses, others for whittling wood and others for cutting plant tissue (Keeley & Toth, 1981; Bunn, 1981; Potts & Shipman, 1981). Thus, we begin to see that in spite of their simplicity these early artefacts had considerable importance in effecting novel adaptations. They are to be understood mainly as meat-cutting tools and as tools for making tools.

From the first appearance of stone artefacts, these occur both scattered over the landscape and in conspicuous localized concentrations which archaeologists call sites. These concentrations are often found to involve quantities of broken animal bones among the artefacts. This has led archaeologists to write into their narratives the early beginning of hunting (or at least, meat eating) and the early adoption of a socio-economic pattern involving 'camps' or 'home bases' and food sharing (e.g. Leakey, 1971;

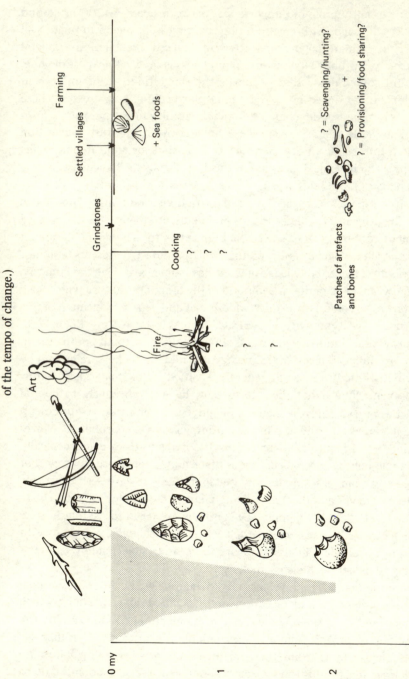
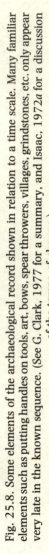

Fig. 25.8. Some elements of the archaeological record shown in relation to a time scale. Many familiar elements such as putting handles on tools, art, bows, spear throwers, villages, grindstones, etc. only appear very late in the known sequence. (See G. Clark, 1977 for a summary, and Isaac, 1972a for a discussion of the tempo of change.)

Isaac, 1978). The validity of these interpretations is currently subject to testing and debate (Binford, 1981; Bunn *et al.*, 1980; Isaac, 1981) (see below).

In Fig. 25.8, notice that the antiquity of control over fire is currently highly uncertain. It goes back at least half a million years, but may go back to one and a half or two million years or more (Gowlett, Harris, Walton & Wood, 1981). Notice also that many material culture attributes of humans appear only in the last 1% to 5% of the record. This wave of innovation occurs in the same time range as the loss of robusticity. This could be taken to mean that many of the familiar accoutrements of being human came only towards the very end of the narrative.

Hominoid fossils of the early and maybe the middle Miocene all come from some sort of tropical forest context. The late Miocene is more complex – some hominoids continued to live in forests, but others, including the ramapithecines which had somewhat hominid-like teeth, seem often to have lived in more open, varied woodland habitats (see Behrensmeyer, 1982 for review with references and Butzer, 1976, 1977). This is interesting, but as we have seen it is quite uncertain whether or not this ramapithecine adaptive radiation is in any way ancestral to the Hominidae.

However this may be, faunal analysis and fossil pollen analyses combine to show that the earliest known fossil specimens of hominids between 4 and 2 mya all derive from strata that were laid down under non-forest conditions. The environments represented are very varied and range from open thorn-veldt grassland (Laetoli and some Transvaal layers) to complex mosaics of grassland, marsh, riverine gallery woods and lake margins (e.g. Hadar, Olduvai, Omo and Koobi Fora, see Jolly, 1978; Bishop, 1978).

This association of hominid fossils with relatively open country has commonly been taken as vindication of Darwin's narrative propositions that our early ancestors left the trees, an idea which has also become enshrined in our folk sense of human evolution. However, one of the surprising twists of discovery in recent years has been the recognition (1) that the hands, feet and shoulders of the early hominids may have been highly adapted for tree climbing (Susman & Creel, 1979; Vrba, 1979) and (2) that early archaeological sites commonly occur where groves of trees would have grown (Isaac, 1972b, 1976). Perhaps bipedalism is yet another example of changing so as to remain the same with the new locomotor pattern being extensively used initially to move between widely spaced patches of trees. Maybe we left the forest a while ago but the trees only much more recently (cf. Romer, 1959; Rodman & McHenry, 1980).

After two million years ago, available evidence allows us to believe in the kind of success story we clearly love for ourselves – expanding geographic distribution, and expanding range of habitats used. Notice though, that the occupation of really extreme environments such as unbroken forests, deserts or tundra can only be documented inside the last 100000 years.

Whether it had any influence or not, the last two and a half million years of geologic time has witnessed global climatic oscillations of increasing amplitude. These involve the so-called ice ages. Following relatively stable, equable conditions in the Miocene and early Pliocene, there have been some 16 or 17 ice ages since the emergence of the genus *Homo* two million years ago (cf. Butzer, 1976; Shackleton, 1982).

Dynamics

Thus it can be seen that over the past decade the outlines of a four million year narrative of human evolution has emerged, and curiosity has begun to switch over to questions about the evolutionary mechanisms involved. Here, I can only touch hastily on aspects of a few selected topics.

One such is the question as to whether human evolution over the past several million years has proceeded by a process of cumulative genetic changes that pervaded populations over wide areas so that all went through evolutionary transformation, or whether successive species of hominids all exhibit stasis, with widespread change being accomplished by species replacement events (Gould & Eldredge, 1977). It should be noted in advance that these alternative models do not seem to me to be entirely mutually exclusive.

Figs 25.9 and 25.10 show data for two relatively simple measurable attributes of hominid fossils plotted against time. Contrary to the view of Cronin, Boaz, Stringer & Rak (1981), Fig. 25.9*b* suggests that both gradualist models and punctuated equilibrium models can equally well be fitted to the available data. The best case for stasis in the record is the taxon *Homo erectus*. It can be argued that the first appearance of this taxon looks like a punctuation event and the taxon lasts a million years. However, a its later end many investigators seem to be reporting mosaic patterns of transition into 'archaic *Homo sapiens*' and this would not be compatible with a clear-cut punctuation event.

Numbers of workers, myself included, have tended to think of the loss of robusticity transition of 30000–50000 years ago as a possible example of a punctuation/genetic replacement event. But this view would seem to be falsified by the new mitochondrial DNA data (Ferris, Wilson & Brown, 1981; Cann, Brown & Wilson, 1982).

Fig. 25.9. (*a*) Endocranial volumes of hominid fossil skulls plotted against a time scale. The degree of uncertainty about age is indicated by the vertical bars. 7 = *Australopithecus afarensis* from Hadar; 6 = *A. africanus*; 5 = *A. robustus* and *A. boisei*; 4 = *Homo habilis*; 3 = *H. erectus* (E. Africa); 2 = *H. erectus* (Java and Lantien); 1 = *H. erectus* (Pekin). Early *H. sapiens* (the range for skulls) P = Petralona, St = Steinheim, S = Saldanha, R = Kabwe (Rhodesia man), Sw = Swanscomb, V = Vertesszöllös. (*b*) The same data (left) fitted to a phyletic gradualist model and (right) to a punctuated model. The species indicated at the right apply to both versions (sources: Holloway, 1981; Day, 1977; Howell, 1978; Tobias, 1981; Cronin *et al.*, 1981). For discussion of relation to body size see Pilbeam & Gould, 1974.

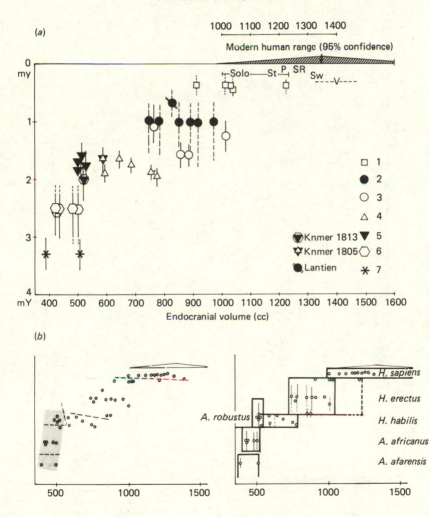

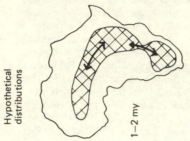
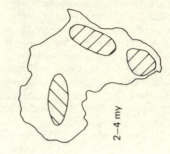

Hypothetical distributions

1–2 my

2–4 my

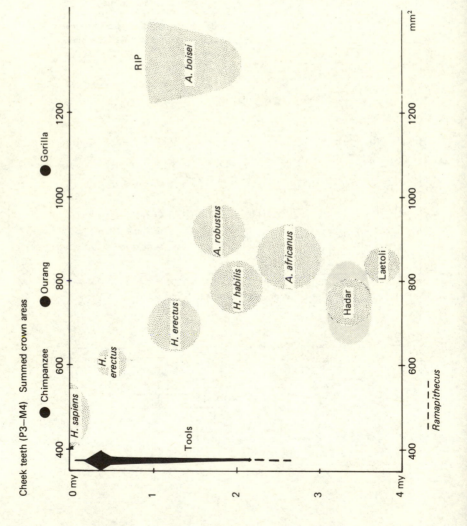

Cheek teeth (P3–M4) Summed crown areas

● Chimpanzee ● Ourang ● Gorilla

● H. sapiens

H. erectus

Tools

H. erectus

A. robustus

H. habilis

A. africanus

Hadar

Laetoli

RIP

A. boisei

mm²

–––– Ramapithecus

Fig. 25.10 also illustrates a possible example within the hominid fossil record of the effects of the breakdown of barriers which had separated trivially differentiated allopatric species. According to one interpretation two species of *Australopithecus* came to have overlapping ranges, and responded by undergoing niche separation and character displacement (Schaffer, 1968; Swedlund, 1974). One of the resultant species or (species complex) is *Australopithecus robustus/boisei* which underwent selection for enlarged body size, and perhaps, following the Jarman–Bell principle, a coarsening of diet. The other became *Homo habilis* and retained moderate body size and took to higher quality foods perhaps acquired in part through the aid of tools. Maybe this is indeed a fairy story, but it is fun and it may turn out to be at least partly true.

The peculiarities of the early hominid megadont phase presumably relates to diet, but what this was continues to baffle us. Scanning electron microscope (SEM) studies of tooth wear by Alan Walker (1981) and others suggest that non-siliceous plant tissues were being consumed – presumably fruits (*sensu lato*) or seeds. But what fruits or seeds? And why such large teeth? These questions call for studies of floristic communities and the feeding opportunities they offer as well as scrutiny of fossils.

A battery of new techniques for palaeodietary studies are being developed, including SEM studies and the analysis for the strontium and ^{13}C composition of old bones. A major onslaught on this fundamental problem seems to be getting underway (Walker, 1981).

As the outlines of the narrative of human evolution have emerged, two particularly intriguing puzzles have emerged with it. *Under what selection pressures did, firstly, the two-legged gait and, secondly, the enlarged brain become adaptive?* The first of these can be rephrased as: Why did ancestral hominids

---

Fig. 25.10. The size (total area in mm^2) of mandibular cheek teeth plotted against a time scale with means for each taxon/sample shown schematically as a stippled zone. Mean values for the closest living relatives of man and for *Ramapithecus* are shown for comparison. If one bears in mind that hominids between 2 and 4 mya were smaller than *H. sapiens* and no bigger than chimpanzees, then it is apparent that they had proportionately large teeth (see Pilbeam & Gould, 1974). *A. boisei* was certainly not as large as a gorilla and was ultra-megadont. The maps at right suggest a scenario of allopatric speciation followed by the breakdown of isolation, range overlap and perhaps character displacement. Some authorities (e.g. Leakey, 1981) would split the Hadar sample between two taxa, others (Johanson & White, 1979) regard it as one highly dimorphic taxon (Sources: Wolpoff, 1975; Tobias, 1981; White, Johnson & Coppens, 1982.)

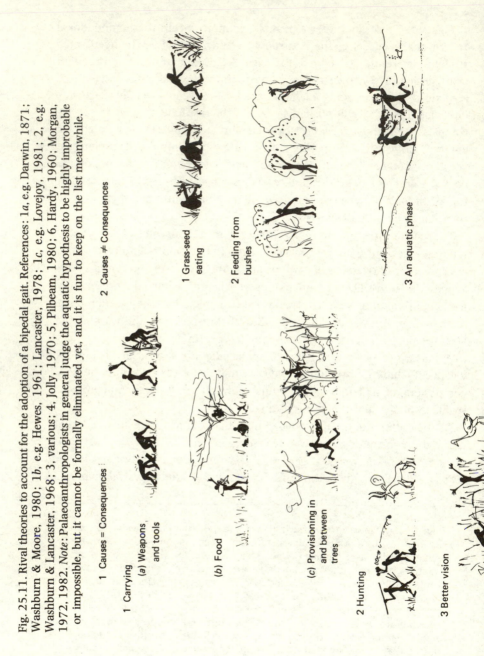

Fig. 25.11. Rival theories to account for the adoption of a bipedal gait. References: 1a, e.g. Darwin, 1871; Washburn & Moore, 1980; 1b, e.g. Hewes, 1961; Lancaster, 1978; 1c, e.g. Lovejoy, 1981; 2, e.g. Washburn & Lancaster, 1968; 3, various; 4, Jolly, 1970; 5, Pilbeam, 1980; 6, Hardy, 1960; Morgan, 1972, 1982. *Note*: Palaeoanthropologists in general judge the aquatic hypothesis to be highly improbable or impossible, but it cannot be formally eliminated yet, and it is fun to keep on the list meanwhile.

1 Causes = Consequences

2 Causes ≠ Consequences

1 Carrying

(a) Weapons and tools

(b) Food

(c) Provisioning in and between trees

2 Hunting

3 Better vision

1 Grass-seed eating

2 Feeding from bushes

3 An aquatic phase

become bipedal when all other primate species which have come to the ground have adopted some or other form of quadrupedal locomotion? Many thinkers on these topics, starting with Darwin, have tended to opt for an all-purpose explanation which might explain both bipedalism and brain enlargement, for instance, tool and weapon carrying. However, since specific evidence for the two evolutionary shifts are separated by at least two million years, it may be wise to uncouple the searches for explanations.

Fig. 25.11 playfully indicates some of the competing explanations which have been or are being discussed in relation to bipedalism.

More fossils, more palaeoenvironmental, and palaeodietary data will certainly help to advance understanding on this question, but it should also be clear that intelligent neontological/ecological work is called for. For instance, do potential feeding niches really exist that would make bipedalism adaptive?

The brain-culture system

We all share in some degree the conviction that our words, our intellect, our consciousness, our aesthetic and moral sense, constitute the quintessential characteristics of being human. Further we associate these qualities directly with the evolutionary enlargement and reorganization of our brains. The issue can be put like this: 'The brain is the organ of culture, and culture is the function of the brain'. The term culture refers to the intricate body of language, craft skills, social custom, traditions and information which humans learn while growing up and living in any human society. (For a good discussion of this, see Geertz, 1973.) Cultural complexity and flexibility of this kind is unknown in any other organism and would be impossible without the hypertrophied brain. It is also hard to make sense of the intricacy of the brain without supposing that the adaptive advantages that have brought it into existence have long involved culture of increasing complexity. However, to keep our topic from becoming dull and predeterministic perhaps we should allow for the possibility that the enlarged brain, like bipedalism, might have been a pre-adaptive development that was favoured by selection for reasons other than culture. This point notwithstanding, for the time being I shall treat the brain and the culture it sustains as likely to have evolved as a single adaptive complex, that is to say as a co-evolution (see Wilson, this volume).

We are rightly impressed with the biological success that seems to have followed from the development of the brain through some critical thresholds, but it must be remembered that enlarged brains require prolonged infant dependency and high quality nutrition (Sacher & Staffeldt, 1974; Martin in Lewin, 1982). Both of these are expensive commodities in the economy of nature. No other lineage has experienced selection producing such an extreme development. The central puzzle to understanding our origins, therefore, remains the problem of figuring out under what novel selective circumstances this trend was initiated, and under what conditions the selection was sustained.

Set out below is a list of some of the distinctive innovations which have been suggested and discussed as prime movers in the initiation of the trend towards elaboration of the brain-culture system.

(1) The use of tools and weapons (e.g. Darwin, 1871; Washburn, 1960; Tobias, 1967, 1981);

(2) Hunting (e.g. Darwin, 1871; Dart, 1925, 1953; Ardrey, 1961; Washburn & Lancaster, 1968);

(3) Gathering (e.g. Zihlman & Tanner, 1979; Tanner, 1981);

(4) Generalized social cooperation with 'autocatalytic' feedback (e.g. Darwin, 1871; Lovejoy, 1981);

(5) Adoption by small-brained hominids of a socio-reproductive system involving food sharing, provisioning and central place foraging (e.g. Hewes, 1961; Washburn, 1965; Isaac, 1978; Lancaster, 1978).

It should be noted that these competing explanations are *not* mutually exclusive, and future research will have to involve subtle assessment of their relative importance at different stages rather than simple Popperian falsification.

It should also be noted that the study of the fossil and archaeological record will not suffice by themselves to distinguish among hypotheses. It is all very well arguing that tool-use was a pivotal development that imposed novel selection pressure, but under what circumstances would tools be adaptive? As I argued in the paper *Casting the Net Wide* (Isaac, 1980) answering this kind of question calls for problem-oriented quantitative field studies of feeding possibilities and foraging strategies.

Over the past 12 years my own research has been focussed first on developing and then on testing the predictions of the so-called 'Food-sharing hypothesis' and its possible bearing on the initiation of selection for larger brain size. I shall briefly indulge myself by discussing aspects of this model and this work.

The first point to be made is that major changes have occurred in human ranging patterns and feeding behaviour (Fig. 25.12). These changes involve the collective acquisition of food, postponement of consumption, transport, and communal consumption at a home base or central place.

These features are so basic in our lives that we take them for granted and very often they do not even appear on lists of contrasts between humans and non-human primates. However, if we could interview a chimpanzee about the behavioural differences separating us, this might well be the item that it found most impressive – 'These humans get food and instead of eating it promptly like any sensible ape, they haul it off and share it with others'.

The food-sharing hypothesis should be renamed the central place foraging hypothesis. It incorporates tools and meat eating. It postulates that at some time before two million years ago, the behaviour of at least one kind of small-brained hominid was modified to include the elements shown

Fig. 25.12. The contrast between human ranging patterns and those of a representative sample of non-human primates (see Isaac, 1980 for sources and discussion). *Note*: Humans are represented by San hunter-gatherers, but the same basic pattern would be found if agriculturalists were represented, or modern city dwellers with offices and supermarkets as the endpoints of radiating movements.

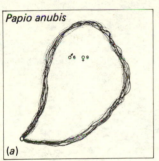

(a)
After Hall (1965)

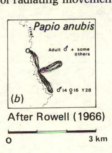

(b)
After Rowell (1966)

0   3 km

(c)
After Schaller (1964)
Fifteen days

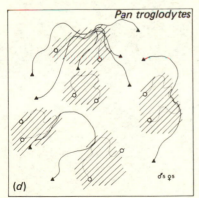

(d)
Invented from Wrangham

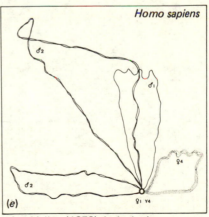

(e)
After Yellen (1972) A single day

Fig. 25.13. (*a*) Two plans of an excavation which appears to be a well preserved early central-place foraging base – the FLK Zirij site at Olduvai (Based on Leakey, 1971). More than 2400 stone artefacts and 40000 bone fragments (right) occur within a radius of 10 m. Some 8% of the identifiable bones show the marks of sharp stone tools (Bunn, 1982) and these include damage due to dismemberment marks and to meat removal. (See also Potts, 1982 and Potts & Shipman, 1981 for detailed information and a more conservative estimate of cut mark damage frequency.) (*b*) A hypothetical model of the processes involved in site formation. The transport of stones and bones (meat) is certain, the transport of plant foods is possible but unconfirmed. Whether sharing occurred at all is harder to judge – and whether it was incidental or 'deliberate' is impossible to tell.

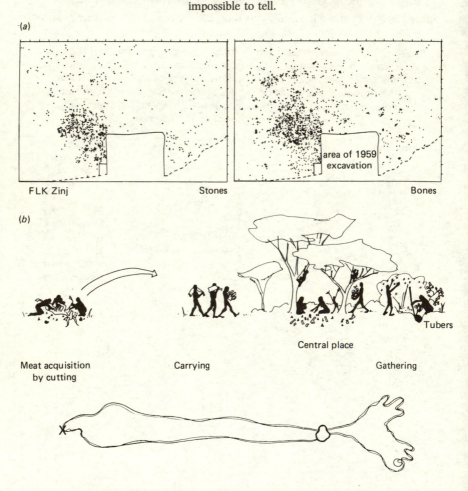

(*a*)

FLK Zinj — Stones

area of 1959 excavation — Bones

(*b*)

Meat acquisition by cutting — Carrying — Central place — Gathering — Tubers in Fig. 25.13, namely the use of tools, the acquisition of meat, perhaps preferentially by males, the transport of portions of that meat to central places where it would be apt to be collectively consumed by members of a social group some of whom, especially females and young, had not participated in its acquisition. At the beginning or at some subsequent stage, female gathering was surely included in the system. Conscious motivation for 'sharing' need *not* have been involved. The model works provided that radiative ranging patterns developed with transport of some food back to the foci of social aggregation.

For me, the interest of the model is not that 'humans' existed 2 mya but that it promises to help explain how the non-human hominids of that time began to be modified into humans. Once food transport was initiated, novel selection pressures would come to bear on (1) ability to communicate about the past, future, and the spatially remote and (2) enhanced abilities to plan complex chains of eventualities and to play what one might call 'social chess' in one's mind. That is, the adoption of food-sharing would have favoured the development of language, social reciprocity and the intellect. Evolutionary strategy models should now be developed to explore the conditions under which food sharing might become an ESS (see Maynard Smith, this volume).

Clearly, part of the nutritional cost of brain enlargement and the costs of prolonged dependency during brain growth with extended learning would be taken care of by the provisioning/nurturing characteristics which in this scenario would already be part of the system.

The model arose as a post-hoc explanation of the existence of concentrated patches of discarded artefacts and of broken-up bones in layers between 1.5 and 2 mya. Having set it up, we have turned around and have been enjoying the sport of trying to knock it down, with the help of fierce critics (e.g. Binford, 1981).

The technicalities of this debate and this research go beyond the scope of this review (see Isaac, 1981, 1982; Bunn *et al.*, 1980; Bunn, 1981; Potts & Shipman, 1981). Suffice it to say that in my view, we have obtained ample confirmation that hominids were indeed acquiring meat through the use of tools and were transporting this to favoured localities where the observed concentrated patches of bones and tools formed. Whether these places were 'home bases' or whether provisioning and/or active food sharing were going on, is harder to judge. My guess now is that in various ways, the behaviour system was less human than I originally envisaged, but that it did involve food transport and de facto, if not purposive, food sharing and provisioning.

The food-sharing model has been widely misunderstood as implying that by two million years ago there existed friendly, cuddly, cooperative human-like hominids. This need not be so. The attractiveness of this model is that it seems entirely feasible for such a behavioural system to come into existence among *non human* hominids that had brains no larger than those of living apes, and it is my strong suspicion that if we had these hominids alive today, we would have to put them in zoos, not in academies.

Clearly, this initial configuration can very readily be plugged into models involving kin-selection, and/or tit-for-tat selection patterns that would provide plausible, if hard-to-test models of the subsequent elaboration of brain–speech–culture–society systems (Fig. 25.14). Amongst other things, the provisioning and division of labour implied by the system would make bonded male–female reproductive modules highly adaptive, if they did not already exist at the outset.

Fig. 25.14. The 'food sharing' or 'central-place-foraging' hypothesis suggests a socio-reproductive milieu in which 'mental and emotional proclivities' could be selected that started to transform non-human hominid systems into human type systems. Communication, reciprocal help and planning ahead would all be favoured (left). Novel behavioural ingredients form structural members in a non-human system. These same elements now envelopped as core-components within highly elaborated, flexible, variable cultural systems (right); see text; Isaac, 1978; Maynard-Smith and E. O. Wilson (this volume) for discussion of the issues and of the modes of selection that are indicated.

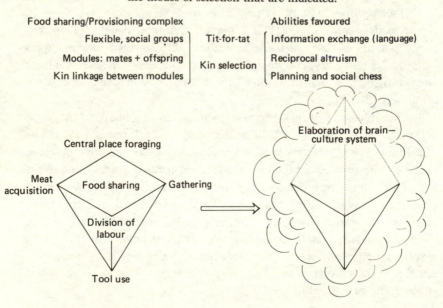

Food sharing/Provisioning complex

Flexible, social groups   Tit-for-tat

Modules: mates + offspring

Kin linkage between modules   Kin selection

Abilities favoured

Information exchange (language)

Reciprocal altruism

Planning and social chess

Central place foraging

Meat acquisition   Food sharing   Gathering

Division of labour

Tool use

Elaboration of brain–culture system

In conclusion

No two people who undertook to review this topic would have tackled it in the same way. I have chosen to stress enquiries focussed on stratified evidence from the past, while other writers would equally legitimately have emphasized the contributions made by studies of biochemistry, ecological dynamics or by comparative behaviour and sociobiology.

Following the pioneer descriptive phases of primate studies, various workers have begun to search out generalizations among non-human primates concerning relationships between food choice, ranging patterns, reproductive strategies and social format (e.g. Clutton-Brock & Harvey, 1977; Wrangham, 1979, 1980; Milton, 1981). The results have not yet been fully assimilated into thinking about human evolution, but already it emerges that humans have distinctive ecological relationships and social configurations that are outside the range of other primate patterns. My hunch would be that this will prove to be connected with the colonization of habitats where potential foods were more patchy and more widely dispersed than is normal for primates. This in turn involved altered diets which focus on two distinctive and quite different things: first on plant foods that yield large numbers of calories per item (e.g. tubers and nuts) and second, significant feeding on the meat of large animals and/or fish. Acquisition of all of these is facilitated by tool use. It is at present uncertain when and by what stages these shifts occurred. Finding out is one of the major challenges that confronts palaeoanthropology.

Relative to other primates, humans have highly distinctive social patterns. In spite of tremendous variation this almost always includes reproductive units involving direct male investment in child rearing, and comprising one male and one or more females. These units are almost invariably integrated as modules into highly variable larger scale social entities. I can see no way of predicting the human pattern from the primate patterns without introducing some novel elements into the mix of variables. One candidate for an influential novelty may well be the significant incorporation of dietary components to which one sex rather than the other had preferential access. Clearly, meat is one such commodity, though it may not be the only one.

Lovejoy (1981) has argued that monogamous pair bonding and food transport preceded meat eating and the formation of bands. However, we need to retain as the alternative hypothesis that pair bonding occurred within multi-male, multi-female social groups and was associated with division of food acquisition labour. Recent examination of relations between mating system, body size and testis size in primates does not support a multi-male social group for *Homo* (Harcourt, Harvey, Larson & Short, 1981; Martin & May, 1981). However, if early ancestral hominids already had mated pair modules within the troop this objection might not apply.

Relating studies of the present to studies of the past will require changes of emphasis. Much of the literature on the stratified record of human evolution is devoted to the taxonomy of individual fossils and to arguments about whether particular ones are on the line or not. The topic is in its own way important, but with major taxa reasonably clearly established the younger generation of scientists is becoming more and more involved in enquiring into relationships between shifts in anatomical configurations and shifts in modes of adaptation. This line of research can be pursued profitably even if we do not know which particular fossils are indeed on the line and which are off it. I would go on to predict that progress with this topic will involve much less narrow focus on fossils. Hominid palaeontologists and archaeologists will need to collaborate in assessing the adaptive significance of technology, subsistence patterns and socio-economic arrangements. For this, the archaeologists will have to give up the artefact typology fixation that has been their equivalent of fossil-philia. Both archaeologists and hominid palaeontologists are also going to have to work closely with ecologists. This has started, e.g. Schaller & Lowther, 1969, Peters & O'Brien, 1981, and J. Sept and A. Vincent (personal communication).

In summary, improvements in knowledge about human evolution require the acquisition of richly diverse classes of information. This includes both stratified evidence from the past and the elucidation of the intricate features of living behavioural and ecological systems. As is normal in science, hypotheses regarding both narrative and mechanisms need to be restlessly formulated, tested and revised. However, as indicated in the introduction, and as amply illustrated in Darwin's own treatment of the topic, the meaning that each of us finds in the growing corpus of secure, tested information, nonetheless remains a humanistic abstraction.

My being in a position to undertake this review stems from an appointment in East Africa, given me in 1961 by the late Louis Leakey. Since then my wife and I have been part of a goodly company of researchers in East Africa during a period of exciting discoveries. Many of my ideas surely derive from this participation. I recognize particularly strong influence from discussions with S. L. Washburn, D. R. Pilbeam, J. D. Clark, Vince Sarich and A. C. Walker. Also from the team that has worked with Richard Leakey and me at Koobi Fora. My wife is a part of the talking, the fieldwork and the laboratory work, and she draws the figures. For this paper, Jeanne Sept has done the light-hearted sketches (Figs. 25.1, 25.3, 25.11, 25.13). Stanley Ambrose encouraged me to think about the material in Fig. 25.10 and he is preparing a paper on character displacement in hominids. I wish to pay tribute to three great scientists, recently deceased, who did much to foster the study of biological and cultural co-evolution–Kenneth Oakley, François Bordes and Charles McBurney. The last-named especially was my mentor during my student days and after.

Further reading

Lancaster, J. (1975). *Primate Behaviour and the Emergence of Human Culture*. New York: Holt, Rinehart & Winston.
Leakey, R. (1981). *The Making of Mankind*. London: Michael Joseph.
Pfeiffer, J. (1978). *The Emergence of Man*. New York: Harper & Row.
Pilbeam, D. R. (1980). Major trends in human evolution. In *Current Argument on Early Man*, ed. L-K. Konigsson, pp. 261–85. Oxford: Pergamon Press.
Reader, J. (1981). *Missing Links*. Boston: Little, Brown & Co.
Washburn, S. L. & Moore, R. (1980). *Ape into Human*. 2nd edn. Boston: Little, Brown & Co.

References

Ardrey, R. (1961). *African Genesis*. London: Collins.
Behrensmeyer, A. K. (1982). The geological context of human evolution. *Annual Review of Earth and Planetary Sciences*, **10**, 39–60.
Behrensmeyer, A. K. & Hill, A. (1981). *Fossils in the Making*. Chicago: Chicago University Press.
Binford, L. R. (1981). *Bones: ancient men and modern myths*. New York: Academic Press.
Bishop, W. W. (1978). *Geological Background to Fossil Man: recent research in the Gregory Rift Valley, East Africa*. Edinburgh: Scottish Academic Press.
Bunn, H. (1981). Archaeological evidence for meat-eating by Plio Pleistocene hominids from Koobi Fora and Olduvai Gorge. *Nature*, **291**, 574–7.
Bunn, H. (1981). Archaeological evidence for meat-eating by Plio-Pleistocene Koobi Fora, Kenya and Olduvai Gorge, Tanzania. In *Proceedings of the 4th International Congress on Archaeo-zoology, London*. Oxford: British Archaeological Record.
Bunn, H., Harris, J. W. K., Isaac, G. Ll., Kaufulu, Z., Kroll, E., Schick, K., Toth, N., & Behrensmeyer, A. K. (1980). FxJj50: an early Pleistocene site in northern Kenya. *World Archaeology*, **12**, 109–36.
Butzer, K. W. (1976). Pleistocene Climates. *Geoscience and Man*, **13**, 27–44.
Butzer, K. W. (1977). Environment, culture and human evolution. *American Scientist*, **65**, 572–84.
Cann, R., Brown, W. M., & Wilson, A. C. (1982). Evolution of Human Mitochondrial DNA: molecular, genetic and anthropological implications. *Proceedings of the 6th International Congress of Human Genetics*, Jerusalem.
Clark, G. (1977). *World Prehistory*, 3rd edn. Cambridge: Cambridge University Press.

Clark, J. D. (1982). New men, strange faces, other minds: an archaeologist's perspective on recent discoveries relating to the origin and spread of modern man. *Proceedings of the British Academy.*

Clutton-Brock, T. H. & Harvey, P. H. (1977). Primate ecology and social organization. *Journal of Zoology,* **183,** 1–39.

Cronin, J., Boaz, N., Stringer, C. & Rak, Y. (1981). Tempo and mode in hominid evolution. *Nature,* **292,** 113–22.

Dart, R. (1925). *Australopithecus africanus:* the man-ape of Southern Africa. *Nature,* **115,** 195–9.

Dart, R. (1953). The predatory transition from ape to man. *International Anthropological and Linguistic Review,* **1,** 201–19.

Darwin, C. (1871). *The Descent of Man and Selection in Relation to Sex.* London: John Murray.

Day, M. (1977). *Guide of Fossil Man.* Chicago: University of Chicago Press.

Ferris, S. D., Wilson, A. C. & Brown, W. M. (1981). Evolutionary tree for apes and humans based on cleavage maps of mitochondrial DNA. *Proceedings of the National Academy of Sciences, USA,* **78,** 2432–6.

Geertz, C. (1973). The growth of culture and the evolution of mind. In *The Interpretation of Cultures,* ed. C. Geertz, pp. 55–83. New York: Basic Books.

Gould, S. J. & Eldredge, N. (1977). Punctuated equilibria: the tempo and mode of evolution reconsidered. *Paleobiology,* **3,** 115–51.

Gowlett, A. J., Harris, J. W. K., Walton, D. & Wood, B. A. (1981). Early archaeological sites, hominid remains and traces of fire from Chesowanja, Kenya. *Nature,* **294,** 125–9.

Hall, R. (1965). Behaviour and Ecology of the wild Patas monkey, *Erythrocebus patas,* in Uganda. *Journal of Zoology,* **148,** 15–87.

Hardy, A. (1960). Was man more aquatic in the past? *New Scientist,* **7,** 642–5.

Harcourt, A. H., Harvey, P. H., Larson, S. G. & Short, R. V. (1981). Testis weight and breeding system in primates. *Nature,* **293,** 55–7.

Hewes, G. (1961). Food transport and the origin of hominid bipedalism. *American Anthropologist,* **63,** 687–710.

Holloway, R. L. (1981). Exploring the dorsal surface of hominoid brain endocasts by stereoplotter and discriminant analysis. *Philosophic Transactions of the Royal Society, London B,* **292,** 155–66.

Howell, F. C. (1978). Hominidae. In *Evolution of African Mammals,* ed. V. J. Maglio & H. B. S. Cooke, pp. 154–248. New York: Academic Press.

Huxley, T. H. (1863). *Evidence as to Man's Place in Nature.* London: Williams & Norgate.

Isaac, G. Ll. (1972a). Chronology and the tempo of cultural change during the Pleistocene. In *Calibration of Hominoid Evolution,* ed. W. W. Bishop & J. A. Miller, pp. 381–430. Edinburgh: Scottish Academic Press.

Isaac, G. Ll. (1972b). Comparative studies of Pleistocene site locations in East Africa. In *Man, Settlement and Urbanism,* ed. P. J. Ucko & G. W. Dimbleby, pp. 165–76. London: Duckworth & Co.

Isaac, G. Ll. (1976). The activities of early African hominids. In *Human Origins: Louis Leakey and the East African Evidence,* ed. G. Ll. Isaac & E. R. McCown, pp. 483–514. Menlo Park: W. A. Benjamin.

Isaac, G. Ll. (1978). Food sharing and human evolution: archaeological evidence from the Plio-Pleistocene of East Africa. *Journal of Anthropological Research,* **34,** 311–25.

Isaac, G. Ll. (1980). Casting the net wide: a review of archaeological evidence for early hominid land-use and ecological relations. In *Current Argument on Early Man,* ed. L-K. Konigsson, pp. 226–53. Oxford: Pergamon Press.

Isaac, G. Ll. (1981). Archaeological tests of alternative models of early hominid behaviour: excavation and experiments. *Philosophical Transactions of the Royal Society, London B*, **292**, 177–88.

Isaac, G. Ll. (1982). Bones in contention. *Proceedings of the 4th International Congress on Archaeo-zoology, London.* Oxford: British Archaeological Record.

Johanson, D. C. & White, T. D. (1979). A systematic assessment of early African hominids. *Science*, **202**, 321–30.

Jolly, C. (1970). The seed eaters: a new model of hominid differentiation based on a baboon analogy. *Man*, **5**, 5–26.

Jolly, C. (1978). *Early Hominids of Africa.* London: Duckworth.

Keeley, L. & Toth, N. (1981). Microwear polishes on early stone tools from Koobi Fora, Kenya. *Nature*, **293**, 464–5.

Lancaster, J. (1978). Carrying and sharing in human evolution. *Human Nature*, **1**, 82–9.

Landau, M. (1981). The Anthropogenic: paleoanthropological writing as a genre of literature. Ph.D. dissertation, Yale University.

Leakey, M. D. (1971). *Olduvai Gorge*, Vol. 3. Cambridge: Cambridge University Press.

Leakey, R. E. (1981). *The Making of Mankind.* London: Michael Joseph.

Lewin, R. (1982). How did humans evolve big brains? *Science*, **216**, 840–1.

Lovejoy, O. (1978). A biomechanical review of the locomotor diversity of early hominids. In *Early Hominids of Africa*, ed. C. Jolly, pp. 403–43. London: Duckworth.

Lovejoy, O. (1981). The origin of man. *Science*, **211**, 341–50.

McHenry, H. & Temerin, L. A. (1979). The evolution of hominid bipedalism: evidence from the fossil record. *Yearbook of Physical Anthropology*, **22**, 105–31.

Martin, R. & May, R. (1981). Outward signs of breeding. *Nature*, **293**, 8–9.

Milton, K. (1981). Distribution patterns of tropical plant foods as an evolutionary stimulus to primate mental development. *American Anthropologist*, **83**, 534–48.

Morgan, E. (1972). *The Descent of Woman.* London: Souvenir Press.

Morgan, E. (1982). *The Aquatic Ape.* London: Souvenir Press.

Oakley, K. P., Campbell, B. C. & Molleson, T. I. (1971–1977). *Catalogue of Fossil Hominids.* London: British Museum (Natural History).

Perper, T. & Schrire, C. (1977). The Nimrod connection: myth and science in the hunting model. *The Chemical Senses and Nutrition*, ed. M. R. Kare, pp. 447–59. New York: Academic Press.

Peters, C. R. & O'Brien, E. M. (1981). The early hominid plant-food niche: insights from an analysis of human, chimpanzee, and baboon plant exploitation in eastern and southern Africa. *Current Anthropology*, **22**, 127–40.

Pilbeam, D. R. (1980). Major trends in human evolution. In *Current Argument on Early Man*, ed. L-K. Konigsson, pp. 261–85. Oxford: Pergamon Press.

Pilbeam, D. R. & Gould, S. J. (1974). Size and scaling in human evolution. *Science*, **186**, 892–901.

Potts, R. (1982). Lower Pleistocene Site Formation and Hominid Activities at Olduvai Gorge, Tanzania. Ph.D. Thesis, Department of Anthropology, Yale University, New Haven.

Potts, R. & Shipman, P. (1981). Cutmarks made by stone tools on bones from Olduvai Gorge, Tanzania. *Nature*, **291**, 577–80.

Propp, V. (1968). *Morphology of the Folktale.* Austin: University of Texas Press.

Reader, J. (1980). *Missing Links.* Boston: Little, Brown & Co.

Rodman, P. & McHenry, H. (1980). Bioenergetics and the origin of bipedalism. *American Journal of Physical Anthroplogy*, **52**, 103–6.

Romer, A. (1959). *The Vertebrate Story*, 4th edn. Chicago: University of Chicago Press.

Rowell, T. (1966). Forest living baboons in Uganda. *Journal of Zoology*, **149**, 344–64.

Sacher, G. A. & Staffeldt, E. F. (1974). Relations of gestation time to brain weight for placental mammals. *American Naturalist*, **108**, 593–614.

Sarich, V. & Cronin, J. (1976). Molecular systematics of the Primates. In *Molecular Anthropology*, ed. M. Goodman, R. E. Tashian & J. H. Tashian, pp. 141–170. New York: Plenum Press.

Schaffer, W. (1968). Character displacement and the evolution of the Hominidae. *The American Naturalist*, **102**, 559–71.

Schaller, G. (1964). *The Mountain Gorilla: Ecology and Behaviour*. Chicago: University of Chicago Press.

Schaller, G. & Lowther, G. (1969). The relevance of carnivore behaviour to the study of early hominids. *Southwestern Journal of Anthropology*, **25**, 307–41.

Shackleton, N. (1982). The deep-sea record of climate variability. *Progress in Oceanography*, **11**, 199–218.

Simons, E. (1972). *Primate Evolution*. New York: Macmillan.

Susman, R. & Creel, N. (1979). Functional and morphological affinities of the subadult hand (OH7) from Olduvai Gorge. *American Journal of Physical Anthropology*, **51**, 311–31.

Swedlund, A. C. (1974). The use of ecological hypotheses in Australopithecine taxonomy. *American Anthropologist*, **76**, 515–29.

Tanner, N. (1981). *On Becoming Human*. Cambridge: Cambridge University Press.

Tobias, P. V. (1967). Cultural hominization among the earliest African Pleistocene hominids. *Proceedings of the Prehistoric Society*, **33**, 367–76.

Tobias, P. V. (1980). *Australopithecus afarensis* and *A. africanus*: critique and alternative hypotheses. *Palaeontologia Africana*, **23**, 1–17.

Tobias, P. V. (1981). The emergence of man in Africa and beyond. *Philosophical Transactions of the Royal Society, London B*, **292**, 43–56.

Trinkaus, E. & Howells, W. W. (1979). The Neanderthals. *Scientific American*, **241**, 118–133.

Vrba, E. (1979). A new study of the scapula of *Australopithecus africanus* from Sterkfontein. *American Journal of Physical Anthropology*, **51**, 117–30.

Walker, A. (1981). Dietary hypotheses and human evolution. *Philosophical Transactions of the Royal Society London B*, **292**, 57–64.

Washburn, S. L. (1960). Tools and human evolution. *Scientific American*, **203**, 62–75.

Washburn, S. L. (1965). An apes eye view of human evolution. In *The Origins of Man*, ed. P. L. DeVore, pp. 89–96. New York: Wenner Gren Foundation.

Washburn, S. L. & Lancaster, C. (1968). Hunting and human evolution. In *Man the Hunter*, ed. R. Lee & I. DeVore, pp. 293–303. Chicago: Aldine.

Washburn, S. L. & Moore, R. (1980). *Ape into Human*, 2nd edn. Boston: Little, Brown & Co.

White, T. D., Johanson, D. C. & Coppens, Y. (1982). Dental remains from the Hadar Formation, Ethiopia: 1974–1977 collections. American Journal of Physical Anthropology, **57**, 545–603.

White, T. D., Johanson, D. C. & Kimbel, W. H. (1981). *Australopithecus africanus*: its phyletic position reconsidered. *South African Journal of Science*, **77**, 445–70.

Wilson, A., Carlson, S. S. & White, T. J. (1977). Biochemical evolution. *Annual Review of Biochemistry*, **46**, 573–639.

Wolpoff, M. (1973). Posterior tooth size, body size and diet in South African gracile australopithecines. *American Journal of Physical Anthropology*, **39**, 375–94.

Wolpoff, M. (1975). Some aspects of human mandibular evolution. In *Determinants of Mandibular Form and Growth*, ed. J. McNamara, pp. 1–64. Ann Arbor: Center for Human Growth and Development.

Wood, B. A. (1981). Tooth size and shape and their relevance to studies of hominid evolution. *Philosophical Transactions of the Royal Society London B*, **292**, 65–76.

Wrangham, R. (1979). On the evolution of ape social systems. *Social Science Information*, **18**, 335–68.

Wrangham, R. (1980). An ecological model of female-bonded primate groups. *Behaviour*, **75**, 262–300.

Yellen, J. (1972). Trip V. Itinerary May 24 – June 9, 1968. Pilot Edition of *Exploring Human Nature*. Cambridge, Mass.: Education Development Center Inc.

Zihlman, A. & Tanner, N. (1979). Gathering and the hominid adaptation. In *Female Hierarchies*, ed. L. Tiger & H. M. Fowler, pp. 163–94. Chicago: Beresford Book Service.

26

Sociobiology and the Darwinian approach to mind and culture

EDWARD O. WILSON

On 3 October 1838, Charles Darwin wrote in his *N* notebook that 'to study Metaphysics, as they have always been studied appears to me to be like puzzling at astronomy without mechanics...Experience shows that the problem of the mind cannot be solved by attacking the citadel itself...the mind is function of body...we must bring some *stable* foundation to argue from...' (in Barrett, 1980). Although Darwin had turned in the right direction, he could do very little with mind and culture during his lifetime for the same reason that he was helpless before the mysteries of heredity: the basic information and modes of thought were lacking to produce the stable foundation which he correctly viewed as essential. Today, one hundred years after his death, we may at last be approaching a sufficient understanding to bring to fruition Darwin's proposal. If so, we can verify another insight, which was entered into the *M* notebook on 16 August 1838: 'Origin of man now proved...Metaphysics must flourish...He who understand baboon would do more toward metaphysics than Locke.'

What will be the outcome of this most problematic, controversial extension of evolutionary theory? It is a common perception that during the 1950s biology replaced physics as the most exciting domain of science. I will make a brash prediction: that by the year 2000 the social sciences, in conjunction with brain studies, will commence to replace biology in the central role. If such an advance is realized, it represents one more step in a progression in which the antidiscipline, that is, the field treating the next level of organization below the one under scrutiny, is partly replaced by the synthetic enterprise to which it gave rigour and impetus (Wilson, 1977). In other words, biology advances the social sciences. Just as physics and chemistry helped to modernize biology and moved it to centre stage during the past thirty years, I believe that biology is about to augment the social sciences greatly and move them to centre stage.

The principal remaining obstacle in this enterprise is the unknown relation between genes and culture. But of course when one uses the phrase 'an unknown relation', he means that it is a puzzle to be solved. In this case the problem is surely one of the most important in all of science, not to mention philosophy, as Darwin perceived 144 years ago. A few writers still speak of a permanent discontinuity between the biological sciences and the social sciences, grounded in epistemology (Eccles, 1980) or at least forced by a fundamental difference in goals (Hampshire, 1978). But others, principally in cognitive science and evolutionary biology, have come to see the gap as a largely unknown evolutionary process, a complicated and fascinating interaction in which culture is generated by biological process while biological traits are simultaneously altered by genetic evolution in response to cultural innovation.

Charles J. Lumsden and I have recently studied this dual evolutionary process, which we call gene–culture coevolution (Lumsden & Wilson, 1981). We have attempted to align a previously independent field of inquiry, cognitive and developmental psychology, with evolutionary biology and particularly sociobiology, and in so doing have constructed an ensemble of models that trace, at times clumsily and imperfectly, behavioural development from the genetic blueprint to the assembly of the nervous system to the learning process – and then back down to the alteration of gene frequencies by natural selection operating within the context of particular cultures. The full sequence covered by our models is referred to as the circuit of gene–culture coevolution. This research is a logical extension of sociobiology, the systematic study of the biological basis of social behaviour and a growing division of evolutionary biology.

The particular problem addressed by the theory is the following. We know that human social behaviour is extremely variable. It is also open-ended, in the sense of being always subject to rapid change due to innovation and importation. Cultural evolution is often characterized as Lamarckian in quality, in other words, dependent on the transmission of acquired characters, and relatively fast; while genetic evolution is Darwinian, that is, dependent on changes in gene frequencies across generations, and slow. But exactly how are these two processes coupled?

The solution to the problem can be found by shifting emphasis from the terminal products, the genetic blueprint and the final cultural product, and concentrating on the developmental procedures that connect them. The reason such analysis has not proceeded more vigorously in the past is that

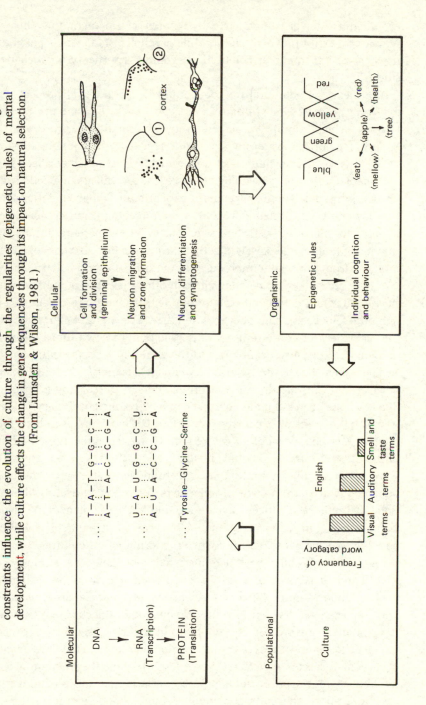

Fig. 26.1. An elementary representation of the circuit of gene-culture coevolution. In this view, biological constraints influence the evolution of culture through the regularities (epigenetic rules) of mental development, while culture affects the change in gene frequencies through its impact on natural selection. (From Lumsden & Wilson, 1981.)

evolutionary biologists have virtually ignored developmental psychology, now a vast field in its own right, while psychologists for their part have not appreciated the great potential of evolutionary theory for their own studies.

The theory of gene–culture coevolution (see Fig. 26.1) proposes the following process. First, human genes affect the way that the mind is formed – which stimuli we perceive, how information is processed, the kinds of memories most easily stored and recalled, the emotions they are most likely to evoke, and so on. These effects, which have been well documented in recent psychological research, are called epigenetic rules. The rules are rooted in the particularities of human biology, and they affect the way culture is formed. For example, outbreeding is much more likely to occur than brother–sister incest because of the apparently innate rule that individuals raised closely together during the first six years of life are inhibited from full sexual intercourse at maturity. Certain colour vocabularies are more likely to be adopted than others because of another rule: the retinal colour cones and certain interneurons within the brain encode light into four basic colours, even when the wavelength of light falling on the eye varies in a continuous manner. The Dani of New Guinea have one of the poorest colour vocabularies in the world, in fact consisting only of two terms, for 'bright' and 'dark' respectively. Eleanor Rosch (1973) took advantage of this fact to conduct an experiment in learning propensity. She gave one group of male volunteers a new colour vocabulary to learn in which the terms were centred on the four basic colours. Another group of men received a vocabulary centred on the wavelengths at the margins of the basic colours. Individuals in the first, 'natural' group learned the words twice as quickly and retained them longer. When given a choice between the two terminologies, Dani men preferred the natural vocabulary. Both of these cases, incest avoidance and the development of colour vocabularies, illustrate nicely how biological constraints in cognition, based on specific genes, can influence the formation of culture.

Epigenetic rules have been demonstrated in virtually every category of cognition and behaviour investigated in such a way as to distinguish choices among stimuli. Examples include odour and taste discrimination, with important effects on the evolution of language and cuisine; preference from infancy onward for certain basic geometric designs over others; phoneme formation; rules of transformational grammar; the development of particular, species-wide facial expressions to denote the emotions of fear, loathing, anger, surprise, and happiness; various other forms of nonverbal communication; the pattern of mother–infant bonding; the method of infant holding by women; fear of strangers (a usually strong response that persists from about six to eighteen months); phobias; and others (see the review by Lumsden & Wilson, 1981). It is clear that during the past twenty years developmental psychologists have come to the edge of a vast array of structural processes in the development of the mind, and an exciting era of experimental research has begun. Most of the ontogenetic patterns occur early enough in life and are sufficiently strongly marked and stereotyped to suggest that they are genetically canalized. It is further true that some degree of heritability has been indicated by twin and pedigree analysis, of varying degrees of sophistication and reliability, in virtually every measurable category of cognitive ability and personality trait thus far studied. Many of these properties form components of the epigenetic rules just cited (Ehrman & Parsons, 1981). By 1980 about 3100 human genes had been distinguished, mostly by biochemical techniques. Of these, 340 were pinpointed to a particular chromosome, with at least one on each of the 23 pairs of chromosomes (McKusick, 1980). Some of the genes and chromosome aberrations affect behaviour in selective ways. A notable example is the major gene recently identified that reduces spatial ability in three standard tests but not in twelve others (Ashton, Polovina & Vandenberg, 1979). It is equally significant that the analysis of complex, multiple-locus systems (polygenes) is well advanced. Recent advances include the calculation of the numbers of chromosomal loci and genes involved in such relatively complicated behaviours as dominance and drug aversive behaviour in mice and the epigenetic rules of colour preference in birds (Thompson & Thoday, 1979).

An important principle of gene–culture coevolutionary theory is that a tabula rasa mind, open to all choices equally and hence totally dependent on the accidents of history, must still have a biological foundation – and a very finely adjusted one at that. The sensory apparatus and brain have to be tuned precisely in order to process all stimuli without bias. Such an effect, which Lumsden and I have called the 'pure cultural transmission' of culture, cannot be achieved merely by removing genetic constraints on cognition and learning. Quite the contrary: it requires a formidable array of homeostatic devices, in order to achieve uniform information processing and hence independence from all but the most generalized set of internal reinforcement mechanisms. And indeed there is no evidence that the human brain works in such an extreme behaviourist manner. Nor does the grain mediate a 'pure genetic transmission' of culture, in which (like the song of the white-crowned sparrow) only a single form of behaviour can be taught and learned. All of the evidence from cognitive studies thus far indicates that human behaviour lies in between these two extremes, that is, in the category we have termed 'gene–culture transmission' of culture. Multiple choices are learned and any one of them can be followed – as for example incest versus outbreeding – but there is an innate predisposition to learn certain ones in preference to others, or else to choose them once they have been acquired.

Let us now review the essential steps in the proposed coevolutionary circuit. Consider for example the case of the avoidance of brother–sister incest, which is based to a substantial degree on an inhibition developed during close domestic association in the first six years of life. Because this epigenetic rule occurs across cultures and is strong enough to defeat countervailing social pressures, it can reasonably be supposed to have a genetic basis. Moreover, those who follow the rule benefit in natural selection. Incest results in higher rates of homozygosity, the more frequent expression of lethal or subvital recessive genes, and hence a greater incidence of hereditary disease and early death among the offspring.

The epigenetic rule thus directs the developing mind to avoid brother–sister incest. The summed preference of members of the society lead to particular cultural patterns, including reinforcing taboos and laws, that prohibit incest. However, because the cultural transmission is of the intermediate, 'gene–culture' form, the preference is not absolute, and scattered individuals in many societies still prefer and may even practise brother–sister incest. The result is some variation among cultures in the frequency of its members who adopt this preference. Consider, for example, groups of 25 individuals. This is the size of many hunter-gatherer bands, the social organization in which mankind has existed throughout most of its history. At any given moment most such bands can be expected to contain no incestuous members. A smaller percentage of the bands will contain one such member, a still smaller percentage will contain two incestuous individuals, and so on. The full array of such fractions, comprising a frequency distribution across all cultures sampled, is called an ethnographic curve. We have devised methods for predicting such curves from a knowledge of two functions: the magnitude of bias in the preference for one cultural choice (such as incest) versus another (outbreeding), and the degree to which the expressed preference of the remainder of the group affects the magnitude of the individual bias.

Two results of general interest emerge from this preliminary analysis. First, it is technically possible to predict patterns of cultural diversity, expressed as the ethnographic curves, from a knowledge of individual cognitive development, and also to perform the reverse: to infer at least some of the principal properties of cognitive development from a knowledge of the pattern of cultural diversity.

The second result of broad interest is that cultural diversity is to be expected even if the underlying cognitive development is rigidly pro-grammed. Cultural anthropologists have commonly argued that the existence of substantial differences among cultures is evidence of the absence of underlying biological influence (Sahlins, 1976; Harris, 1981). But this conclusion is entirely wrong. Cultural diversity *per se* is evidence neither for nor against such control. Rather, what matters is the *pattern* of the diversity. As biological bias is increased toward one choice as opposed to another in the course of genetic evolution, the mode of the ethnographic curve can be expected to shift in that direction. And as the influence of peer activity is increased – this influence itself may well be biologically determined – there will be a tendency for the ethnographic curve to change from a unimodal to a multimodal form. By examining the pattern of cultural diversity in an explicit form such as the ethnographic curve, the nature of the underlying epigenetic rules can be partially inferred.

A recurrent working hypothesis of gene–culture coevolutionary theory is that the epigenetic rules are shaped by natural selection over many generations. Returning to the brother–sister incest case for illustration, we note that individuals who conform to the aversion leave more offspring. As a result, genes underwriting the avoidance of incest remain at a high level in the population. Consequently the predisposition is sustained as one of the epigenetic rules. In general, the rules leading to higher rates of survival and reproduction tend to increase in the population. Thus the assembly rules of the mind build up during evolution, element by element.

In the genetic models, the tabula rasa brain, in which the mind is created solely by the circumstances of history, proves to be a very improbable outcome in the evolution of any conceivable intelligent species. Even if a species somehow managed to begin with such a brain it would soon evolve in the direction of structural and biased epigenetic rules. And as the evolution proceeds, small changes in the degree of bias can be expected to result commonly in much greater changes in the final cultural product. For example, a barely detectable innate bias toward the use of body adornment, if combined with a moderate sensitivity to peer usage, would result in most or all members using such adornment in all societies.

Finally, a detectable amount of genetic evolution in the brain and mind can occur within only thirty or forty generations, or very roughly a thousand years. If correct, this still purely theoretical conclusion implies that epigenetic rules and mental traits might have continued to evolve into historical times. The conventional view, that such biological evolution ceased tens of thousands of years ago and human change has consisted entirely of cultural evolution since then, may be incorrect.

In closing, I want to take note of the familiar lament that science and technology have created not just a cornucopia but terrible dangers as well. What is meant by science in this case is of course the physical sciences and to a lesser degree the biological sciences. But the solution is not, as a few modern Luddites have suggested, the curtailment of science itself, including sociobiology and the social sciences. Quite the opposite: the solution is to make every effort to extend new scientific procedures into the deeper reaches of human nature in order to provide solutions to those residual problems that continue to defy simple economic and technological solution. The peculiar clockwork of the human mind, not scientific knowledge itself, is the source of the danger.

If evolutionary theory can be successfully extended to the assembly of the mind and the creation of cultural diversity, the result may well rank as the completion of the Darwinian revolution. Whether the particular scheme summarized here can contribute substantially to that end remains to be seen, but I hope at the very least I have been able to express why I believe that the social sciences will eventually be fused with biology. No natural boundary appears to exist between the natural and social sciences. Their blend zone, a mysterious and sometimes prohibited domain, offers a great immediate potential for scientific discovery in the postulational-deductive and experimental tradition of the natural sciences.

Further reading

Sociobiology is defined as the systematic study of the biological basis of all forms of social behaviour. It is based to a substantial degree on population biology, including the ecology and genetics of populations, and thereby stresses not only behaviour but also the demography and structure of whole societies. Most sociobiology is concerned with animals, but of course the limited extensions that have been made to human beings are both important and liable to unusual amounts of controversy. The following books can be consulted for a general introduction to the subject and collections of recent research articles:

Wilson, E. O. (1975). *Sociobiology: The New Synthesis*. Cambridge, Mass.: Belknap Press of Harvard University Press.

Barash, D. P. (1981). *Behavior and Sociobiology*, 2nd edn. New York: Elsevier North Holland.

Chagnon, N. A. & Irons, W. (ed.) (1979). *Evolutionary Biology and Human Social Behavior: An Anthropological Perspective*. North Scituate, Mass.: Duxbury Press.

Markl, H. (ed.) (1980). *Evolution of Social Behavior: Hypotheses and Empirical Tests*. Life Sciences Research Report 18. Dahlem Workshops: Verlag Chemie.

The four principal journals in the field are *Behavioral Ecology and Sociobiology* (New York: Springer Verlag), *Insectes Sociaux* (Paris: Masson), *Journal of Social and Biological Structures* (New York: Academic Press) and *Ethology and Sociobiology* (New York: Elsevier North Holland). The last two are concerned principally with human behaviour.

References

Ashton, G. C., Polovina, J. J. & Vandenberg, S. G. (1979). Segregation analysis of family data for 15 tests of cognitive ability. *Behavior Genetics*, 9,329–47.

Barrett, P. H. (1980). *Metaphysics, Materialism, and the Evolution of Mind: Early Writings of Charles Darwin*. Chicago: University of Chicago Press.

Eccles, J. (1980). *The Human Psyche*. New York: Springer International.

Ehrman, L. & Parsons, P. A. (1981). *Behavior, Genetics, and Evolution*. New York: McGraw-Hill.

Hampshire, S. (1978). The illusion of sociobiology. *New York Review of Books* (12 October 1978).

Harris, M. (1981). *Cultural Materialism: The Struggle for a Science of Culture*. New York: Random House.

Lumsden, C. J. & Wilson, E. O. (1981). *Genes, Mind, and Culture*. Cambridge, Mass.: Harvard University Press.

McKusick, V. (1980). The anatomy of the human genome. *Journal of Heredity*, 71, 270–91.

Rosch, E. (1973). Natural categories. *Cognitive Psychology*, 4, 328–50.

Sahlins, M. (1976). *The Use and Abuse of Biology*. Ann Arbor, Mich.: University of Michigan Press.

Thompson, J. N., Jr. & Thoday, J. M. (eds) (1979). *Quantitative Genetic Variation*. New York: Academic Press.

Wilson, E. O. (1977). Biology and the social sciences. *Daedalus*, 106, 127–40.

27

Evolution, ethics, and the representation problem

BERNARD WILLIAMS

This paper is concerned with culture and with evolution, but not with cultural evolution. It discusses the relations between biological evolution and the areas of human culture which may broadly be called 'ethical'. The concept of *cultural* evolution is problematical, and there are notorious difficulties about applying the notions of evolution and natural selection to cultural development; in particular, the ends served by various cultural developments are themselves defined by culture, as are the 'choices' to which Wilson refers in his paper (this volume). That area, however, is not the concern of the present discussion.

There are two kinds of connection between evolutionary theory and ethics: one normative, and one explanatory. There is also a connection between these two, to which I shall come later. The first of these is older than the second, and has acquired a bad name; indeed, it acquired it fairly early, in some part from the monumental and unappealing system of Herbert Spencer. In fact, as John Burrow has shown (Burrow, 1966) a lot of this material ante-dated *The Origin of Species*; the concept of 'the survival of the fittest' (Spencer's own phrase) was already implicit in earlier sociological work which Spencer derived from Malthus. Darwin himself had little sympathy for these ideas and not much, personally, for Spencer, though he did once say – I quote Burrow (p. 182) – 'in a moment of enthusiasm...that Spencer's *Principles of Biology* made him feel that he "is about a dozen times my superior", and thought that Spencer might one day be regarded as the equal of Descartes and Leibniz, rather spoiling the effect by adding, "about whom, however, I know very little"'.

The bad normative applications of evolutionary theory to ethics which were made by Spencer and others also, of course, involved a lot of bad evolutionary theory: if normative lessons could be drawn from Darwinian

theory, there is certainly no reason why they should take the form
suggested by Social Darwinists. However, there is in addition a standard
objection which holds that no such lessons can be drawn at all, at least
in any directly logical way, since any project of deriving ethical content
from premisses of evolutionary theory commits the 'naturalistic fallacy',
an error which is today often equated with that of trying to derive *ought*
from *is*.*

Interesting questions about 'naturalism' in ethics in fact go beyond these
purely logical issues. Naturalism in a broader sense consists in the attempt
to lay down certain fundamental aspects of the good life for man on the
basis of considerations of human nature. If this project fails, it is not for
purely logical reasons; it will rather be for the more interesting reason that
the right sort of truths do not exist about human nature. I shall come back
to this wider question at the end of this paper.

The point about *ought* and *is*, so far as it goes, does have some force.
It can be put in the following way. Suppose that considerations of
evolutionary theory show that certain behaviour is in some sense appropriate for human beings. Either human beings can diverge from this
pattern, or they cannot. If they can, then the biological considerations are
not going to show that they ought not to; while, if they cannot so diverge,
then there is no question of *ought*. This argument seems to me sound so
far as it goes, but it does not go very far.

Implicit in this last argument is another logical relation which is more
interesting for this question than that between *ought* and *is*: the relation,
that is to say, between *ought* and *can*. This relationship underlies some
important negative arguments which by citing certain claims to the effect
that human beings cannot, as they may suppose, live in a certain way, lead
to the conclusion that certain ethical goals or ideals are unrealistic and
should be revised. By arguments of this kind, biological or similar
arguments could coherently yield *constraints* on social goals, personal
ideals, possible institutions and so forth.

To say that human beings *cannot* do certain things is, of course, an
extremely vague form of statement. At one extreme, it may mean that the
world will not contain an example of any single human being doing that
thing; at the other end, it may merely mean that if a group of human beings

---

* It is perhaps worth remarking that when G. E. Moore introduced this term (in Moore,
1903) he applied it to what he regarded as a mistake with respect to the notion of
goodness. The extension of it to the prohibition on 'deriving *ought* from *is*'
rests – clearly in the case of R. M. Hare (e.g. in Hare, 1952) and less clearly in the case
of other writers – on a particular philosophical doctrine, that *good* is to be explained in
terms of *ought*.

adopt a norm requiring that behaviour, the norm will often be broken, its observance will give rise to a good deal of anxiety, those who comply without anxiety to the norm will be unusual in other respects, and so forth. This vagueness will not matter so long as one is clear about the level at which the formula of '*ought* implies *can*' is being applied: thus the latter, and weaker, kind of 'cannot' would be enough to provide a strong argument against the behaviour being made into a norm for a human society, but it would not be enough if the question concerned the adoption of a personal ideal in an individual case. Here, as so often, it is a centrally important question, who is supposedly being addressed by a given piece of ethical discourse. Granted that one is clear about that, the fact that relevant statements of what human beings can and cannot do come in various strengths is not so important. What is vitally important is the difficulty of knowing which of them, relevant to difficult ethical issues, are on biological grounds in fact true.

In this area there is an important connection, which I mentioned before, between what I called the normative and the explanatory interest. If some biological constraint can rule out, or make unrealistic, some normative practice or institution, then knowledge of it may not only encourage us to decline that practice if it is suggested, but may also contribute an explanation of why human communities do not in general display that practice or institution. Might biological considerations then go further and explain the human adoption of other practices, which are conformable to biological constraints?

This raises a general question which is central to these areas, and which I shall call *the representation problem*. It is a problem which comes up at various points in considering the relations between biology and human practices, and may be put in the following way: *how is a phenotypic character which would present itself in other species as a behavioural tendency represented in a species which has a culture, language and conceptual thought?* It may be said that in some cases, at least, such a tendency will show up in that species merely as itself – that is to say, as a merely biological character of that species. But, in fact, virtually no behavioural tendency which constitutes genuine action can just show up in a cultural context 'as itself'. Where there is culture, it affects everything, and we should reject the crude view that culture is applied to an animal in a way which leaves its other characteristics unmodified. (Related to that view is the naive assumption of certain sociobiologists that sociobiology should expect to be more closely related to social anthropology than to other social sciences, because the 'primitive' peoples studied by social anthropology are nearer to nature than human beings who live in large industrialized societies.) None of this is to deny that there is a biological basis for elements in human behaviour which are culturally affected, moulded and elaborated. It is not to deny that some culturally elaborated behaviour can usefully be explained from a biological perspective. It is simply to recall the fact that almost all human behaviour, at least that which deserves the name of 'action', is in fact culturally moulded and elaborated.

In accepting that there is a representation problem, I reject two views according to which there would be no such problem. First is a simple reductionist view, which would neglect the way in which culture not only shapes but constitutes the vast mass of human behaviour. When ancient Greek thought first discovered the opposition of 'nature' and 'convention', it also discovered that an essential part of human nature is to live by convention. The study of human nature *is*, in good part, the study of human conventions, and that is what it is from the strictest ethological point of view. That is how *this* species is. It is a claim additional to this, but one which I also believe to be true, that human conventions, at least beyond a certain state of elaboration, can be understood only with the help of history, and that the social sciences accordingly have an essential historical base. To pursue the question of whether that is so, would go beyond the limits of the present discussion, but it is worth bearing in mind, when the relations are discussed of biology to the social sciences, that an essential social science is likely to prove to be history.

The second point of view which is excluded by taking seriously the representation problem is one which I am disposed, perhaps unfairly, to call 'the Wittgensteinian cop-out'. This is a view implicit in the idea that the central concept for gaining insight into human activities is that of a 'language game'. Since 'language' in this formulation is regarded both as the key to human convention, and also as something which human beings possess and animals do not, the phrase itself implies the lack of interesting explanatory or constraining connections between human and animal behaviour. It suggests an autonomy of the human, under a defining idea of linguistic and conceptual consciousness, which tends to put a stop to any interesting questions of the biological kind before they even start. It therefore does not give any help even in the areas, such as sex and hunger, where we most obviously need means of describing the relations between culture and the biological.

The feature of human culture and human activities that gives rise to the representation problem is above all that human communities embody *norms*, and it is this notion that I shall principally discuss. However, there are other ways, as well, of picking out differences between human activities and those of other animals. One is the very general feature that humans possess conceptual and reflexive consciousness; this, and the very large philosophical problems introduced by those three terms, I shall happily leave on one side. Another distinction between human and animal behaviour is that considerations of *motive* are appropriate to the assessment of human action. This is a matter that is worth some brief discussion, since it is closely connected with the fuss that has been made about the application of the term 'altruism' to animal as to human behaviour.

In other animals there is behaviour which benefits another individual, and moreover there is behaviour the end of which is to benefit another individual, in a sense of 'end' which requires a lot of work to make clear, but which is uncontentiously illustrated by behaviour the end of which is that the animal should take in food. In the human case, many more layers can be added, and other distinctions drawn. There are questions of intention, where this concerns what thoughts produce the action, and what features of the action are, relative to that thought, accidental. There are questions of underlying desire. Some actions which benefit others come from the desire just to benefit that particular person, while others flow from some more general disposition, while the desire to benefit a particular person or group may be accompanied by a variety of other desires, for instance to extract goodwill from them, or the possibility of a reward.

The cultural and psychological elaboration of these various motives of course raises difficulties for any simple relation of them to the biological. Some of those difficulties arise just from the general problem of applying biological models to a species which engages in intentional thought; to that extent there is no *special* problem about altruism and morality. People think that there is a special barrier here to the application of biological models, I believe, because they take 'altruism', in a 'properly moral' sense, to refer to some quite peculiarly pure motive, such as the intention to benefit others derived from impartial reflection on their interests and associated with no other desire whatsoever. But it is extremely unreasonable to suppose that all (perhaps any) human beings act from that motivation, either, and if morality is to be a generally human phenomenon, it is simply a mistake to equate it from the beginning with such exigently Kantian formulations, and it is a mistake even from the point of view of the human sciences. It is no doubt true that a biological perspective will make one more suspicious of extremely intellectualist or, again, very purist views of morality; but equally, so will a reasonable historical and psychological understanding of morality.

It is the notion of a *norm* that perhaps gives rise to the central representation problem. The main point is condensed in the question raised by Pat Bateson in his paper (this volume), about the relation between an inhibition and a prohibition. The most, it seems, that a genetically acquired character could yield would be an inhibition against behaviours of a certain kind; what relation could that have to a socially sanctioned prohibition? Indeed, if the inhibition exists, what *need* could there be for such a prohibition? If the prohibitory norm is to be part of the 'extended phenotype' of the species, how could we conceive, starting from an inhibition, that this should come about?

This is a central example of the problem, but it is not the only example even with respect to norms, and it will be helpful to distinguish various things that fall under the general heading of a 'norm'. Not everything that falls under this heading is a sanctioned prohibition. We can distinguish various items; I will represent them as *stacked*, in a way which is typical, but not by any means universal.

(1) Behaviour which is normal. This does not just mean 'frequent': exceptions are perceived as 'odd', but are not necessarily disapproved of, sanctioned, etc.

(2) (1) together with an institution. This can be applied to the case of marriage, where there will of course be usually sanctions of varying degrees against behaviours that threaten marriage, and sexual activity outside marriage may be disapproved of, but this does not imply that merely not engaging in marriage is disapproved of, nor that an unmarried condition is sanctioned.

(3) Behaviour which lies outside (1) and (2), and to which in addition there may be strong personal disinclination: e.g. homosexuality as regarded in enlightened circles.

(4) (3) together with rejection and disapproval of the deviant behaviour: e.g. homosexuality as regarded in less enlightened circles.

(5) (4) together with sanctions personal or legal: e.g. homosexuality in the least enlightened circles.

Among the cases in which the options are not stacked like this, is that in which the sanctions and disapproval exist against behaviour which is in fact frequent and not the subject, perhaps, of any deep personal disinclination; this, at the limit, is pure humbug, like the old school-master's attitude to masturbation.

(1) to (3) of course raise some difficulties for a biological approach, particularly with regard to institutions, and an adequate treatment of the representation problem will deal with all these levels. The question of

inhibitions and prohibitions arises most clearly at levels (4) and (5). There is, moreover, a specially paradoxical version of it which arises from certain cases in which not only does extra conceptual content have to be introduced to characterize the human prohibition, but also the introduction of that content stands in conflict with the proposed biological explanation of it.

A clear example of this arises with the famous example of the incest taboo, which has been discussed by Bateson (this volume). There are of course many incest taboos, that is to say, prohibitions on sexual relations between persons of various degrees of familial relation, and some of these are hardly even candidates for biological explanation. Moreover, there may well be some very severe doubts about the application of the biological model even to the favourite cases. The present discussion, is not, however, concerned with the factual merits of these explanations, but only with the shape that they take.

In other species, there are behavioural drives the function of which is to avoid inbreeding. Such a drive, however, has to be operationalized in some other way, since the animals do not have any direct knowledge of the matters relevant to inbreeding: the inhibition against mating has to be triggered by the recognition of or reaction to some property adequately correlated with the kin relationship, such as being an individual with which the animal has been brought up. It is *this* inhibition that is allegedly displayed, in the well-known case, by those brought up in the kibbutz. But we have not yet reached any incest taboo. There are no sanctions against marrying those that one is brought up with (as such); the sanction is against marriages which would constitute close in-breeding. The conceptual content of the prohibition is thus different from the content that occurs in the description of the inhibition. It indeed relates to the suggested *function* of that inhibition, but that fact will not explain how the prohibition which is explicitly against in-breeding will have arisen. It certainly does not represent a mere 'raising to consciousness' of the inhibition. It can have come about, in fact, only given human knowledge of relevant facts – presumably, of the ill-effects of in-breeding. But once that is an essential step in the explanation, we no longer need the biological element in the explanation (of the prohibition, that is to say, rather than of the inhibition). It turns out that we have to appeal in any case to something like a rational collective agency, directed towards avoiding recognized and agreed evils, and that *already* provides an adequate explanation – a fairly traditional one – of the incest prohibition.

A similar paradox can arise with other norms supposedly based at a biological level, but there are cases that avoid it. Consider for instance the 'double standard' in sexual morality, traced by Symons (1979) to the disparity between ovum and sperm. This account, though it applies much more widely, is essentially the same as an explanation of these social phenomena which goes back at least to Hume, who accounted for 'the artificial virtues of chastity and modesty in women' by referring to the naturally greater disposition of males to protect children that they believe to be their own. Here again, there may be serious doubts about relevant anthropological facts, but the present point concerns the principle of the explanation, which involves an important difference from the incest case. Here, it *is* natural to think in terms of the institutionalization of a disposition which could be displayed in a simpler form pre-culturally. The conceptual content required in this case to describe the institution, though it involves a great deal of cultural elaboration, does not display the same kind of break between the pre-cultural and the cultural as is found in the incest case; and the biological pattern of explanation could recognizably run through such ideas as human beings finding certain institutions 'natural', which does not require any appeal to a rational collective agency to understand the basic biological idea, as is damagingly the case with the incest example. In fact, an explanation which went back to a biologically grounded disposition could in this case precisely avoid the invocation of rational collective agency, which is rather an intellectualist embarrassment to the story as Hume (1738–40) tells it. None of this implies that even if such biological elements did play some role in explaining these institutions, the institutions would then be necessary or unchangeable – even if the explanation were true, this could still be a case in which becoming conscious of their rationale was a help in changing them.

In one of the two cases we have considered – incest – the prohibition is paradoxically related to pre-cultural dispositions: it expresses their function, but not their content. In the case just considered, social institutions could in principle be an expression of a pre-cultural disposition. In other cases, again, the existence of norms seems to be a *substitute* for a pre-cultural disposition. This might well be so with the control of aggression and of self-seeking behaviour; I shall make one or two remarks about this question without pursuing it at length.

In the work of Maynard Smith and others (see, for example, Maynard Smith, this volume) games theory is applied to explaining selection for certain genetically based patterns of behaviour. Games theory can equally be applied to characterizing human norms which are instituted against aggression and other non-cooperative behaviour. (Ullmann-Margalit (1977) gives a recent analysis, though the outlines of the idea that sanctioned norms can represent a solution to the Prisoners' Dilemma can be found in Hobbes.) The principles of the two applications of games theory are in many ways the same but their results point, in a sense, in opposite directions. If sanctioned norms are necessary in the human case, or socialization into rule-observing behaviour, this must be because constraints on human responses in these areas are not, or not significantly, genetically based. Granted structures of the Prisoners' Dilemma type, cooperative behaviour can be secured only granted a certain level of assurance, and the need of norms (in particular of sanctioned norms) to produce that assurance shows that the assurance cannot be adequately delivered by genetically based signals. It is very tempting to suppose that the lack of any such reliable signals, and the perilously low level of security often reached in human communities, must be connected with a high level of conceptual and, in particular, predictive thought, and also an associated capacity for deceit. This perhaps gives a special force to the Voltairean remark about the function of language being to conceal thought.

The previous remarks have raised some questions about the relations between human norms and possible underlying dispositions determined at a biological level. They represent some aspects of what I have called the representation problem, and it is only through further investigation of that problem, and by becoming clearer about how the various kinds of norm *could* relate to our biological inheritance, that we can come to see much about what biological constraints there might be, beyond the obvious ones, on social and ethical arrangements. I do not believe it to be excluded *a priori* that there could be some, and I do not believe that very much is to be achieved by very general assertions or denials of the possibility. What is needed is more detailed analysis, not only anthropological but philosophical, of the demands that any explanations of this sort would have to meet.

It will be needed, above all, if we are to be able to read the historical record. It is only if we can read that record that we can discover some very important biological characteristics of human beings, since (to repeat an earlier point) it is through convention, convention that has a history, that human nature is expressed. It is not merely that without a hold on the representation problem we cannot discover the relevant content in the historical record; without understanding that problem, we cannot adequately control the idea that there is any relevant content at all. If a biologically grounded disposition showed up simply in the form of what human beings could not or would not do, then there would be no real problem of alternative behaviours. The alternatives will simply be absent from the record, and it is unlikely that anyone, except as the most extreme perversity, would want to undertake them. This, of course, is the area in which the *is/ought* argument scores its clear but uninteresting success. What is much more interesting, I have already suggested, is the idea that there could be patterns of behaviour which human beings are entirely capable of wanting and indeed, on an individual or limited scale, of achieving, but which for biological reasons are bound to be psychologically costly, or confined to a small group of otherwise unusual individuals, or otherwise bound to fail as *general* social institutions. To understand how this could be involves some understanding of the representation problem, and to decide that any given pattern of behaviour has this character of being, as one might put it, 'biologically discouraged' requires one to be able to read the historical record. It hardly needs emphasizing that on any question that is interesting, such as social roles of the sexes, we would have to be able to read the historical record better than we now can in order to arrive at any strong conclusions about what is biologically discouraged.

I come back finally to what I mentioned at the beginning of this chapter as the area of 'naturalism' more broadly conceived: that is to say, the question of founding human ethics on considerations of human nature, in some way which goes beyond merely respecting the limits, biological or other, on what human beings are able to do. This is the project of thinking out, from what human beings are like, how they might best and most appropriately live.

Such a project continues to attract some philosophers. Its attractions are obvious. It does not, in any obvious way, require any supernatural warrant, while it is less arbitrary or relativistic than other secular ways of looking at the content of morality. It seems to offer some promise of being both well-founded and contentful.

It seems to me that a correct understanding of human evolution is very relevant to projects of this kind, but that the effect of that understanding is largely discouraging to them. This is for two different kinds of reason. The first is a reason at a more particular and factual level and is correspondingly more sensitive than the other to changes in hypotheses about the emergence of human beings. It is simply that the most plausible stories now available about that evolution, including its very recent date and also certain considerations about the physical characteristics of the species, suggests that human beings are to some degree a mess, and that the rapid and immense development of symbolic and cultural capacities has left man as a being for which no form of life is likely to prove entirely satisfactory, either individually or socially. Many of course have come to that conclusion before, and those who have tried to reach a naturalistic morality which transcends it have had to read the historical record, or read beyond the historical record, in ways which seek to reveal a partly hidden human nature which is waiting to be realized or perfected. The evolutionary story, to the extent that it can now be understood (and to the much more modest extent to which I understand it myself) seems to me to give some support to the view that in this respect the historical story means much what it looks as though it means.

The second and more general reason lies not in the particular ways in which human beings may have evolved, but simply in the fact that they have evolved, and by natural selection. The idea of a naturalistic ethics was born of a deeply teleological outlook, and its best expression, in many ways, is still to be found in Aristotle's philosophy, a philosophy according to which there is inherent in each natural kind of thing an appropriate way for things of that kind to behave. On that view it must be man's deepest desire – need? – purpose? – satisfaction? – to live in the way that is in this objective sense appropriate to him (the fact that modern words break up into these alternatives expresses the modern break-up of Aristotle's view). Other naturalistic views, Marxist and some which indeed call themselves 'evolutionary', have often proclaimed themselves free from any such picture, but it is basically very hard for them to avoid some appeal to an implicit teleology, an order in relation to which there would be an existence which would satisfy all the most basic human needs at once. The first and hardest lesson of Darwinism, that there is no such teleology at all, and that there is no orchestral score provided from anywhere according to which human beings have a special part to play, still has to find its way fully into ethical thought.

Further reading

Ruse (1979), particularly ch. 9, gives a good critical account of some main issues, with references to the literature. Flew (1967) and Quinton (1966) forcefully state the Naturalistic Fallacy arguments. A lot of Ruse's criticism is directed against Wilson (1975); Wilson (1978) expresses more moderate, but also less distinctive, views. Midgley (1978) is not always accurate on evolutionary biology, but makes many suggestive points over a wide field. Stent (1978) contains discussions which touch on most of the issues.

Flew, A. G. N. (1967). *Evolutionary Ethics*. London: Macmillan.

Midgley, Mary (1978). *Beast and Man*. Hassocks: Harvester.

Quinton, A. M. (1966). Ethics and the Theory of Evolution. In *Biology and Personality*, ed. I. T. Ramsey. Oxford: Blackwell.

Ruse, Michael (1979). *Sociobiology: Sense or Nonsense?* Dordrecht: Reidel.

Stent, G. S. (ed.) (1978). *Morality as a Biological Phenomenon*. Dahlem Conference Life Sciences Research Report 9. Berlin: Dahlem.

Wilson, E. O. (1975). *Sociobiology: the New Synthesis.* Cambridge, Mass.: Belknap.
Wilson, E. O. (1978). *On Human Nature.* Cambridge, Mass.: Harvard University Press.

References

Burrow, J. W. (1966). *Evolution and Society.* Cambridge: Cambridge University Press.
Hare, R. M. (1952). *The Language of Morals.* Oxford: Oxford University Press.
Hume, David (1738–40). *A Treatise of Human Nature,* 3 vols. London: John Noon & Thomas Longman.
Moore, G. E. (1903). *Principia Ethica.* Cambridge: Cambridge University Press.
Symons, D. (1979). The evolution of human sexuality. Oxford: Oxford University Press.
Ullmann-Margalit, Edna (1977). *The Emergence of Norms.* Oxford: Oxford University Press.

EPILOGUE

28

The mysterious case of Charles Darwin*

JOHN PASSMORE

The place of Charles Darwin in the pantheon of Great Thinkers is, one might easily suppose, wholly assured. After all, we are commemorating his death a century later, in a multiplicity of volumes, at a multiplicity of conferences. Few scientists are thus honoured.

Naturally enough, much of this commemoration has been left to historians of science, dedicated to exploring the origins and influence of Darwin's theories. The present volume is an exception. True enough, the opening papers fall into that category. But the general title *Evolution from Molecules to Men* makes no explicit reference to Darwin. The authors, for the most part, are active scientists. If, naturally enough, they usually begin with a bow in Darwin's direction and an appropriate quotation, diligently sought out, they are not, in general, talking *about* Darwin. Rather, they are presenting us with independent investigations conducted in a Darwinian spirit, testimonies to the fact that Darwin is still a living force. No scientist could ask for more than this a century after his death. There is something very special about a scientist who is commemorated in that way.

Nor is that all. Darwin's innovations constitute, so Mayr tells us, the greatest intellectual revolution of all time; we all of us, not only scientists but everybody except a few reactionary dissentients, see the world and man's place in it in a Darwinian way. That makes of Darwin not only a great scientist but, as few scientists have been, the source of a profound transformation in human culture. Indeed, in so far as he inspired E. O. Wilson's sociobiology, one might think of him, even, as compelling us to reconsider what culture amounts to, the degree to which it can be

---

* This epilogue was prepared in the course of listening to the Conference at which the papers in this volume were delivered. I have not read the papers. That explains the indirectness of the references.

sharply sundered from nature. We may well regard some of the applications of Darwinism to ethics or to society as wholly deplorable. Williams, for one, sharply criticizes E. O. Wilson, as not being able to bridge the gap between a norm-less nature and a norm-governed culture, even if Bateson seeks to fill that gap. But we can scarcely deny Darwin's fecundity, or the degree to which Darwinian ideas have penetrated our modes of thought.

Then why describe Darwin's as a strange case? Well, consider this fact. It is widely agreed both that there is bound to be another commemoration of Darwin in a century's time and that he would still, at that time, be a centre of controversy. This is very peculiar, on the face of it. We don't have conferences on Newton at which scientists argue about whether Newton was right or wrong. Something of this sort can happen, to be sure, at conferences celebrating a great philosopher, but not at the commemoration of a great scientist. For the most part, such commemorations are pious celebrations, principally in the hands of historians and philosophers of science.

The explanation, the reply might come, is just that Darwin so disturbed the traditional picture of man's place in nature. Anyone who revolutionizes not just a particular area of science, where his ideas will finally be absorbed to form part of impersonal scientific knowledge, but our entire notion of ourselves must expect to be the centre of lengthy and bitter controversies. This we can grant. The attacks of creationists need neither surprise nor disturb us; of themselves, they do nothing to suggest that Darwin's is a strange case.

Certainly, the contents of the present volume will do nothing to assuage the hostility of theologians. Quite the contrary. When theologians have sought to absorb Darwin into their world-view rather than wholly rejecting him, they have commonly argued that the work of a Great Designer can be seen as clearly in an evolutionary pattern culminating in *Homo sapiens* as in the simultaneous once and for all creation of separate species. But, as Jacob has observed, it is more and more apparent that evolution proceeds by tinkering, pottering about, *bricolage*, rather than like an engineer working from a blue-print. In Hodge's language, natural selection is 'opportunistic'. There is no sign in its operations of a pre-conceived plan.

The most far-reaching objections to Darwin, however, come not from fundamentalists disturbed in their devotions but from serious philosophers and serious scientists. Creationists may quote them for their own purposes but they are not themselves creationists. Darwin's reputation as a scientist has been subject to extraordinary, indeed unparalleled, fluctuations. In a history of biology written in the nineteen-twenties he was relegated to obscurity. The work of the geneticists, incorporated in the Haldane synthesis, seemed for a time to secure his reputation. But the attacks continue and have recently intensified. We are told not only, which is sufficiently serious, that the Darwinian programme of explaining macro-evolution by microevolution cannot be carried through, since natural selection can explain only variations, not species-change, but even that the theory of evolution is nothing more than a pleasant fiction, the latest version of ancient myths.

The best we can say of Darwin's so-called 'theory', Popper has argued – in a quite different spirit but with equally damaging results – is that it is a successful metaphysical research programme. Others have described it, what is even worse, as a series of tautologies, telling us no more than that those species which are capable of reproducing themselves will succeed in doing so or that mutations which cannot survive will die out.

The strange case of Charles Darwin, then, consists in the fact that a century after his death, it is still disputed whether *The Origin of Species* really counts as a scientific achievement. The mere fact that we freely use the word 'Darwinism' is in itself significant. It suggests that Darwin ought to be grouped with those two other controversial figures, Marx and Freud, as the author of a *Weltanschauung* rather than as a scientist. Who would speak of 'Einsteinism' or 'Planckism', as one does of Darwinism, Marxism, Freudianism?

How far will the present volume succeed in dispelling such doubts? First for microevolution and macroevolution. On this point, Darwin is explicitly defended by Dawkins and Ayala, implicitly defended whenever micro-evolutionary theories were used to throw light on macroevolutionary problems. Yet the change of tone in the volume is notable, as the professional studies of molecular and bacterial evolution are succeeded by the quasi-philosophical discussions of whole organisms. The mood of expert confidence is dispelled. Disappointment is openly expressed that so much progress has been made at the molecular level, so little at the macro level, whether in relation to embryology or the evolution of animal behaviour.

This, Lewontin suggests, is because Darwin used a nineteenth-century methodology, for which the organism was passive in relation to its environment. What is really needed is a twentieth-century methodology, for which the organism *creates* the environment. It is very surprising to hear the ideas of those paradigmatic nineteenth-century figures, Hegel and Marx, resuscitated as a twentieth-century methodology. What underlies these objections, however, is a dissatisfaction with the present condition of macro theory; the philosophical thrashings about flow from that dissatisfaction.

Such dissatisfactions are not unique. They are echoed, in their general form, at meetings where the successes of neurophysiology are sadly contrasted with the failures of psychology and sociology, or the successes of microeconomics with the failures of macroeconomics. The resemblance is not an accidental one. In each case the micro enquiry can be safely conducted at a level which ignores the complexities arising out of historical circumstances. This does not mean that it is useless in macro inquiries. Maynard Smith, for example, shows us how the micro-investigations of game theory can be used to distinguish between stable and unstable behavioural strategies.

Nevertheless, and granting that the ingenuity of investigators in making such applications will continue to astonish us, it is sensible to ask what can reasonably be expected from microtheories, whether in the way of explanation or prediction. Excessive enthusiasm can generate excessive disillusionment. One cannot reasonably expect that the work of the ethologist will ever be rendered unnecessary by molecular biology, any more than advances in neurophysiology, or in psychology, could ever make it unnecessary to ask whether traffic in Japan keeps to the left or the right. Or so the 'whole organism' theorists are substantially arguing. Molecular biology is often interpreted in a preformationist way, as if behaviour were a kind of unfolding. Not unnaturally, those who are accustomed to look at animals in their natural habitat, subject to a variety of pressures, find that approach wholly unsatisfactory. Gould looks for formal factors as a way of explaining not only – as molecular biology does so successfully – how replication and mutation occur but why some species persist so unchanged over a million years, in what their peculiar 'success' consists. The test-tube takeover is not universally welcomed.

To turn now to the charge that Darwinism is at best a metaphysical research programme, at worst an idle story or an empty tautology, one thing this volume makes plain is just how successful a research programme it has been, suggesting both problems and modes of seeking their solution. Most of the authors would want to insist, nevertheless, that the research fruitfulness of Darwinism derives precisely from its being a successful theory, and more specifically a successful *scientific* theory, even if Hull throws some doubts on the familiar contrast between the scientific and the non-scientific. They are not prepared to regard *The Origin of Species* as purely and simply a research programme, let alone as a *metaphysical* research programme; they think of it as a theory which sought to explain why life has evolved as it has done, in the circumstances existing here on earth.

As for the view that Darwin's central propositions are tautologous, this is explicitly rejected by Hodge and Eigen, both of whom seek to formulate the theory of natural selection in a non-tautologous manner. Indeed, here unlike Hodge, Eigen sets out to show that natural selection can be mathematically formulated as a law in exactly the same sense as Newton's laws. Dawkins, however, claims universality for Darwin, in a way which raises the question whether this universality is of the type of physical laws – which operate, given appropriate boundary conditions, as much in other galaxies as on this galaxy – or whether Dawkins is saying that it is *logically* inconceivable that evolution could operate in some different way in a different planet, that natural selection is implicit in the very definition of evolution, so that the tautology objection would reassert itself. It is generally granted that biologists sometimes express themselves in a way which would suggest that every biological difference *must* contribute to survival, merely in virtue of existing, so that 'surviving' and 'being adapted to the environment' are tautologically related. But this way of formulating Darwinism, as we shall be further suggesting, is now widely rejected as misrepresenting the actual situation.

Some other contemporary criticisms of Darwin are scarcely ventilated in the present volume. Surprisingly little is said about the debate between punctuated equilibrium and gradualism, except for the passing observation by Dawkins that Darwin himself had thought that a punctuated view was more suitable for his purposes but incompatible with the facts. Nor is there any considerable discussion of what survival is at issue, whether of the species, the organism, the gene, or DNA. No one challenges either, except in passing, the cladistic-inspired doctrine that systematics is *real* biological science and evolution nothing more than a beguiling story. (That objection, I suspect, is the ideology of a proletarian revolt, with systematists taking advantage of cracks in the Darwinian ranks to assume what they regard as their rightful central status, where they have been for so long despised and rejected. (I have heard biologists say: 'I don't call systematics *biology*!')) Isaacs comes nearest to their position, even if from a somewhat different, literary-anthropological, standpoint, in suggesting that Darwinism has assumed the form of a classical myth in which through a series of terrible vicissitudes, the hero – *Homo sapiens* – finally emerges, triumphant. Or, looked at somewhat differently, that it has occupied a societal niche, left empty by the decline of Christianity, as a creation-story. These suggestions will not unnaturally arouse indignation. Darwinism, the reply will come, is serious science, not mythology. But one has only to look at the popular presentations of Darwin, from the nineteenth century onwards, to recognize that he has often been read in this way, sometimes by biologists themselves. Much of the material presented in this volume is specifically directed against such interpretations, fully recognizing the importance of stating Darwinism precisely if one is to see it as a science replacing myth, rather than as a rival myth.

The spirit of the more strictly scientific papers is, however, the best testimony to the fact that Darwin was a scientist, subject to the rigorous criticism characteristic of science, not a religious sage, whose every word is sacred. (Contrast it in this respect with papers on Marx by Marxists or on Freud by Freudians.) Darwin's errors, Darwin's ignorance – these are as much insisted upon as his knowledge, his insights. Clarke on bacteria, Woese on primordial forms, consider in detail what Darwin had only glimpsed. Shapiro, Bodmer and Jeffreys go far beyond Darwin's imaginings, in a way which helps to resolve problems he had no way of settling. To such papers, the philosopher can only refer with a respect not unmixed with envy. The extension of evolution, at the hands of Phillips and Jacob, into the realm of the molecule not only broadens Darwin's theory but raises questions about its range. Are we to think of Darwinism as a general theory of development, having application not only to organisms but to all natural differentiations which arise over time, a theory about the origins of increasing complexity? If concepts like 'natural selection' apply to all such changes, is this also true of concepts like 'inclusive fitness'? Or does biological evolution still have its peculiarities? Is there a risk that too broad a theory will collapse into emptiness?

Darwin would have admired these technical papers, as clearing up problems about replication and mutation which he had been unable to solve, as confirming his suspicion that the study of bacteria would be of great significance for evolutionary theory, as establishing continuities between molecular change and the development of organisms. But Darwin, as Mayr insists, was not only a great theorist, he was also a great naturalist. He loved his plants and his animals as Clarke loves her bacteria and Phillips his proteins. He wanted to know not only how stability could be secured and yet variations made possible through the mechanisms of genetic transmission but also why particular mutations occurred, and survived, at particular places and times. He would have been delighted by Hallam's work on tectonics, Clutton-Brock's on sexual selection, and even more, perhaps, by Harper's sward studies, although – or especially because – they would have given him ground for much further reflection and reconsideration.

Reference to Harper's paper leads us on to an issue which in some measure now divides biologists. Can one properly use such phrases as DNA 'junk'? One man's junk, of course, can be another man's treasured possession, whether he be archaeologist or sculptor or antique dealer. But the view that DNA can be 'junk' in the sense of finding no genetic expression provokes considerable opposition. DNA, many biologists feel, is 'there' to be a mechanism of replication.

Is this objection merely a religious prejudice, a relic of the belief that everything in the universe must have a place, a purpose, a function? Or of the doctrine of perfect adaptation expressed by Paley, once Darwin's favourite reading and a name still sometimes invoked? Well, it might rather be nothing more than a heuristic principle, that it is unwise, perhaps obscurantist, to dismiss the question: 'what is this DNA doing?' One cannot suppose, as Clarke emphasizes, that it is on stand-by for future emergencies but it might still have present effects which have not yet been discerned.

Yet it is certainly important to dispel the suggestion that this *must* be so. The good housekeeping view – 'a place for everything and everything in its place' – which found classical expression in the doctrine of 'the Great Chain of Being' is still firmly ensconced in our culture. Lovejoy once suggested that the theory of evolution was nothing but the great chain of being temporalized. The papers in this volume may help to persuade us otherwise, to expel the last traces of Paley from Darwinism. As Burkhardt emphasizes, the world is an untidy place; it is far from being true that all is for the best in the best of all biologically possible worlds, even if some species have found so suitable a niche that they could complacently occupy it over very long periods of time. Not every variation is either advantageous or disadvantageous; it may persist simply because it costs nothing and would take energy to get rid of, as, under similar circumstances, books often remain on our shelves. Human beings do not necessarily possess the genetic code which would be most suitable for their present needs and circumstances. Both their grandeur and their misery may partly stem from that fact. Biologists need no longer feel obliged, in the manner of mediaeval theologians, to 'explain' every biological variation in functional terms. It is enough to show how it could come about, how it could survive, in a world in which parsimony is not the ruling principle. Only by the removal of such metaphysical accretions will the true scientific accomplishment of Charles Darwin emerge. Or so this volume suggests.

Index

Page numbers in italic type refer to figures and tables

577

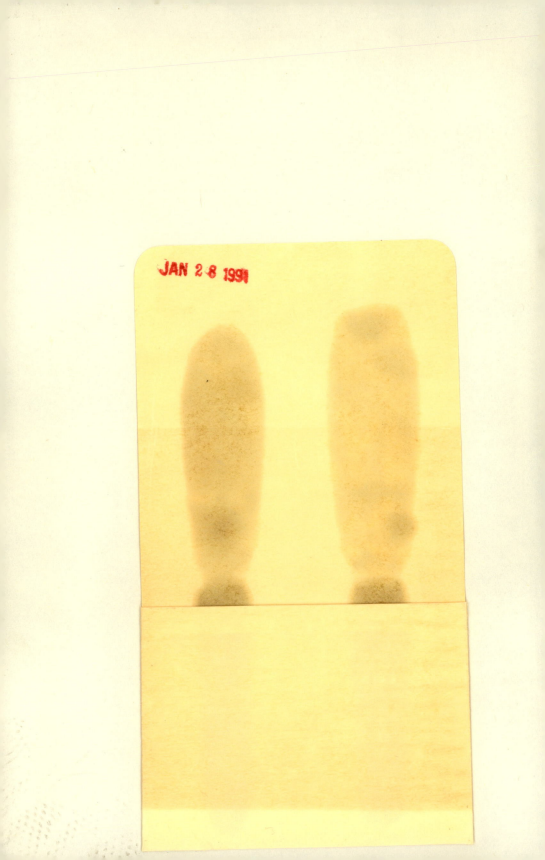